Free Radicals
in Biology and Medicine

The Cover Photograph

Copper-zinc superoxide dismutase. Electrostatic potential is mapped on to the enzyme's molecular surface to show the highly positive potential around the active site channel, allowing guidance of the negatively charged O_2^- substrate into the active site. The molecular surface is shown as dots coloured by electrostatic potential (red highly negative, yellow negative, green neutral, cyan positive, blue highly positive). The active-site copper ion is represented as a purple sphere.

Computer rendering by Elizabeth D. Getzoff and John A. Tainer of the Research Institute, Scripps Clinic (see also *Nature* [1983] **306**, 287), to whom we are very grateful.

Free Radicals in Biology and Medicine

Second Edition

BARRY HALLIWELL
Professor of Medical Biochemistry,
University of London King's College

and

JOHN M. C. GUTTERIDGE
Senior Scientist,
National Institute for Biological
Standards and Control
and
Visiting Professor,
Oklahoma Medical Research Foundation

CLARENDON PRESS · OXFORD
1989

Oxford University Press, Walton Street, Oxford OX2 6DP
Oxford   New York   Toronto
Delhi   Bombay   Calcutta   Madras   Karachi
Petaling Jaya   Singapore   Hong Kong   Tokyo
Nairobi   Dar es Salaam   Cape Town
Melbourne   Auckland
and associated companies in
Berlin   Ibadan

Oxford is a trade mark of Oxford University Press

Published in the United States
by Oxford University Press, New York

First edition 1985
Second edition 1989

British Library Cataloguing in Publication Data
Halliwell, Barry
Free radicals in biology and medicine.—2nd ed.
1. Organisms. Free radicals
I. Title II. Gutteridge, John M. C.
574.19'282
ISBN 0–19–855294–7 ✓
ISBN 0–19–855291–2 (pbk.)

Library of Congress Cataloging in Publication Data
Halliwell, Barry.
Free radicals in biology and medicine / Barry Halliwell and John
M. C. Gutteridge. — 2nd ed.
p.   cm.
Includes bibliographies and index.
1. Free radicals (Chemistry)—Physiological effect.
2. Superoxide—Physiological effect.   3. Pathology. Molecular.
I. Gutteridge, John M. C.   II. Title.
[DNLM: 1. Biology.   2. Free Radicals.   3. Medicine. QP 527
H191f]
RB170.H35   1989 612'.01524 — dc 19   89–3131
ISBN 0–19–855294–7
ISBN 0–19–855291–2 (pbk.)

Typeset by Latimer Trend & Company Ltd
Printed in Great Britain
by Bookcraft (Bath) Ltd
Midsomer Norton, Avon.

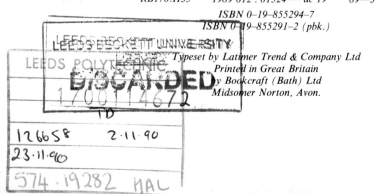

Preface to the second edition

The explosive growth of interest in free radicals, and the enormous amount of research work undertaken since 1984, has necessitated the writing of a second edition of this book after only three years. During the extensive re-writing necessary, we have sometimes felt like painters of the Forth bridge. Production of the revised edition has been helped by information and critical comments provided by the following scientists, to whom we are very grateful. Any remaining errors are the entire responsibility of the authors, however.

B. N. Ames
B. Anderton
B. M. Babior
D. R. Blake
J. M. Braughler
L. Breimer
G. Burton
R. Cammack
C. J. Chesterton
T. Connors
F. Corongiu
J. T. Curnutte
C: Dahlgren
A. T. Diplock
E. A. Dratz
H. Esterbauer
I. Fridovich
E. Getzoff

D. J. Hockley
R. L. Hoult
V. Kagan
D. Leake
T. Lindahl
M. Matsuo
M. J. Mitchinson
D. P. R. Muller
H. J. Okamoto
Philips Analytical
E. A. Porta
W. A. Pryor
C. Rice-Evans
G. Rotilio
L. L. Smith
T. F. Slater
Y. Sugiura
S. P. Wolff
R. L. Willson

London
October 1988

B.H.
J.M.C.G.

Preface to the first edition

The importance of radical reactions in radiation damage, food preservation, combustion, and in the rubber and paint industry, has been known for many years to people in the respective fields, but it has rarely been appreciated by biologists and clinicians. The interest in radicals shown by the latter groups has been raised recently by the discovery of the importance of radical reactions in normal body chemistry and in the mode of action of many toxins. The discoveries of hypoxic cell sensitizers that potentiate radiation-induced radical damage to cancerous tumours, of the enzyme superoxide dismutase, and of the mechanism of action of such toxins as paraquat and carbon tetrachloride provide major examples of this importance.

Any expanding field attracts the charlatans, such as those who make money out of proposing that consuming radical scavengers will make you live for ever or that taking tablets containing superoxide dismutase will enhance your health and sex life. In evaluating these and other less-obviously silly claims, it is useful to understand the basic chemistry of radical reactions.

This book is aimed mainly at biologists and clincans. It assumes virtually no knowledge of chemistry and attempts to lead the reader as painlessly as possible into an understanding of what free radicals are, how they are generated, and how they can react. Having established this basis, the role of radical reactions in several biological systems is critically evaluated in the hope that the careful techniques needed to *prove* their importance will become more widely used. We believe that free-radical chemists should also find these latter chapters useful.

London   B.H.
                                                                   J.M.C.G.

Acknowledgements

We are very grateful to the following scientists and publishers who have granted permission to reproduce material or otherwise supplied us with figures:

Dr R. W. Hardy and the American Society of Plant Physiologists (Fig. 1.1); Dr O. R. Brown and Academic Press (Fig. 1.2); Dr H. B. Michaels (Fig. 1.3); Dr J. D. Balentine (Fig. 1.4); Dr G. L. Huber and Springer-Verlag (Fig. 1.5); Dr C. L. Greenstock and Pergamon Press (Fig. 2.1 part A); Professor R. L. Willson (Fig. 2.1 part B); Dr M. Dizdaroglu (Fig. 2.2); Dr B. Franck and Verlag Chemie (Fig. 2.8); Professor B. Chance and Academic Press (Fig. 3.3); Professors B. Chance, H. Sies and the American Physiological Society (Figs. 3.1 and 3.6); Professor G. Rotilio and Elsevier/North-Holland (Figs. 3.8 and 6.9); Professor L. Hurley and Elsevier/North-Holland (Fig. 3.10); Dr David Hockley (Fig. 4.4); Professor H. Esterbauer and A. R. Liss Inc (Fig. 4.8); Professor P. Hochstein and Academic Press (Fig. 4.9); Professor W. A. Pryor and Ann Arbor Science Publishers (Figs. 4.12 and 6.6); Professor L. L. Smith (Fig. 4.10); Dr A. A. Noronha-Dutra (Fig. 4.11); Philips Analytical (Fig. 4.13); Professor F. Corongiu and Elsevier/North-Holland (Fig. 4.14); Dr D. P. R. Muller (Fig. 4.19); Professor E. A. Dratz and Elsevier/North-Holland (Fig. 4.20); Dr C. Rice-Evans (Fig. 4.23); Dr J. M. Braughler (Fig. 4.24); Professor H. Okamoto (Fig. 6.4); Drs B. H. Anderton and J. P. Brion (Fig. 6.15); Professor A. W. Segal (Fig. 7.5); Professor W. G. Hocking (Fig. 7.6); Dr W. Dawson (Fig. 7.7); Dr C. Dahlgren (Fig. 7.9); Professor R. S. Sohal and Elsevier/North-Holland (Fig. 8.5); Dr C. Verdone-Smith, Dr H. E. Enesco and Pergamon Press (Fig. 8.6); Dr M. J. Mitchinson (Fig. 8.7); Professor D. Armstrong and Elsevier/North-Holland (Figs. 8.8 and 8.9); Dr D. Hockley (Fig. 8.11).

J.M.C.G. is indebted to his wife Pushpa, and children Samantha and Mark, for their encouragement and support during the preparation of this book. B.H. is grateful to Emma Seale and Yvonne D'Souza-Rauto for their invaluable help with typing.

Contents

Contents

1

Oxygen is poisonous—an introduction to oxygen toxicity and free radicals

The diatomic oxygen molecules in the Earth's atmosphere are themselves 'free radicals' and major promoters of radical reactions in living cells. It is therefore appropriate to begin by making some general comments about oxygen, after which the nature and definition of radicals will be considered.

1.1 Oxygen and the Earth

Except for those organisms that are especially adapted to live under anaerobic conditions, all animals and plants require oxygen for the efficient production of energy. Free oxygen appeared in the Earth's atmosphere in significant amounts about 2×10^9 years ago, probably due to the evolution of oxygen-evolving photosynthetic organisms. The appearance of oxygen must have been accompanied by the appearance of a layer of ozone (O_3) in the high atmosphere, and the absorption of damaging solar ultraviolet radiation by oxygen and ozone probably permitted the evolution of more complex terrestrial organisms. Oxygen is now the most prevalent element in the Earth's crust (atomic abundance 53.8 per cent) and the percentage of oxygen in the atmosphere has risen to 21 per cent in dry air. If the barometric pressure of dry air at sea level is 760 mm Hg (1 mm Hg = 1 torr), the partial pressure of oxygen would thus be about 159 mm Hg.

Oxygen is also found dissolved in seas, lakes, rivers, and other bodies of water; the oxygen content of surface water is generally in equilibrium with the atmosphere. The solubility of oxygen in sea-water at 10 °C corresponds to a concentration of $0.284 \, \text{mmol} \, l^{-1}$, and decreases at higher temperatures (e.g. $0.212 \, \text{mmol} \, l^{-1}$ at 25 °C). Oxygen is more soluble in fresh water, e.g. for distilled water: $0.258 \, \text{mmol} \, l^{-1}$ at 25 °C, $0.355 \, \text{mmol} \, l^{-1}$ at 10 °C. Of course, the oxygen concentration within living cells will depend on how far the oxygen has to move in order to get to them as well as on how quickly they consume it. For example, the O_2 tension in human venous blood is only 40 mm Hg (about $53 \, \mu\text{mol} \, l^{-1} \, O_2$), about 25 per cent of ambient. Within some eukaryotic cells, e.g. heart or liver, there is an oxygen gradient, decreasing in concentration from the cell membrane to the oxygen-consuming mitochondria. Oxygen is seven-to-eight times more soluble in organic solvents than in water, a point worth bearing in mind when considering oxidative damage to the hydrophobic interior of biological membranes (Chapter 4).

As the oxygen content of the atmosphere rose, it also exposed living matter to oxygen toxicity: oxidations in the cell harmful to the organism and in some cases lethal. There was considerable pressure upon organisms to evolve protective mechanisms against oxygen toxicity, or to retreat to environments that the oxygen did not penetrate. Studies of present-day anaerobes show us what must have happened to the numerous primitive species that failed to adapt and were lost during evolution.

1.2   Oxygen and anaerobic organisms

The term 'anaerobic organism' covers a wide range of biological variation. There are 'strict' anaerobes such as the bacteria *Treponema denticola* and several *Clostridia* that will grow in the laboratory only if oxygen is virtually absent. 'Moderate' anaerobes can grow in atmospheres up to about 10 per cent O_2 (e.g. *Bacteroides fragilis* or *Clostridium novyi* Type A), whereas microaerophiles, such as *Campylobacter jejuni* (a major cause of diarrhoea in humans) and *Treponema pallidum* (the causative agent of syphilis), require a low concentration of oxygen for growth but cannot tolerate 21 per cent O_2. Even 'strict anaerobes' display a wide spectrum of oxygen tolerance. Some are killed by even a brief exposure to oxygen whereas in others oxygen inhibits growth but does not kill the cells, e.g. *Methanobacterium AZ* ceases growth at 0.01 ppm O_2 but survives exposure for several days to 7 ppm dissolved O_2, equivalent to an atmospheric concentration of 20 per cent.

Any environment that can develop a low enough oxygen concentration can harbour anaerobes. For example, in the human mouth, strict anaerobes can be cultured from pockets in the gums, from decaying teeth, and from the deeper layers of dental plaque; whereas less strict anaerobes and microaerophiles can be found in the more superficial layers of plaque on the teeth. The human colon (over 90 per cent of faecal bacteria are anaerobes), rotting material, polluted waters, and gangrenous wounds all provide places for anaerobic bacteria to thrive. Indeed, the treatment of gas gangrene due to *Clostridial* infections by exposure of the patient to pure oxygen at high pressure is based on the known sensitivity of these anaerobes to oxygen. As discussed below, however, such treatment is not without problems.

The damaging effects of oxygen on strict anaerobes seem to be due to the oxidation of essential cellular components. Anaerobes thrive in reducing environments and, by oxidizing such constituents as NAD(P)H, thiols, iron–sulphur proteins, and pteridines, the oxygen can 'drain away' the reducing equivalents that are needed for biosynthetic reactions within the cell. Some enzymes in anaerobes are inactivated by oxygen, e.g. the nitrogenase enzyme of *C. pasteurianum* is inactivated due to the oxidation of essential components at its active site. This enzyme, which catalyses reduction of nitrogen to ammonia, is essential for the survival of the organism in environments poor in nitrogen compounds. Indeed, all nitrogenase enzymes are inactivated by oxygen to some extent, but not all nitrogen-fixing species are strict anaerobes. Indeed, a study of nitrogen-fixing organisms has shown a variety of ways around this problem. *C. pasteurianum*, as we have seen, adopts a simple solution and keeps away from oxygen. Several aerobic, nitrogen-fixing (and other) bacteria surround themselves with a thick slime capsule to restrict the entry of oxygen; some cyanobacteria locate their nitrogenase in specialized, thick-walled, oxygen-resistant cells known as 'heterocysts'. In the root nodules of leguminous plants an oxygen-binding protein, leghaemoglobin, is present to control the free oxygen concentration and to prevent the nitrogen-fixing bacteroids of the nodule from being damaged. The nitrogen-fixing aerobe *Azotobacter* has one of the highest respiration rates of any microorganism, which may serve to consume all the oxygen entering the cell and prevent it reaching the nitrogenase. The photosynthetic cyanobacterium *Gloeocapsa* contains both nitrogenase and an oxygen-evolving photosynthetic apparatus within the same cell, but its life cycle is such that nitrogenase is only highly active when the rate of photosynthesis is low.

Anaerobes can teach us a great deal about the evolution of protective mechanisms against oxygen toxicity and we shall consider them again when reviewing the various protective mechanisms thought to exist (Chapter 3).

1.3 Oxygen and aerobes

Oxygen supplied at concentrations greater than those in normal air has long been known to be toxic to plants, animals, and to aerobic bacteria such as *Escherichia coli*. Indeed, studies of bacterial chemotaxis to O_2 (*aerotaxis*) show that several strains swim away from regions of high O_2 concentration. Plots of the logarithm of survival time against the logarithm of the oxygen pressure have shown inverse, approximately linear, relationships, for protozoa, mice, fish, rats, rabbits, and insects. Indeed, there is considerable evidence that even 21 per cent O_2 has slowly-manifested damaging effects. Figure 1.1 shows one example of oxygen effects: the dry matter accumulation in the leaves of soybean plants is actually increased if they are placed in subnormal oxygen concentrations. All plant tissues are damaged at oxygen concentrations above normal; there is an inhibition of chloroplast development, decrease in seed viability and root growth, membrane damage, and an eventual shrivelling and dropping-off of leaves. Green plants produce oxygen during photosynthesis and can expose themselves and their surroundings to

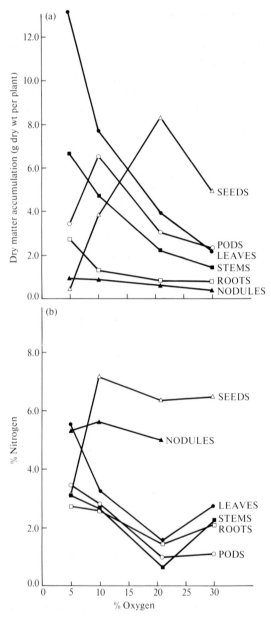

Fig. 1.1.   Dry matter accumulation and nitrogen content of various parts of soybean plants grown in chambers containing different percentages of atmospheric oxygen. Seed production is less sensitive to oxygen than other parts of the plant. From Quebedeaux, B., Havelka, U. D., Livak, K. L., and Hardy. R. W. F., *Plant physiol.* **56,** 761–4 (1975), with permission.

it, e.g. oxygen bubbles from some aquatic plants have been reported to interfere with the breeding of mosquitoes.

The growth of *E. coli* and other aerobic bacteria is slowed by exposure to pure oxygen at 1 atmosphere pressure; Fig. 1.2 shows that exposure of this organism to high-pressure oxygen causes immediate growth inhibition. Oxygen enhances the damaging effects of ionizing radiation both to bacteria and to animal cells in culture; Fig. 1.3 shows this effect for cultured Chinese hamster ovary cells exposed to X-rays. As will be discussed in Chapter 2, the effects of oxygen and those of ionizing radiation on organisms have some similarities.

The toxicity of O_2 to animals, including man, has been of interest in relation to diving, underwater swimming, and escape from submarines, and, more recently, in the use of oxygen in the treatment of cancer, gas gangrene, multiple sclerosis, and lung diseases, and in the design of the gas supply in spacecraft. Rises in the oxygen partial pressure to which an organism is subjected can be due not only to an increase in the percentage of oxygen in

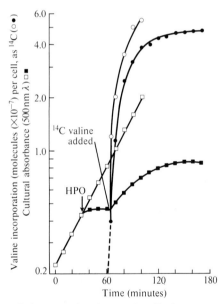

Fig. 1.2. Inhibition of the growth of *E. coli* cells by exposure to high-pressure oxygen. The growth medium was mineral salts, glucose, and amino acids (no valine) at 37°C. At the point marked HPO the atmosphere was changed from air to one of 80 per cent O_2 at 5 atm total pressure. At the point indicated valine was added and growth was restored. Closed symbols: HPO experiment; open symbols: normal air control. From Brown, O. R. and Yein, F., *Biochemical and Biophysical Research Communications*, **85**, 1219–21 (1978), with permission.

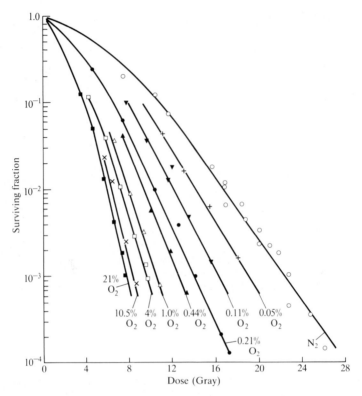

Fig. 1.3.   The 'oxygen effect' in exposing cultured Chinese hamster ovary cells to ionizing radiation. Figure by courtesy of Dr H. B. Michaels.

the air but also, as in diving, to an increase in the total pressure. High-pressure oxygen frequently causes acute central nervous system toxicity, producing convulsions. Figure 1.4 shows a rat in this sorry state. Oxygen at 1 atmosphere pressure does not usually produce such convulsions, but oxygen concentrations of 50 per cent or above, corresponding to an inspired partial pressure of 360 mm Hg, gradually damage the lungs. Exposure of humans to pure oxygen at 1 atmosphere pressure for as little as 6 hours causes chest soreness, cough, and a sore throat in a few people; further exposure leads in all cases to damage to the alveoli of the lungs. This is manifested at first as an increased thickness of the air–blood barrier caused by oedema. Further oxygen exposure causes the death of alveolar epithelial cells and an eventual laying down of inelastic fibrous material in the lungs. Such damage can never be repaired. Figure 1.5 shows the gradual development of pulmonary oxygen toxicity as seen on chest X-rays. Recent clinical and experimental observations suggest that oxygen may worsen lung damage caused by other means even at concentrations thought to be 'safe'.

Fig. 1.4. An adult female Sprague–Dawley rat with convulsive paralysis, especially of the forelimbs, induced by exposure to pure oxygen at 5 atm pressure. Photograph by courtesy of Dr J. D. Balentine.

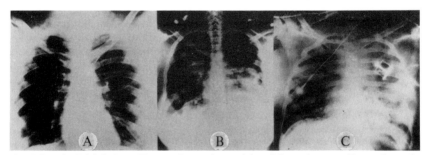

Fig. 1.5. A: Initial chest X-ray of a patient with mild respiratory discomfort after administration of oxygen for a non-pulmonary condition. There are no visible significant abnormalities. B: Further exposure causing X-ray-visible damage with diffuse, irregular pulmonary densities of various sizes in both lungs. C: Late radiological manifestations of pulmonary oxygen toxicity with extension and joining-up of the lesions. Radiological manifestations are due to fluid accumulation (oedema), atelectasis, and accumulation of cellular debris in alveolar spaces and in the terminal airways. There is laboured, gasping breathing often accompanied by frothy, bloody sputum. The damaged lungs cannot absorb sufficient oxygen for the body, resulting in cyanosis. From Huber, G. L. and Drath, D. B., Chapter 14 in *Oxygen and living processes, an interdisciplinary approach* (Gilbert, D. L., Ed.), Springer-Verlag, New York, 1981, with permission.

Other tissues do not escape damage when animals are exposed to high oxygen concentrations, however. The form of blindness known as retrolental fibroplasia (from the Latin for 'formation of fibrous tissue behind the lens') arose abruptly in the early 1940s among infants born prematurely, and quickly became widespread. Not until 1954 was it realized that this disease is associated with the use of high oxygen concentrations in incubators for premature babies, and more careful control of oxygen use has greatly decreased its incidence. Elevated O_2 appears to inhibit the growth of retinal blood vessels. On return to a normal atmosphere there is an excessive regrowth of the vessels, which sometimes occurs to an extent that causes detachment of the retina and subsequent blindness. The new vessels lack structural integrity and often bleed.

Table 1.1 lists some typical effects of oxygen on other animal tissues. High oxygen concentration also cause a general 'stress reaction' in animals, which stimulates the action of some endocrine glands. Removal of, for example, the thyroid gland decreases the toxic effects of O_2 in some animals whereas administration of thyroxine, cortisone, or adrenalin often makes them worse. Exposure of pregnant animals to elevated oxygen concentrations has been reported to increase the incidence of fetal abnormalities.

The damaging effects of oxygen on aerobic organisms vary considerably with the type of organism used, its age, physiological state, and diet. Different tissues of an animal are affected in different ways. For example, the effective oxygen concentration in the swim-bladder of the rat-tail fish at a depth of 3000 m is 2500 times greater than ambient, yet the bladder remains undamaged. The fish as a whole cannot tolerate anywhere approaching this oxygen concentration, and so its swim-bladder must be specially protected. Cold-blooded animals, such as turtles and crocodiles, are relatively resistant to oxygen toxicity at low environmental temperatures, but become more sensitive at higher temperatures. Young rats are more resistant to oxygen than are adult rats: adult humans are less sensitive than are adult rats. Oxygen toxicity is also influenced by the presence in the diet of varying amounts of vitamins A, E, and C, heavy metals, anti-oxidants (now added to many human foods), and polyunsaturated fatty acids. For example, rats fed on a fat-free test diet supplemented with cod-liver oil could tolerate pure O_2 much better than if the supplement consisted of coconut oil. In rats, an elevated blood glucose concentration has been reported to delay the onset of convulsions caused by hyperbaric oxygen.

1.4   What causes the toxic effects of oxygen?

Perhaps the earliest suggestion made to explain oxygen toxicity was that oxygen inhibits cellular enzymes. Indeed, direct inhibition by oxygen ac-

Table 1.1. Some typical effects on animal tissues of exposure to high oxygen concentrations

Species used	Nature of exposure	Organ examined	Results found
Adult, male rats	Pure O_2 at 5 atm for 75 min	Heart	Mitochondrial swelling followed by damage to myofibrils
Cats	Pure O_2 at 8 atm for 50 min	Kidney	Swelling of tubules, glomerular abnormalities
Rats	Pure O_2 at 0.33 atm for 3 days	Liver	Mitochondrial damage
Monkeys	Pure O_2 at 0.5 atm for up to 22 days	Liver	Proliferation and abnormality of smooth endoplasmic reticulum, decrease in glycogen content
Male hamsters	70% O_2 for 3–4 weeks	Testes	Degeneration of semeniferous epithelium, cessation of sperm production
Humans	'Hyperbaric oxygen therapy'	Ear	Haemorrhages of inner ear, deafness
Guinea pigs	70% O_2 at 1 atm for 6–36 days	Bone-marrow	Inhibition of erythroid cell development

counts for the loss of nitrogenase activity in oxygen-exposed *C. pasteurianum* (see above) and in a few other cases. Figure 1.2 shows that the inhibition of growth observed on exposing *E. coli* to high-pressure oxygen can be relieved by adding the amino acid valine to the culture medium, apparently because its synthesis is impaired due to a rapid inhibition of the enzyme dihydroxy-acid dehydratase in the metabolic pathway leading to valine. Even when valine is supplied, however, growth soon ceases because of a slower inhibition of other cellular enzymes, and supplementation of the culture medium with niacin and thiamin can then permit further growth. The onset of oxygen-induced convulsions in animals is correlated with a decrease in the cerebral content of the neurotransmitter GABA (γ-aminobutyric acid), perhaps because of an inhibition of the enzyme glutamate decarboxylase (glutamate $\rightarrow$ GABA + CO_2) by oxygen. In neither of these cases, however, has it been shown that the enzyme inhibition *in vivo* is due to oxygen itself rather than, say, to an increased production of oxygen radicals (see below).

Perhaps the best example of the direct effect of oxygen on aerobes comes from green plants. During photosynthesis, illuminated green plants fix carbon dioxide (CO_2) into sugars by a complex metabolic pathway known as the Calvin cycle. The first enzyme in this pathway, ribulose bisphosphate carboxylase, combines carbon dioxide with a five-carbon sugar (ribulose 1,5-bisphosphate) to produce two molecules of phosphoglyceric acid (Chapter 5). Oxygen is an inhibitor of this reaction competitive with carbon dioxide, and so at elevated oxygen concentrations there is less carbon dioxide fixation and less plant growth. This simple mechanism accounts for part, but not all, of the decreased leaf-growth at elevated oxygen concentrations (Fig. 1.1).

In general, however, the rates of enzyme inactivation by oxygen in aerobic cells are too slow and too limited in extent to account for the rate at which toxic effects develop and many enzymes are totally unaffected by O_2 at all. This led Rebecca Gershman and Daniel L. Gilbert, in the USA, to propose, in 1954, that most of the damaging effects of oxygen could be attributed to the formation of free oxygen radicals. Let us now consider exactly what 'free radicals' are. (In order to understand the discussion in the next section, it is essential to appreciate clearly what is meant by such chemical terms as 'covalent bond', 'Pauli principle', 'atomic orbital', 'antibonding molecular orbital', 'spin quantum number', 'Hund's rule', and 'transition metal'. Readers requiring explanation of such terms are advised to consult Appendix I before reading further in this chapter.)

1.5 What is a free radical?

The term 'radical' is often used loosely in chemistry to refer to various groups of atoms that behave as a unit, such as the carbonate radical (CO_3^{2-}), nitrate radical (NO_3^-), and the methyl radical (CH_3-). We shall avoid this use and define a 'free radical' as follows: *a free radical is any species capable of independent existence that contains one or more unpaired electrons.* (An unpaired electron is one that occupies an atomic or molecular orbital by itself.)

The presence of one or more unpaired electrons causes the species to be attracted slightly to a magnetic field (i.e. to be *paramagnetic*), and sometimes makes the species highly reactive. Consideration of the above broad definition shows that there are many free radicals in chemistry and biology (e.g. the hydrogen atom; see Appendix I). Radicals can be formed by the loss of a single electron from a non-radical, or by the gain of a single electron by a non-radical. They can easily be formed when a covalent bond is broken if one electron from each of the pair shared remains with each atom, a process known as *homolytic fission*. The energy required to dissociate the covalent bond can be provided by heat, electromagnetic radiation, or other means as will be discussed further in subsequent chapters. Many covalent bonds only dissociate at high temperatures, e.g. 450–600 °C is often required to rupture C–C, C–H, or C–O bonds. Many studies of radical reactions have been carried out in the gas phase at high temperatures; combustion is well known to chemists as a free-radical process.

If A and B are two atoms covalently bonded ($\overset{x}{\underset{x}{}}$ representing the electron pair), homolytic fission can be written as:

$$A_x^x B \rightarrow \dot{A}^x + B_x.$$

A^x is an A-radical, often written as A˙, and B_x is a B-radical (B˙). Homolytic fission of one covalent bond in the water molecule will yield a hydrogen radical (H˙) and a hydroxyl radical (usually written as OH˙ but sometimes as ˙OH, the latter emphasizing the location of the unpaired electron on oxygen). The opposite of homolysis (homolytic fission) is *heterolytic fission* in which one atom receives both electrons when a covalent bond breaks, i.e.

$$A_x^x B \rightarrow A_x^{x-} + B^+.$$

The extra electron gives A a negative charge and B is left with a positive charge. Heterolytic fission of water gives the hydrogen ion H^+ and the hydroxide ion OH^-. In fact, pure water is very slightly ionized in this way and contains 10^{-7} moles per litre each of H^+ and OH^- ions at 25°C. (A mole of a substance is x grams of it, where x is its relative atomic or molecular mass.)

Let us now look at some atoms and molecules of biological importance to see how far they fit into our definition of radicals.

1.5.1 *Oxygen and its derivatives*

Inspection of Fig. 1.6 shows that the oxygen molecule, as it occurs naturally, certainly qualifies as a radical: it has two unpaired electrons each located in a different π^* antibonding orbital. These two electrons have the same spin quantum number (or, as is often written, they have *parallel spins*). This is the most stable state, or *ground state*, of oxygen. Oxygen is a good oxidizing agent, the basic definitions being:

> *Oxidation*: loss of electrons by an atom or molecule (e.g. the conversion of a sodium atom to the ion Na^+).
> *Reduction*: gain of electrons by an atom or molecule (e.g. the conversion of a chlorine atom to the ion Cl^-).

An oxidizing agent, therefore, is good at absorbing electrons from the molecule it oxidizes (as is chlorine) whereas a reducing agent (such as sodium) is an electron donor. (These definitions are simplified, but sufficient for our purpose.) If oxygen attempts to oxidize another atom or molecule by accepting a pair of electrons from it, both of these electrons must be of antiparallel spin so as to fit in to the vacant spaces in the π^* orbitals (Fig. 1.6). A pair of electrons in an atomic or molecular orbital would not meet this criterion, however, since they would have opposite spins in accordance with Pauli's principle. This imposes a restriction on electron transfer which tends to make oxygen accept its electrons one at a time, and contributes the fact that oxygen reacts sluggishly with many non-radicals. Theoretically, the complex organic compounds of the human body should immediately com-

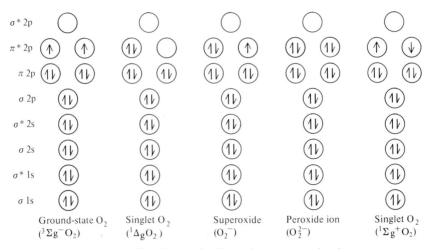

Fig. 1.6. Bonding in the diatomic oxygen molecule.

bust in the oxygen of the air (as occult magazines, such as *The Unexplained*, occasionally claim that people have done) but the spin restriction and other factors slow this down, fortunately!

More reactive forms of oxygen, known as *singlet oxygens*, can be generated by an input of energy (see Chapter 2 for details of how this is done). The $^1\Delta g O_2$ state (Fig. 1.6) has an energy 22.4 kcal above the ground state. The $^1\Sigma g^+ O_2$ state is even more reactive, 37.5 kcal above the ground state. By our definition, $^1\Delta g O_2$ is not a radical; there are no unpaired electrons. In both forms of singlet oxygen the spin restriction is removed and so the oxidizing ability is greatly increased.

If a single electron is added to the ground-state O_2 molecule, it must enter one of the π^* antibonding orbitals (Fig. 1.6). The product is the *superoxide radical* O_2^-. With only one unpaired electron, superoxide is actually less of a radical than is O_2 itself, despite its name and so the authors write it as O_2^- rather than $O_2^{\cdot-}$ (since we do not write oxygen as $O_2^{\cdot}$). The properties of O_2^- are considered in the next chapter. Addition of one more electron will give O_2^{2-}, the *peroxide ion* which, as may be seen from Fig. 1.6, is not a radical. Since the extra electrons in O_2^- and O_2^{2-} are entering antibonding orbitals, the strength of the oxygen–oxygen bond is decreasing (see Appendix I for an explanation of this). In ground-state O_2 the atoms are effectively bonded by two covalent bonds, but in O_2^- only by one-and-a-half (there is an extra electron in an antibonding orbital), and in O_2^{2-} by one bond only. Hence the oxygen–oxygen bond in O_2^{2-} is quite weak. Addition of another two electrons to O_2^{2-} would eliminate the bond entirely since they would go into the σ^*2p orbitals, so giving $2O^{2-}$ species. Usually in biological systems the two-electron reduction product of oxygen is hydrogen peroxide (H_2O_2), and the four-electron product, water. To summarize:

$$O_2 \xrightarrow[\text{reduction}]{\text{one-electron}} O_2^-$$

$$O_2 \xrightarrow[\substack{\text{reduction}\\ \text{(plus 2H}^+)}]{\text{two-electron}} H_2O_2 \text{ (protonated form of } O_2^{2-})$$

$$O_2 \xrightarrow[\substack{\text{reduction}\\ \text{(plus 4H}^+)}]{\text{four-electron}} 2H_2O \text{ (protonated form of } O^{2-})$$

Hydrogen peroxide is a pale-blue covalent viscous liquid, boiling point 150°C. It mixes readily with water and acts as an oxidizing agent, which gives it some antibacterial properties. In fact, the weak antiseptic activity of honey, which has been used in wound treatment since ancient times, is probably due to the formation of hydrogen peroxide by enzymes contained within it.

Since the O–O bond is relatively weak (see above), hydrogen peroxide decomposes easily: homolytic fission would give the hydroxyl radical

$$H_2O_2 \xrightarrow{energy} 2OH^{\cdot}$$

The structure of hydrogen peroxide is shown in Fig. 1.7. Although a covalent molecule, it does ionize at highly alkaline pH values (way above the biological range, however)

$$H_2O_2 \rightarrow 2H^+ + O_2^{2-}$$

1.5.2   *Ozone and oxides of nitrogen*

Ozone (O_3), a pale-blue gas, is an important shield for solar radiation in the higher reaches of the atmosphere. It is produced by the photodissociation (i.e. splitting-up caused by light energy) of molecular O_2 into oxygen atoms, which then react with oxygen molecules

$$O_2 \xrightarrow{solar\ energy} 2O$$

$$O_2 + O \rightarrow O_3$$

The two oxygen–oxygen bonds in the ozone molecule are of equal length and intermediate in nature between those of an oxygen–oxygen single bond and a double bond. Ozone has an unpleasant smell and severely damages the lungs. It is a much more powerful oxidizing agent than ground-state oxygen, i.e. it is only helpful to us in its proper place. Significant amounts of ozone can form in the lower atmosphere in urban air as a result of a series of complex photochemical events resulting from pollution. The ozone molecule is not a radical and is diamagnetic (weakly repelled by a magnetic field), although the

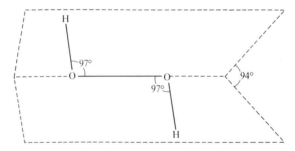

Fig. 1.7.   Structure of hydrogen peroxide.

damage that it causes is often mediated by free-radical production (Chapter 4).

Recently there has been great concern over the use of fluorinated hydro-carbons in, for example, aerosol sprays because they may help to deplete the ozone layer in the upper atmosphere. Photodissociation of gases such as CF_2Cl_2 and $CFCl_3$ produces chlorine atoms in the atmosphere, which cause a breakdown of ozone. Two of the oxides of nitrogen, nitric oxide (NO) and nitrogen dioxide (NO_2), can also deplete ozone. Nitric oxide is a colourless gas, a weak reducing agent (i.e. it tends to donate electrons to other molecules), and it reacts with oxygen to give nitrogen dioxide. Nitrogen dioxide is a dense, brown poisonous gas and a powerful oxidizing agent. Sources of nitrogen oxides in the higher atmosphere include solar flares and supersonic aircraft exhausts. They are also both found in cigarette smoke and since both NO and NO_2 molecules possess odd numbers of electrons, they fall into our definition of free radicals. Further consideration of their noxious effects can be found in Chapter 6.

1.5.3 Transition metals

All the metals in the first row of the d-block in the Periodic Table contain unpaired electrons and can thus qualify as radicals, with the sole exception of zinc (see Appendix I for further explanation if necessary). Copper does not really fit the definition of a transition element since its 3d-orbitals are full, but it readily forms the Cu^{2+} ion by loss of two electrons, one from the 4s- and one from a 3d-orbital. This leaves an unpaired electron. Many transition elements are of great biological importance (Table 1.2) and so it is worth-while examining their properties.

The transition elements are all metals. Their most important feature from a radical point of view is their variable valency, which allows them to undergo changes in oxidation state involving one electron. For example, iron has two common valencies in which the electronic configurations are as follows (see Appendix I for an explanation of this notation, if necessary):

Table 1.2. Biological importance of some d-block elements

Metal	Biochemical significance
Copper (Cu)	Essential in human diet. Required for enzymes such as superoxide dismutase, cytochrome oxidase, lysine oxidase, dopamine-β-hydroxylase, and caeruloplasmin. About 80 mg Cu in adult human body. Highest concentrations in liver and brain. Overall blood content in males 0.106 mg/100 ml. Toxic in excess
Zinc (Zn)	Non-transition element, fixed valency of 2. Suggested that it sometimes inhibits iron-dependent radical reactions by displacing iron from its binding site. Essential in human diet; found in RNA polymerase, carbonic anhydrase, superoxide dismutase. Plasma zinc approx. 0.112 mg/100 ml. Toxic in excess
Vanadium (V)	Essential in animals but requirement in man not yet established. Accumulated in large amounts in some tunicates. Involved in cholesterol metabolism. Inhibits strongly the ATPase enzyme which exchanges Na^+ and K^+ ions across cell membranes: this might be a physiological regulatory mechanism
Chromium (Cr)	Probably essential in diet, involved in regulation of glucose metabolism. Normal serum Cr is 1–5 ng/ml (1 nanogram (ng) = 10^{-9} g)
Manganese (Mn)	Essential in animals, probably in man as well (normal blood level 9 μg/ml). Needed for mitochondrial superoxide dismutases, also activates a number of hydrolase and carboxylase enzymes. Free and total Mn contents in liver cells of fed rats estimated as about 0.71 and 34 nmol ml^{-1} of cell water, respectively. (1 nanomole = 10^{-9} moles)
Iron (Fe)	Essential in human diet: deficiency causes simple anaemia. Most abundant transition metal in humans. Normal serum iron in males approx. 0.127 mg/100 ml mostly bound to the protein transferrin. Regulation of iron content of the body is done by regulation of iron uptake in the gut. Needed for haemoglobin, myoglobin, cytochromes, several enzymes, and non-haem-iron proteins (see Chapter 2)
Cobalt (Co)	Essential as a component of vitamin B_{12} but little else known.
Nickel (Ni)	Probably essential in animals, requirement in man not yet established. Found in urease in plant cells and in several bacterial enzymes, such as hydrogenases and carbon monoxide dehydrogenase
Molybdenum (Mo)	Essential in trace amounts for some flavin metalloenzymes, e.g. xanthine dehydrogenase, nitrogenase, sulphite oxidase, nitrate reductase

The existence of a ferryl (iron IV) species has also been suggested. In solution in the presence of air, the iron(III) state is the most stable, whereas iron(II) salts are weakly reducing and ferryl compounds are powerful oxidizing agents. If a solution of an iron(II) salt, e.g. 'ferrous sulphate' ($FeSO_4$), is left exposed to the air it slowly oxidizes to the iron(III) state. This is a one-electron oxidation, and oxygen dissolved in the solution is reduced to the superoxide radical O_2^- (Fig. 1.6).

$$Fe^{2+} + O_2 \rightleftharpoons Fe^{2+} - O_2 \leftrightarrow Fe^{3+} - O_2^- \rightleftharpoons Fe^{3+} + O_2^-$$

intermediate complexes

Copper has two common valencies, copper(I) and copper(II), formerly known as 'cuprous' and 'cupric':

Copper atom   Ar   3d [↑↓ ↑↓ ↑↓ ↑↓ ↑↓]   4s [↑]

Copper (I) ion (not a radical)   Ar   [↑↓ ↑↓ ↑↓ ↑↓ ↑↓]   [ ]

Copper (II) ion   Ar   [↑↓ ↑↓ ↑↓ ↑↓ ↑]   [ ]

Again, the one-electron difference between these valency states allows copper to take part in radical reactions. Under appropriate conditions, for example, copper salts can both accept electrons from, and donate electrons to, the superoxide radical O_2^-,

$$Cu^{2+} + O_2^- \rightarrow Cu^+ + O_2$$

$$Cu^+ + O_2^- \rightarrow Cu^{2+} + O_2^{2-}$$

$$O_2^{2-} + 2H^+ \rightarrow H_2O_2$$

Net reaction:

$$O_2^- + O_2^- + 2H^+ \rightarrow H_2O_2 + O_2$$

The copper salt, by changing its valency, is causing the net combination of two O_2^- radicals and two H^+ ions to form H_2O_2 and O_2. The copper salt is acting as a *catalyst*: it remains unchanged in amount and chemical nature at the end of the reaction whilst speeding it up enormously. The variable valency of transition metals helps them to be effective catalysts of many reactions involving oxidation and reduction and they are used for this purpose at the active sites of many enzymes catalysing such reactions. The radical reactions they promote can overcome the spin restriction on direct reaction of oxygen with non-radical species.

Manganese has a most stable valency state in aqueous solution of Mn^{2+}; more oxidized species such as Mn(III), Mn(IV) and Mn(VII) also exist. Again radical reactions are possible, e.g. Mn^{2+} and O_2^-:

$$Mn^{2+} + O_2^- + 2H^+ \rightarrow Mn^{3+} + H_2O_2$$

Zinc, by contrast, has only one valency (Zn^{2+}) and does not promote radical reactions. It has been suggested that zinc may inhibit some radical reactions *in vivo* by displacing other transition metal ions such as iron from the binding sites at which they are promoting such reactions.

Fenton reaction

A mixture of hydrogen peroxide and an iron(II) salt reacts with many organic molecules, as was first observed by Fenton in 1894. The reactivity is most likely due to formation of the hydroxyl radical:

$$Fe^{2+} + H_2O_2 \rightarrow \text{intermediate complex} \rightarrow Fe^{3+} + OH^\cdot + OH^-$$

Traces of Fe^{3+} might be able to react further with H_2O_2, although this is very slow at physiological pH:

$$Fe^{3+} + H_2O_2 \rightarrow \text{intermediate complex} \rightarrow Fe^{2+} + O_2^- + 2H^+$$

Even more reactions are possible:

$$OH^\cdot + H_2O_2 \rightarrow H_2O + H^+ + O_2^-$$
$$O_2^- + Fe^{3+} \rightarrow Fe^{2+} + O_2$$
$$OH^\cdot + Fe^{2+} \rightarrow Fe^{3+} + HO^-$$

Thus this simple mixture of an iron salt and hydrogen peroxide, which can almost certainly form in biological systems under certain circumstances (Chapter 2), can provoke a whole series of radical reactions. The overall sum of these, unless some other reagent is added, is an iron-catalysed decomposition of hydrogen peroxide.

$$2H_2O_2 \xrightarrow[\text{catalyst}]{\text{Fe-salt}} O_2 + 2H_2O.$$

Other reagents present can react with the various radicals and change the mechanism of the reaction.

Copper(I) salts are thought to react with H_2O_2 to make hydroxyl radicals:

$$Cu^+ + H_2O_2 \rightarrow Cu^{2+} + OH^{\cdot} + OH^-$$

Another possibility is that a Cu(III) species, a powerful oxidizing agent, can be generated by reaction of Cu^+ with H_2O_2.

1.5.4 *Other free radicals*

Unpaired electrons can be associated with a wide range of other atoms. For example, thiol (R–SH) compounds oxidize in the presence of transition metal ions to form, among other products, *thiyl radicals*, $RS^{\cdot}$.

$$RSH + Cu^{2+} \rightarrow RS^{\cdot} + Cu^+ + H^+$$

Thiyl radicals are *sulphur-centred* radicals that have considerable reactivity. They can combine with O_2:

$$RS^{\cdot} + O_2 \rightarrow RSO_2^{\cdot}$$

oxidize NADH to another radical, $NAD^{\cdot}$

$$RS^{\cdot} + NADH \rightarrow RS^- + NAD^{\cdot} + H^+$$

and oxidize ascorbic acid. Indeed, thiols oxidizing in the presence of iron or copper ions generate a wide range of radicals, including $OH^{\cdot}$, $O_2^{\cdot-}$ and $RS^{\cdot}$. The toxic actions of sporidesmin and diphenyl disulphide may involve thiyl radicals (Chapter 6).

Thiyl radicals can also be formed by the homolytic fission of disulphide bonds in proteins:

$$cys\text{-}S\text{–}S\text{-}cys \rightarrow cys\text{-}S^{\cdot} + {}^{\cdot}S\text{-}cys$$

Human finger-nails are composed largely of α-keratin, a protein rich in disulphide bonds. Chandra and Symons in England have found that ESR signals (for an explanation of this technique see Chapter 2, Section 2.5.1) characteristic of sulphur-centred radicals are produced by repeated cutting of finger-nails. Indeed, grinding of proteins, especially at low temperatures, is a well-established way of generating free radicals.

Carbon-centred radicals can be formed in many biological systems, e.g. during the metabolism of carbon tetrachloride (CCl_4) by liver microsomes (Chapter 6) the trichloromethyl radical is formed:

$$CCl_4 \xrightarrow[\text{P-450 system}]{\text{cytochrome}} {}^{\cdot}CCl_3 + Cl^-$$

trichloromethyl
radical

Carbon-centred radicals often react rapidly with O_2 to give *peroxyl radicals*, e.g. for $^{\cdot}CCl_3$

$$^{\cdot}CCl_3 + O_2 \rightarrow {}^{\cdot}O_2CCl_3$$

trichloromethylperoxyl radical

Nitrogen-centred radicals also exist, e.g. the phenyldiazine radical ($C_6H_5N = N^{\cdot}$) is formed during oxidation of phenylhydrazine by erythrocytes (Chapter 6). *Phosphorus-centred* radicals have been described in the chemical literature.

In the next chapters, we shall examine the biological significance of some inorganic radicals in detail.

1.6   Further reading

Asmus, K. D. (1983). Sulphur-centered free radicals. In *Radioprotectors and anticarcinogens*, p. 23. Academic Press, New York.

Balentine, J. D. (1978). Experimental pathology of oxygen toxicity. In *Oxygen and physiological function* (F. F. Jobsis, Ed.). Professional Information Library, Dallas, Texas.

Balentine, J. D. (1982). *Pathology of oxygen toxicity*. Academic Press, New York.

Chandra, H. and Symons, M. C. R. (1987). Sulphur radicals formed by cutting α-keratin. *Nature* **328**, 833.

Clark, J. M. (1988). Pulmonary limits of oxygen tolerance in Man. *Exp. Lung Res.* **14**, 897.

Deneke, S. M. and Fanburg, B. L. (1980). Normobaric oxygen toxicity of the lung. *New Engl. J. Med.* **303**, 76.

Dulka, J. J. and Risby, T. H. (1976). Ultratrace metals in some environmental and biological systems. *Anal. Chem.* **48**, 640A.

Gallon, J. R. (1981). The oxygen-sensitivity of nitrogenase: a problem for biochemists and micro-organisms. *Trends Biochem. Sci.* January 1981, 19.

Gilbert, D. L. (Ed.) (1981). *Oxygen and living processes: an inter-disciplinary approach*. Springer, New York.

Gutteridge, J. M. C., Westermarck, T., and Halliwell, B. (1985). Oxygen radical damage in biological systems. In *Free radicals, aging and degenerative diseases*, p. 99. (J. E. Johnson Jr. *et al.*, Eds). A. R. Liss Inc, New York.

Halliwell, B. (1981). Free radicals, oxygen toxicity and ageing. In *Age pigments*, p. 1. (R. S. Sohal, Ed.). Elsevier/North-Holland Biomedical Press, Amsterdam.

Haugaard, N. (1968). Cellular mechanisms of oxygen toxicity. *Physiol. Rev.* **48,** 229.

Jones, R. R. (1987). Ozone depletion and cancer risk. *Lancet*, **ii,** 443.

Krieg, N. R. and Hoffman, P. S. (1986). Microaerophily and oxygen toxicity. *Ann. Rev. Microbiol.* **40,** 107.

Mönig, J. *et al.* (1987). On the reaction of molecular oxygen with thiyl radicals: a re-examination. *Intl. J. Radiat. Biol.* **52,** 589.

Morris, J. G. (1976). Oxygen and the obligate anaerobe. *J. Appl. Bacteriol.* **40,** 229.

Morris, J. G. (1979). Oxygen and growth of the oral bacteria. In *Saliva and dental caries* (I. Kleinberg *et al.*, Eds). Information Retrieval Ltd., New York.

Prasad, A. S. (Ed.) (1978). *Trace elements and iron in human metabolism.* Plenum, USA.

Rajagopalan, K. V. (1987). Molybdenum—an essential trace element. *Nutr. Rev.* **45,** 321.

Saez, G. *et al.* (1982). The production of free radicals during the autoxidation of cysteine and their effect on isolated rat hepatocytes. *Biochim. Biophys. Acta* **719,** 24.

Shioi, J., Dang, C. V., and Taylor, B. L. (1987). Oxygen as attractant and repellent in bacterial chemotaxis. *J. Bacteriol.* **169,** 3118.

Simons, T. J. B. (1979). Vanadate. A new tool for biologists. *Nature* **281,** 337.

Walling, C. (1975). Fenton's reagent revisited. *J. Am. Chem. Soc.* **8,** 125.

Walsh, C. T. and Orme-Johnson, W. H. (1987). Nickel enzymes. *Biochemistry* **26,** 4901.

Willson, R. L. (1978). Iron, zinc, free radicals and oxygen in tissue disorders and cancer control. In *Iron metabolism*, p. 331. CIBA Foundation Symposium. Elsevier, Amsterdam.

Wood, P. M. (1988). The potential diagram for oxygen at pH 7. *Biochem. J.* **253,** 287.

2

The chemistry of oxygen radicals and other oxygen-derived species

Before we can really understand what oxygen radicals and related species can and cannot do in biological systems, we must first look at their chemical properties.

2.1 Reaction rates and rate constants

In describing the properties and reactions of radicals, frequent reference will be made to the rates at which reactions proceed, so it is worthwhile beginning by establishing clearly how rates are usually expressed. The rate of a reaction can be measured either by following the loss of the starting materials (*reactants*), or by following the formation of the products. Reaction rate is then simply defined as the amount of product formed in unit time, or as the amount of reactant used up in unit time. Time is usually quoted in seconds and the amounts in *moles*, one mole of a substance being its molecular weight (relative molecular mass is a more precise term) expressed in grams. One mole of any substance contains the same number of molecules, that number being *Avogadro's number*, numerically equal to the enormous figure of 6.023×10^{23}.

The rate of a reaction will obviously depend on the concentration of reactants present. To take a simple case, suppose one mole of a substance A in solution in a volume of one litre is reacting to form another substance B:

$$A \to B$$

Suppose further that after 1 second, 0.01 moles of A has been converted into B. The reaction rate (R) can then be expressed either as 0.01 moles of B formed in 1 litre in 1 second ($R = 0.01$ mol l^{-1} s^{-1} for B), or as 0.01 moles of A used up in 1 litre in 1 second ($R = 0.01$ mol l^{-1} s^{-1} for A).

If the concentration of A is doubled, it is likely that the rate of reaction will double. The exact mathematical relationship between the rate of a reaction and the concentration of the reactants is known as the *rate law*. In this case R is proportional to the concentration of A, expressed as moles of A per litre of solution (mol l^{-1}). This is mathematically equivalent to saying that R is equal to the concentration of A multiplied by a constant, the *rate constant* for the reaction, i.e. the rate law is:

$$R = k_1[A]$$

where k_1 is the rate constant at the temperature of the experiment, and [A] means the concentration of A in moles per litre. Once the reaction has started, A is used up, [A] falls and so R will fall. Hence rate measurements are always made in the first few seconds of a reaction so that the concentration of reactants has not changed significantly from that originally present (so-called *initial rate measurements*). Rate constants, and hence rates of reactions, increase as temperature is raised.

In the rate law:

$$R = k_1[A]$$

the rate of the reaction depends only on the first power of the concentration of A; another way of saying this is that the reaction is *first order with respect to* A. If the rate law had been found by experiment to be:

$$R = k_2[A]^2$$

(i.e. the rate is proportional to the square of the concentration of A), the reaction would be *second order with respect to* A, and if:

$$R = k_3[A]^3$$

then it would be *third order with respect to* A. k_1 would be called a *first-order rate constant* with units of s^{-1}, k_2, a *second-order rate constant* with units of 1 s^{-1} mol^{-1} (sometimes written as M^{-1} s^{-1} where M (molar) is another way of writing moles per litre), and k_3 is a *third-order rate constant*. It is these rate constants that are usually published in the scientific literature.

Now consider another simple reaction in which there are two reactants, e.g.

$$A + B \rightarrow products$$

This type of equation often represents the reaction of a radical A with some other material B and it usually follows the rate law:

$$R = k_2[A][B]$$

where k_2 is a second-order rate constant. The reaction is first order with respect to A, first order with respect to B and *second order overall*. The constant k_2 has the units $1\,s^{-1}\,mol^{-1}\,(M^{-1}\,s^{-1})$.

As an example of the information that can be gleaned from published rate constants, let us look at the formation of hydroxyl radicals from H_2O_2 in the presence of either Fe^{2+} or Cu^+ ions, as described in Chapter 1. The published approximate second-order rate constants are:

$$H_2O_2 + Fe^{2+} \rightarrow Fe^{3+} + OH^- + OH^{\cdot} \qquad k_2 = 76\,1\,mol^{-1}\,s^{-1}$$
$$H_2O_2 + Cu^+ \rightarrow Cu^{2+} + OH^- + OH^{\cdot} \qquad k_2 = 4{\cdot}7 \times 10^3\,1\,mol^{-1}\,s^{-1}$$

If equal concentrations of H_2O_2 are mixed with equal concentrations of Fe^{2+} or Cu^+, the initial rate of hydroxyl radical formation in the latter case will be greater by a factor of $4{\cdot}7 \times 10^3/76$, i.e. a factor of 61·8. Values of the rate constant can be applied to see how fast a reaction might occur under biological conditions. As discussed later in this chapter, the concentrations of hydrogen peroxide and Fe^{2+} in, say, liver cells are likely to be low, in the range of μmoles per litre ($10^{-6}\,mol\,l^{-1}$) or less. If $1\,\mu mol\,l^{-1}$ of H_2O_2 comes into contact with $1\,\mu mol\,l^{-1}$ of Fe^{2+}, how much $OH^{\cdot}$ radical will be formed? The rate law is:

$$R = k_2[H_2O_2]\,[Fe^{2+}]$$
$$= 76(10^{-6})\,(10^{-6}) = 7.6 \times 10^{-11}\,mol\,l^{-1}\,s^{-1}$$

This seems a tiny figure, but remember that 1 mole of a substance contains 6.023×10^{23} molecules. Hence the *number* of hydroxyl radicals formed per litre per second is 4.58×10^{13}—much more impressive! If the cell volume is 10^{-12}–10^{-11} litre (average volumes for a liver cell) this still means 46 to 458 hydroxyl radicals formed per cell every second. Of course, as the reaction proceeds both Fe(II) and H_2O_2 will be used up and the rate of $OH^{\cdot}$ production will fall unless they are continuously replenished. Thus, even reactions with low rate-constants (such as the Fenton reaction) can be biologically important.

2.2   Measurement of reaction rates for radical reactions

Many radical reactions proceed extremely quickly and so special techniques are required to measure their rates. Two techniques have commonly been used, *stopped flow* and *pulse radiolysis*, and they are considered briefly below.

2.2.1 *Pulse radiolysis*

In this technique, the compound to be studied is placed in a reaction cell in solution and a radical is formed directly in the cell by a 'pulse' of ionizing radiation, e.g. from a linear accelerator. By the appropriate choice of experimental conditions, specific radicals can be generated and their reactions followed over a microsecond (10^{-6} s) time-scale or even longer. Since many radicals absorb light at different wavelengths than their parent compound, the progress of the reaction is followed by changes in the light absorbance of the cell components, usually displayed on an oscilloscope. Figure 2.1(a) shows an outline of the apparatus. Radiation sources useful for pulse radiolysis generally provide pulses of electrons in the energy range 1–30 mega-electron-volts. Exposure of water to these produces ionization and excitation within 10^{-16} second:

$$2H_2O \rightarrow H_2O^+ + e^- + H_2O^*$$

where e^- represents an electron and H_2O^* an excited water molecule. Such excited molecules undergo homolytic fission in 10^{-14}–10^{-13} second to give hydrogen atoms (which can equally well be called hydrogen radicals since they contain one unpaired electron) and hydroxyl radicals:

$$H_2O^* \rightarrow H^\cdot + OH^\cdot$$

Within the same time-scale H_2O^+ also reacts to give $OH^\cdot$

$$H_2O^+ + H_2O \rightarrow H_3O^+ + OH^\cdot$$

The electrons become surrounded by clusters of water molecules within 10^{-12}–10^{-11} second. These hydrated electrons can be written as e^-_{aq} where 'aq' is an abbreviation for 'aqueous'. Hence three different radicals are produced on 'pulsing' an aqueous solution: $H^\cdot$, $OH^\cdot$, and e^-_{aq}. Alterations of pH, and addition of various compounds, can 'select out' a particular radical for further study. For example, if the aqueous solution is saturated with nitrous oxide (N_2O) gas before pulsing, e^-_{aq} are removed by the reactions:

$$e^-_{aq} + N_2O \rightarrow N_2 + O^-$$
$$O^- + H^+ \rightarrow OH^\cdot$$

and converted into hydroxyl radicals. By contrast, if the solution is saturated with oxygen gas and also contains 0.1 mole per litre of sodium formate (an ionic solid containing sodium ions and formate ions, $HCOO^-$) the following reactions occur to produce the superoxide radical:

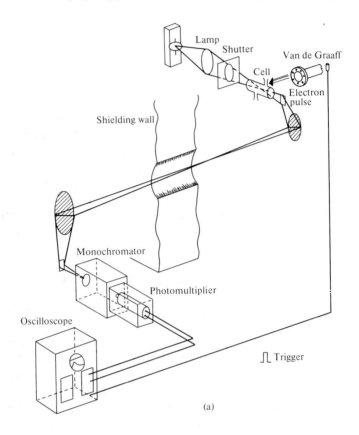

(a)

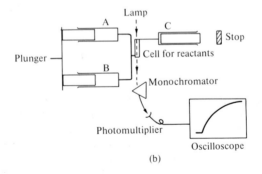

(b)

$$e^-_{aq} + O_2 \rightarrow O_2^-$$
$$H^\cdot + HCOO^- \rightarrow H_2 + CO_2^{\cdot-}$$
$$OH^\cdot + HCOO^- \rightarrow H_2O + CO_2^{\cdot-}$$
$$CO_2^{\cdot-} + O_2 \rightarrow CO_2 + O_2^-$$

Thus relatively 'clean' sources of $OH^\cdot$ or O_2^- can be produced and studies of their reactions with various compounds can be made. This technique has been especially useful in investigating reactions of $OH^\cdot$ and O_2^- with biological molecules. Usually the reaction is observed directly by following the rise in light absorbance of a reaction product, or the loss of absorbance of a reactant. If this is not possible, a 'competition method' may be used. For example, $OH^\cdot$ reacts with thiocyanate ion (CNS^-) to give a strongly absorbing product and the rate constant for this reaction is known. If another compound (X) that reacts with $OH^\cdot$ is added, then it will intercept some of the $OH^\cdot$, and the absorbance change due to the CNS^- reaction will be smaller. Knowing the concentrations of CNS^- and X, and the above rate constant, the rate constant for the reaction between X and $OH^\cdot$ can easily be calculated. The dye ABTS is also well suited for such 'pulse competition' studies.

The pulse radiolysis method can be used to study other radicals as well. For example, removal of one electron from ascorbic acid (vitamin C) produces an ascorbate radical. The absorption spectrum and properties of this radical can be observed by generating it in a pulse radiolysis apparatus, either by reducing dehydroascorbic acid (ascorbate that has lost two electrons) with e^-_{aq} or by oxidizing ascorbate with the $OH^\cdot$ radical:

$$OH^\cdot + \text{vit C} \rightarrow \text{vit C}^{\cdot-} + OH^-$$

or with another oxidizing radical such as $Br_2^{\cdot-}$, formed by adding bromide (Br^-) ions.

$$OH^\cdot + 2Br^- \rightarrow OH^- + Br_2^{\cdot-}$$

Fig. 2.1. Techniques for measuring fast reactions. Light is shone through a reaction cell and analysed for changes in light absorbance by the monochromator and photomultiplier. The results are displayed on an oscilloscope. In pulse radiolysis (part (a)) reaction is started by generating a radical in the reaction cell by means of a burst of ionizing radiation, e.g. from a Van de Graaff accelerator. In stopped flow (part (b)) the reactants are contained in separate syringes (A and B) and they enter the reaction vessel when the plunger is pressed, eventually passing into the collecting syringe C. Figure (a) is taken from *Progress in reaction kinetics 11* (1982), pp. 73–135, with permission. Figure (b) by courtesy of Professor R. L. Willson.

2.2.2   *Stopped-flow methods*

Stopped-flow methods are often used where the rates of radical reactions are too slow to be measured conveniently by pulse radiolysis, yet too fast to be measured by standard biochemical techniques. Solutions of the compounds to be reacted are contained in syringes (A and B in Fig. 2.1(b)). To initiate the reaction the plungers are pushed so that the syringe contents are forced simultaneously into a quartz cell, where they mix, and then into a collecting syringe (C in Fig. 2.1(b)). Experimental conditions are chosen so that, while the solutions are flowing, any given volume of mixed solution does not stay in the cell long enough for significant reaction to occur. When the plunger of the collecting syringe strikes the 'stop', the flow halts abruptly and the reaction within the cell proceeds to completion. Its absorbance changes can be measured and recorded on an oscilloscope and so reaction rates can be calculated. For example, a solution of superoxide ion (as its potassium salt $K^+O_2^-$) in an organic solvent can be mixed with a compound in aqueous solution and the rate of reaction measured (see Section 2.9 for further details).

2.3   Reactions of the hydroxyl radical

Rate constants for OH˙ reactions have mainly been determined by pulse radiolysis methods. Inspection of Table 2.1 shows that this radical reacts with extremely high rate-constants with almost every type of molecule found in living cells: sugars, aminoacids, phospholipids (e.g. lecithin), nucleotides, and organic acids. Indeed, it is one of the most reactive chemical species known. For example it is pointless to try to demonstrate hydroxyl radical reactions *in vitro* in solutions containing Tris buffer, since OH˙ attacks this buffer rapidly and a Tris-derived radical is produced. Reactions of OH˙ can be classified into three main types: *hydrogen abstraction, addition,* and *electron transfer*. These reactions illustrate an important principle of radical chemistry: *reaction of a free radical with a non-radical species produces a different free radical, which may be more or less reactive than the original radical*. Radicals produced by reactions with OH˙ are usually less reactive, however, since OH˙ is such an aggressive species.

As an example of hydrogen abstraction, consider the reaction of OH˙ with alcohols. The OH˙ 'pulls off' a hydrogen atom (H) and combines with it to form water, leaving behind an unpaired electron on the carbon atom, e.g. for the alcohol ethanol:

$$H-\underset{\underset{H}{|}}{\overset{\overset{H}{|}}{C}}-\underset{\underset{H}{|}}{\overset{\overset{H}{|}}{C}}-O-H + OH^{\cdot} \rightarrow H-\underset{\underset{H}{|}}{\overset{\overset{H}{|}}{C}}-\underset{\underset{H}{|}}{\overset{}{C}}^{\cdot}-O-H + H_2O$$

hydroxyethyl radical

or for methanol:

$$CH_3OH + OH^{\cdot} \rightarrow {\cdot}CH_2OH + H_2O$$
hydroxymethyl radical

Further reactions of the carbon radical can then occur, e.g. reaction with oxygen to give peroxyl radicals (Chapter 1, Section 1.5.4):

$${\cdot}CH_2OH + O_2 \rightarrow {\cdot}O_2CH_2OH$$
peroxyl radical

or the joining-up of two radicals to form a non-radical product, the two unpaired electrons between them forming a covalent bond:

$$CH_3\overset{\cdot}{C}HOH + CH_3\overset{\cdot}{C}HOH \rightarrow CH_3CHOH$$
$$|$$
$$CH_3CHOH$$

The reaction of OH· with lecithin (also known as phosphatidylcholine), an important phospholipid found in biological membranes, is of the hydrogen-abstraction type, and the carbon radicals left behind undergo a series of reactions leading to membrane damage (this is discussed further in Chapter 4). Attack of OH· on a sugar such as deoxyribose, found in DNA, produces a huge variety of different products, some of which have been shown to be mutagenic in bacterial test systems.

The reaction of OH· with aromatic ring structures can proceed by addition, and similar reactions occur with the purine and pyrimidine bases present in DNA and RNA. For example OH· can add on across a double bond in the pyrimidine base thymine:

Table 2.1.   Typical second-order rate constants for reactions of the hydroxyl radical

Compound tested	pH	Rate constant ($M^{-1} s^{-1}$)	Compound tested	pH	Rate constant ($M^{-1} s^{-1}$)
Carbonate ion, CO_3^{2-}	10.7	2.0×10^8	Glycylglycine	2	7.8×10^7
Bicarbonate ion, HCO_3^-	6.5	1.0×10^7	Glycyltyrosine	2	5.6×10^9
Fe^{2+}	2.1	2.5×10^8	Guanine	—	1.0×10^{10}
H_2O_2	7	4.5×10^7	Haemoglobin	—	3.6×10^{10}
Adenine	7.4	3.0×10^9	Histidine	6–7	3.0×10^9
Adenosine	7.7	2.5×10^9	Hydroxyproline	2	2.1×10^8
AMP	5.4	1.8×10^9	Lactate ion	9	4.8×10^9
Arginine	7	2.1×10^9	Lecithin	—	5.0×10^8
Ascorbic acid	1	7.2×10^9	Mannitol	7	2.7×10^9
Benzene	7	3.2×10^9	Methanol	7	4.7×10^8
Benzoic acid	3	4.3×10^9	Methionine	7	5.1×10^9
Butan-1-ol (*n*-butanol)	7	2.2×10^9	Nicotinic acid	—	6.3×10^8
Catalase	—	2.6×10^{11}	Phenol	7	4.2×10^9
Citric acid	1	3.0×10^7	Phenylalanine	6	3.5×10^9
Cysteine	1	7.9×10^9	Propan-1-ol	7	1.5×10^9
Cystine	2	3.2×10^9	Pyridoxal phosphate	—	1.6×10^9
Cytidine	2	2.0×10^9	Ribonuclease	—	1.9×10^{10}
Cytosine	7	2.9×10^9	Ribose	7	1.2×10^9
Deoxyguanylic acid	7	4.1×10^9	Serum albumin	—	2.3×10^{10}
Deoxyribose	7.4	3.1×10^9	Thiourea	7	4.7×10^9
Ethanol	7	7.2×10^8	Thymine	7	3.1×10^9
Glucose	7	1.0×10^9	Tryptophan	6	8.5×10^9
Glutamic acid	2	7.9×10^7	Uracil	7	3.1×10^9
Glutathione	1	8.8×10^9	Urea	9	$< 7.0 \times 10^5$

Values are mostly taken from the compilation by Anbar, M. and Neta, P. (1967). *Intl. J. Appl. Radiat. and Isotopes*, **18**, 493–523. Value for mannitol is from Hoey *et al.* (1988). *Free Radical Res. Commun.* **4**, 259–63.

The thymine radical then undergoes a series of further reactions, including reaction with oxygen to give a fairly reactive thymine peroxyl radical. Thus hydroxyl radical severely damages the bases and sugars of DNA, and also induces strand breakage. If damage is so extensive that it cannot be repaired, the cell may die. Even survivable damage can result in mutation (Chapter 8, Section 8.8.1).

Hydroxyl radicals take part in electron-transfer reactions with inorganic and organic compounds, e.g. with the chloride ion:

$$Cl^- + OH^{\cdot} \rightarrow Cl^{\cdot} + OH^-$$
$$Cl^{\cdot} + Cl^- \rightarrow Cl_2^-$$

It is clear from Table 2.1 that the reactivity of $OH^{\cdot}$ radicals is so great that, if they are formed in living systems, they will react immediately with whatever biological molecule is in their vicinity, producing secondary radicals of variable reactivity. For example, their reaction with carbonate ion (CO_3^{2-}) produces carbonate radicals $(CO_3^{\cdot-})$, which are powerful oxidizing agents. Let us see how $OH^{\cdot}$ radicals could be produced.

2.4 Production of hydroxyl radicals in living systems

2.4.1 *Ionizing radiation*

Since the major constituent of living cells is water, exposure of them to ionizing radiation such as X-rays or γ-rays will result in hydroxyl radical production, as described previously. Hydroxyl radicals are responsible for a large part of the damage done to cellular DNA and to membranes by ionizing radiation. Single- and double-strand breaks in DNA are considered to be very important damaging events, especially as double-strand breaks cannot be repaired by the cell. Oxygen, normally present in most biological systems, aggravates the damage done by ionizing radiation. Reasons for this 'oxygen effect' are explored in Chapter 6.

Detailed studies by French, German, British, and American scientists have characterized many of the products formed by attack of $OH^{\cdot}$ upon DNA. For example, Fig. 2.2 shows the separation of modified bases from irradiated DNA. Calf thymus DNA was exposed to γ-rays in solutions saturated with N_2O. The $OH^{\cdot}$ formed attacked the DNA which was then hydrolysed by acid and the modified bases were converted into volatile products (by *trimethylsilylation*) to enable their separation by gas chromatography. The peaks were identified by mass spectrometry.

In mammalian cells, DNA strand breakage can lead to a drop in the intracellular concentration of the coenzyme NAD^+. This occurs because a

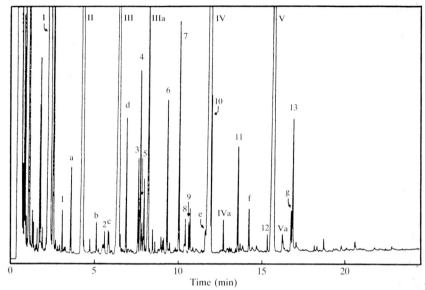

Fig. 2.2. Gas chromatogram of a trimethylsilylated acid-hydrolysate of DNA γ-irradiated in N₂O-saturated aqueous solution at a dose of 330 Gy. Column, fused silica capillary (12.0 m) coated with cross-linked 5 per cent phenylmethylsilicone phase. Temperature program, 100 to 250 °C at a rate of 7°C/min after 3 min at 100°C. *Peak identification*: I, phosphoric acid; 1, uracil; II, thymine; 2, 5,6-dihydrothymine; III, cytosine; d, 5-methylcytosine; 3, 5-hydroxy-5,6-dihydrothymine; 4, 5-hydroxyuracil; 5, 5-hydroxy-5,6,dihydrouracil; IIIa, cytosine; 6, 5-hydroxycytosine; 7 and 8, *cis-* and *trans-*thymine glycol; 9, 5,6-dihydroxyuracil; IV, adenine; 10, 4,6-diamino-5-formamidopyrimidine; IVa, adenine; 11, 8-hydroxyadenine; 12, 2,6-diamino-4-hydroxy-5-formamidopyrimidine; V, guanine; Va, guanine; 13, 8-hydroxyguanine. Some compounds appear twice because they derivatize in different ways. Diagram by courtesy of Dr M. Dizdaroglu.

chromatin-bound enzyme, poly(ADP-ribose) synthetase, splits the NAD⁺ molecule and transfers the ADP-ribose part of it on to nuclear proteins, as shown at the top of p. 33.

The formation of poly(ADP-ribose) usually promotes the repair of damaged DNA. However, excessive activation of the enzyme, due to numerous DNA strand breaks, can so deplete intracellular NAD⁺ pools that the cell is killed. This effect has been termed the 'suicide response'; since DNA repair is not completely efficient, a cell with extensively-damaged DNA may destroy itself so that unwanted mutations do not arise. Poly(ADP-ribose) synthetase can be inhibited by several reagents, such as *theophylline*, *theobromine*, and *3-aminobenzamide*. These diminish the falls in NAD⁺ seen in cells with

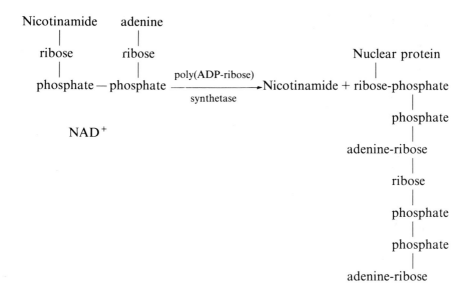

extensive DNA damage due to irradiation or to treatment with drugs that induce DNA strand breaks.

2.4.2   *Ultrasound, lithotripsy, and freeze-drying*

Ultrasonication of aqueous solutions has been shown to produce both OH˙ radicals and hydrogen atoms. The 'sonochemistry' of aqueous solutions can be likened to radiation chemistry. Whether any radical production during the use of ultrasound in medical diagnostic imaging has biological effects remains to be determined, although formation of radicals during ultrasonication of human amniotic fluid *in vitro* has been demonstrated. Kidney stones are now often treated by shattering them using the technique of *extracorporeal shock-wave lithotripsy*. The high-energy shock waves delivered by the lithotripter have been shown to produce radicals *in vitro*, but there is no evidence as yet for any harm produced by such radicals *in vivo*.

The processes of freezing, drying, and freeze-drying (*lyophilization*) can generate radicals capable of damaging bacterial cells, an effect particularly dependent on the concentration of oxygen present.

2.4.3   *Reaction of metal ions with hydrogen peroxide*

As discussed in Section 2.10, hydrogen peroxide is formed in many aerobic cells. Thus the reaction of iron ions with H_2O_2 is a feasible source of OH˙ *in vivo*, provided that such ions are present. A similar comment can be made about Cu^+ ions, which may react with H_2O_2 to form $OH^{\cdot}$ and/or the oxidizing species Cu(III).

Iron distribution: is iron available for the Fenton reaction?

Although the presence of iron in blood was not discovered until the eighteenth century, treatment of human disease with iron is said to date back to 2735 BC in China and 1500 BC in Europe. Records show that in 1681 a liquor, made from iron filings, sugar, and wine, was used successfully by the English physician Sydenham to treat a malady later shown to be caused by iron deficiency. By 1832, a lowered iron content in the blood of anaemic patients had been observed and iron salts were introduced for oral treatment of the disease. These are, of course, still widely available throughout the world for the medically-prescribed or self-prescribed treatment of anaemia. Indeed, many foods and beverages are fortified with iron salts or elemental iron from which, it has been estimated, some 10–15 per cent of our total iron intake is derived.

An average adult human male contains about 4.5 g of iron, absorbs about 1 mg of iron per day from the diet and excretes approximately the same amount when in iron-balance. Since the total plasma iron turnover is some 35 mg per day, extremely efficient mechanisms of iron preservation must exist in the body. Slight disturbances of iron metabolism will readily lead either to iron-deficiency or to iron-overload. It has been estimated that more than 500 million people in the world are iron-deficient and several million are iron-overloaded. There are no specific physiological mechanisms for iron excretion; loss occurs by the turnover of intestinal epithelial cells, in sweat, faeces, urine, and by menstrual bleeding in women.

About two-thirds of body iron is found in haemoglobin, with smaller amounts in myoglobin, various enzymes, and the transport protein transferrin. Iron not required for these is stored as ferritin and haemosiderin. Ferritin consists of a protein shell surrounding an iron core that holds up to 4500 ions of iron per molecule of protein. Iron enters ferritin as Fe^{2+} which becomes oxidized by the protein to Fe^{3+} and deposited in the interior. Similarly, iron can be removed from ferritin, as Fe^{2+}, by the action of a number of biological reducing agents, including cysteine, reduced flavins, and ascorbate. Ferritin can be converted in lysosomes into an insoluble product known as haemosiderin. It is currently thought that the conversion is achieved by proteolytic attack, although this remains to be proved. O'Connell *et al.* in England have produced data suggesting that free-radical reactions stimulated by ferritin iron could be involved in the formation of haemosiderin from ferritin.

Myoglobin and haemoglobin from animal tissues represent the chief food source of haem iron for non-vegetarians. The greatest percentage of dietary iron is non-haem iron, however, present in the Fe^{3+} state. Haem is absorbed as such and the iron removed from it in the intestinal mucosal cells, but other forms of iron require solubilization and reduction to the Fe^{2+} state to aid absorption. The hydrochloric acid in the stomach achieves solubilization and dietary vitamin C (ascorbic acid, a reducing agent) aids absorption. The most active sites of iron absorption are the duodenum and upper jejenum. Not all the dietary iron is taken up into the intestinal cells (see below) and not all the iron taken up is transferred into the circulation. Some is stored in ferritin within the mucosal cells and is eventually lost again when these cells exfoliate from the mucosal surface during normal turnover.

That fraction of iron taken up by the gut that is transferred to the circulation enters the plasma bound to the protein *transferrin*, which functions as a carrier molecule. Transferrin is a glycoprotein and each molecule has two separate binding sites to which Fe^{3+} attaches extremely tightly at physiological pH. Tight binding requires the presence at each site of an anion, usually bicarbonate (HCO_3^-). Under normal conditions the transferrin present in the human bloodstream is only about 30 per cent loaded with iron on average, so that the amount of free iron salts available in the blood plasma would be expected to be virtually zero, a result confirmed by experiment (see page 42). A similar protein to transferrin, known as *lactoferrin*, is found in some body fluids, and in milk, and is produced by phagocytic cells. Lactoferrin also binds 2 moles of Fe^{3+} per mole of protein, but retains its iron at a much lower pH than transferrin.

Iron from transferrin must enter the various cells of the body for use in synthesizing iron enzymes and proteins. The current view is that transferrin is taken into iron-requiring cells by receptor-mediated endocytosis, so that it enters the cytoplasm in a vacuole. The contents of the vacuole are then acidified. This aids the release of iron from transferrin; the iron released probably chelates to various cellular constituents such as citrate, ATP, GTP, or other phosphate esters. The iron-free transferrin (apotransferrin) is then ejected from the cell whilst the small pool of 'non-protein-bound iron' can be used in the synthesis of iron proteins. For example, mitochondria have been reported to take up iron salts rapidly, e.g. for incorporation into haem and non-haem-iron proteins. Mitochondria may contain small 'pools' of chelated iron salts in the matrix, although it is fair to say that the size and the chemical nature of the pool of 'non-protein-bound iron' in cells and organelles is ill-defined. Indeed, Crichton and Charloteaux-Waters (1987) have described it as 'appearing somewhat like the Loch Ness Monster, only to disappear from view before its presence or nature can be confirmed'.

This small pool of non-protein bound iron, presumably attached to such biological metal-binding agents as ATP, ADP, GTP, or citrate, could conceivably provide iron for the Fenton reaction, since complexes of these substances with Fe^{2+} have been shown to react with H_2O_2 to form $OH^·$ *in vitro*. Indeed, the authors have argued that minimizing the amount of non-protein-bound transition metal ions in biological systems is an important part of antioxidant defence (Chapter 3).

Are iron proteins 'Fenton catalysts'?

Does iron bound to proteins react with H_2O_2 to form $OH^.$? This has been an area of considerable controversy in the literature. In order to answer the question properly, it is important to understand what is happening to the protein in the reaction mixture. Thus the authors find that iron correctly bound to the two iron-binding sites of lactoferrin or transferrin is not available to generate $OH^.$ at pH 7.4. However, if the proteins have been incorrectly loaded with iron, it can become detached from the protein during incubation, and the resulting iron ions *will* promote $OH^.$ formation. Transferrin will also release iron, in the presence of chelating agents such as ATP, ADP, and citrate, at acidic pH values. Incubation of ferritin with reducing agents (including ascorbic acid and superoxide radical) causes the reductive release of Fe^{2+} from the protein, and this Fe^{2+} can then go on to form $OH^.$. The authors have shown that treatment of human haemoglobin, cardiac myoglobin, or the protein leghaemoglobin from root nodules of soybean, with a molar excess of H_2O_2 causes, in all cases, haem degradation. The breaking-down of haem seems to release iron ions that can then react with H_2O_2 to give $OH^.$. Thus iron proteins can lead to $OH^.$ generation if conditions in the reaction mixture (or *in vivo*) are such that iron can be released from the protein. If iron ions attached to a protein did react with H_2O_2 to form $OH^.$, the $OH^.$ would be expected to attack components of the protein itself rather than escaping from the protein.

Indeed, any $OH^.$ formed *in vivo* will combine at or very close to its site of formation. Thus the molecular nature of the damage done to cells and tissues by H_2O_2 may depend on the precise location of metal ion complexes that can convert H_2O_2 into $OH^.$, i.e. the sites to which 'catalytic' metal ions are bound will tend to be preferentially attacked. Scientists in Brazil and in the USA have studied the toxicity of H_2O_2 to several mammalian cells, and have found that an early damaging event is the formation of DNA strand breaks, accompanied by activation of poly (ADP-ribose) synthetase and NAD^+ depletion (Section 2.4.1), often to an extent that depresses cellular ATP synthesis. Pure H_2O_2 reacts slowly, if at all, with purified DNA *in vitro*. It is therefore possible that DNA *in vivo* contains transition metal ions bound to it. If sufficient H_2O_2 added to the cells survives to reach the nucleus, then a 'site-specific' formation of $OH^.$ by these metal ions can occur, giving DNA damage. Another possibility is that the H_2O_2 treatment of the cells causes mobilization of transition metal ions (e.g. from a vacuole within the cell) that quickly bind to DNA.

Iron overload

The amount of iron in the human body is determined by the amount entering the circulation from the gastrointestinal tract. Once iron has entered the body, there is no obvious physiological mechanism for disposing of it (unless one counts menstrual bleeding in women). The consequences of inability to dispose of excess iron are seen in iron overload. Acute iron overload is caused by the ingestion of large quantities of iron salts, usually by children who eat iron tablets prescribed for their parents. There is vomiting, gastrointestinal bleeding, and peripheral vascular collapse, followed by severe shock. Since the introduction of desferrioxamine therapy (see page 38), the mortality rate from acute iron poisoning has fallen. Large quantities of iron salts, particularly in the Fe^{2+} state, can degrade the protective layer of gastrointestinal mucus and attack the cells underneath to cause erosion of the gastric mucosa.

A slower-developing iron overload of dietary origin is seen in the Bantu tribe of Africa, who drink acidic beer out of iron pots, and also in patients suffering from an inherited disease (*idiopathic haemochromatosis*) in which much more dietary iron than usual is absorbed by the gut. The metabolic abnormality causing the increased iron uptake is unknown. The time taken for clinically-significant iron overload to develop in haemochromatosis is often 40 or more years and depends to some extent on the diet of the patient. The iron-binding capacity of transferrin in the serum of iron-overloaded patients is often exceeded, so that non-transferrin-bound iron is present. This iron may exist loosely-bound to serum albumin, to citrate, and to other low-molecular-mass chelating agents. The liver attempts to remove this non-transferrin-bound iron by rapidly taking it up from the blood, and so the liver becomes iron-overloaded. The pathology consequent upon iron overload in haemochromatosis includes liver damage, sometimes leading to hepatoma (liver cancer), weakness and malaise, weight loss, skin pigmentation, diabetes (since pancreatic β-cells are damaged), and cardiac malfunctions. It is also interesting to note the association of haemochromatosis with an arthritis-like syndrome in view of suggestions concerning the role of iron in this disease (Chapter 8; Section 8.3.2).

Iron overload can also result from medical treatment of other diseases. For example, the thalassaemias (named from a Greek word meaning 'the sea') are inborn conditions in which the rate of synthesis of one of the haemoglobin chains is diminished, the prefix a- or β-thalassaemia being used to identify the chain that is synthesized abnormally slowly. Thalassaemia major and minor refer to the homozygous and heterozygous states, respectively. Untreated patients with thalassaemia major die of anaemia in infancy, but can be kept alive by regular blood transfusions. Since each unit of blood contains about 200 mg of iron, the patients eventually become overloaded with iron, leading to saturation of transferrin and often the appearance of non-transferrin-bound iron in the blood. Iron accumulates, especially in the liver and spleen. Some iron is present in the heart, but this organ appears to be very sensitive to it so that, as in iron overload secondary to haemochromatosis, many thalassaemic patients treated by blood transfusion suffer cardiac malfunctions and liver damage. Similar problems arise in the treatment of other chronic anaemias by transfusion.

It has been shown by the authors that the non-transferrin bound iron found in the sera of iron-overloaded haemochromatosis patients will stimulate lipid peroxidation (Chapter 4) and the formation of OH˙ radicals from H_2O_2 in experiments *in vitro*, which suggests that the pathology of iron overload is related to increased radical formation *in vivo*. The sensitivity of the heart to even a small loading with iron could then be explained by its relatively poor protection against oxygen radicals: cardiac catalase activity is low and activities of superoxide dismutase and glutathione peroxidase are only moderate (Chapter 3). Other examples of the sensitivity of the heart to radical damage include the cardiomyopathy induced by the anti-tumour drug doxorubicin, which is known to stimulate oxygen radical production (Chapter 8, Section 8.8.4) and the heart lesions seen in Keshan disease, in which lack of dietary selenium causes a drop in tissue glutathione peroxidase activities (Chapter 4, Section 4.7.5). It has even been suggested that the lower incidence of heart disease in women as compared to men is related to their lower body iron stores. The pancreatic β-cells are also sensitive to free radicals (Chapter 6, Section 6.2).

Some evidence consistent with this 'radical hypothesis' of damage induced by iron overload, has been obtained. An additional mechanism of damage is the labilization of lysosomal membranes caused by excessive formation of haemosiderin within them. Of course, lysosomal damage might be a consequence of iron-stimulated lipid peroxidation (Chapter 4). However, present evidence, although suggestive, certainly does not rigorously prove that free-radical reactions are a major cause of the pathology of iron overload.

Chelating agents
Treatment of iron-overload resulting from idiopathic haemochromatosis is usually by blood-letting (*phlebotomy* or *venesection*), whereas chelating agents are administered to transfused thalassaemic patients in an effort to slow the accumulation of iron in the body. Children with β-thalassaemia major in Britain were first given the chelating agent desferrioxamine B in 1962, and it has been successful in prolonging their lifespan. Desferrioxamine B, a very powerful chelator of Fe^{3+}, is isolated from *Streptomyces pilosus*. Desferrioxamine and its Fe^{3+}-complex (*ferrioxamine*) are rapidly excreted, in both urine and bile, so removing iron from the body. Table 2.2 shows the structure of desferrioxamine. Large doses of desferrioxamine are required, it cannot be given by mouth (requiring subcutaneous or intravenous infusion), and it penetrates only slowly into several cell types. Hence there is considerable interest in the development of new chelating agents without the disadvantages of desferrioxamine; candidates include rhodotorulic acid, 1,2-dimethyl-3-hydroxypyrid-4-one, PIH, and 2,3-dihydroxybenzoate (Table

Table 2.2. Chelating agents

Name	Formula	Comments
Penicillamine	$\mathrm{HS{-}C(CH_3)_2}$ $\mathrm{H_2N{-}CHCOOH}$	Useful in promoting urinary excretion of copper salts. Also binds iron(II).
EDTA (ethylenediamine-tetraacetic acid), ion	$^-\mathrm{OOCCH_2}$ and $\mathrm{CH_2COO}^-$ on $\mathrm{N(CH_2)_2N}$; $^-\mathrm{OOCCH_2}$ and $\mathrm{CH_2COO}^-$	Chelates several metal ions. Copper-EDTA chelates usually less active than free copper ions in radical reactions, whereas chelates of EDTA with Fe(II) or Fe(III) still react with H_2O_2 or superoxide.
DETAPAC (diethylenetri aminepentaacetic acid), ion	$^-\mathrm{OOCCH_2}$, $\mathrm{CH_2COO}^-$ on $\mathrm{N(CH_2)_2N(CH_2)_2N}$; $^-\mathrm{OOCCH_2}$, $\mathrm{CH_2COO}^-$, $\mathrm{CH_2COO}^-$	Chelates several metal ions other than iron and copper. Not much used clinically as has several toxic side effects including magnesium depletion.
Rhodotorulic acid	[structure]	Both this and desferrioxamine (below) are examples of *siderophores*, i.e. compounds produced by some micro-organisms in order to chelate iron from the growth medium and bring it into the cell. Both bind iron as Fe(III).
2,3-Dihydroxybenzoic acid, ion	[aromatic ring with COO^-, OH, OH]	Has given promising results in treatment of thalassaemias. Can be given orally.

Table 2.2.—*continued*

Name	Formula	Comments
Desferrioxamine B		A linear molecule that 'bends round' to complex Fe^{3+} with six oxygen ligands, forming a bright-red complex known as ferrioxamine. Commercially available (CIBA) as 'desferal'—desferrioxamine **B** methanesulphonate (mol. wt. 657).
PIH (pyridoxal isonicotinoyl hydrazone)		Shown to be an effective iron chelator in animals. Can be given orally.
o-Phenanthroline (1,10-phenanthroline)		Good chelator of Cu^{2+} ions. Also binds Fe^{2+}. Can prevent H_2O_2-mediated damage to DNA in some isolated mammalian cells. However, a Cu^{2+}-phenanthroline complex can *stimulate* DNA degradation and is the basis of the phenanthroline assay.

1,2-Dimethyl-3-hydroxy-pyrid-4-one

A member of a series of chelators, the 1-alkyl-3-hydroxy-2-methylpyrid-4-ones, which can be given orally and have shown promising preliminary results in the treatment of thalassaemia.

2.2). Iron ions bound to desferrioxamine are usually poorly active, or inactive, in promoting iron-dependent radical reactions, and similar properties must be shown by any newly-introduced chelating agent. It is interesting to note that the chelating agent EDTA, commonly used by biochemists, does not prevent the reaction of iron ions with H_2O_2 or with oxygen radicals (Table 2.2). Reactions of iron chelates with oxidants are discussed further in Chapter 3, Section 3.5.4.

Some of the problems of an iron-overloaded state, such as increased iron deposition in the liver, associated lysosomal damage, and (probably) increased free-radical reactions, can arise as side-effects of other disease states. Excessive alcohol consumption promotes liver damage and iron accumulation. Golden *et al.* in Jamaica have suggested that an iron-overload syndrome, coupled with a lack of dietary antioxidants (Chapter 3), contributes to the pathology of the malnutrition disease kwashiorkor.

The bleomycin assay
The authors have developed an assay which is the first attempt to measure the availability in body fluids of iron complexes that can accelerate free-radical reactions such as the formation of $OH^\cdot$ from H_2O_2. It is based on the fact that the antibiotic bleomycin requires the presence of iron salts in order to degrade DNA (Chapter 8, Section 8.8.4). If other reagents are present in excess, the extent of the DNA degradation is proportional to the amount of iron in a system that is available to be bound by bleomycin (Fig. 2.3). Bleomycin has a fairly low affinity for iron, and it cannot remove iron ions at pH7.4 from pure iron proteins (e.g. lactoferrin, transferrin, ferritin, haemoglobin, catalase). Hence no bleomycin-detectable iron is present in freshly-prepared serum or plasma from normal humans, because most iron is bound to transferrin, and some to ferritin (Table 2.3). Bleomycin-detectable iron is,

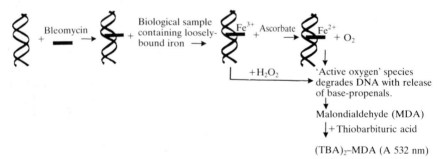

Fig. 2.3. Outline of the assay for measuring bleomycin-detectable iron. The products of DNA degradation by bleomycin include base propenals that react, on heating with thiobarbituric acid (TBA) at low pH, to form a pink TBA-malondialdehyde adduct (Chapter 8, Section 8.8.4).

Table 2.3.   Bleomycin-detectable iron in human extracellular fluids

Fluid tested	Iron detected in assay μmoles per litre
Serum	
Normal (49)	0
Rheumatoid arthritis (21)	0
Osteoarthritis (10)	0
Haemochromatosis (7) (iron-overloaded)	4.3 ± 6.7
Sweat fluid	
Arm fluid (12)	0.72 ± 2.5
Trunk fluid (10)	4.63 ± 2.9

Results are given as mean $\pm$ SD for the number of samples quoted in parentheses. See Gutteridge and Halliwell, *Life Chem. Rep.* **4**, 113–42 (1987) for a full review of these results.

however, present in some sweat samples and in the plasma from iron-overloaded patients (Table 2.3). Bleomycin-detectable iron presumably represents iron ions bound to low-molecular-mass chelating agents such as citrate, or loosely bound to certain proteins, such as albumin. In haemochromatosis patients with iron overload, the plasma concentration of bleomycin-detectable iron does not correlate with total plasma iron, but does correlate with plasma ferritin. Several studies by the authors upon human body fluids show that the iron measurable in the bleomycin assay does seem to be iron that is available to stimulate radical reactions, although the chemistry of the assay itself is extremely complex. Thus the redox-cycling of bleomycin itself produces radicals (Chapter 8, Section 8.8.4) that can release metal ions from metalloproteins under certain circumstances.

Because of the high sensitivity of the bleomycin assay, it easily detects the traces of iron contamination present in most biological reagents (Table 2.4 shows some typical examples of this contamination, obtained in the authors' laboratories). Some iron contamination can be removed by treating reagents with *Chelex* resin, which binds both iron and copper salts. However, Chelex treatment usually leaves the solution with an alkaline pH value and this limits the treatment of most biological buffers. An alternative and more effective procedure for the removal of iron ions from biological buffers is to use the high affinity of transferrin and related proteins for Fe(III) ions, since most iron ion contamination is present in the Fe(III) state. Thus if one places sealed dialysis tubing containing transferrin or ovotransferrin (the iron-binding protein of egg white, sometimes called *conalbumin*, which is much

Table 2.4.   Iron contamination of laboratory reagents

Solution analysed	Iron concentration ($\mu mol\, l^{-1}$)
5.8 M Hydrochloric acid	1–2
'Saline buffer' (67.5 mM Na_2HPO_4 + 4 mM KCl adjusted to pH 7.4 with HCl)	10
Old saline buffer (stored in laboratory for several weeks in a flask covered with 'parafilm')	18
50 mM EDTA	8
0.5 M Sodium formate	9
0.5 M Urea	6
0.5 M Thiourea	3
20 mM Ascorbic acid	4

Solutions were made up in double-distilled water that itself contained no iron. The iron content was measured by atomic absorption analysis (data from Wong *et al. J. Inorg. Biochem.* **14**, 127–34 [1981]). The presence of these traces of iron is of great significance in studies of lipid peroxidation and of the iron-catalysed Haber–Weiss reaction (Chapter 3).

cheaper than transferrin) into a buffer, the Fe(III) contamination is gradually removed. The pH range at which this can be achieved is limited by the affinity of transferrins for iron, which declines at acidic pH values.

Copper availability

Reaction of Cu^+ ions with H_2O_2 appears to generate $OH^{\cdot}$, and/or a reactive Cu(III) species. Is copper available to promote such reactions *in vivo*?

An average adult human contains about 80 mg of copper. It is absorbed from the diet in the stomach or upper small intestine, probably as complexes with amino acids (such as histidine) or small peptides. Transport to the liver is aided by binding to albumin, which has one high-affinity binding site for copper (except in dogs), plus several weaker binding sites. In the liver, copper is incorporated into the protein caeruloplasmin, which has a relative molecular mass of around 134 000 with 6 or 7 copper ions per molecule. Six of these copper ions are tightly bound and only released at low pH in the presence of a reducing agent. Caeruloplasmin may be able to donate copper within cells for incorporation into other copper proteins, such as superoxide dismutase (Chapter 3) and cytochrome oxidase. This 'copper donor' role of caeruloplasmin is often referred to as a 'copper transport' function. However, caeruloplasmin does not function in the same way as transferrin; caeruloplasmin has to be irreversibly degraded to release copper from it, whereas apotransferrin is released from cells to bind more iron.

Normal human plasma contains 200–400 mg/l of caeruloplasmin, comprising 90 per cent or more of total plasma copper. The rest seems to consist of copper ions bound to albumin, to histidine, or to small peptides. The actual size of this 'non-caeruloplasmin copper pool' *in vivo* is uncertain because chromatography, dialysis or even low-temperature storage of plasma or serum can lead to release of copper from caeruloplasmin, and such techniques have frequently been used to study non-caeruloplasmin copper.

In vitro, caeruloplasmin catalyses the oxidation of a wide range of polyamine and polyphenol substrates, including bioamines, but it is uncertain if these activities have biological significance. Caeruloplasmin also has a 'ferroxidase' activity, catalysing oxidation of Fe^{2+} to Fe^{3+} with simultaneous reduction of O_2 to H_2O. This may aid the binding of iron onto transferrin (which binds Fe^{3+}). Non-enzymic oxidation of Fe^{2+} by O_2 is quite fast at pH 7.4 but it produces oxygen radicals (Section 1.5.3), whereas caeruloplasmin-catalysed oxidation does not. Indeed, caeruloplasmin has antioxidant properties, in part because its ferroxidase activity depresses Fe^{2+}-dependent radical reactions, such as formation of OH˙ from H_2O_2 (Chapter 4, Section 4.7.6). The ferroxidase activity of caeruloplasmin is inhibited by azide, and is often called 'ferroxidase I', to distinguish it from other ferroxidases that have been described.

Thus caeruloplasmin-bound copper does not accelerate radical reactions. However, copper ions attached to albumin or to histidine can still react with H_2O_2 to form highly-reactive species, OH˙ and/or Cu(III). The reactive oxidant appears to attack the ligand to which the copper is bound and is not released into free solution. This is a further illustration of the principle that the damage done by OH˙ generated from H_2O_2 *in vivo* will depend on the location of transition metal complexes. Japanese and Israeli scientists have shown that the binding of copper ions to DNA, viruses, and proteins (such as albumin or the enzyme penicillinase), can lead to site-specific damage when these systems are exposed to H_2O_2. In this context, the authors have argued that the ability of albumin to bind copper ions may be a biologically-significant protective mechanism. The presence of very high albumin concentrations in plasma may prevent copper ions from binding to more important sites, e.g. on cell membranes. If H_2O_2 and O_2^- are produced in plasma (e.g. by activated phagocytic cells; Chapter 7), OH˙ and/or Cu(III) may be produced on the albumin surface and damage the protein, but so much albumin is present and its turnover is so rapid that this is unlikely to lead to significant biological consequences.

Copper overload
The toxicity of excess copper is well illustrated by Wilson's disease, an inherited metabolic defect characterized by low concentrations of caeruloplasmin in the blood. Copper is deposited in the liver, kidney, cornea, and brain, causing damage that leads to lack of co-ordination, tremors, and

progressive mental retardation. It seems likely that copper-stimulated free radical reactions are involved in the pathology of Wilson's disease, although this has not been experimentally proven. Treatment involves a copper-restricted diet and use of chelating agents such as penicillamine (Table 2.2) that promote copper excretion. Oral administration of zinc salts may also help to prevent copper accumulation, by interfering with the intestinal absorption of copper. Zinc (Zn^{2+}) might also compete with copper ions for binding to target sites that could be damaged by free radicals.

The phenanthroline assay

The chelating agent, 1,10-phenanthroline (*o*-phenanthroline, Table 2.2) degrades DNA in the presence of copper ions, oxygen, and a suitable reducing agent. Degradation results in the release of a product from DNA that reacts, upon heating with thiobarbituric acid at acidic pH, to form a pink chromogen. The authors have made this reaction the basis of a technique to detect and measure 'available' copper in biological fluids, i.e. copper that can be complexed by phenanthroline. All reagents are carefully treated with the metal-binding resin Chelex to remove contaminating copper ions, and azide is added to inactivate catalase which might arise from cell lysis, since catalase inhibits the DNA degradation. Copper, complexed by phenanthroline added to biological fluids, is reduced by the thiol mercaptoethanol and the resulting damage to DNA is measured and is proportional to the amount of copper available to phenanthroline. Phenanthroline-detectable copper has been found in some sweat samples and in cerebrospinal fluid. The 'phenanthroline' assay detects copper bound to the high-affinity site of albumin and to histidine, but not to caeruloplasmin. The precise significance of 'phenanthroline-detectable copper' in catalysing radical reactions *in vivo* is still under investigation. An interesting observation is that cerebrospinal fluid from patients with Parkinson's disease has a higher phenanthroline-detectable copper level than normal, although there is no rise in the iron or manganese content of the fluid.

Degradation of DNA by the Cu-phenanthroline complex involves reduction of Cu^{2+} to Cu^+ by the reducing agent, followed by its reaction with H_2O_2 to form OH$^.$ or a Cu(III) species such as cupryl, $Cu(OH)_2^+$, that immediately attacks the adjacent DNA.

2.4.4   *Formation of hydroxyl radicals from ozone*

Ozone (O_3), an important component of photochemical air pollution, is a powerful oxidizing agent that can attack many biological molecules. For example, cysteine, methionine, and histidine in proteins can be oxidized. Tyrosine residues in proteins can be cross-linked, after oxidizing their —OH groups, to give O,O'-dityrosine. Oxidation of polyunsaturated fatty acids by ozone can lead to lipid peroxidation (Chapter 4, Section 4.4.2). There is also evidence that ozone can produce OH˙ radicals in aqueous solution, which could contribute to toxicity. Hydroxyl radical formation is favoured at alkaline pH values, but might occur to some extent at physiological pH. It has been reported that exposure of mice to 0.45 ppm of O_3 led to an increase in poly(ADP-ribose) synthetase activity in lung tissue, suggestive of increased DNA damage *in vivo* (Section 2.4.1).

2.4.5 *Ethanol metabolism*

Most adult humans ingest ethanol to improve their lifestyle and even in teetotallers it is produced in small amounts by the gut flora. Ethanol absorbed from the gut is metabolized in the liver, mainly by the action of an alcohol dehydrogenase enzyme that oxidizes it to ethanal (acetaldehyde). A minor contribution to ethanol metabolism is made by a 'microsomal ethanol oxidizing system' (MEOS), so called because it is found in the microsomal pellet upon subcellular fractionation of liver homogenates. When liver cells are disrupted, the plasma membrane and endoplasmic reticulum are torn up and the pieces form membrane vesicles which can be sedimented by high-speed centrifugation. Thus the microsomal fraction so obtained is a hetero-geneous collection of membrane vesicles from different parts of the cell; and this should be borne in mind when interpreting the results of experiments upon this fraction. We have seen more than one paper in which the authors spoke of microsomes as if they were a discrete subcellular organelle.

   The MEOS system probably originates from the endoplasmic reticulum, which contains systems that generate hydrogen peroxide (Section 2.10). There is evidence that *some* (exactly how much is not clear) of the ethanol is oxidized by hydroxyl radicals generated from this hydrogen peroxide. Since the action of MEOS *in vitro* is partially inhibited by the iron chelator desferrioxamine, the OH˙ may be produced by a Fenton reaction using iron bound to the microsomal membranes. A specific ethanol-inducible form of cytochrome P-450 (Section 3.5.1) also participates in ethanol oxidation by microsomal fractions.

2.5 Detection of hydroxyl radicals in biological systems

2.5.1 *Electron spin resonance and spin-trapping*

Electron spin resonance (ESR) is a technique that can be applied to free radicals, since it detects the presence of unpaired electrons. An unpaired electron has a spin of either $+\frac{1}{2}$ or $-\frac{1}{2}$ and behaves as a small magnet. If it is exposed to an external magnetic field, it can align itself either parallel or antiparallel (in opposition) to that field, and thus can have two possible energy levels. If electromagnetic radiation of the correct energy is applied, it will be absorbed and used to move the electron from the lower energy level to the upper one. Thus an absorption spectrum is obtained, usually in the micro-wave region of the electromagnetic spectrum. For reasons that need not concern us, ESR spectrometers are set up to display *first-derivative spectra*, which show not the absorbance but the *rate of change of absorbance*, i.e. a point on the derivative curve corresponds to the gradient (slope) at the equivalent point on the absorption plot.

The condition to obtain an absorbance is:

$$\Delta E = g\beta H$$

where ΔE is the energy gap between the two energy levels of the electron, H is the applied magnetic field and β a constant known as the *Bohr magneton*. The value of g (the '*splitting factor*') for a free electron is 2.00232 and nearly all biologically important radicals have values close to this. Thus, if this equation is obeyed, an absorption spectrum results. For a single electron this can be crudely represented as:

but, if presented as its first derivative (as ESR machines do) it will appear as:

A number of atomic nuclei, such as those of hydrogen and nitrogen, also behave like small magnets and will align either parallel or antiparallel to the applied magnetic field. Thus in a hydrogen atom the single unpaired electron will actually see two different magnetic fields: the one applied plus that from the nucleus, or the one applied minus that from the nucleus. Thus there will be two energy absorptions and the single line becomes a doublet, i.e.

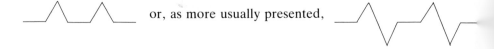

or, as more usually presented,

If the unpaired electron 'sees' two hydrogen nuclei, each can be aligned in the same way with the applied field, in opposite ways, or one in the same way and one opposite, i.e.

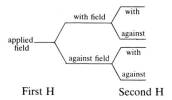

giving a three-line spectrum in which the intensities are in the ratio 1:2:1, i.e.

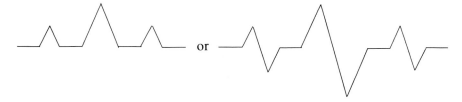

In the carbon-centred methyl radical ($^{\cdot}CH_3$ or $CH_3^{\cdot}$) the unpaired electron on the carbon can 'see' three hydrogen nuclei and the ESR spectrum contains four lines. Remembering that the field of each hydrogen nucleus can align with or against the applied magnetic field, a 'tree' diagram like that above can be used to predict the spectrum, i.e.

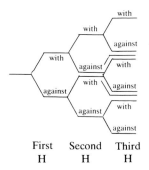

which therefore consists of four lines with intensity ratios 1:3:3:1, or:

The number of lines in the ESR spectrum of a radical is called the *hyperfine structure* and is often very large in complicated radicals containing many nuclei. A radical can be identified from its ESR spectrum by looking at the g value, hyperfine structure, and line shape.

ESR is a very sensitive method and can detect radicals at concentrations as low as $10^{-10}\,mol\,l^{-1}$, provided that they stay around long enough to be measured. For very unstable radicals a number of techniques are available to detect their presence. One can use flow systems whereby the radicals are continuously generated in the spectrometer so as to maintain a steady-state concentration (see Fig. 2.1(b) for the principles of this). Another approach is to generate the radical in a frozen transparent solid matrix which prevents it from colliding with other species and undergoing reaction. Such a 'rapid-freezing' technique was used in 1969 by P. F. Knowles and others in England to identify the superoxide radical being produced in an enzyme-catalysed reaction (xanthine oxidase catalysing the oxidation of xanthine) by observing its ESR spectrum. By allowing the matrix to warm up, reactions of the radical can then be observed.

Another approach is *spin trapping*. A highly reactive radical, difficult to observe by normal ESR, is allowed to react with a compound to produce a long-lived radical. Reaction of nitroso compounds (R·NO) with radicals often produces nitroxide radicals that have a long lifetime

$$R-N{=}O \;\;+\;\; R'^{\textbf{·}} \;\longrightarrow\; \begin{matrix} R \\ \diagdown \\ \diagup \\ R' \end{matrix} N-O^{\textbf{·}}$$

R symbolizes 'rest of molecule'	very reactive radical	nitroxide radical (fairly stable)

Nitrones (R$=$N$^+$$-O^-$) also produce nitroxide radicals in a similar reaction.

Spin-trapping methods have often been used to detect the presence of superoxide and hydroxyl radicals in biological systems, and also the formation of organic radicals during lipid peroxidation (this is discussed further in Chapter 4). Table 2.5 shows some of the trapping molecules that have been used, DMPO being especially popular. The 'ideal' trap should react rapidly and specifically with the radical one wants to study, to produce a product that is stable and has a highly characteristic ESR spectrum. It is also worth noting that if a biological process is dependent on, say, hydroxyl radicals, then addition of a trap that reacts with such radicals will inhibit the process to an extent that depends on how much trap is added and on the rate constant for reaction. For example, the spin-trap DMPO decreases ethanol

Table 2.5. A selection of the 'spin-traps' that have been used in biological systems

Name	Abbreviation	Structure
tert-Nitrosobutane (nitroso-*tert* butane)	tNB (NtB)	$H_3C-\overset{\overset{\textstyle CH_3}{\mid}}{\underset{\underset{\textstyle CH_3}{\mid}}{C}}-N{=}O$
a-Phenyl-*tert*-butylnitrone	PBN	
5,5-Dimethylpyrroline-*N*-oxide	DMPO	
tert-Butylnitrosobenzene	BNB	
a-(4-Pyridyl 1-oxide)-*N*-*tert*-butylnitrone	4-POBN	

oxidation by the MEOS system as it scavenges OH⋅. None of the spin-traps at present in use is ideal, although better ones are being developed. For example, DMPO reacts with both OH⋅ and O_2^- radicals to form products with different ESR spectra (Fig. 2.4). The second-order rate constants for the reactions are very different, however, at approximately $10\,M^{-1}s^{-1}$ for O_2^- and 3.4×10^9 for OH⋅. Thus OH⋅ is trapped much more quickly than is O_2^-. Unfortunately, the product of reaction of DMPO with O_2^- (DMPO—OOH) is unstable and decomposes to give DMPO—OH, the same product that is

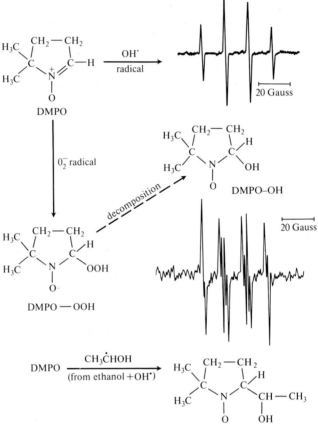

Fig. 2.4. Reactions of the spin-trap DMPO with superoxide, hydroxyl, and ethanol radicals. The ESR spectra of the DMPO—OH and DMPO—OOH adducts are shown diagrammatically to illustrate the difference between them.

given by direct reaction with OH˙ (Fig. 2.4). Even worse, decomposition of the DMPO—OOH adduct has been claimed to release some OH˙ radical, i.e. addition of DMPO to a system producing only O_2^- might *cause* small amounts of OH˙ to be formed.

There are solutions to these problems. If the OH˙ signal observed in a system is really due to formation of OH˙ and its reaction with DMPO, then the addition of excess ethanol should *completely* abolish the signal, since ethanol scavenges OH˙. Further, reaction of ethanol with OH˙ produces a hydroxyethyl radical (Fig. 2.4) that reacts with DMPO to give another adduct with a different ESR spectrum. This spectrum should be observed as the OH˙ signal is lost. If, however, the signal arose from decomposition of a DMPO—OOH adduct, then ethanol will not completely abolish it, since it does not scavenge O_2^-. Dimethylsulphoxide may be used instead of ethanol: its reaction with OH' produces a methyl (CH_3) radical that forms a characteristic adduct with DMPO ($DMPO-CH_3$).

Another problem is that the radical adducts of several spin-traps in current use are easily reduced by cellular reducing agents, such as ascorbic acid, to give 'ESR-silent' species (i.e. products that no longer give an ESR signal). This limits the application of such spin-traps to biological systems. Much work is being carried out to develop improved spin-traps.

2.5.2 *Aromatic hydroxylation*

The oxidation of aromatic compounds (i.e. compounds containing benzene rings: Appendix I, Section A.2.4) by metal ion–H_2O_2 mixtures has been known for over 80 years, and, since the pioneering work of Merz and Waters in 1949, has generated an enormous chemical literature. The reactions are very complex. In the case of benzene itself, there is a fast addition of OH' to the aromatic ring structure to give a *hydroxycyclohexadienyl* radical:

Two such radicals can join together to give a dimer, that can lose water to form *biphenyl*:

Hydroxycyclohexadienyl radical can also be oxidized to phenol, e.g.

The presence of O_2, Fe^{3+}, or Cu^{2+} tends to increase the yield of hydroxylated products.

If substituted benzenes are attacked by OH', reactions become even more complex. For example, using aromatic acids, decarboxylation reactions (loss of the carboxyl, $-COOH$, group) are favoured at low pH values in the absence of metal ions such as Cu^{2+} or Fe^{3+}, whereas hydroxylated product formation is favoured if such metal ions are present. Hence under physiologically relevant conditions (pH 7.4, metal ions and O_2 present), hydroxylation will be the predominant reaction pathway observed. Sagone and others in the USA used decarboxylation of benzoic acid, [benzoic acid structure with COOH], labelled with ^{14}C in the carboxyl group, as an assay for generation of OH˙ in biochemical systems. The assay is very sensitive, since even small amounts of $^{14}CO_2$ can be trapped in alkaline solutions and accurately measured by scintillation counting. However, as explained above, decarboxylation is likely to be a very minor reaction pathway under physiological conditions and measurement of hydroxylated products seems more satisfactory in principle. Hydroxylated products can be separated by gas–liquid chromatography (GLC) or high-performance liquid chromatography (HPLC) and measured by electrochemical detection, fluorescence spectra, or simple 'colour' reactions (e.g. Table 2.6).

HPLC separation combined with highly-sensitive electrochemical detection has been used by the authors' laboratories and others to measure the products formed by attack of OH˙ on aromatic compounds, as a highly-sensitive assay to measure the generation of OH˙ by cells, organelles, and

Table 2.6.   Formation of hydroxyl radicals during the oxidation of xanthine by xanthine oxidase

Reagent added	Rate constant for reaction with OH˙ ($M^{-1} s^{-1}$)	Amount of hydroxylated products formed (nmoles)	Percentage inhibition of hydroxylation
None	—	102	—
Mannitol (5 mM)	2.7×10^9	64	37
Sodium formate (5 mM)	3.7×10^9	49	52
Thiourea (5 mM)	4.7×10^9	24	76
Urea (5 mM)	$< 7.0 \times 10^5$	102	0

Experiments were carried out at pH 7.4; for further details see Halliwell, B. *FEBS Lett.* **92**, 321–6 (1978). Hydroxyl radicals were detected by measuring the formation of hydroxylated products from 2-hydroxybenzoic acid, [structure with COOH and OH]. The amount of OH˙ reacting with this compound can be decreased by adding other reagents that react with OH˙. The more quickly these other reagents react with OH˙, the more inhibition of the formation of hydroxylated products will they produce.

perfused organs. For example, the attack of OH˙ on salicylic acid (2-hydroxybenzoic acid) produces two dihydroxylated products (2,3- and 2,5-dihydroxybenzoates), together with a small amount of catechol, by decarboxylation (Fig. 2.5). Attack of OH˙ upon phenylalanine produces three dihydroxylated products: 2-hydroxyphenylalanine (*o*-tyrosine), 3-hydroxy-phenylalanine (*m*-tyrosine), and 4-hydroxyphenylalanine (*p*-tyrosine).

If an aromatic compound reacts with OH˙ to form a specific set of hydroxylated products that can be accurately measured in body fluids or tissue extracts, and one or more of these products is not identical to enzyme-

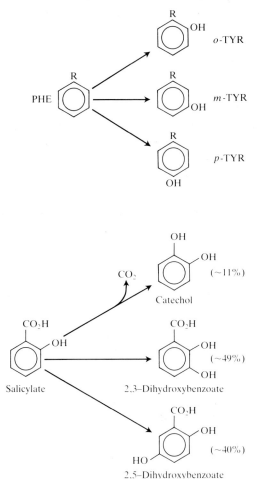

Fig. 2.5. Products of the reaction of hydroxyl radical with aromatic compounds. PHE (phenylalanine); *o*-TYR (*ortho*-tyrosine); *m*-TYR (*meta*-tyrosine); *p*-TYR (*para*-tyrosine).

produced hydroxylated products, then formation of the 'unnatural' products could *conceivably* be used to assess OH˙ radical formation *in vivo*. This assumes that the aromatic detector molecule is present at the sites of OH˙ radical generation at concentrations sufficient to compete with any other molecules that might scavenge OH˙ and that any unnatural hydroxylated product is not immediately metabolized. For example, dihydroxybenzoates are rapidly metabolized in perfused liver, so that salicylate is not a good detector of OH˙ in this organ system. Both salicylate and phenylalanine might conceivably be used in this way as 'probes' for the generation of OH˙ *in vivo*. Thus 2,3-dihydroxybenzoate is formed by attack of OH˙ on salicylate (Fig. 2.5), yet has not been reported as an enzyme-produced metabolite of salicylate in humans. Normal metabolism of phenylalanine by the enzyme phenylalanine hydroxylase generates *p*-tyrosine, but attack of OH˙ also produces *o*- and *m*-tyrosines (Fig. 2.5). There is now considerable interest in attempts to use aromatic compounds to detect OH˙ generated *in vivo*, but it is probably too early as yet to evaluate the usefulness of the technique. Aromatic hydroxylation is certainly a sensitive method of detecting OH˙ *in vitro*.

2.5.3   *Measurement of products of radical attack upon DNA*

DNA seems to be an important cellular target for attack by radicals *in vivo*, such as OH˙ generated by ionizing radiation or formed from H_2O_2 by site-specific reactions involving metal ions (Sections 2.4.1 and 2.4.3). Attack of OH˙ upon DNA produces a wide range of chemical changes (Fig. 2.2). These include conversion of thymine residues into thymine glycol and 5-hydroxy-methyluracil, and of guanine residues into 8-hydroxyguanine (Figs 2.2 and 2.6). Methods have been developed for the measurement of these products. For example, Dizdaroglu in the USA (Fig. 2.2) and others have identified multiple products of OH˙ attack in DNA subjected to ionizing radiation. Kasai, Nishimura and others in Japan detected 8-hydroxyguanine in DNA obtained from the livers of mice that had been subjected to γ-irradiation. Ames' group in the USA reported that thymine glycol and thymidine glycol (thymine glycol still attached to deoxyribose) are excreted in human urine, and might serve as an index of radical attack upon DNA *in vivo*. 5-hydroxymethyluracil can also be measured in rat and human urine. Ames *et al.* reported that normal humans excrete about 100 nmoles (a nanomole is 10^{-9} moles) of thymine glycol, thymidine glycol and hydroxymethyluracil every day. If it is assumed that this all results from the repair of oxidative damage to cellular DNA (i.e. ignoring products from gut bacteria or from DNA ingested in food), it can be calculated that there are, on average, 10^3 'oxidative hits' per day on the DNA within each of the 6×10^{13} cells in the body. Since there are many other products of radical attack on DNA (Fig. 2.2), the true figure might be greater.

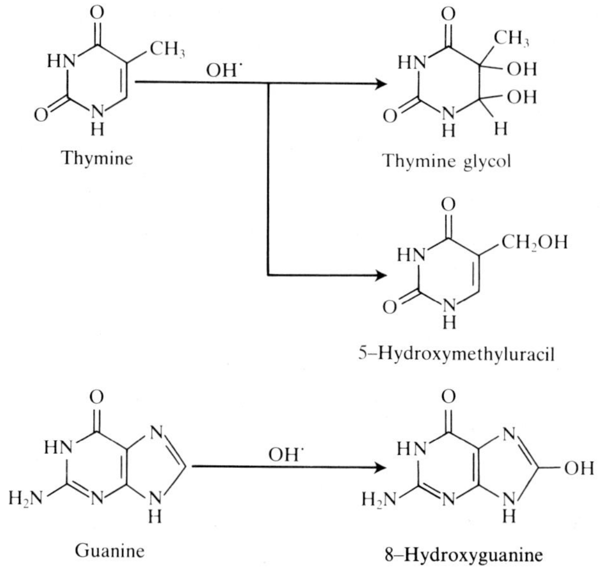

Fig. 2.6. Two products formed by attack of OH˙ radicals upon DNA. A wide range of other products is also produced (Fig. 2.2), including products of OH˙ attack on deoxyribose. Indeed, irradiation of aqueous thymine solutions generates over 30 different products.

2.5.4 Conversion of methional and related compounds into ethene (ethylene) gas (C_2H_4)

This assay method, developed in Fridovich's laboratory in the USA, was the first used to show formation of hydroxyl radicals in a biological system. Methional, which has the structure $CH_3-S-CH_2-CH_2-CHO$, reacts with hydroxyl radicals in a very complex series of reactions to form ethene $(H_2C = CH_2)$ gas, which may be measured by GLC. The amino acid methionine

$$CH_3-S-CH_2-CH_2-\underset{\underset{NH_3}{\overset{\displaystyle |}{}}}{\overset{+}{C}}H-COO^-$$

and the compound KTBA (2-keto- 4-methylthiobutanoic acid),

$$CH_3{\cdot}S{\cdot}CH_2{\cdot}CH_2{\cdot}\overset{\displaystyle O}{\overset{\displaystyle \|}{C}}{\cdot}COOH$$

also release ethene

upon exposure to hydroxyl radicals. Ethene formation from these compounds is not specific as a test for the hydroxyl radical since there are several other oxidizing radicals that can convert them into ethene. However, Fridovich's group showed that formation of ethene in the system that they studied was inhibited by compounds that react with the hydroxyl radical in the way expected from their known rate-constants (Table 2.1). If the ethene formation is due to OH˙, then addition of, say, sodium formate at a concentration that competes with methional for the OH˙ should decrease ethene formation. Addition of an equal concentration of ethanol should inhibit less, because it is less reactive with OH˙ than is formate. Thiourea should inhibit more, and urea, which reacts only slowly with OH˙ (Table 2.1) should have little inhibitory effect. If the ethene formation is inhibited by OH˙ scavengers to extents that can be correlated with known rate-constants for reactions of these compounds with OH˙ (Table 2.1) then it can probably be attributed to OH˙. (Table 2.6 shows the principle of the use of these compounds in a different method of detecting OH˙). It is essential to use a wide range of different OH˙ 'scavengers' in such experiments. If one or more scavengers does not inhibit in the way expected, then the oxidizing species is not OH˙ produced in free solution. However, scavengers often do not protect against damage by OH˙ generated in site-specific reactions (Section 2.4.3), since OH˙ formed at the site of the bound transition metal ion immediately attacks the binding molecule and scavengers cannot gain easy access to it. This is further discussed in Chapter 3, Section 3.5.4.

2.5.5   *Other methods*

Hydroxyl radical reacts with many other compounds to give products that can be easily detected. It is essential that 'scavenger' experiments (Table 2.6) be performed to show whether or not the radical producing the product is indeed OH˙ or something else. Table 2.7 summarizes some of these other methods, several of which have been applied to biological systems with varying degrees of success.

2.6   Production of singlet oxygen

We saw in Chapter 1 that two singlet states of oxygen exist in which the spin restriction that slows reaction of this molecule with non-radicals is removed, resulting in increased reactivity. The $^1\Sigma g^+$ state of oxygen is extremely energetic and rapidly decays to the $^1\Delta g$ state, so only the latter need usually be considered in biological systems. Hence references to just 'singlet oxygen' below can be taken to refer to this state. Although $^1\Delta g O_2$ is not a free radical (Chapter 1) it can be formed in some radical reactions and can trigger off others, so its chemistry is worth discussing.

Methods for detection of hydroxyl radicals

Method	Principle of method	Comments
Bleaching of *p*-nitroso dimethylaniline (PNDA) N=O—⟨ ⟩—N(CH₃)₂	PNDA reacts rapidly with OH˙ but not with O_2^- or singlet O_2. Reaction is accompanied by loss of the yellow colour ('bleaching') (see Bors *et al.*, *Eur. J. Biochem.* **95**, 621–7 (1979)).	Use in biological systems has found many bleaching reactions not inhibited by compounds that react with OH˙; suggesting that it is bleached non-specifically by many biological molecules. We do not recommend it for use.
Deoxyribose method	Reaction of OH˙ with deoxyribose produces products that, when heated with thiobarbituric acid at low pH, give a colour (see Halliwell and Gutteridge, *FEBS Lett.* **128**, 343–52 [1981]).	Highly sensitive method, but inhibition studies (Table 2.6) must be done to prove that the reacting species is OH˙. Can be used to determine approximate rate constants for reactions of OH˙ (Halliwell *et al.*, **165**, 215–19 [1987]).
Tryptophan method	Reaction of OH˙ with tryptophan produces a characteristic set of products (see Singh *et al.*, *Bull. Eur. Physiopath. Resp.* **17**, 31–41 [1981]).	Tryptophan also reacts with singlet O_2 but the products are different.
Dimethylsulphoxide (DMSO) method	OH˙ radicals react with DMSO, CH₃—S—CH₃ to give CH₃—S—CH₃ (OH˙) which decomposes to give, among other products, methane gas, measured by gas–liquid chromatography (see Klein *et al.*, *Biochemistry* **20**, 6006–12 [1981]).	Inhibition studies (Table 2.6) must be done to prove that the reacting species is OH˙. Methanal (H·CHO) is also produced and can be measured instead. Babbs *et al.* in the USA have suggested that oxidation of DMSO to methanesulphinic acid might be a means of detecting OH˙ formation *in vivo*.
Benzoate fluorescence	Reaction of benzoic acid with OH˙ gives 3- and 4-hydroxybenzoates, which are highly fluorescent at 407 nm when excited at 305 nm	Highly-sensitive method. See Gutteridge, *Biochem. J.* **243**, 709–14 (1987).

One system that can be used to generate singlet oxygen in the laboratory is a mixture of hydrogen peroxide and the hypochlorite ion, OCl^-, which is formed when chlorine gas is passed into cold alkaline solution:

$$Cl_2 + OH^- \rightarrow Cl^- + OCl^- + H^+$$
$$OCl^- + H_2O_2 \rightarrow Cl^- + H_2O + O_2(singlet).$$

The singlet oxygen arises from the hydrogen peroxide molecules. This reaction might be biologically relevant, since OCl^- can be formed by the enzyme myeloperoxidase during phagocytosis (Chapter 7). The decomposition of the compound potassium peroxochromate, K_3CrO_8, in aqueous solution also produces singlet oxygen. Its usefulness as a source of this species is restricted, however, since it also generates O_2^- and $OH^{\cdot}$ radicals as it decomposes.

Singlet oxygen is most often generated in the laboratory by *photosensitization reactions*. If certain molecules are illuminated with light of a given wavelength they absorb it and the energy raises the molecule into an 'excited state'. The excitation energy can then be transferred onto an adjacent oxygen molecule, converting it to the singlet state whilst the photosensitizer molecule returns to the ground state. Popular sensitizers of singlet oxygen formation in the laboratory include the dyes *acridine orange, methylene blue, rose bengal,* and *toluidine blue*; but many compounds found *in vivo* are also effective, such as the water-soluble vitamin riboflavin and its derivatives FMN (flavin mononucleotide) and FAD (flavin adenine dinucleotide), chlorophylls a and b, the bile pigment bilirubin, retinal, and various porphyrins, both free and bound to proteins (Fig. 2.7). The singlet oxygen produced on illumination of these substances with light of the correct wavelength can react with other molecules present or it can attack the photosensitizer molecule itself. The chemical changes thereby produced are known as *photodynamic effects*. Hence illuminated solutions of flavins lose their orange colour and chlorophylls their green colour as they are attacked; this is called *photobleaching*. Reactions of this type can cause the dyes in your clothes or curtains to fade when exposed to sunlight. Indeed, exposure of cells in culture to high-intensity visible light causes damage, especially to the mitochondria, which are rich in haem proteins and flavin-containing proteins. Haem-containing enzymes such as catalase are also inactivated. Some of the media used for growth of cell cultures contain added riboflavin, which makes these effects worse since the fluorescent lighting found in many laboratories is quite

---

Fig. 2.7. Compounds that can sensitize formation of singlet O_2 when illuminated with light of the correct wavelength. (a) *Psoralen*. Psoralen derivatives are used in treating certain skin diseases. (b) The basic *porphyrin* nucleus. R_1–R_8 are different side chains (c) *Bilirubin*. (d) *Riboflavin*. (e) *Chlorophyll a*. (f) *Retinal*.

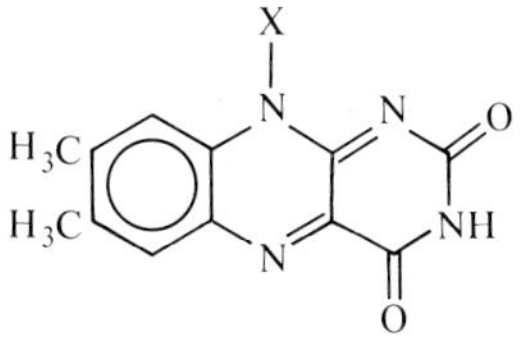

Riboflavin: X is $\quad -CH_2(CHOH)_3CH_2OH$

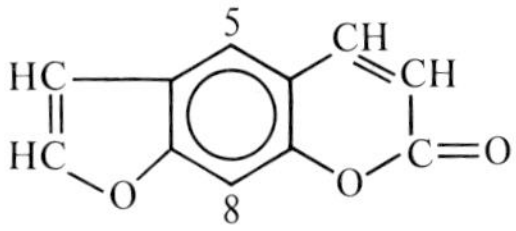

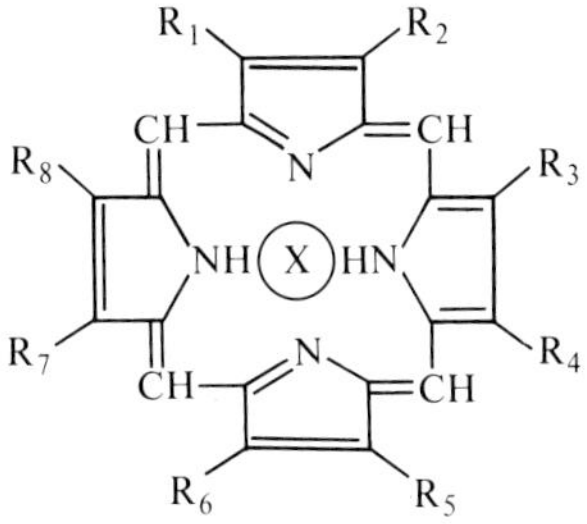

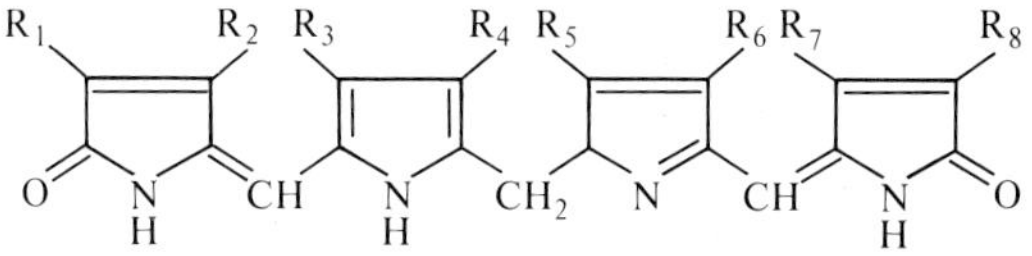

FMN: $\quad$ X is $\quad -CH_2(CHOH)_3CH_2O\overset{\displaystyle O^-}{\underset{\displaystyle OH}{\overset{|}{\underset{|}{P}}}}=O$

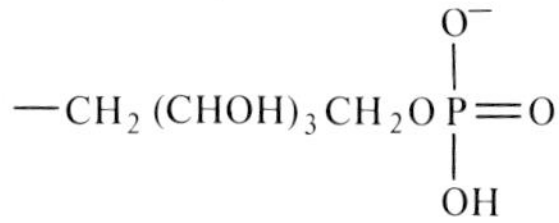

FAD: $\quad$ X is $\quad -CH_2O\overset{\displaystyle O}{\overset{\|}{\underset{\displaystyle O^-}{\underset{|}{P}}}}-O-\overset{\displaystyle O}{\overset{\|}{\underset{\displaystyle O^-}{\underset{|}{P}}}}-O-CH_2 \quad$ Adenine ring

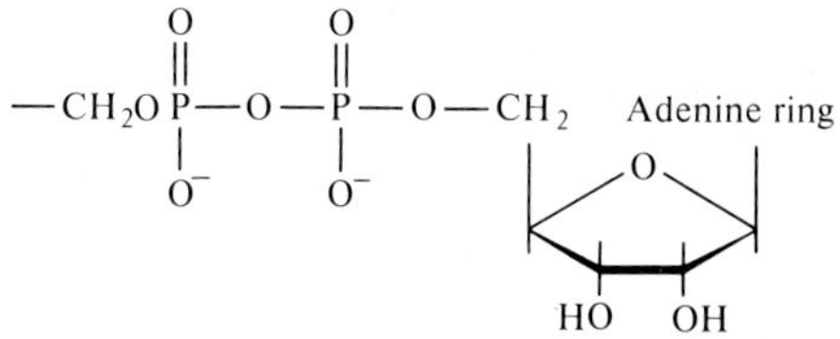

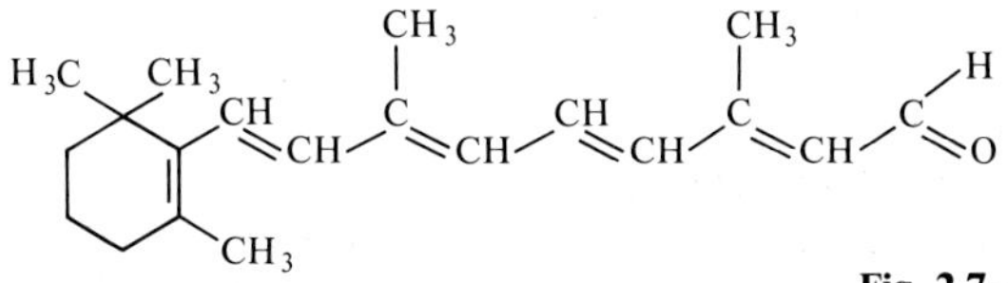

Fig. 2.7—*continued overleaf*

intense enough to cause flavin-sensitized singlet oxygen formation. Not all photosensitization damage need arise by means of singlet oxygen production, however, since the excited state of the photosensitizer can often cause damage itself. Such damage is said to proceed by a *type I mechanism* as opposed to that caused by singlet oxygen production, described as a *type II mechanism*. Both mechanisms may operate simultaneously and the relative importance of each type of damage will depend on the nature of the molecule present, the efficiency of energy transfer to oxygen, and on the oxygen concentration (remember that oxygen concentrates in membranes, where many sensitizers are located). In addition, illumination of several porphyrins and acridine dyes in aqueous solution has been reported to produce hydroxyl radical, and sometimes superoxide.

It is possible to distinguish between type I and type II mechanisms of damage to a biological system by immobilizing the sensitizer on a solid support, and allowing the singlet oxygen produced to diffuse a short distance to the target. Thus the biological target does not come into contact with the excited state of the photosensitizer. For example, experiments of this type have shown that externally-generated singlet O_2 is highly cytotoxic to

Salmonella, but is poorly mutagenic (if mutagenic at all) for this organism.

Photosensitization reactions, probably involving singlet oxygen, are important in many biological situations. The most obvious is the chloroplasts of higher plants, which contain chlorophylls and a high oxygen concentration—the problems they face are discussed in detail in Chapter 5. The retina of the eye is obviously exposed to light and its rod cells contain the pigment retinal (Fig. 2.7) bound to a protein known as 'opsin', the whole being called *rhodopsin* or 'visual purple'. Retinal can sensitize singlet oxygen formation and thus damage itself. Indeed, visual cells are known to be injured by prolonged exposure to bright light. The outer segment membranes of the retinal rods contain a large number of highly unsaturated lipids (i.e. containing a lot of carbon–carbon double bonds, $\diagup C = C \diagdown$) which could readily be attacked by singlet oxygen. It has also been suggested that the lens of the human eye contains sensitizers of singlet oxygen formation. Indeed, singlet oxygen-mediated oxidation of lens proteins *in vitro* gives products similar to those that accumulate in the lens during ageing (Chapter 5). Illumination of milk or milk products causes development of 'off-flavours' as the riboflavin present photosensitizes the degradation of milk proteins and lipids.

Several diseases can lead to excessive singlet oxygen formation. For example, the *porphyrias* are diseases caused by defects in the metabolism of haem and other porphyrins (Fig. 2.7). In these diseases, porphyrins are excreted in the urine and often accumulate in the skin, exposure of which to light causes damage leading to unpleasant eruptions, scarring, and thickening (Fig. 2.8). The severity of the damage depends on the exact structure of the porphyrin accumulated and thus will differ in different diseases. Illumination of some porphyrins has also been reported to produce the hydroxyl radical.

Sometimes photochemical effects are made use of in medicine, as in *photodynamic therapy* in the treatment of 'cold sores' and genital sores caused by the virus *Herpes simplex*. The lesions are painted with a dye such as neutral red or proflavin, which enters and binds to the DNA of the virus. Subsequent illumination destroys the DNA, prevents viral replication and allows the lesion to heal. This therapy is not without problems, however. One unfortunate male patient with genital herpes had the whole of his penis painted with a red dye and exposed to incandescent light. This helped deter the herpes virus, but soon afterwards skin cancer developed elsewhere on his penis. A better-established use of photodynamic therapy is in treatment of the jaundice often developed by premature infants soon after birth. The yellow colour is due to the accumulation in the skin of the pigment *bilirubin* (Fig. 2.7). The bulk of bilirubin is derived from the breakdown of the haemoglobin in worn-out red blood cells, and from the destruction of other haem proteins. It travels in the blood tightly bound to serum albumin, which

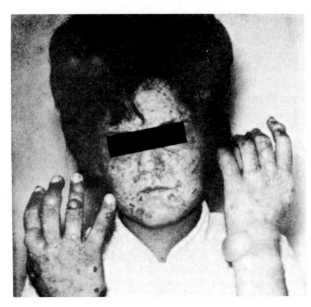

Fig. 2.8.   Skin and tissue damage in a porphyria patient. The porphyrins accumulated in porphyrias can be found in all body organs. Among other symptoms this is recognizable by deep-red staining of the tooth roots and the teeth themselves. The photosensitizing actions of porphyrins cause skin blistering and swelling.   From Franck, B., *Angew-Chem. Int. Ed.* **21,** 343 (1982), with permission.

can bind two moles of bilirubin per mole of protein. The liver takes up the bilirubin and converts it into a water-soluble product by the action of the enzyme *glucuronyl transferase*, which catalyses the reaction:

bilirubin + 2UDP-glucuronic acid → bilirubin diglucuronide + 2UDP.

The diglucuronide is disposed of by excretion into the bile. In premature babies sufficient glucuronyl transferase is often not present in the liver so that the lipid-soluble bilirubin accumulates in the blood, and deposits in tissues with a high lipid content, such as the brain, where it can cause irreversible damage. The jaundice can be reduced by careful exposure of the babies to blue light from a sunlamp, whereupon the pigment deposited in the skin sensitizes its own destruction in a reaction that involves singlet oxygen. In addition to this, and probably more important in quantitative terms, is a light-induced rearrangement of the structure of bilirubin to give water-soluble products that can be excreted. This rearrangement is called *photoisomerization*.

A third example of the application of photosensitization reactions in medicine is the use of psoralens (Fig. 2.7) in the treatment of skin diseases such as psoriasis. The treatment consists of the combined application of ultraviolet light in the wavelength range 320–400 nm (sometimes known as UVA) and a psoralen, and is often referred to as PUVA therapy (psoralen ultraviolet A). Psoralens are a class of compounds produced by plants and they are powerful photosensitizers. There is considerable debate about the safety of PUVA therapy, e.g. does it increase the risk of skin cancer?

It has been observed that certain porphyrins are taken up by cancerous tumours. After injection of a porphyrin derivative known as 'HPD' (which stands for haematoporphyrin derivative), fluorescent products are strongly retained by tumour tissues and this can be used to detect the presence of the tumour by observing the fluorescence. Irradiation with light of a wavelength absorbed by HPD can damage the tumour. Such reactions are of potential use in cancer chemotherapy, especially for skin cancer and for some forms of lung cancer, in which the tumour can be illuminated with light from a fibre-optic bronchoscope. Both hydroxyl radicals and singlet oxygen have been suggested to cause HPD-mediated tumour damage.

Photosensitization reactions are also of importance in veterinary medicine. Chlorophyll digestion in ruminants forms the pigment *phylloerythrin*, which is absorbed from the gut and excreted in the bile. Liver damage or malfunction can allow it to enter the bloodstream and deposit in the tissues. When the animals are exposed to sunlight the pigment in the skin causes damage that results in reddening and swelling. The fungal product *sporidesmin* achieves a similar effect by inhibiting bile production. (The toxicity of this substance is discussed further in Chapter 6, Section 6.15.) The buck-wheat plant synthesizes a compound that sensitizes singlet oxygen formation and its consumption by animals must be avoided. The same is true of the *St. John's wort* plant which produces the sensitizer *hypericin*. Japanese scientists have reported a light-induced dermatitis in some people who have ingested large quantities of tablets made from the alga *Chlorella,* and have attributed the damage to a product derived from chlorophyll.

Several drugs have been shown to sensitize singlet O_2 formation *in vitro*, including some phenothiazines (used as tranquillizers), tetracycline anti-biotics, and the compound benoxaprofen. This was introduced as an anti-inflammatory agent (Chapter 8, Section 8.3.4), but it is now rarely used because it can cause acute phototoxicity in patients. Some constituents of cosmetics, such as musk ambrette and musk xylene, can also act as photosensitizers.

2.7   Reactions of singlet oxygen

Singlet oxygen can interact with other molecules in essentially two ways: it can either combine chemically with them, or else it can transfer its excitation energy to them, returning to the ground state while the molecule enters an excited state. The latter phenomenon is known as *quenching*. Sometimes both can happen (Fig. 2.9). The best studied chemical reactions of singlet oxygen are those involving compounds that contain carbon–carbon double covalent bonds, C=C . Such bonds are present in many biological molecules, such as carotenes, chlorophyll, and the fatty-acid side-chains present in membrane lipids (Chapter 4). Compounds containing two double bonds separated by a single bond (known as *conjugated double bonds*) often react to give *endoperoxides*.

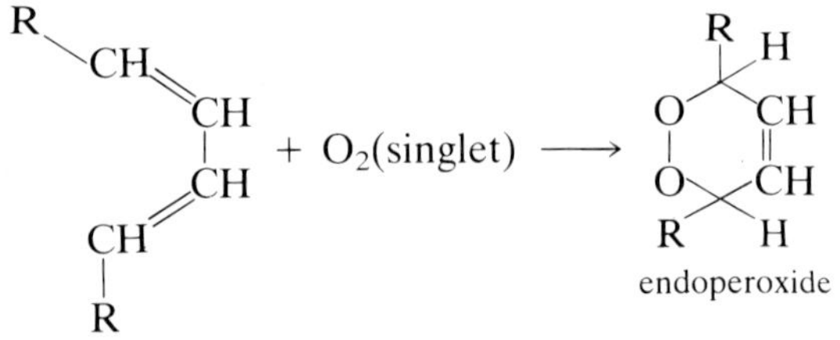

endoperoxide

Diphenylisobenzofuran reacts in this way (Fig. 2.9). If one double bond is present, the ene-reaction can occur—the singlet oxygen adds on and the double bond shifts to a different position:

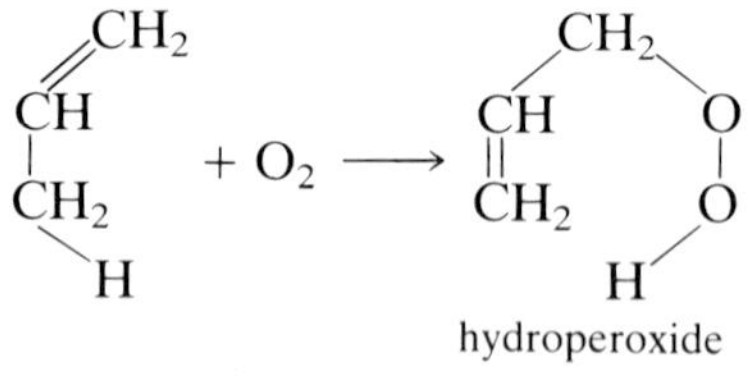

hydroperoxide

This is of importance in lipid peroxidation (Chapter 4).

If there is an electron-donating atom, such as N or S, adjacent to the double bond, singlet oxygen may react by *dioxetane* formation. Dioxetanes are unstable and decompose to give compounds containing the carbonyl group, C=O e.g.

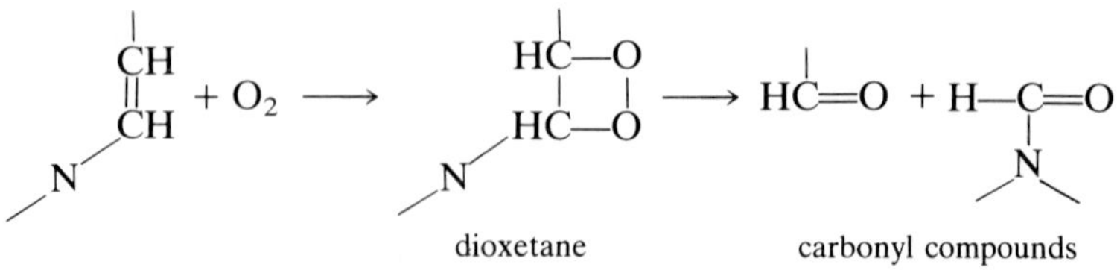

dioxetane   carbonyl compounds

Tryptophan can react in this way (Figure 2.9).

Compound	Structure	Type of reaction
DABCO (1,4-diazabicyclooctane)	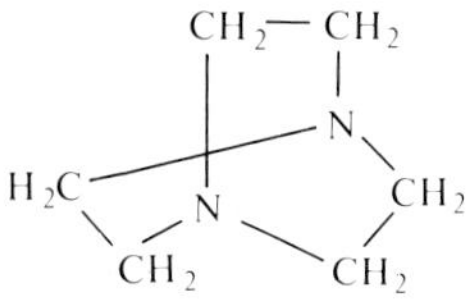	Quenching only. Fairly unreactive – about 50 mM needed in aqueous solution to quench half the $^1O_2$ formed
Azide ion	N_3^-	Mostly quenching. Much more effective than DABCO but inhibits many enzymes and scavenges $OH^\cdot$, thus not generally useful in biological systems
α-Tocopherol (vitamin E)	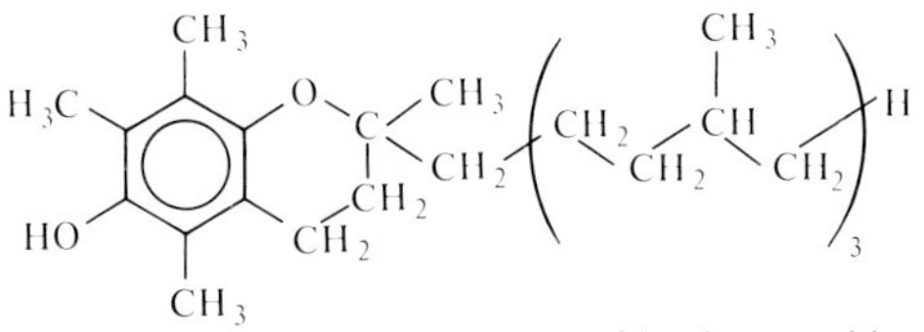	Mostly quenching but some chemical reaction to give 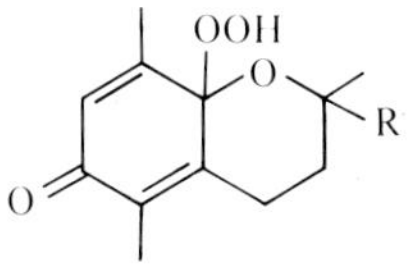 which decomposes to various products including α-tocopherylquinone (R is the side chain) 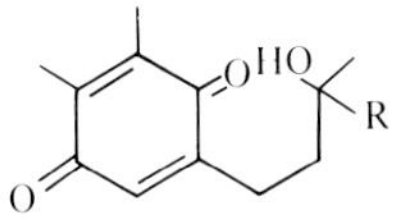
Phenols	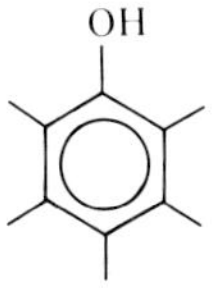	Both quenching and chemical reaction often of the type 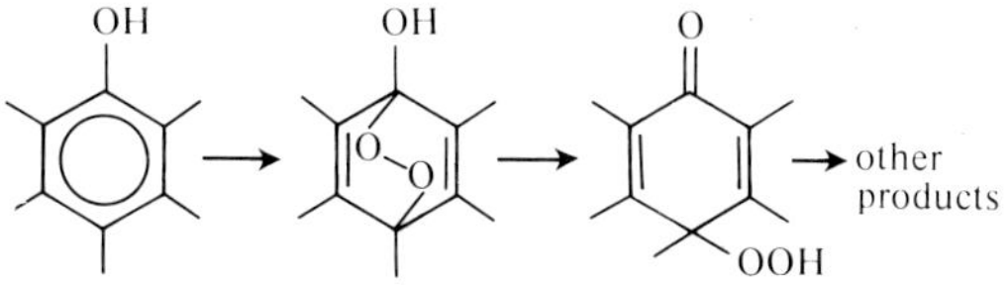 → other products
Bilirubin	(see fig. 2.7)	Mostly quenching but some chemical reaction

Fig. 2.9—*continued overleaf*

Compound	Structure	Type of Reaction
DNA	–	Complex mixture of products from all types of reaction with purine and pyrimidine bases, e.g. uracil hydroperoxide formation and subsequent decomposition
Cholesterol	C_8H_{17} HO	Major product is the $5\alpha-$ hydroperoxide C_8H_{17} HO OOH
β-Carotene	H_3C CH_3 CH_3 CH_3 H_3C H_3C H_3C CH_3 H_3C CH_3	Mostly quenching, some (very complex!) chemical reactions
Diphenylisobenzofuran (other furans react similarly)	O	Endoperoxide formation, accompanied by loss of the light absorption at 415 nm C_6H_5 O O C_6H_5
Tryptophan	$CH_2 CH COO^-$ NH_3^+ N H	Reacts by several mechanisms, including an ene-reaction to give initially OOH R N (R is the side-chain) and also formation of a dioxetane R O O N H which decomposes to N-formylkynurenine O NH_3^+ ‖	$CCH_2 CH COO^-$ NH CHO

Compound	Structure	Type of reaction

Methionine

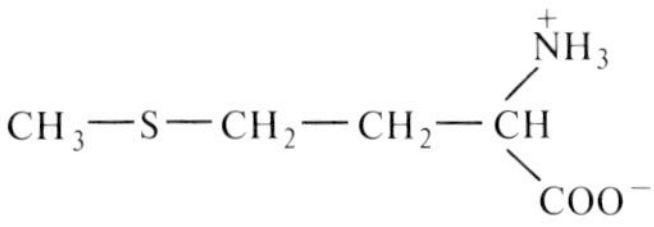

$$CH_3 - S - CH_2 - CH_2 - \overset{\overset{+}{N}H_3}{\underset{COO^-}{CH}}$$

Forms methionine sulphoxide

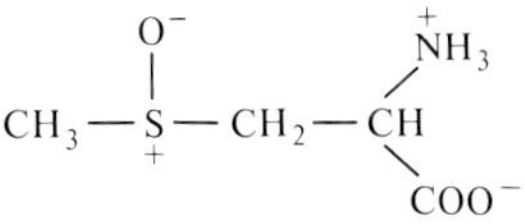

$$CH_3 - \overset{O^-}{\underset{+}{S}} - CH_2 - \overset{\overset{+}{N}H_3}{\underset{COO^-}{CH}}$$

Cysteine

$$HS - CH_2 - \overset{\overset{+}{N}H_3}{\underset{COO^-}{CH}} \quad (R - SH)$$

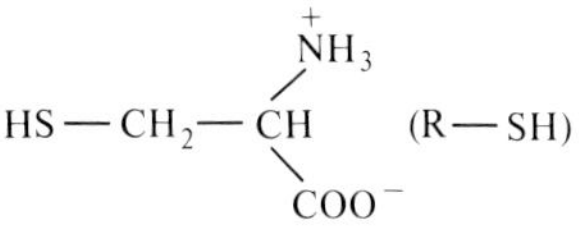

Reaction not well characterized: both disulphides $(R - S - S - R)$ and sulphonic acids $(R - SO_3H)$ are produced

Histidine

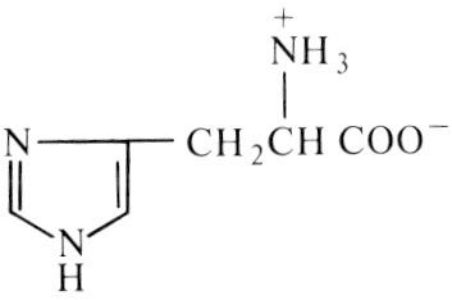

$$\overset{\overset{+}{N}H_3}{\underset{}{\;}}$$
N-----CH$_2$CH COO$^-$ (imidazole ring with N, N, H)

Probably reacts first to give an endoperoxide

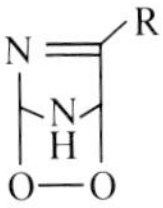

(endoperoxide structure with N=R, N, H, O—O)

which decomposes to a complex mixture of products

NADPH

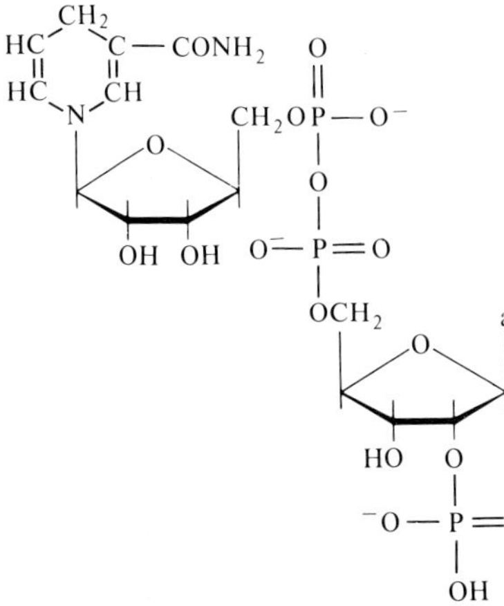

(structure of NADPH with nicotinamide ring: CH$_2$, HC, HC, C—CONH$_2$, N, CH; ribose rings with OH OH; phosphate groups O—P, O—P=O, OCH$_2$; adenine; HO; O—P=O; OH)

Reacts rapidly with singlet oxygen, being converted into NADP$^+$,

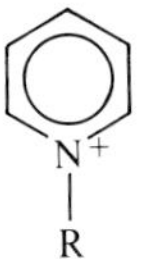

(pyridinium ring with N$^+$—R)

where R is the rest of the molecule

Fig. 2.9. Compounds that react with singlet oxygen. In a structure such as that of cholesterol, each 'angle' represents a carbon atom, e.g. ⬡ is an abbreviation

for (cyclohexane ring: CH$_2$, CH$_2$, CH$_2$, CH$_2$, CH$_2$, CH$_2$) and ⬡ an abbreviation for (cyclohexene ring: CH$_2$, CH$_2$, CH$_2$, CH, CH$_2$, CH)

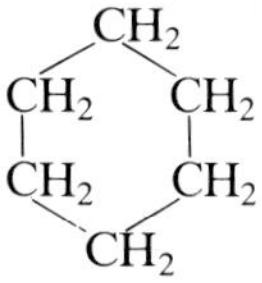

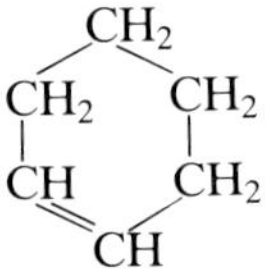

Figure 2.9 shows the reactions of singlet oxygen with several biologically important molecules. Damage to proteins by singlet oxygen is often due to oxidation of essential methionine, tryptophan, histidine, or cysteine residues. The reaction mechanisms are extremely complicated. Histidine has the highest second-order rate constant for reaction with singlet oxygen at $10^8 \, M^{-1} s^{-1}$, tryptophan 3×10^7, and methionine $1.7 \times 10^7 \, M^{-1} s^{-1}$. Photosensitization by dyes bound to the active site of enzymes has often been used to identify amino acid residues essential for the catalytic activity, it being especially useful in investigating the role of histidine residues. The histidine is destroyed at a controlled rate by careful illumination, and the effect on enzyme activity is measured.

2.8   Detection of singlet oxygen in biological systems

2.8.1   *Use of scavengers*

In distinguishing between damage caused by type I or by type II mechanisms in photosensitization reactions, or in investigating the role of singlet oxygen in other biological systems, scientists have often relied on the use of 'singlet-oxygen scavengers' and quenchers. Popular compounds have included DABCO, diphenylisobenzofuran, histidine, and azide (Fig. 2.9). Addition of these should inhibit a reaction dependent on singlet oxygen. If, when added at high concentrations, none of them inhibits the reaction under study, one may conclude that singlet oxygen is not required for it to proceed. If they do inhibit, this does *not* prove a role for singlet oxygen since all these compounds react with hydroxyl radical, often with a greater rate constant than the reaction with singlet oxygen. For example, azide reacts with OH˙ to give a reactive azide radical, N_3˙

$$N_3^- + OH˙ \rightarrow N_3˙ + OH^-$$

Fortunately, the products of reaction of cholesterol and tryptophan with singlet oxygen are different from those obtained on reaction with OH˙, so isolation and characterization of them allows distinction. This is hard work, however, and the use of singlet-oxygen scavengers, like that of spin-traps, cannot be recommended to the chemically-naïve.

2.8.2   *Deuterium oxide*

Another approach, particularly employed in studying photodynamic effects, is the use of deuterium oxide. The lifetime of singlet oxygen is longer in D_2O than in H_2O by a factor of ten or fifteen. Thus if a reaction in aqueous solution is dependent on singlet oxygen, carrying it out in D_2O instead should greatly potentiate the reaction. Theoretically a type I photodynamic reaction should be unaffected.

2.8.3 *Light emission*

Light emission (*luminescence*) has also been investigated. As singlet oxygen decays back to the ground state, some of the energy is emitted as light. The light from individual singlet oxygen molecules appears in the infra-red, but a so-called *dimol luminescence* in which two singlet oxygen molecules co-operate also occurs and produces light at 634 and 703 nm. Many other chemical reactions produce light, however (e.g. the treatment of several haem proteins with hydrogen peroxide), and so the mere production of light cannot be taken to imply singlet oxygen formation unless the exact spectrum has been measured. Light emission has been detected from activated phagocytes (Chapter 7), Fenton's reaction, and peroxidizing lipids (Chapter 4); but it can never be assumed that the light originates from singlet O_2 unless the spectrum of emitted light has been determined and shown to be that expected for singlet O_2. If, however, singlet oxygen is formed it might well excite adjacent molecules, which could emit light of different wavelengths as they return to the ground state. The decomposition of dioxetanes can produce carbonyl compounds in excited states, which then emit light as they return to the ground state.

Light emission is measured by an electronic device known as a photomultiplier, either in a liquid scintillation counter, or in a device specifically designed to measure luminescence (a *luminometer*). Following earlier work in Soviet laboratories, scientists in the USA and West Germany have carried out detailed studies on the light emitted by oxidant-generating systems *in vitro*, and by tissues and whole organs. It seems that the 'low level chemiluminescence' that can be detected from such systems (but not seen by the naked eye because the photon count is far too low to stimulate the retina) provides useful information on reactions involving oxidants, especially as measurement of the light emitted can be performed continuously, and it is a non-invasive method. For example, light emission by tissues and organs usually increases significantly if the surrounding O_2 concentration is raised. At present it is not often possible to identify conclusively the precise chemical reactions that are producing the light, although lipid peroxidation (Chapter 4, Section 4.6.6) often seems to contribute.

2.9 Reactions of the superoxide radical

The 'free-radical' hypothesis of Gershman and Gilbert (Chapter 1) was extended by J. M. McCord and I. Fridovich in the USA into a 'superoxide theory of oxygen toxicity' following their discovery in 1968 of an enzyme, superoxide dismutase, which specifically catalyses removal of the superoxide radical O_2^- (Chapter 3). This theory proposes that O_2^- formation is a major factor in oxygen toxicity and that superoxide dismutase enzymes constitute an essential defence against it. The current status of this theory is evaluated in detail in the next chapter, but here we shall lay the background to it by discussing the chemistry of the superoxide radical.

Superoxide chemistry differs greatly according to whether reactions are carried out in aqueous solution or in organic solvents, in many of which it is very stable. Various methods are available for producing superoxide for chemical studies. Firstly, oxygen may be reduced electrochemically in an appropriate electrolytic cell in the presence of an organic solvent such as dimethylsulphoxide or acetonitrile. Secondly, tetramethylammonium super-oxide, an ionic salt of formula $(CH_3)_4N^+O_2^-$, can be dissolved in a number of organic solvents. Thirdly, if potassium metal is burned in oxygen the pale-yellow ionic compound potassium superoxide, $K^+O_2^-$, is obtained. This is slightly soluble in organic solvents and its solubility can be increased by the addition of compounds called *crown ethers*. Essentially these are cyclic compounds, a hole in the centre of which can bind K^+. They are very soluble in organic solvents and so 'drag into solution' the central K^+ ion together with its associated O_2^-. A popular crown ether in KO_2 experiments is dicyclohexyl-18-crown-6 (Fig. 2.10). The reaction with O_2^- of another compound added to the organic solvent can be observed, or the O_2^--containing organic solvents can be mixed with an aqueous solution of the compound, often in stopped-flow experiments (Fig. 2.1). Lastly, O_2^- may be generated by pulse radiolysis of aqueous solutions as already described. Its presence may be detected by observing its UV-absorption spectrum (maximal absorption around 245 nm) or its ESR spectrum at low temperatures. Superoxide dissolved in the above organic solvents is very stable if water is kept away, but in aqueous solution it disappears rapidly. It has been

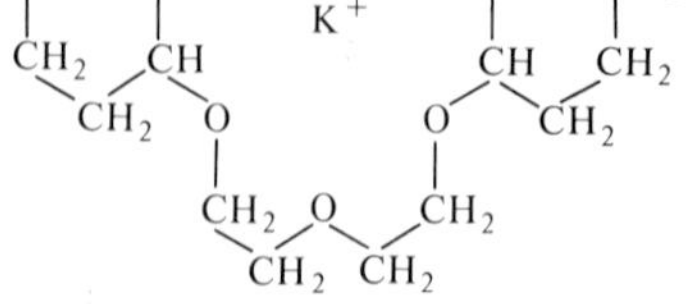

Fig. 2.10.   Structure of dicyclohexyl-18-crown-6. The K^+ ion fits into the central hole.

suggested that the surface of the planet Mars holds oxygen as a superoxide complex which is stable in the absence of water.

One of the reactions of superoxide in aqueous solution is to act as a *base*; a base being defined as an acceptor of protons (H^+ ions). When O_2^- accepts a proton it forms the *hydroperoxyl radical* ($HO_2^\cdot$). $HO_2^\cdot$ can dissociate to release H^+ ions again, i.e. it can act as a supplier of protons (an *acid*). Hence when O_2^- and H^+ ions are mixed together, an equilibrium is set up:

$$HO_2^\cdot \rightleftharpoons H^+ + O_2^-$$

The pH of a solution is defined as $-\log_{10}[H^+]$, the square brackets being used to denote concentration of H^+ in moles per litre. The pH at which the acid $HO_2^\cdot$) and base (O_2^-) forms are present in equal concentrations is known as the pK_a, approximately equal to 4.8 in this system. Since pH is $-\log_{10}[H^+]$, a pH of 4.8 corresponds to $[H^+]$ equal to $1.58 \times 10^{-5}\,mol\,l^{-1}$. pH and pK_a are related by the *Henderson–Hasselbalch equation*:

$$pH = pK_a + \log_{10}\frac{[\text{base}]}{[\text{acid}]}$$

or, in this case:

$$pH = 4.8 + \log_{10}\frac{[O_2^-]}{[HO_2^\cdot]}$$

At a pH of 3.8, $[O_2^-]/[HO_2^\cdot]$ is therefore $1/10$, whereas at pH 5.8 it is $10/1$. As the pH of most body tissues and fluids is in the range 6.4–7.5, then obviously the ratio of $[O_2^-]/[HO_2^\cdot]$ will be very large, e.g. $100/1$ at pH 6.8. Hence any O_2^- generated will remain almost entirely in this form rather than becoming protonated.

The reason for the disappearance of O_2^- in aqueous solution is the so-called *dismutation reaction*. The overall reaction may be represented as:

$$O_2^- + O_2^- + 2H^+ \rightarrow H_2O_2 + O_2$$

but in fact the rate constant for this reaction as written is virtually zero ($< 0.3\,M^{-1}\,s^{-1}$). By contrast, the reaction:

$$HO_2^\cdot + O_2^- + H^+ \rightarrow H_2O_2 + O_2$$

has $k_2 = 8 \times 10^7\,M^{-1}\,s^{-1}$, and the reaction:

$$HO_2^\cdot + HO_2^\cdot \rightarrow H_2O_2 + O_2$$

has $k_2 = 8 \times 10^5\,M^{-1}\,s^{-1}$

Dismutation is thus most rapid at the acidic pH values needed to protonate O_2^- and will become slower as the pH rises (i.e. becomes more alkaline) and the concentration of $HO_2^{.}$ in equilibrium with a given concentration of O_2^- decreases. For example, it may be calculated that in aqueous solution the dismutation reaction will have an overall rate constant of about $10^2 \, M^{-1} s^{-1}$ at pH 11 and about $5 \times 10^5 \, M^{-1} s^{-1}$ at pH 7.0. Any reaction undergone by O_2^- in aqueous solution will be in competition with this dismutation reaction, and it also follows that a system generating O_2^- must be producing H_2O_2.

As well as acting as a weak base, O_2^- in aqueous solution is a reducing agent, i.e. a donor of electrons. For example, it reduces cytochrome c, a haem protein. The iron at the centre of the haem ring is reduced from the Fe^{3+} to the Fe^{2+} state:

$$\text{cyt c} \, (Fe^{3+}) + O_2^- \rightarrow O_2 + \text{cyt c} \, (Fe^{2+})$$

The copper protein plastocyanin (Chapter 5) is also reduced:

$$\text{plastocyanin} \, (Cu^{2+}) + O_2^- \rightarrow O_2 + \text{plastocyanin} \, (Cu^+)$$

Superoxide reduces the yellow dye nitro-blue tetrazolium (NBT^{2+}) to produce a blue product called formazan, although the reaction mechanism is complex (Fig. 2.11). It can also both oxidize and reduce Fe^{2+}:

$$Fe^{3+} + O_2^- \rightleftarrows (Fe^{3+}-O_2^- \leftrightarrow Fe^{2+}-O_2) \rightleftarrows Fe^{2+} + O_2$$

intermediate complexes

$$2H^+ + Fe^{2+} + O_2^- \rightarrow Fe^{3+} + H_2O_2$$

The rates of these two reactions depend on the experimental conditions and on the molecule to which the iron ions are attached in the solution being studied. Complexes of Fe^{3+} with the chelating agent EDTA (Table 2.2) are still reduced by O_2^-:

$$Fe^{3+}-EDTA + O_2^- \rightleftarrows \text{(intermediate complexes)} \rightleftarrows Fe^{2+}-EDTA + O_2$$
$$k_2 = 1.3 \times 10^6 \, M^{-1} s^{-1} \text{ at pH 7}$$

whereas complexes of Fe^{3+} with DETAPAC or desferrioxamine (Table 2.2) are reduced much more slowly, if at all.

Superoxide in aqueous solution is also a weak oxidizing agent (electron acceptor). For example, it oxidizes Fe^{2+} (see above) and ascorbic acid. The second-order rate constant for the reaction of ascorbic acid with O_2^- has

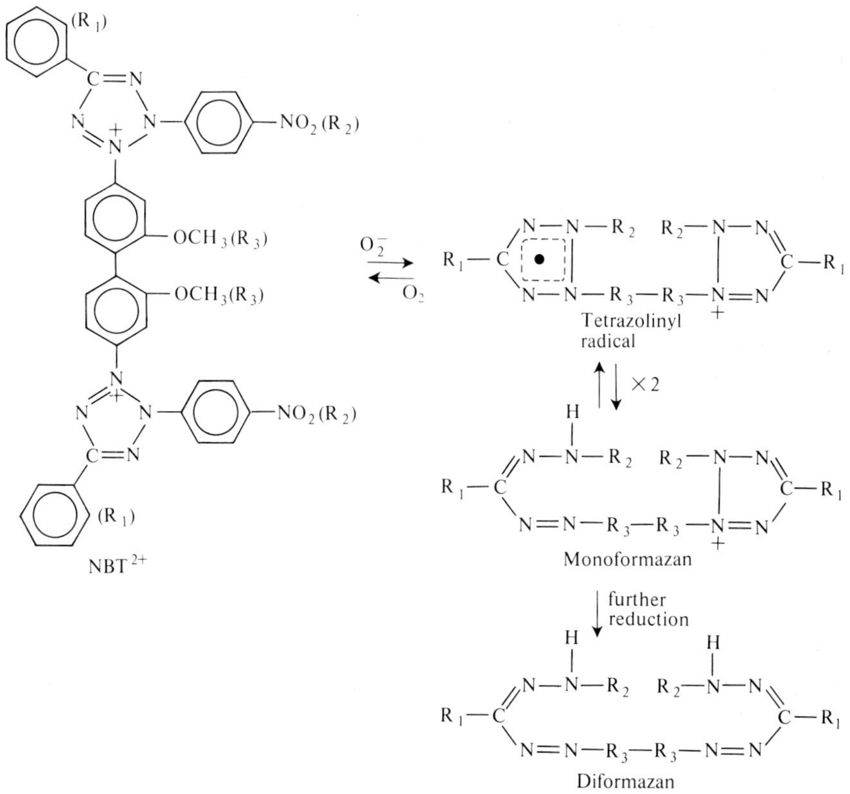

Fig. 2.11. Reduction of NBT^{2+} by superoxide radical. The detailed reaction mechanism is shown. It is probable that the tetrazolinyl radical can react with O$_2$ to form O$_2^-$, i.e. the initial reaction is reversible. Hence high O$_2$ concentrations depress formazan production from O$_2^-$ plus NBT^{2+}.

been quoted as $2.7 \times 10^5 \, \text{M}^{-1} \text{s}^{-1}$ at pH 7.4. Superoxide does not oxidize NADPH or NADH at measurable rates. However, it can interact with NADH bound to the active site of the enzyme lactate dehydrogenase (but not other dehydrogenases) to form an NAD· radical (Table 2.8):

$$\text{enzyme} - \text{NADH} + \text{O}_2^- + \text{H}^+ \rightarrow \text{enzyme} - \text{NAD}^\cdot + \text{H}_2\text{O}_2$$

Superoxide may slowly oxidize compounds containing the thiol group; its reaction with glutathione (GSH) is discussed further in Chapter 3 (Section 3.2.2).

The protonated form of O$_2^-$ (HO$_2^\cdot$) is a more powerful reducing agent and oxidant than is O$_2^-$ itself, although not much HO$_2^\cdot$ will be present at

Table 2.8. Rates of reaction of superoxide radical with various compounds in aqueous solution

Compound tested	pH	Second-order rate constant $(M^{-1} s^{-1})$	Reaction
Cholesterol, membrane lipids, pyruvate, aspartate, histidine, methionine, tryptophan, tyrosine, H_2O_2	8–10	< 1	Essentially none.
Cytochrome c (Fe^{3+})	8.5	2.6×10^5	Reduction to Fe^{2+} form.
Tetranitromethane	—	2×10^9	Reduction $C(NO_2)_4 + O_2^- \rightarrow C(NO_2)_3^- + NO_2 + O_2.$
Benzoquinone (and other quinones)	7	1×10^9	Reduction; an equilibrium is established (position depends on the quinone)
Catechol (and other diphenols)	7	1×10^9	Oxidation, to establish an equilibrium
Ascorbic acid	7.4	2.7×10^5	Semidehydroascorbate radical (Chapter 3) produced.

	pH	Rate constant	
NAD(P)H	7.4	<1	Essentially none.
NADH at active site of lactate dehydrogenase (other dehydrogenases not affected)	7.5	1×10^5	NAD$\cdot$ radical formed which reduces O_2 to O_2^- (forming NAD$^+$) and starts a chain reaction.
Plastocyanin	7.7	1×10^6	Reduction of Cu^{2+} at active site to Cu^+
Bilirubin	8.3	2.3×10^4	Bleaching; mechanism not characterized.
Nitrobluetetrazolium	7–11	6×10^4	NBT^{2+} reduced to a tetrazolium radical, two of which combine together to give a blue dye (monoformazan) and NBT^{2+}. For a detailed mechanism see Fig. 2.11

physiological pH. For example, unlike O_2^-, $HO_2^{\cdot}$ will directly oxidize NADH ($k_2 = 1.8 \times 10^5 \, M^{-1} s^{-1}$). It reduces cytochrome c with a rate constant of 2×10^6 as compared with O_2^- at 2.6×10^5 (Table 2.8). NADH bound at the active site of the enzyme glyceraldehyde-3-phosphate dehydrogenase is oxidized more rapidly by $HO_2^{\cdot}$ ($k_2 = 2 \times 10^7 \, M^{-1} s^{-1}$), although this enzyme, unlike lactate dehydrogenase (see above), does not promote reaction with O_2^- with bound NADH.

When superoxide is dissolved in organic solvents its ability to act as a base and as a reducing agent are increased. For example, it can reduce dissolved sulphur dioxide (SO_2) gas in organic solvents but not in aqueous solution:

$$SO_2 + O_2^- \rightarrow O_2 + SO_2^-$$

Also, if protons are not readily available, then dismutation is prevented and the O_2^- stays around much longer. Further, it gains the ability to act as a *nucleophile*, a reagent that is attracted to centres of positive charge in a molecule. Consider, for example, an ester molecule of general formula:

$$\begin{array}{c} O^{\delta-} \\ \parallel \\ R-C-O-R' \\ {\scriptstyle \delta+} \end{array}$$

where R and R' are hydrocarbon groups. Since oxygen is more electronegative than carbon, the carbonyl group is slightly polarized (Appendix I). O_2^- will be attracted to the $\delta+$ charge and will attack the molecule. Reactions have been suggested to occur as follows, although some scientists have expressed doubts about this mechanism:

$$\begin{array}{ccc}
\begin{array}{c} O \\ \parallel \\ R-C-O-R' \\ \uparrow \\ O_2^- \end{array}
& \longrightarrow &
\begin{array}{c} O^- \\ \mid \\ R-C-O-R' \\ \mid \\ O_2^{\cdot} \end{array}
\quad \longrightarrow \quad
\begin{array}{c} O \\ \parallel \\ R-C + R'-O^- \\ \mid \\ O_2^{\cdot} \end{array}
\end{array}$$

$$\begin{array}{c} O \\ \parallel \\ R-C-OO^{\cdot} + O_2^- \end{array} \longrightarrow \begin{array}{c} O \\ \parallel \\ R-C-OO^- + O_2 \end{array}$$

Superoxide can displace chloride ion from chlorinated hydrocarbons such as chloroform (trichloromethane, $CHCl_3$), tetrachloromethane (carbon tetrachloride, CCl_4), hexachlorobenzene (C_6Cl_6), and from some polychlorobiphenyls, important environmental toxins. For example, in the case of CCl_4 the reaction:

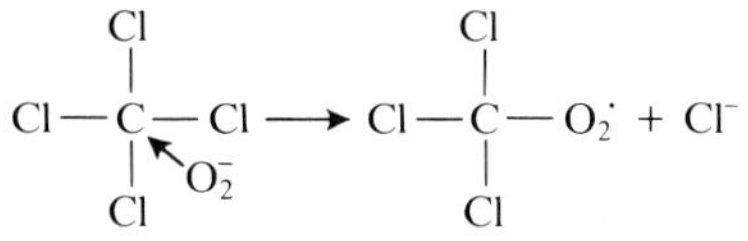

$$\underset{\underset{\text{Cl}}{|}}{\overset{\overset{\text{Cl}}{|}}{\text{Cl}-\text{C}-\text{Cl}}} \xrightarrow{\overset{}{\text{O}_2^-}} \underset{\underset{\text{Cl}}{|}}{\overset{\overset{\text{Cl}}{|}}{\text{Cl}-\text{C}-\text{O}_2^{\cdot}}} + \text{Cl}^-$$

peroxyl radical

is followed by further displacements.

The nucleophilicity of O_2^- in aqueous solution is, by contrast, very low, in part because of competition by the dismutation reaction.

The oxidizing capacity of O_2^- in organic solvents is only seen with compounds that can donate H^+ ions, such as ascorbate, catechol, and a-tocopherol (Table 2.8). Tocopherol (Ht) is slowly oxidized by O_2^- in organic solvents to give tocopheryl radical ($t^{\cdot}$); reactions suggested to account for this include:

$O_2^- + Ht \rightarrow t^- + HO_2^{\cdot}$ (deprotonation by O_2^- giving tocopherol ion)
$HO_2^{\cdot} + Ht \rightarrow H_2O_2 + t^{\cdot}$ (oxidation of tocopherol by $HO_2^{\cdot}$)
$O_2 + t^- \rightarrow O_2^- + t^{\cdot}$
$2t^{\cdot} \rightarrow dimer \rightarrow other\ products$

Similar proton transfers occur in the oxidation of other substances by O_2^- dissolved in organic solvents. In aqueous solution, however, O_2^- probably does not directly react with a-tocopherol, any oxidation being mediated by $HO_2^{\cdot}$ in equilibrium with O_2^- in the presence of water.

In the next chapter we shall examine the superoxide theory of oxygen toxicity in relation to O_2^- chemistry.

2.10   Hydrogen peroxide in biological systems

Any biological system generating O_2^- will produce hydrogen peroxide by the dismutation reaction unless, of course, all the O_2^- is intercepted by some other molecule (e.g. a high concentration of cytochrome c). Hydrogen peroxide production, probably mainly via O_2^-, has been frequently observed from mitochondria and microsomes *in vitro*; and the amount produced increases as the surrounding oxygen concentration is raised. There are also several enzymes that produce hydrogen peroxide without the intermediacy of free O_2^- radical. These include glycollate oxidase, D-aminoacid oxidase, and urate oxidase. Professor Britton Chance's group in the USA studied the rate of hydrogen peroxide production in the isolated perfused rat liver by observing the spectral intermediates of the enzyme catalase (Chapter 3) and arrived at a figure of 82 nmol (82×10^{-9} moles) of H_2O_2 produced per minute per gram of liver in normally-fed animals. Inclusion of glycollate or urate in the perfusion medium increased this rate, as the above oxidases become active. Much less of this hydrogen peroxide arose from the endoplasmic reticulum than would be expected from the rate at which microsomes produce hydrogen peroxide *in vitro*: remember that microsomes are artefacts of subcellular fractionation, and the membrane rearrangements undergone by them during homogenization and centrifugation may increase their production of O_2^- and hence hydrogen peroxide (this is discussed further in Chapter 3). Because the liver has effective mechanisms for disposing of hydrogen peroxide, they estimated the steady-state H_2O_2 concentration as being in the range 10^{-7}–10^{-9} moles per litre, hydrogen peroxide being continuously generated and destroyed. In other animal cells, with less effective H_2O_2-removal mechanisms, more hydrogen peroxide is found, e.g. the lens of the human eye contains 10–$25\,\mu\mathrm{mol\,l^{-1}}$ H_2O_2, and rabbit spermatozoa actually release hydrogen peroxide into the surrounding medium, again probably from O_2^-. Many bacteria and mycoplasmas (these are the smallest known organisms that exist independently and differ from bacteria in not having a cell wall) also release hydrogen peroxide into their surroundings, as do some blue-green algae upon illumination. Hydrogen peroxide is known to be produced during photosynthesis (Chapter 5) and during phagocytosis (Chapter 7), and hydrogen peroxide vapour has been detected in human expired air. The breathing of pure oxygen by experimental subjects increases the amount of hydrogen peroxide exhaled. Both H_2O_2 and O_2^- are produced by photochemical reactions in sea-water. Indeed, H_2O_2 at concentrations of 10^{-5}–$10^{-8}\,\mathrm{M}$ has been measured in rainwater, sea-water and river-water.

Hydrogen peroxide is a weak oxidizing agent and can inactivate a few enzymes directly, usually by oxidation of essential thiol (—SH) groups. Glyceraldehyde-3-phosphate dehydrogenase, an enzyme of the glycolytic pathway, is inactivated by H_2O_2 in this way. Thus exposure of cells to large doses of H_2O_2 can lead to ATP depletion by inhibition of glycolysis. Spinach chloroplast fructose bisphosphatase is similarly inactivated by H_2O_2 (Chapter 5). H_2O_2 is also capable of non-enzymically oxidizing certain keto-acids such as pyruvate (CH_3COCOO^-). Indeed, high concentrations of pyruvate added to culture media can protect cells against H_2O_2 by scavenging it. H_2O_2 at high concentrations is often used as a disinfectant. Some bacterial strains are very sensitive to it, and some animal cells grown in culture (e.g. human fibroblasts) can be damaged by hydrogen peroxide added to the culture medium at concentrations in the micromolar range. Hydrogen peroxide produced by, for example, *Mycoplasma pneumoniae*, attacks epithelial cells in the trachea.

Hydrogen peroxide can cross cell membranes rapidly whereas O_2^- usually cannot. Once inside the cell, H_2O_2 can probably react with Fe^{2+}, and possibly Cu^+, ions to form the hydroxyl radical and this may be the origin of many of its toxic effects, as discussed in Section 2.4.3. Evidence supporting OH· as an important mediator of H_2O_2 toxicity comes from several observations. For example, the killing of bacterial spores by hydrogen peroxide can be related to their content of transition-metal ions. The killing of *Staphylococcus aureus* cells by hydrogen peroxide was much more efficient if they had been grown in medium with a high iron content, so increasing the intracellular amount of iron. Supplying the cells with dimethylsulphoxide, which penetrates into the cell and reacts with OH·, decreased the toxicity, and the production of methane (Table 2.7) could be detected. Hydroxyl-radical scavengers that could not enter the cells, such as mannitol, had little effect. Hydrogen peroxide also enhances the damaging effect of near-ultraviolet radiation on bacteria and viruses—usually the damage seen in the presence of both agents is much greater than that done by each agent alone (a *synergistic effect*). It is possible that the UV light can cause homolytic fission of the hydrogen peroxide and hence increase the production of OH·

$$H_2O_2 \xrightarrow[\text{energy}]{} 2OH·$$

The damage appears to affect DNA especially, causing single-strand breaks and DNA-protein cross-links.

Let us now go on and look in more detail at the biological importance of some of the reactions discussed above.

2.11 Further reading

Andrae, U., Singh, J., and Siegle-Skylakakis, K. (1985). Pyruvate and related α-ketoacids protect mammalian cells in culture against hydrogen peroxide-induced cytotoxicity. *Toxicol. Lett.* **28**, 93.

Anon (1977). Photodye herpes therapy—Cassandra confirmed? *J. Am. Med. Assoc.* **238**, 133.

Aruoma, O. I. and Halliwell, B. (1987). Superoxide-dependent and ascorbate-dependent formation of hydroxyl radicals from hydrogen peroxide in the presence of iron. Are lactoferrin and transferrin promoters of hydroxyl radical generation? *Biochem. J.* **241**, 273.

Bielski, B. H. J. *et al.* (1985). Reactivity of HO_2/O_2^- radicals in aqueous solution. *J. Phys. Chem. Ref. Data* **14**, 1041.

Borg, D. C. and Schaich, K. M. (1984). Cytotoxicity from coupled redox cycling of autoxidizing xenobiotics and metals. *Isr. J. Chem.* **24**, 38.

Bors, W., Saran, M., and Michel, C. (1982). Assays of oxygen radicals. Methods and mechanisms. In *Superoxide dismutase* (L. W. Oberley, Ed), vol. II, p. 31. CRC Press, Florida, USA.

Boveris, A., Cadenas, E., and Chance, B. (1981). Ultraweak chemiluminescence: a sensitive assay for oxidative radical reactions. *Fed. Proc.* **40**, 195.

Brewer, G. J. *et al.* (1987). Treatment of Wilson's disease with zinc. Prevention of reaccumulation of hepatic copper. *J. Lab. Clin. Med.* **109**, 526.

Britigan, B. E. *et al.* (1987). Detection of the production of oxygen-centred free radicals by human neutrophils using spin trapping techniques: a critical perspective. *J. Leukocyte Biol.* **41**, 349.

Britton, R. S., Bacon, B. R., and Recknagel, R. O. (1987). Lipid peroxidation and associated hepatic organelle dysfunction in iron overload. *Chem. Phys. Lipids* **45**, 207.

Brodie, A. E. and Reed, D. J. (1987). Reversible oxidation of glyceraldehyde 3-phosphate dehydrogenase thiols in human lung carcinoma cells by hydrogen peroxide. *Biochem. Biophys. Res. Commun.* **148**, 120.

Butler, J. and Halliwell, B. (1982). Reaction of iron-EDTA chelates with the superoxide radical. *Arch. Biochem. Biophys.* **218**, 174.

Carson, D. A. *et al.* (1986). DNA strand breaks, NAD metabolism and programmed cell death. *Exp. Cell. Res.* **164**, 273.

Cederbaum, A. I. and Dicker, E. (1983). Inhibition of microsomal oxidation of alcohols and of hydroxyl radical scavenging agents by the iron-chelating agent desferrioxamine. *Biochem. J.* **210**, 107.

Chan, P. C. and Bielski, B. H. J. (1980). Glyceraldehyde-3-phosphate dehydrogenase-catalysed chain oxidation of reduced nicotinamide adenine dinucleotide by per-hydroxyl radicals. *J. Biol. Chem.* **255**, 874.

Chang, L. Y. L. and Packer, L. (1979). Damage to hepatocytes by visible light. *FEBS Lett.* **97**, 124.

Clare, D. A. *et al.* (1984). Effect of molecular oxygen on detection of superoxide radical with nitroblue tetrazolium and on activity stains for catalase. *Anal. Biochem.* **140**, 532.

Clare, N. J. (1955). Photosensitisation in animals. *Adv. Vet. Sci. Comp. Med.* **2**, 182.

Crichton, R. R. and Charloteaux-Waters, M. (1987). Iron transport and storage. *Eur. J. Biochem.* **164**, 485.

Cross, C. E. *et al.* (1987). Oxygen radicals and human disease. *Ann. Intern. Med.* **107**, 526.

Crum, L. A. *et al.* (1987). Free radical production in amniotic fluid and blood plasma by medical ultrasound. *J. Ultrasound Med.* **6**, 643.

Dahl, T. A., Midden, R. W., and Hartman, P. E. (1988). Pure exogenous singlet oxygen: nonmutagenicity in bacteria. *Mutat Res.* **201**, 127.

Dizdaroglu, M. *et al.* (1987). Ionizing-radiation-induced damage in the DNA of cultured human cells. Identification of 8,5 cyclo-2'-deoxyguanosine. *Biochem. J.* **241**, 929.

Dodge, A. D. and Knox, J. P. (1986). Photosensitisers from plants. *Pestic. Sci.* **17**, 579.

Duran, N. (1982). Singlet oxygen in biological processes. In *Chemical and biochemical generation of excited states*, p. 345. Academic Press, New York.

Ebert, M. *et al.* (Eds) (1965). *Pulse radiolysis*. Academic Press, London and New York.

Ekström, G. *et al.* (1987). Cytochrome P-450 dependent ethanol oxidation. Kinetic isotope effects and absence of stereoselectivity. *Biochemistry* **26**, 7348.

Ennever, J. F. *et al.* (1987). Rapid clearance of a structural isomer of bilirubin during phototherapy. *J. Clin. Invest.* **79**, 1674.

Feierabend, J. and Engel, S. (1986). Photoinactivation of catalase *in vitro* and in leaves. *Arch. Biochem. Biophys.* **251**, 567.

Foote, C. S. (1979). Detection of singlet oxygen in complex systems: a critique. In *Biochemical and clinical aspects of oxygen* (W. S. Caughey, Ed.) p. 603. Academic Press, New York.

Foote, C. S. (1981). Photo-oxidation of biological model compounds. In *Oxygen and oxy-radicals in chemistry and biology* (M. A. J. Rodgers and E. L. Powers, Eds.) p. 425. Academic Press, New York.

Golden, M. N. H. (1987). Free radicals in the pathogenesis of Kwashiorkor. *Proc. Nutr. Soc.* **46**, 53.

Gutteridge, J. M. C. and Halliwell, B. (1987). Radical-promoting loosely-bound iron in biological fluids and the bleomycin assay. *Life Chem. Rep.* **4**, 113.

Gutteridge, J. M. C. and Stocks, J. (1981). Caeruloplasmin: physiological and pathological perspectives. *CRC Crit. Rev. Clin. Lab. Sci.* **14**, 257.

Gutteridge, J. M. C. *et al.* (1985). The behaviour of caeruloplasmin in stored human extracellular fluids in relation to ferroxidase II activity, lipid peroxidation and phenanthroline-detectable copper. *Biochem. J.* **230**, 517.

Hall, R. D. *et al.* (1987). Near-infrared detection of singlet molecular oxygen produced by photosensitization with promazine and chlorpromazine. *Photochem. Photobiol.* **46**, 295.

Halliwell, B. and Grootveld, M. (1987). The measurement of free radical reactions in humans. Some thoughts for future experimentation. *FEBS Lett.* **213**, 9.

Halliwell, B. and Gutteridge, J. M. C. (1986). Oxygen free radicals and iron in relation to biology and medicine: some problems and concepts. *Arch. Biochem. Biophys.* **246**, 501.

Harrison, P. M. and Hoare, R. J. (1980). *Metals in biochemistry.* Chapman and Hall, London.

Hassan, T. and Khan, A. U. (1986). Phototoxicity of the tetracyclines: photosensitized emission of singlet delta oxygen. *Proc. Natl. Acad. Sci. USA* **83**, 4604.

Hershko, C. and Peto, T. E. A. (1987). Non-transferrin plasma iron. *Br. J. Haematol.* **66**, 149.

Hoigne, J. and Bader, H. (1975). Ozonation of water: role of hydroxyl radicals as oxidising intermediates. *Science* **190**, 782.

Hussain, M. Z. *et al.* (1985). Stimulation of poly(ADP-ribose) synthetase activity in the lungs of mice exposed to a low level of ozone. *Arch. Biochem. Biophys.* **241**, 477.

Hyslop, P. A. *et al.* (1988). Mechanisms of oxidant-mediated cell injury. The glycolytic and mitochondrial pathways of ADP phosphorylation are major intracellular targets inactivated by hydrogen peroxide. *J. Biol. Chem.* **263**, 1665.

Johnson, G. R. A. and Nazhat, N. B. (1987). Kinetics and mechanism of the reaction of the *bis*(1,10-phenanthroline) copper(I) ion with hydrogen peroxide in aqueous solution. *J. Am. Chem. Soc.* **109**, 1990.

Kasai, H. and Nishimura, S. (1986). Hydroxylation of guanine in nucleosides and DNA at the C-8 position by heated glucose and oxygen radical-forming agents. *Environ. Health Perspect.* **67**, 111.

Knowles, P. F. *et al.* (1969). ESR evidence for enzymic reduction of oxygen to a free radical: the superoxide ion. *Biochem. J.* **111**, 53.

Kontoghiorghes, G. *et al.* (1987). Effective chelation of iron in thalassaemia with the oral chelator 1,2-dimethyl-3-hydroxypyrid-4-one. *Br. Med. J.* **295**, 1509.

Korycka-Dahl, M. and Richardson, T. (1977). Photogeneration of superoxide anion in serum of bovine milk and in model systems containing riboflavin and amino acids. *J. Dairy Sci.* **61**, 400.

Kwon, B. M. and Foote, C. S. (1988). Chemistry of singlet oxygen. 50. Hydroperoxide intermediates in the photoxygenation of ascorbic acid. *J. Am. Chem. Soc.* **110**, 6582.

Martin, J. P. and Logsdon, N. (1987). The role of oxygen radicals in dye-mediated photodynamic effects in *E. coli*. B. *J. Biol. Chem.* **262**, 7213.

Mason, R. P. *et al.* (1985). Free radical metabolites of toxic chemicals. *Environ. Health Perspect.* **64**, 3.

Mello Filho, A. C., Hoffmann, M. E., and Meneghini, R. (1984). Cell killing and DNA damage by hydrogen peroxide are mediated by intracellular iron. *Biochem. J.* **218**, 273.

Modell, B. *et al.* (1982). Survival and desferrioxamine in thalassaemia major. *Br. Med. J.* **284**, 1081.

Morgan, T. R. *et al.* (1988). Free radical production by high-energy shock waves—comparison with ionizing radiation. *J. Urol.* **139**, 186.

Nordmann, R., Ribière, C., and Rovach, H. (1987). Involvement of iron and iron-catalysed free radical production in ethanol metabolism and toxicity. *Enzyme* **37**, 57.

O'Connell, M. J., Baum, H., and Peters, T. J. (1986). Haemosiderin-like properties of free-radical-modified ferritin. *Biochem. J.* **240**, 297.

Oshino, N., Jamieson, D., and Chance, B. (1975). The properties of hydrogen peroxide production under hyperoxic and hypoxic conditions of perfused rat liver. *Biochem. J.* **145**, 53.

Pall, H. S. *et al.* (1987). Raised cerebrospinal-fluid copper concentration in Parkinson's disease. *Lancet* **ii**, 238.

Pathak, M. A. and Joshi, P. C. (1984). Production of active oxygen species ($^1O_2$ and O_2^-) by psoralens and ultraviolet radiation. *Biochim. Biophys. Acta* **798**, 115.

Puppo, A. and Halliwell, B. (1988). Formation of hydroxyl radicals in biological systems. Does myoglobin stimulate hydroxyl radical formation from hydrogen peroxide? *Free Radical Res. Commun.* **4**, 415.

Rice-Evans, C. and Halliwell, B. (Eds) (1988). *Free Radicals, Methodology and Concepts*. Richelieu Press, London.

Riesz, P., Berdahl, D., and Christman, C. L. (1985). Free radical generation by ultrasound in aqueous and nonaqueous solution. *Environ. Health Perspect.* **64**, 233.

Rosen, G. M. and Turner, M. J. (1988). Synthesis of spin traps specific for hydroxyl radical. *J. Med. Chem.* **31**, 428.

Rush, J. D. and Bielski, B. H. J. (1985). Pulse radiolytic studies of the reaction of HO_2/O_2^- with Fe(II)/Fe(III) ions. The reactivity of HO_2/O_2^- with ferric ions and its implication on the occurrence of the Haber–Weiss reaction. *J. Phys. Chem.* **89**, 5062.

Sawyer, D. T. and Valentine, J. S. (1981). How super is superoxide? *Acc. Chem. Res.* **14,** 393.

Schafer, A. I. *et al.* (1981). Clinical consequences of acquired transfusional iron overload in adults. *New Engl. J. Med.* **304,** 319.

Scholes, G. (1983). Radiation effects on DNA. *Br. J. Radiol.* **56,** 221.

Schraufstatter, I. U. *et al.* (1986). Oxidant injury of cells. *J. Clin. Invest.* **77,** 1312.

Sies, H. (1987). Intact organ spectrophotometry and single photon counting. *Arch. Toxicol.* **60,** 138.

Sugimoto, H., Matsumoto, S., and Sawyer, D. T. (1987). Oxygenation of polychloro aromatic hydrocarbons by a superoxide ion in aprotic media. *J. Am. Chem. Soc.* **109,** 8081.

Teschke, R. and Gellert, J. (1986). Hepatic microsomal ethanol-oxidizing system (MEOS): metabolic aspects and clinical implications. *Alcoholism Clin. Exp. Res.* **10,** 20S.

Turrens, J. F., Freeman, B. A., and Crapo, J. D. (1982). Hyperoxia increases H_2O_2 release by lung mitochondria and microsomes. *Arch. Biochem. Biophys.* **217,** 411.

Wardman, P. (1978). Application of pulse radiolysis methods to study the reactions and structure of biomolecules. *Rep. Prog. Phys.* **41,** 259.

Zigler, J. S. and Goosey, J. D. (1981). Photosensitized oxidation in the ocular lens: evidence for photosensitizers endogenous to the human lens. *Photochem. Photobiol.* **33,** 869.

3

Protection against oxidants in biological systems: the superoxide theory of oxygen toxicity

3.1 Protection by enzymes

3.1.1 *Protection against hydrogen peroxide by catalase and peroxidases*

We saw in Chapter 2 that hydrogen peroxide is damaging in living systems, often because it can give rise to the formation of $OH^{\cdot}$ radicals. It is therefore biologically advantageous for cells to control the amount of hydrogen peroxide that is allowed to accumulate.

Two types of enzyme exist to remove hydrogen peroxide within cells. They are *the catalases*, which catalyse the reaction:

$$2H_2O_2 \rightarrow 2H_2O + O_2$$

and *the peroxidases*, which bring about the general reaction:

$$SH_2 + H_2O_2 \rightarrow S + 2H_2O$$

in which SH_2 is a substrate that becomes oxidized. The oxygen produced by catalase is ground-state: no singlet oxygen can be detected.

Catalase
Most aerobic cells contain catalase activity, although a few do not, such as the bacterium *Bacillus popilliae*, *Mycoplasma pneumoniae*, the green alga *Euglena*, several parasitic helminths (e.g. the liver fluke), and the blue-green alga *Gloeocapsa*. A few anaerobic bacteria, such as *Propionibacterium shermanii* also contain catalase, but most do not. In animals catalase is present in all major body organs, being especially concentrated in liver and erythrocytes. The brain, heart, and skeletal muscle contain only low amounts, however, although the activity does vary between muscles and even in different regions of the same muscle (Table 3.1).

Most purified catalases have been shown to consist of four protein subunits, each of which contains a haem (Fe(III)—protoporphyrin) group

Table 3.1. Catalase and glutathione peroxidase activities in normal human tissues

Tissue		Catalase activity (mg^{-1} protein)	Glutathione peroxidase activity (mg^{-1} protein)
Liver	A	1300	190
	B	1500	120
Erythrocytes	A	990	19
	B	1300	19
Kidney cortex	A	430	140
	B	110	87
Adrenal gland	B	300	120
Kidney medulla	A	700	90
	B	220	73
Spleen	A	56	50
Lymph node	A	120	160
Pancreas	A	100	43
	B	120	110
Lung	A	210	53
	B	180	54
Heart	A	54	69
Skeletal muscle	A	36	38
	B	25	22
Brain grey-matter	A	11	71
	B	3	66
Brain white-matter	A	20	76
Adipose tissue	A	270	77
	B	560	89

Data were abstracted from Marklund *et al.* (1982). *Cancer Res.* **42**, 1955–61. Glutathione peroxidase was assayed with a hydroperoxide substrate (see text and Chapter 4). Results are expressed as enzyme activity per milligram (10^{-3} g) of protein. Two individuals were used, denoted A and B, as sources of tissue samples.

bound to its active site. Each subunit also usually contains one molecule of NADPH bound to it, which helps to stabilize the enzyme. Dissociation of catalase into its subunits, which easily occurs on storage, freeze-drying, or exposure of the enzyme to acid or alkali, causes loss of catalase activity. The three-dimensional structures of catalase from beef liver and from the fungus *Penicillium vitale* have been determined by X-ray crystallography.

The catalase reaction mechanism may be written as follows:

$$\text{catalase}-\text{Fe(III)} + H_2O_2 \xrightarrow{k_1} \text{compound I}$$

$$\text{compound I} + H_2O_2 \xrightarrow{k_2} \text{catalase}-\text{Fe(III)} + H_2O + O_2$$

For rat liver catalase, the two second-order rate constants, k_1 and k_2, have values of $1.7 \times 10^7 \, \mathrm{M}^{-1}\mathrm{s}^{-1}$ and $2.6 \times 10^7 \, \mathrm{M}^{-1}\mathrm{s}^{-1}$, respectively. Formation of compound I leads to characteristic changes in the absorption spectrum of the molecule (Fig. 3.1). The exact structure of compound I is uncertain—the iron is oxidized to a nominal valency of Fe(V) but the extensive charge-delocalization in haem rings (Appendix I) makes description of the exact structure difficult. It is probably intermediate in structure between a ferric peroxide $(\mathrm{Fe(III)-HOOH})$ and $\mathrm{Fe(V)}{=}\mathrm{O}$. It is very difficult to saturate catalase with hydrogen peroxide—its maximal velocity (V_{max}) for the destruction of hydrogen peroxide is enormous. However, the above equations show that complete removal of hydrogen peroxide requires the impact of two molecules of hydrogen peroxide upon a single active site, which becomes less likely as hydrogen peroxide concentrations fall. The amount of compound I present in a mixture of catalase and hydrogen peroxide depends on the concentrations of catalase and hydrogen peroxide and on the rate constants k_1 and k_2. It may be calculated that at all reasonable concentrations, the rate of removal of hydrogen peroxide is given by the equation:

$$\text{moles } H_2O_2 \text{ used } 1^{-1}\mathrm{s}^{-1} = 2k_2[H_2O_2]\,[\text{compound I}]$$

$$= 2k_1[H_2O_2]\,[\text{free catalase}].$$

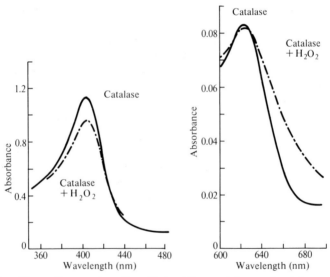

Fig. 3.1. Absorption spectra of purified rat liver catalase and catalase compound I. The large absorbance of catalase around 400 nm is known as the *Soret band*, and is seen with all haem proteins. An absorption spectrum is simply a plot of the amount of light absorbed by the protein as a function of the wavelength of that light.

If the concentration of hydrogen peroxide is fixed, the initial rate of removal of it will be proportional to the concentration of catalase present and hence will be higher in liver than in, say, brain or heart. Similarly, for a given concentration of catalase, the initial rate of hydrogen peroxide removal will be proportional to the hydrogen peroxide concentration. As a result, the specific activities (μmoles H_2O_2 decomposed per min per mg protein) quoted by manufacturers for their catalase preparations are meaningless unless they describe exactly how the assay was done. Hydrogen peroxide decomposition can be followed by the loss of its light absorbance at 240 nm, or by measuring the release of oxygen by using an oxygen electrode.

Catalase activity can be inhibited by azide or cyanide, but these inhibit many other enzymes. A more useful inhibitor is *aminotriazole* (Fig. 3.2) which inhibits catalase activity when fed to animals or plant tissues. Its inhibitory action is exerted on compound I, so it will only inhibit catalase if hydrogen peroxide is present to allow generation of this intermediate. This H_2O_2-dependent inhibition of catalase by aminotriazole has been used *in vitro* to measure low rates of hydrogen peroxide production by various biological systems (e.g. wood-rotting fungi and *Mycoplasma*) and the fact that aminotriazole inhibits catalase when it is fed to whole animals or plants indicates that hydrogen peroxide must be being produced *in vivo*.

Catalase is also capable of bringing about certain peroxidase-type reactions in the presence of a steady supply of hydrogen peroxide, which allows formation of compound I. Compound I will oxidize the alcohols methanol (CH_3OH) and ethanol (CH_3CH_2OH) to their corresponding aldehydes HCHO (formaldehyde or methanal) and CH_3CHO (acetaldehyde or ethanal); but propanols or butanols are much poorer substrates. In spinach leaves, formic acid (HCOOH, methanoic acid) is oxidized to carbon dioxide by the peroxidase action of compound I. It can also oxidize nitrite ion (NO_2^-) into nitrate (NO_3^-) *in vitro* and it has been suggested that it can oxidize elemental mercury (Hg) absorbed into the human body to form Hg^{2+} ions. The presence of peroxidatic substrates for catalase *in vivo* will decrease the concentration of compound I, causing more free catalase to be formed, and so this is yet another variable that must be considered in assessing how quickly hydrogen peroxide is removed. The separated catalase subunits show little catalase activity, but have peroxidase activity on a wider range of

Fig. 3.2. Structure of aminotriazole. The full name of this compound is 3-amino-1,2,4-triazole.

substrates, including NADH. This has no physiological significance, but is of interest in considering the active site chemistry.

The drug cyanamide ($H_2N—C \equiv N$) is converted *in vivo* into a product that inhibits the enzyme aldehyde dehydrogenase, which oxidizes ethanal (acetaldehyde) generated by ethanol metabolism (Chapter 2, Section 2.4.5). Consumption of ethanol after taking cyanamide causes ethanal accumulation, with unpleasant symptoms, and so cyanamide is used as an alcohol deterrent. It has been suggested that catalase is responsible for oxidizing cyanamide to the product that inhibits aldehyde dehydrogenase.

The catalase activity of animal and plant tissues is largely located in subcellular organelles bounded by a single membrane and known as *peroxisomes*. Although a significant proportion of the catalase activity detected in homogenates of animal and plant tissues is found not to be bound to organelles, this could be, in part or in whole, due to the rupture of fragile peroxisomes during the homogenization (as subcellular fractionation techniques have improved, the amount of soluble catalase observed has decreased). However, some non-peroxisomal catalase may occur in the livers of a few animals, such as guinea-pigs.

Peroxisomes also contain some of the cellular H_2O_2-generating enzymes, such as glycollate oxidase (see legend to Fig. 3.3), urate oxidase and the flavoprotein dehydrogenases involved in the β-oxidation of fatty acids (a metabolic pathway that operates in both mitochondria and peroxisomes in animal tissues). Mitochondria (at least in liver), chloroplasts, and the endoplasmic reticulum contain little, if any, catalase activity, so any hydrogen peroxide they generate *in vivo* cannot be disposed of in this way. It has been reported, however, that rat heart mitochondria contain some catalase activity in the matrix. One must realize that isolated subcellular fractions, especially microsomes, may be heavily *contaminated* with catalase activity and this can confuse experimental results. For example, the MEOS system (Chapter 2, Section 2.4.5) was argued by some workers to be merely due to the peroxidase action of contaminating catalase on ethanol using hydrogen peroxide produced by the microsomal fraction, although further work has shown this not to be the case. Some ethanol can be oxidized by catalase *in vivo*, however (see Fig 3.3), although this would occur in the peroxisomes.

Britton Chance in the USA has pioneered the direct observation of the absorption spectrum of catalase compound I in perfused animal organs, or organs *in situ*, as a means of assessing the rate of intracellular H_2O_2-production. Light is shone through a portion of, say, the liver or kidney, and the transmitted light analysed. The intracellular concentrations of hydrogen peroxide in the liver, quoted in Chapter 2, were obtained by this method. Fig. 3.3 shows a typical experimental result. An alternative approach has been to measure the rate at which radioactively-labelled (^{14}C) methanol and formate are oxidized to $^{14}CO_2$ by organs or tissues, based on the assumption that these

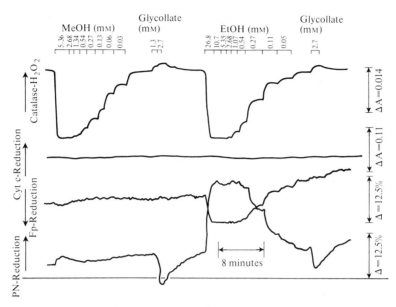

Fig. 3.3. Production of hydrogen peroxide in perfused rat liver. Rat liver was perfused with a bicarbonate–saline solution containing 2 mM L-lactate and 0.3 mM pyruvate at 30 °C. Light was shone through a liver lobe and the concentration of catalase compound I measured by dual-wavelength spectrophotometry. There is a steady concentration of compound I from which the endogenous rate of H_2O_2 production can be calculated. Inclusion of methanol in the perfusion medium reduces compound I concentration because it is a substrate for the peroxidase action of catalase. Ethanol has a similar effect, but it also causes an increased reduction of pyridine nucleotides (PN) since it is a substrate for the alcohol dehydrogenases which convert NAD^+ to NADH (Chapter 2). There is also some reduction of flavoproteins (Fp). Infusion of glycollate raises the steady-state concentration of compound I because it is oxidized in the peroxisomes to form hydrogen peroxide by glycollate oxidase:

$$glycollate + O_2 \rightarrow glyoxylate + H_2O_2$$

Key: PN, pyridine nucleotides; Fp, flavoproteins; cyt-c, cytochrome c. Data from Oshino and Chance, *Arch. Biochem. Biophys.* **154**, 117–31 (1973), with permission.

oxidations are entirely due to the peroxidase action of catalase. It must be checked that this is true before these methods can be used, e.g. spinach leaves contain an NAD^+-dependent formate dehydrogenase activity which catalyses the oxidation of formate to carbon dioxide, in addition to the catalase activity.

Manganese-containing catalases

Several micro-organisms, such as some strains of pediococci, streptococci, and lactobacilli, contain 'pseudocatalase', an H_2O_2-degrading enzyme that is insensitive to inhibition by azide or cyanide and does not contain haem. Purification of an enzyme of this type from *Lactobacillus plantarum* showed it to have a relative molecular mass of around 172 000 and to consist of six subunits, each of which contains one Mn^{3+} ion. A similar enzyme from *Thermoleophilum album* contained only four subunits, however. It is thought that the Mn^{3+} becomes oxidized to a Mn(V) state by H_2O_2, so that the H_2O_2 is decomposed by a reaction mechanism similar to that for haem catalase:

$$\text{catalase-Mn}^{3+} + H_2O_2 + 2H^+ \rightarrow \text{catalase-Mn(V)} + 2H_2O$$

$$\text{catalase-Mn(V)} + H_2O_2 \rightarrow \text{catalase-Mn}^{3+} + 2H^+ + O_2$$

The Mn-catalase of *L. plantarum* has been shown to be important in protecting this organism against H_2O_2.

Glutathione peroxidase

The enzyme glutathione peroxidase was discovered in animal tissues in 1957 by G. C. Mills in the USA. It is not generally present in higher plants or bacteria, although it has been reported in some algae and fungi. Its substrate is the low-molecular-weight thiol compound glutathione, which is found in animals, plants, and some bacteria (e.g. in *E. coli* but not in most anaerobic bacteria) at concentrations that are often in the millimolar (10^{-3} moles per litre) range (Table 3.2). The structure of glutathione is shown in Fig. 3.4. Most free glutathione *in vivo* is present as GSH rather than GSSG, but up to one-third of the total cellular glutathione may be present as 'mixed' disulphides with other compounds that contain —SH groups, such as cysteine, coenzyme A, and the —SH of the cysteine residues of several proteins. If R—SH is used to represent these other molecules, then the mixed disulphides have the general formula:

$$\text{glu—cys—gly}$$
$$|$$
$$\text{S—S—R}$$

Glutathione peroxidase catalyses the oxidation of GSH to GSSG at the expense of hydrogen peroxide,

$$H_2O_2 + 2GSH \rightarrow GSSG + 2H_2O$$

It is found at high activity in liver, moderate activity in heart, lung, and brain, and low activity in muscle (Tables 3.1 and 3.2). The enzyme

Table 3.2. Presence of glutathione and enzymes using it in different organisms

System studied	[GSH]	Ratio GSH/GSSG	Glutathione peroxidase activity	Glutathione reductase activity
Spinach chloroplasts	3.0 mM	> 10/1	Absent	High
Rat tissues				
liver	7–8 mM	> 10/1	High	High
erythrocyte	2 mM	> 10/1	Moderate	Moderate
heart	2 mM	> 10/1	Moderate	Moderate
lung	2 mM	> 10/1	Moderate	Moderate
lens	6–10 mM	> 10/1	Moderate	Moderate
spleen	4–5 mM	> 10/1	—	—
kidney	4 mM	> 10/1	Moderate	Moderate
brain	2 mM	> 10/1	Moderate	Moderate
skeletal muscle	1 mM	> 10/1	Low	Low
blood plasma	0.02–0.03 mM	~ 5/1	Low	Low
adipose tissue	3.2 µg per 10^6 cells	> 100/1	Low	Low
Human tissue				
liver	4 µmol g^{-1} wet weight	> 10/1	High	High
lens	6–10 mM	> 10/1	Moderate	Moderate
erythrocytes	240 µg ml^{-1} blood	> 10/1	Moderate	Moderate
N. crassa	20 µmol g^{-1} dry weight	150/1	Absent	Moderate
E. coli				
aerobically-grown	27 µmol g^{-1}	> 10/1	Absent	—
anaerobically-grown	7 µmol g^{-1}			

Whenever possible, concentrations are expressed as millimoles per litre (mM) but these cannot always be calculated from published data. These GSH values are not to be taken too literally, since they (i) decrease with age in animals, (ii) are different at different times of day in animals and at different points of the growth cycle in bacteria and fungi, (iii) in liver, fall on starvation, and (iv) will vary in the different cell types present in animal organs. Consumption of ethanol also decreases liver GSH concentrations.

is specific for GSH as a hydrogen donor but will accept other peroxides as well as hydrogen peroxide (Chapter 4). It is made up of four protein subunits, each of which contains one atom of the element selenium at its active site. Selenium is in group VI of the Periodic Table (see Appendix I) and has properties intermediate between those of a metal and a non-metal. It is probably present at the active site as seleno-cysteine, the amino acid cysteine in which the normal sulphur atom has been replaced by a selenium atom (R—SeH instead of R—SH, where R is

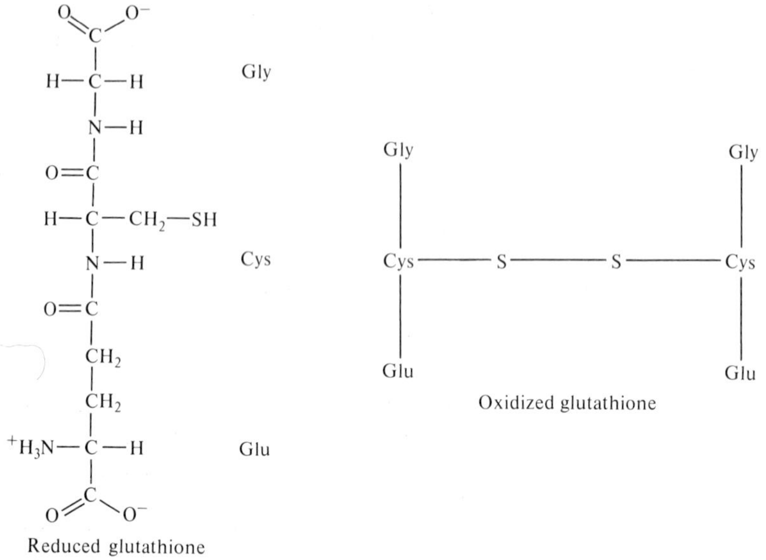

Fig. 3.4. Structure of glutathione. Glutathione is a simple tripeptide (glutamic acid–cysteine–glycine) in its reduced form, usually abbreviated to GSH. In the oxidized form, GSSG, two GSH molecules join together as the —SH groups of cysteine are oxidized to form a disulphide bridge, —S—S.

$$\overset{+}{N}H_3$$
$$|$$

—CH$_2$—CH—COO$^-$). The GSH apparently reduces the selenium and the reduced form of the enzyme then reacts with hydrogen peroxide with an approximate rate constant of $5 \times 10^7\,\mathrm{M^{-1}\,s^{-1}}$. Trace amounts of selenium are required in animal diets, although it is toxic in excess, and several of the symptoms of selenium deficiency can be explained by the resulting lack of glutathione peroxidase. It must not be assumed that this is the only biochemical role performed by selenium, however (see Chapter 4 for a further discussion).

The ratios of GSH/GSSG in normal cells are kept high (Table 3.2) so there must be a mechanism for reducing GSSG back to GSH. This is achieved by *glutathione reductase* enzymes, which catalyse the reaction:

$$\text{GSSG} + \text{NADPH} + \text{H}^+ \rightarrow 2\text{GSH} + \text{NADP}^+$$

Glutathione reductases can also catalyse reduction of certain 'mixed disulphides', such as that between GSH and coenzyme A. The NADPH required is mainly provided in animal tissues by a complex metabolic pathway known as the *oxidative pentose phosphate pathway*. The first enzyme in this pathway is *glucose 6-phosphate dehydrogenase*,

glucose 6-phosphate + NADP$^+$ → 6-phosphogluconate

$$+ \text{NADPH} + \text{H}^+$$

followed by *6-phosphogluconate dehydrogenase*,

6-phosphogluconate + NADP$^+$ → CO_2 + NADPH

$$+ \text{H}^+ + \text{ribulose 5-phosphate.}$$

The rate at which the pentose phosphate pathway operates is controlled by the supply of NADP$^+$ to glucose 6-phosphate dehydrogenase. As glutathione reductase operates and lowers the NADPH/NADP$^+$ ratio, the pentose phosphate pathway speeds up to replace the NADPH.

Glutathione reductases contain two protein subunits, each with the flavin FAD (Chapter 2) at its active site. Apparently the NADPH reduces the FAD, which then passes its electrons onto a disulphide bridge ($-S-S-$) between two cysteine residues in the protein. The two $-SH$ groups so formed then interact with GSSG and reduce it to 2GSH, re-forming the protein disulphide.

Cerami and others in the USA have shown that 'glutathione reductases' from certain trypanosomes, such as *Crithidia fasciculata* or *Trypanosoma brucei*, require a GSH derivative for their action. This co-factor consists of glutathione covalently linked to the polyamine spermidine, and it has been called *trypanothione* (Fig. 3.5). It has been suggested that inhibitors of trypanothione-dependent removal of H_2O_2 might be therapeutically useful in treating human tropical diseases caused by parasitic trypanosomes, e.g. African sleeping sickness, *Chagas' disease* and leishmaniasis.

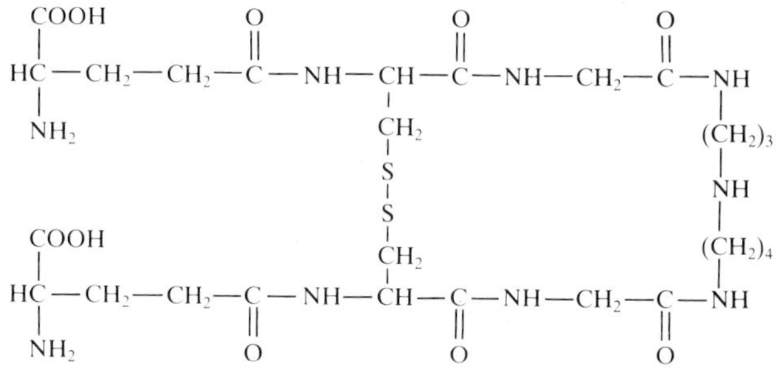

Fig. 3.5. Structure of trypanothione. Trypanothione consists of glutathione (Fig. 3.4) covalently bonded to the polyamine spermidine.

Co-operation between catalase and glutathione peroxidase in animal tissues in the removal of hydrogen peroxide

Brain and spermatozoa contain little catalase activity but more glutathione peroxidase, so the question as to which enzyme is more important in removing hydrogen peroxide *in vivo* is easily answered. A number of animal tissues contain both enzymes, however, so how do they co-operate with each other? Mammalian erythrocytes (red blood cells) contain no subcellular organelles, and both catalase and glutathione peroxidase enzymes float around in the cell sap, although it is possible that some catalase might be attached to the inside of the erythrocyte membrane. The normal low rate of production of hydrogen peroxide in these cells (via superoxide dismutase— see Fig. 3.16) seems to be mainly dealt with by glutathione peroxidase. Indeed, humans suffering from an inborn defect in the catalase gene which produces an unstable mutant enzyme and so decreases erythrocyte catalase activities, show no life-threatening harmful effects. If the concentration of hydrogen peroxide is raised, e.g. by supplying these cells with a drug that increases intracellular H_2O_2 generation, then catalase becomes more important.

Mammalian erythrocytes operate the pentose phosphate pathway in order to provide NADPH for glutathione reduction. However, over a hundred million people, principally in tropical and Mediterranean areas, have an inborn defect in the gene coding for glucose 6-phosphate dehydrogenase, so that its activity in erythrocytes is reduced below normal. This deficiency does cause some damage to the erythrocyte membranes, but it is not usually severe enough to cause clinical symptoms unless the rate of H_2O_2-production in erythrocytes is increased, e.g. by certain drugs. If the rate of H_2O_2-production exceeds the capacity of the enzyme to generate NADPH, then GSH/GSSG ratios fall, and glutathione peroxidase stops working, leading to destruction of the red blood cells (*haemolysis*), anaemia, and jaundice (Chapter 2) due to the excessive degradation of released haemoglobin. The survival of the defective gene in such large numbers in tropical areas has often been suggested to occur because erythrocytes with lowered dehydrogenase activity are resistant to the presence of malarial parasites within them. This could possibly be because the malarial parasite uses NADPH from its host erythrocyte to maintain its own internal GSH concentration, and so cannot live within cells deficient in NADPH generation. Indeed, as a result of suggestions made by I. A. Clark in Australia, there is now considerable interest in the role of free-radical generators in the eradication of malarial parasites (see Chapter 6 for further discussion). An inborn deficiency of erythrocyte glutathione peroxidase or in the synthesis of GSH itself is much less common, but can result in severe haemolysis.

Liver contains high concentrations of both catalase and glutathione peroxidase. Whereas catalase is largely or entirely in the peroxisomes, the latter enzyme is found mainly in the cytosol but also in the matrix of mitochondria (about one-third of the total in rats but rather less in humans). The distribution of GSH is similar. Thus hydrogen peroxide produced by, say, glycollate oxidase and urate oxidase in the peroxisomes is largely disposed of by catalase, whereas hydrogen peroxide arising from mitochondria, the endoplasmic reticulum, or soluble (cytosolic) enzymes such as superoxide dismutase (Section 3.1.3) is acted upon by the peroxidase. The capacity of the glutathione system to cope in other tissues depends on the activity of peroxidase, glutathione reductase, and the pentose phosphate pathway enzymes. The glutathione content of tissues also varies at different times of day (Table 3.2). In lung, eye, and muscle the capacity of the system is restricted. For example, inhibition of catalase present in the eyes of rabbits by feeding aminotriazole to the animals caused the concentration of hydrogen peroxide in the aqueous humour of the eye to rise from about 0.06 to 0.15 mM even though glutathione reductase or peroxidase activities were unaffected. The glutathione system here cannot cope with the extra load caused by the loss of catalase activity. Feeding young rabbits with aminotriazole can cause cataracts to develop (cataract is defined as clinically-significant loss of lens transparency), perhaps due to oxidation of lens proteins (see Chapter 5).

The rate of operation of the glutathione peroxidase system *in vivo* can be assessed in a number of ways. One approach has been to measure the pentose phosphate pathway activity by supplying $[1\text{-}^{14}C]$-labelled glucose to the tissue and measuring the release of radioactive $^{14}CO_2$ in the 6-phosphogluconate dehydrogenase reaction (see above). An increased pathway activity has been observed upon exposing isolated perfused rat lung, ox retina, or erythrocytes to elevated oxygen concentrations, presumably as more NADPH is consumed by glutathione reductase as it deals with increased GSSG production from glutathione peroxidase. An alternative approach has been to measure GSSG release: if cells are treated with chemical reagents that oxidize internal GSH to GSSG (such as *diamide*) they rapidly eject the GSSG into the surrounding medium. In the whole liver, GSSG is released into the bile. This rate of release of GSSG in perfused organs can be taken as a measure of glutathione peroxidase activity if glucose is omitted from the perfusing medium, so that NADPH cannot be produced by the pentose phosphate pathway for glutathione reductase activity. Exposure of isolated perfused liver and lung to elevated oxygen concentrations causes a rapid increase in GSSG release. It must be noted that glutathione peroxidase acts on organic hydroperoxides in addition to hydrogen peroxide (Chapter 4) and so the increased GSSG release cannot be entirely attributed to the latter molecule. Figure 3.6 shows the rate of GSSG release when hydrogen peroxide is infused into a perfused rat liver—the saturation of the effect is probably related to the increased action of catalase at the higher H_2O_2-concentrations. Inclusion

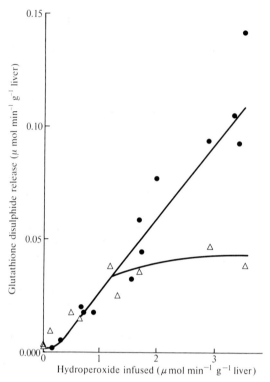

Fig. 3.6. Release of GSSG from the isolated perfused liver during infusion of hydrogen peroxide. Hydrogen peroxide was included in the perfusion medium at the concentrations stated (open triangles). The closed circles show the results when an organic hydroperoxide was used (see Chapter 4 for discussion).

of glycollate in the perfusing medium at physiological concentrations causes only a small increase in GSSG release, indicating that hydrogen peroxide generated by glycollate oxidase in the peroxisomes is largely disposed of by catalase, as expected. If aminotriazole is used to inhibit catalase, glycollate infusion then does cause a marked increase in GSSG release.

Cytochrome c peroxidase
The haem-containing enzyme cytochrome c peroxidase is found between the inner and outer membrane of yeast mitochondria, which contain no catalase or glutathione peroxidases. It is also found in some bacteria. Cytochrome c peroxidase reacts rapidly with hydrogen peroxide to form a stable enzyme–substrate complex that has an absorption maximum at 419 nm, whereas that of the free enzyme is at 407 nm. On addition of reduced cytochrome c [Fe(II)] it is oxidized to the Fe(III) form and the enzyme returns to its resting state:

$$\text{enzyme} + H_2O_2 \rightarrow \text{complex} (\lambda_{max} \text{ 419 nm})$$

$$\text{complex} + 2\,\text{cyt-c}(Fe^{2+}) \rightarrow \text{enzyme} + 2\,\text{cyt-c}(Fe^{3+}) + 2OH^-$$

Spectrophotometric measurement of the intermediate complex has been used as a highly accurate and sensitive method for measuring rates of H_2O_2-formation. This method has been applied to protozoa and to isolated mitochondrial and peroxisomal fractions in the laboratory.

NADH peroxidase and oxidase
Several bacteria, such as *Lactobacillus casei* or *Streptococcus faecalis*, contain a peroxidase which uses hydrogen peroxide to oxidize NADH into NAD$^+$. An aerotolerant mutant of the anaerobe *Clostridium perfringens* was reported to have gained such an NADH peroxidase. This enzyme should not be confused with the *NADH oxidase* enzyme found in some bacteria, in which NADH is oxidized to NAD$^+$, and oxygen simultaneously reduced to water. Several aerotolerant anaerobic bacteria synthesize NADH oxidase on exposure to oxygen; by its action it reduces oxygen to water and so removes it from the environment of the bacteria. Not until this enzyme has been overwhelmed by too much oxygen can irreversible damage occur. The NADH oxidase is useful because it allows the cell to survive exposure to limited amounts of oxygen, but the NADH it requires must be provided by metabolic pathways such as glycolysis, as is also true for NADH peroxidase. Drainage of the cell's reducing equivalents away from biosynthetic reactions into the NADH peroxidase or oxidase reactions could contribute to the growth-inhibitory effects of oxygen. This is better than cell death however.

'Non-specific' peroxidases
Plants and bacteria often harbour haem-containing peroxidases (Fig. 3.7) that are capable of acting on a very wide range of substrates. They are usually assayed in cell extracts by using artificial substrates, which they oxidize in the presence of hydrogen peroxide to give coloured products. Such artificial substrates include guaiacol (produced by certain millipedes), benzidine, and *o*-dianisidine. Often the substrates of these peroxidases *in vivo* have not been identified. 'Non-specific' peroxidases have also been found in a few animal systems. For example, *lactoperoxidase* is found in milk and saliva. It may function to regulate the growth of some strains of bacteria since, among many other substrates, it can oxidize thiocyanate (SCN$^-$) ions, which are found in both milk and saliva, into hypothiocyanite (OSCN$^-$), which is very toxic to some bacterial strains, including *Escherichia coli*, *Streptococci*, and *Salmonella typhimurium*. Lactoperoxidase may be one of the factors in milk that protects babies against infections of the gastrointestinal tract. The hydrogen peroxide that it needs to oxidize SCN$^-$ seems to be derived from

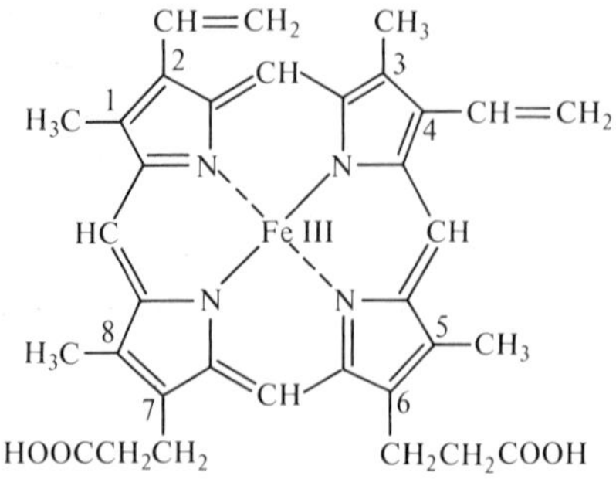

Fig. 3.7. Most peroxidases contain the structure shown, iron(III) protoporphyrin IX, at their active site. Of those that do not, e.g. myeloperoxidase, most contain similar porphyrin structures.

some of the other strains of bacteria present, which excrete hydrogen peroxide into their surroundings. *Myeloperoxidase*, another 'non-specific' peroxidase, is found in phagocytic cells and will be further considered in Chapter 7. *Thyroid peroxidase* is found in the thyroid gland and probably serves to oxidize iodide ion (I^-) into iodine and to attach it to the thyroid hormones. A *uterine peroxidase* has also been described. Its function is unknown and it is possible that the enzyme is produced by eosinophils (Chapter 7) present within the tissue. In general, however, 'non-specific' peroxidases are not widely distributed in animal tissues.

Perhaps the most-studied non-specific peroxidase in biology is *horseradish peroxidase*, obtained from the roots of the horseradish plant (*Armoracia lapathifolia*). Several different forms of the enzyme exist, each containing bound carbohydrate and calcium ions, but they all have very broad substrate specificity. For example, they will oxidize guaiacol, pyrogallol, CN^- ion, NADH, thiol compounds, phenols, and the plant hormone, indoleacetic acid (auxin). Oxidations by horseradish peroxidase, and probably by most other plant peroxidases, can in almost all cases be represented by the following series of reactions, in which SH_2 is the substrate:

$$\text{peroxidase} + H_2O_2 \rightarrow \text{compound I}$$
$$\text{compound I} + SH_2 \rightarrow SH^{\cdot} + \text{compound II}$$
$$\text{compound II} + SH_2 \rightarrow SH^{\cdot} + \text{peroxidase}$$

The iron in the haem ring of 'resting' peroxidase is in the oxidation state Fe(III) (Fig. 3.7). Hydrogen peroxide removes two electrons to give compound I; the exact structure of this is unknown but it probably contains iron in the Fe(IV) oxidation state (perhaps as Fe(IV)=O), the extra oxidizing capacity being located elsewhere in the active site. The two electrons are replaced in two one-electron steps, in each of which a substrate molecule forms a radical, SH. Compound II is the intermediate state of the enzyme.

The substrate-derived radicals (SH) then usually undergo a disproportionation reaction, one reducing the other to SH_2 and simultaneously being itself oxidized to S (*disproportionation* is any reaction in which one atom or molecule is reduced and an identical atom or molecule oxidized).

$$SH + SH \rightarrow S + SH_2$$

However, radicals of some substrates can react with oxygen and reduce it to superoxide:

$$SH + O_2 \rightarrow S + O_2^- + H^+$$

Superoxide can give H_2O_2 by the dismutation reaction (Chapter 2)

$$O_2^- + O_2^- + 2H^+ \rightarrow H_2O_2 + O_2$$

Thus, when horseradish peroxidase is oxidizing a substrate whose SH radical reduces O_2 to O_2^-, only catalytic quantities of H_2O_2 need be added to cause oxidation of the substrate. For example, oxidation of NADH by horseradish peroxidase occurs without addition of H_2O_2, since traces of H_2O_2 are always present in NADH solutions. Oxidation produces the NAD radical, which can reduce O_2 to O_2^-, i.e.

$$\text{Peroxidase} + H_2O_2 \rightarrow \text{compound I}$$
$$\text{compound I} + \text{NADH} \rightarrow \text{compound II} + \text{NAD} + H_2O$$
$$\text{compound II} + \text{NADH} \rightarrow \text{peroxidase} + \text{NAD} + H_2O$$
$$2\text{NAD} + 2O_2 \rightarrow 2\text{NAD}^+ + 2O_2^-$$
$$2O_2^- + 2H^+ \rightarrow H_2O_2 + O_2 \text{ (dismutation)}$$

overall reaction: $2\text{NADH} + O_2 + 2H^+ \rightarrow 2\text{NAD}^+ + 2H_2O$

NAD radicals can also join together to give an NAD dimer:

$$\text{NAD} + \text{NAD} \rightarrow (\text{NAD})_2$$

but this reaction is slower (k_2 3×10^7 $M^{-1}s^{-1}$) than the reaction of NAD with O_2 (k_2 1.0×10^9 $M^{-1}s^{-1}$). Superoxide radical formed by reaction of NAD with O_2 can combine with horseradish peroxidase to generate a form of the enzyme known as *oxyperoxidase*, or *compound III*:

$$\text{enzyme-Fe}^{3+} + O_2^- \rightarrow \text{enzyme (Fe}^{2+}-O_2)$$

Oxyperoxidase oxidizes NADH very slowly, and so its accumulation during the reaction slows down the overall rate of NADH oxidation.

NADH oxidation is one example of the so-called *oxidase reactions* of peroxidase, as compared to its 'normal' reactions in which equal amounts of SH_2 and H_2O_2 must be provided and no O_2 is consumed. Oxidase reactions occur when the SH˙ radicals (e.g. NAD˙) can reduce O_2 directly.

Horseradish peroxidase and lactoperoxidase can oxidize GSH into a thiyl radical (Chapter 1) in the presence of H_2O_2

$$H_2O_2 + 2GSH \rightarrow 2H_2O + 2GS˙$$

The glutathione thiyl radical can then participate in several reactions that result in O_2 uptake, e.g.

$$GS˙ + GS^- \xrightarrow{\hspace{2cm}} G\dot{S}\bar{S}G$$

(ionized
form of
GSH)

$$G\dot{S}\bar{S}G + O_2 \xrightarrow{\hspace{2cm}} GSSG + O_2^-$$

superoxide

$$GS˙ + O_2 \xrightarrow{\hspace{2cm}} GSO_2˙$$

The exact reactions taking place depend upon the assay conditions and the enzyme used. Horseradish peroxidase also oxidizes the amino-acid cysteine to a thiyl radical, and a similar series of reactions occurs.

Another example of an oxidase reaction of peroxidase is its oxidation of 2-nitropropane $(CH_3CH(NO_2)CH_3)$, a compound which has been widely used as a solvent and as a component of inks, paints and varnishes (USA production in 1982 was estimated as 30 million pounds). Again, peroxidase initiates nitropropane oxidation to generate intermediate radicals that reduce O_2 to O_2^-. This then participates in the continued, non-enzymic oxidation of the nitropropane.

Lignification of plant cell walls involves the polymerization of a number of phenols derived from the aromatic amino acid phenylalanine,

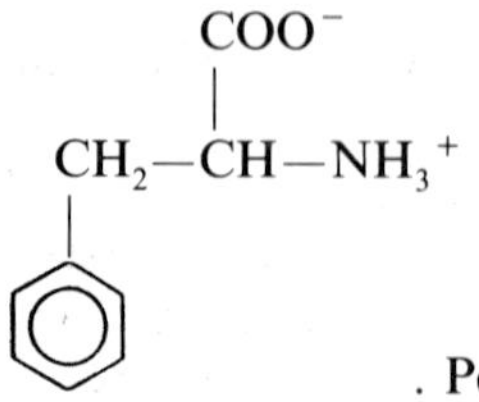

COO⁻
|
$CH_2-CH-NH_3^+$
|

. Peroxidase bound to the cell walls functions to oxidize

these phenols into *phenoxyl radicals* which polymerize to form the lignin. One source of the hydrogen peroxide required for this oxidation may be the simultaneous oxidation by peroxidase of NADH generated by a malate dehydrogenase enzyme, also bound to the cell walls. Plant peroxidases are involved in the degradation of the hormone, indoleacetic acid (auxin), and thus in the regulation of plant growth. This reaction is much more complicated than the usual peroxidase reactions, however.

Apart from these cases, the identity of the *in vivo* substrates of plant and bacterial 'non-specific' peroxidases is unknown, making it very difficult to assess their contribution to H_2O_2-removal *in vivo*. A low-level luminescence (Chapter 2) has been observed from root and stem tissues of a wide variety of plants, and has been suggested to arise in part from reactions carried out by peroxidase. For example, the oxidation of auxin can lead to the formation of a dioxetane intermediate (Chapter 2) which decomposes to a carbonyl compound in an excited state. As this decays to the ground state, light is emitted. It can also sensitize the formation of singlet oxygen as it decays (see Chapter 2):

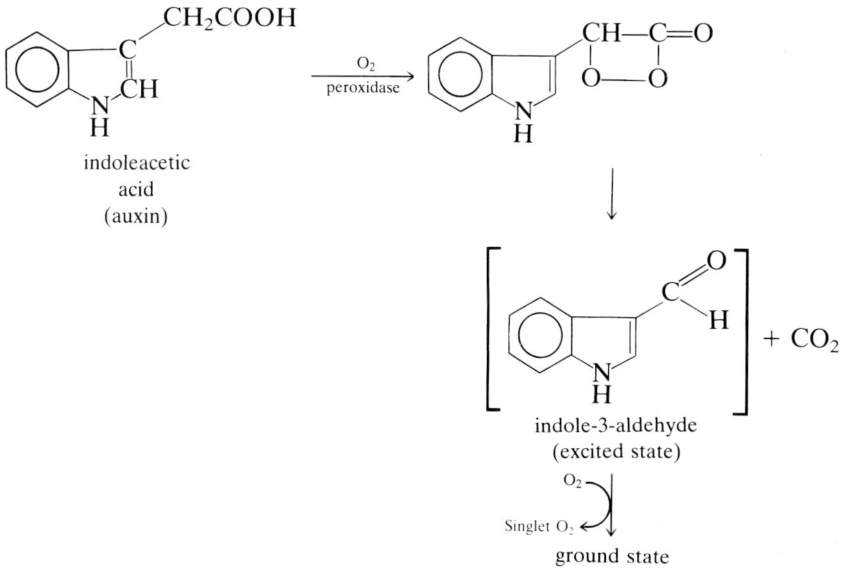

indoleacetic
acid
(auxin)

indole-3-aldehyde
(excited state)

$+ CO_2$

ground state

Indole-3-aldehyde is not the only product of the action of peroxidase on auxin, however.

Horseradish peroxidase is much used in the laboratory as a method of measuring H_2O_2-production, often employing as its substrate the compound *scopoletin*. Scopoletin emits light (fluoresces) at 450 nm when illuminated with light at 360 nm, but the product of its oxidation by peroxidase does not fluoresce. After suitable calibration, this loss of fluorescence can be used to measure the rate of H_2O_2-production in a system. Of course, if that system contains other substrates for the peroxidase that can compete with scopoletin, an underestimate of H_2O_2-production will result. For example, this method cannot be applied to measure H_2O_2-production in chloroplasts since they contain large amounts of ascorbic acid, which is oxidized by the peroxidase. Thiol compounds also interfere with peroxidase-based assays for H_2O_2.

Table 3.3 summarizes the various methods that have been used to measure H_2O_2-production by biological systems.

Chloroperoxidase and bromoperoxidase
Chloroperoxidase is a non-specific peroxidase first isolated from the fungus *Caldariomyces fumago*. It catalyses the usual peroxidase reactions, but, in addition, can catalyse introduction of halogen atoms onto a wide range of substrates in the presence of hydrogen peroxide and the halide ions, chloride (Cl^-), bromide (Br^-), or iodide (I^-). If SH is the substrate and X^- the halide, these may be written:

$$SH + X^- + H_2O_2 + H^+ \rightarrow SX + 2H_2O$$

Many marine organisms are rich in halogenated compounds and similar enzymes have been isolated from several of these, such as the purple bleeder sponge and the tropical marine sponge. In some cases only one halide can act as a substrate, e.g. a bromoperoxidase has been isolated from several marine organisms. Bromoperoxidases from the marine brown alga *Ascophyllum nodosum* and from the lichen *Xanthoria parietina* have been reported to contain vanadium as a prosthetic group instead of haem.

Ascorbate peroxidase
The chloroplasts of higher plants and the green alga *Euglena* contain no catalase, glutathione peroxidase, or 'non-specific' peroxidase activities, but they do contain high activities of an ascorbate peroxidase enzyme, which catalyses the overall reaction

$$ascorbate + H_2O_2 \rightarrow 2H_2O + dehydroascorbate$$

The enzyme purified from *Euglena* is a haem protein inhibited by cyanide and azide and it is likely that the chloroplast enzyme is similar. Disposal of hydrogen peroxide by this enzyme is probably one reason why chloroplasts contain a high internal concentration of ascorbic acid. An ascorbate peroxidase activity has recently been reported in *Trypanosoma cruzi*, which also lacks catalase activity.

Table 3.3. Methods for detecting H_2O_2-production in biological systems: a summary

Method	Principle of the method	Systems to which it has been applied
Oxidation of dimethylthiourea to a dioxide product	DMTU is oxidized by $OH^{\cdot}$, hypochlorous acid, and H_2O_2, but only H_2O_2 gives the dioxide product	Enzymes, neutrophils. See Curtis *et al.*, *Proc. Natl. Acad. Sci. USA*, **85**, 3422–5 (1988).
Observation of intracellular catalase compound I, or oxidation of [^{14}C]-methanol or [^{14}C]-formate.	See Section 3.1.1	Bacteria, liver (perfused and *in situ*), organ slices or homogenates.
Cytochrome c peroxidase	See Section 3.1.1	Animal and plant mitochondria, protozoa, peroxisomes, microsomes.
Horseradish peroxidase + scopoletin	Section 3.1.1	Animal and plant mitochondria, sub-mitochondrial particles, phagocytes, protozoa, microsomes.
O_2-electrode method	Add large excess of catalase and measure release of oxygen: $2H_2O_2 \rightarrow 2H_2O + O_2$	Only useful if little catalase present to start with. Used to study H_2O_2-removal in chloroplasts.
Catalase inhibition	If a reaction requires H_2O_2, then it should be inhibited by catalase	Catalase very slow at destroying low concentrations of H_2O_2, so a large amount must be added. Often used to investigate the role of H_2O_2 in radical reactions.
GSSG release	Section 3.1.1	Perfused organs.
Aminotriazole inhibition of catalase	Section 3.1.1	Fungi, various bacteria, *Mycoplasma*.

Miscellaneous peroxidase activities

It has long been known that myoglobin, haemoglobin, and a complex of haemoglobin with the haemoglobin-binding proteins in plasma (*haptoglobins*) display peroxidase activities *in vitro* using hydrogen peroxide and a suitable electron donor. No physiological peroxidase role has ever been ascribed to these proteins, but the peroxidase properties of haemoglobin are widely used as the basis of a diagnostic test for gastrointestinal bleeding (*faecal occult blood test*).

3.1.2   *Superoxide dismutase*

The copper–zinc enzymes
In 1938, T. Mann and D. Keilin in England described a blue-green protein containing copper that they had isolated from bovine blood. They called it *haemocuprein*. In 1953, a similar protein was isolated from horse liver and named *hepatocuprein*. Other proteins of this type were later isolated, such as *cerebrocuprein* from brain. In 1970, it was discovered that the erythrocyte protein contains zinc as well as copper. No enzymic function was detected in any of these proteins, so it was often suggested that they served as metal stores. However, in 1968 the work of J. M. McCord and I. Fridovich in the USA showed that the erythrocyte protein is able to remove catalytically the superoxide radical and thus they identified its function as a *superoxide dismutase* enzyme. Despite an intensive search, no other substrate on which superoxide dismutase enzymes act catalytically has been discovered, i.e. we may regard them as specific for the superoxide radical.

Copper–zinc-containing superoxide dismutases (CuZnSODs) are highly stable enzymes and thus easily isolated. In purifying this enzyme from erythrocytes, the cells are lysed and haemoglobin removed by treatment with chloroform and ethanol, followed by centrifugation. The enzyme actually enters the organic phase, from which it can be precipitated out by addition of cold propanone (acetone) and then further purified by ion-exchange chromatography. Not many enzymes will tolerate these procedures. Copper–zinc superoxide dismutases are also quite resistant to heating, to attack by proteases, and to denaturation by such reagents as guanidinium chloride, sodium dodecyl sulphate (SDS), or urea.

Subsequent studies have shown that CuZnSODs are found in virtually all eukaryotic cells such as yeasts, plants and animals (Table 3.4) but not generally in prokaryotic cells such as bacteria or blue-green algae. The first exception to this rule to be discovered, in A. M. Michelson's laboratory in France, is the luminescent bacterium *Photobacterium leiognathi*, which contains a CuZnSOD. This organism exists in a symbiotic relationship with the ponyfish, occupying a special gland and imparting a characteristic luminescence to the fish. Comparison of the amino-acid composition of the bacterial enzyme with that of higher organisms shows that it is closely related to fish CuZnSOD enzymes. This might be taken to mean that the bacterium obtained the gene for its CuZnSOD by gene transfer from its host fish, although recent sequencing studies have cast some doubt on this suggestion.

Table 3.4. Some systems from which copper–zinc SOD has been purified

Mammalian tissues	Plant tissues
Bovine erythrocytes and retina	*Neurospora crassa*
Human erythrocytes	*Fusarium oxysporum*
Spermatozoa (especially high activity in donkey semen)	Green peas
	Maize seeds
Rat liver	Wheat-germ
Bovine liver	Spinach chloroplasts
Horse liver	Yeast (*Saccharomyces cerevisiae*)
Bovine milk ⎱ (present at low activity in	Pea seedlings*
Human milk ⎰ milk)	Corn seedlings*
Pig liver	Tomatoes
	Cucumber*
	Green peppers*
	Lens esculenta

Fish	Other organisms
Shark (*Prionace glauca*)*	Fruit-fly (*Drosophila melanogaster*)
Cuttlefish (*Sepia officinalis*)*	*Photobacterium leiognathi*
Ponyfish	Chicken liver
Snapper	*Caulobacter crescentus*
Sea bass	*Trichinella spiralis*
Croaker	Housefly (*Musca domestica*)
Merlin	
Trout	
Swordfish	

Unless indicated by a star (*), the enzyme was purified to homogeneity and shown to contain two subunits. All purified enzymes have one ion of copper and one ion of zinc at each active site.
   This list is being added to constantly, perhaps in the hope of finding an enzyme different from the norm.

The free-living (non-symbiotic) bacterium *Caulobacter crescentus* CB15 also contains a CuZnSOD, although studies of its amino-acid sequence have shown that it is not closely related to eukaryotic CuZnSODs.

All the CuZnSOD enzymes so far isolated from eukaryotic cells have relative molecular masses around 32 000 and contain two protein subunits, each of which bears an active site containing one copper ion and one zinc ion (Table 3.4).

For all CuZnSODs the reaction catalysed is the same—the dismutation reaction of O_2^- is greatly accelerated (Fig. 3.8)

$$O_2^- + O_2^- + 2H^+ \rightarrow H_2O_2 + O_2 \text{ (ground-state)}$$

Whereas the overall rate constant for the uncatalysed dismutation reaction depends strongly on the pH of the solution (Chapter 2) and is about

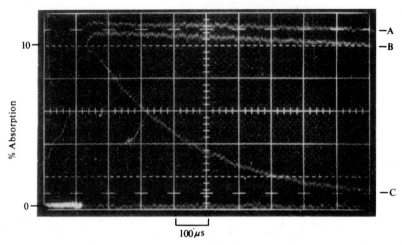

Fig. 3.8. The catalytic action of superoxide dismutase as demonstrated by pulse radiolysis. The oscilloscope traces show the decay at pH 8.8 of O_2^- radical (initial concentration 32 μmoles per litre) as followed by the loss of its absorbance at 250 nm. Trace A, spontaneous dismutation of O_2^-; C, plus 2 μmoles per litre of SOD; B as C but SOD boiled for 5 min to destroy enzyme activity. Data from G. Rotilio *et al.*, *Biochim. Biophys. Acta* **268**, 605–9 (1972), with permission.

$5 \times 10^5 \, M^{-1} s^-$ at physiological pH, the reaction catalysed by bovine erythrocyte CuZnSOD is relatively independent of pH in the range 5.3–9.5, the rate constant for reaction of O_2^- with the active site being about $1.6 \times 10^9 \, M^{-1} s^{-1}$. Cyanide is an extremely powerful inhibitor of CuZnSODs. These enzymes are also inactivated on prolonged incubation with the compound *diethyldithiocarbamate* $(CH_3CH_2)_2N-\overset{\parallel}{\underset{S}{C}}-SH$, which binds to the copper at the active sites and removes this metal from the enzyme. Diethyldithiocarbamate has been used to inhibit CuZnSOD activity in isolated erythrocytes, intestinal cells, and in whole animals, although caution should be exercised in its use since it inhibits a number of other enzymes as well. For example, when 1.5 g of diethyldithiocarbamate per kilogram of body weight was injected into mice, the SOD activity, 3 hours later, of blood had decreased by 86 per cent, that of liver by 71 per cent, and that of brain by 48 per cent.

The copper ions appear to function in the dismutation reaction by undergoing alternate oxidation and reduction, i.e.

$$Enzyme-Cu^{2+} + O_2^- \rightarrow E-Cu^+ + O_2$$
$$E-Cu^+ + O_2^- + 2H^+ \rightarrow E-Cu^{2+} + H_2O_2$$

net reaction: $O_2^- + O_2^- + 2H^+ \rightarrow H_2O_2 + O_2$

However, at least one other mechanism of action is also feasible, in which the first O_2^- does not reduce the copper ion, but forms a complex with it.

The Zn^{2+} does not function in the catalytic cycle but it appears to stabilize the enzyme—this conclusion is drawn from experiments in which the metals are removed from the active sites and replaced either singly or together. Indeed, the regaining of SOD activity on addition of copper ions to the metal-free CuZnSOD enzyme (apoenzyme) has been suggested as a simple method for the measurement of trace amounts of copper. In general, ions of other transition metals, such as Mn^{2+}, cannot replace the copper to yield a functional enzyme, but cobalt, mercury, or cadmium ions can replace Zn(II) in increasing enzyme stability. If the Cu^{2+} is replaced by cobalt ions (Co^{2+}), however, the enzyme can still apparently catalyse O_2^- dismutation, although with a rate constant of only $4.8 \times 10^6 \, M^{-1} s^{-1}$.

The complete amino-acid sequences of CuZnSODs from several plants and animals have been studied and they are all very similar. The three-dimensional structure of the bovine enzyme has been elucidated by X-ray crystallography. Each subunit is composed primarily of eight antiparallel strands of β-pleated sheet structure that form a flattened cylinder, plus three external loops. The copper ion is held at the active site by interaction with the nitrogens in the imidazole ring structures (Fig. 3.9) of four histidine residues (numbers 44, 46, 61, and 118 in the amino-acid sequence); whereas the zinc ion is bridged to the copper by interaction with the imidazole of histidine 61 and it also interacts with histidines 69 and 78 and the carboxyl ($-COO^-$) group of aspartate 81. Histidine 61, which interacts with both metals, may be involved in supplying the protons needed for the dismutation reaction. Most of the surface of each protein subunit is negatively charged, repelling O_2^-, except for positively-charged 'tracks' that lead into the active site (see cover photograph). A similar arrangement probably exists in the manganese and iron SODs (see below). Hence O_2^- approaching any other part of a subunit seems to be 'guided' into the active site. Chemical modifications of these

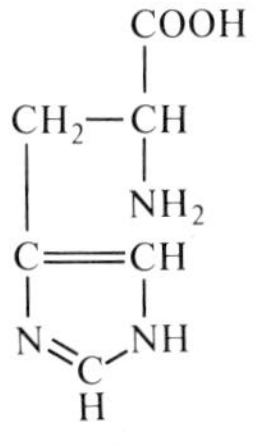

$$
\begin{array}{c}
COOH \\
| \\
CH_2-CH \\
| \quad\quad | \\
\quad\quad NH_2 \\
C=\!\!=CH \\
| \quad\quad | \\
N_{\diagdown C \diagup} NH \\
| \\
H
\end{array}
$$

Fig. 3.9. Structure of the amino-acid histidine. The ring structure is known as the imidazole ring and contains two nitrogen atoms. Each has five electrons in its outermost shell (see Appendix I), three of which are being used in covalent bonding. The remaining two constitute a lone pair (see Appendix I) and can interact with metal ions as explained in the text.

positively charged amino-acid side-chains markedly decrease enzyme activity. Although the two active sites on the enzyme are some distance from each other, the separated SOD subunits themselves catalyse O_2^- dismutation only slowly, if at all.

It is possible to visualize SOD enzymes after electrophoresis on polyacrylamide gels, as explained in the legend to Fig. 3.10. Inhibition by cyanide ion (CN^-) can be used to identify CuZnSOD enzymes. Electrophoresis of some tissue extracts, or even of purified SOD enzymes, has sometimes shown the presence of multiple bands, e.g. cow liver shows seven bands of CuZnSOD activity. Caution must be exercised in attributing such multiple bands to the presence of SOD isoenzymes since they might arise by attack on the SOD protein by proteolytic enzymes present in the extract. Storage of purified *Neurospora crassa* CuZnSOD at low temperatures causes it to show multiple bands on subsequent electrophoresis. If, for example, metal ions were lost

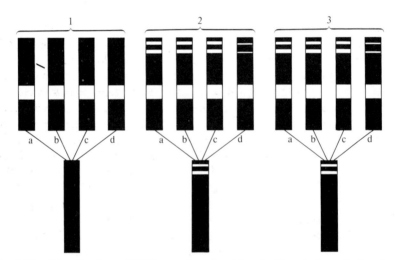

Fig. 3.10. Visualization of SOD on polyacylamide gels. Protein is applied to the gel and electrophoresis carried out. The gel is then soaked in a solution of nitro-blue tetrazolium and exposed to a O_2^--generating system. O_2^- reduces NBT to the blue-coloured formazan (Chapter 2) and so the gel turns blue except at the points where SOD activity is located. The enzyme quickly removes the O_2^-, formazan production is prevented, and so a white 'achromatic zone' is detected (for further details see C. Beauchamp and I. Fridovich (1971). *Anal. Biochem.* **44**, 276–80). The figure shows the pattern obtained in polyacrylamide gel electrophoresis of extracts from (a) brain, (b) heart, (c) liver, and (d) lung of (1) rat, (2) mouse, (3) chicken. Upper panel, no cyanide added. Lower panel, 2 mM CN^- ions present, which inhibit CuZnSODs. Figure taken from De Rosa *et al.*, *Biochim. Biophys. Acta* **566**, 32–9 (1979), with permission. (The enzyme activity *tetrazolium oxidase*, extensively studied by geneticists, is identical with SOD.)

from the enzyme this should increase its net negative charge. However, purification and analysis of the enzymes has proved the existence of two slightly different forms of CuZnSOD in wheat seeds, and isoenzymes have also been shown to exist in several other organisms (e.g. *Drosophila*). Indeed, an electrophoretic variant of human CuZnSOD known as SOD-2 has been found to occur in Northern Sweden and Northern Finland. Most of the population is homozygous for 'normal' SOD (SOD-1), but there are some heterozygotes with both SOD-1 and SOD-2, and a very few SOD-2 homozygotes. SOD-2 is rarely detected in other populations, except in one of the Orkney islands; perhaps the Vikings may have introduced the SOD-2 gene on one of their rampages. SOD-2 has also been found in some Mormons in Utah, USA; between 1850 and 1905, more than 30 000 Mormon converts left Scandinavia for the USA. The SOD-2 enzyme has a slightly lower activity than SOD-1, but SOD-2 homozygotes appear to suffer no ill-effects from this.

The manganese enzymes
The superoxide dismutase first isolated from *E. coli* proved to be entirely unlike the CuZn enzymes. It was pink rather than blue-green, was not inhibited by cyanide or diethyldithiocarbamate, had a relative molecular mass of 40 000 rather than 32 000, was destroyed by treatment with chloroform plus ethanol (and hence did not survive the typical purification methods for CuZnSOD) and contained manganese at its active site, this being in the Mn(III) state in the 'resting' enzyme. MnSODs catalyse exactly the same reaction as do the CuZnSODs. At pH 7.0 the rate constants for the two enzymes are similar, but unlike CuZnSODs the rate constant for MnSODs decreases at alkaline pH (e.g. for the *E. coli* enzyme, k_2 at pH 7.8 is 1.8×10^9, but $0.33 \times 10^9 \, \text{M}^{-1} \text{s}^{-1}$ at pH 10.2). Thus assays of tissues at high pH for SOD activity underestimate the amount of MnSOD present in relation to CuZnSOD. Manganese SODs are also much more labile to denaturation by heat or chemicals, such as detergents, than are CuZnSODs.

Manganese SOD has since been detected in several other bacteria and also in extracts of animal and plant tissues. For example, the CN^--insensitive SOD bands shown in Fig. 3.10 are attributable to the Mn-enzyme. The activity of the MnSOD in relation to CuZnSOD depends on the tissue and on the species (Fig. 3.10). Mammalian erythrocytes contain no MnSOD, it is about 10 per cent of total SOD activity in rat liver but much more than this in human liver. In a normal growth medium, the fungus *Dactylium dendroides* contains 80 per cent of its SOD activity as CuZnSOD, and 20 per cent as MnSOD; but if its copper supply is restricted, more MnSOD is synthesized to maintain the total cellular SOD activity constant. Similarly, a rise in CuZnSOD activity has been observed in the liver of chickens fed on a Mn-restricted diet.

Subcellular fractionation studies upon the liver of rats have shown that most of the CuZnSOD is in the cytosol of the cell, with some activity being present in lysosomes and possibly between the inner and outer mitochondrial membranes and in the nucleus. The MnSOD is located in the mitochondrial matrix. Mitochondria isolated from chicken liver, the fungus *N. crassa*, maize seedlings, and spinach leaves also contain MnSOD in the matrix and possibly have CuZnSOD in the intermembrane space. In human and baboon liver, however, there is some MnSOD outside the mitochondria as well as in the matrix, and the 'extra' MnSOD synthesized by copper-depleted *D. dendroides* appears in the cytosol.

MnSODs have been purified from a number of sources (Table 3.5). All of the MnSODs from higher organisms contain four protein subunits and usually have 0.5 or 1.0 ions of Mn per subunit. Most, but not all, of the bacterial enzymes have two subunits. Removal of the manganese from the active site causes loss of catalytic activity, and it cannot in general be replaced by any other transition-metal ion, including iron, to yield a functional enzyme. Detailed mechanistic studies of MnSODs from *E. coli* and *Bacillus stearothermophilus* show that the manganese undergoes changes of its valency state during catalysis, as would be expected, but the exact nature of these has not been elucidated; it is not a simple two-stage mechanism.

The amino-acid sequences of all MnSODs, whether from animals, plants, or bacteria, are extremely similar to each other and unrelated to those of the CuZnSODs. This is consistent with the *endosymbiotic theory* for the origin of mitochondria, which suggests that they evolved as a symbiosis between a primitive eukaryote (with CuZnSOD) and a prokaryote (with MnSOD) that eventually became incorporated into the eukaryotic cytoplasm, wrapped in a membrane (the outer mitochondrial membrane).

The iron enzymes

From the bacterium *E. coli*, three SOD enzymes can be purified, one of which is the MnSOD already described. A second is an iron-containing enzyme (FeSOD, for short), and similar enzymes were found subsequently in several other bacteria and in algae. The third *E. coli* SOD is a hybrid enzyme containing subunits of the manganese enzyme and of the iron enzyme in the same dimeric molecule. In *E. coli*, both FeSOD and MnSOD are found in the cell matrix—an early report that the iron enzyme is located in the periplasmic space (i.e. in-between the cell wall and cell membrane) has been retracted, although it is still widely quoted.

Iron-containing superoxide dismutases usually contain two protein subunits, although at least two tetrameric enzymes exist, one from *Mycobacterium tuberculosis* and one from *Methanobacterium bryantii* (Table 3.6). The dimeric enzymes usually contain one or two ions of iron per molecule of

Table 3.5. Some organisms from which manganese SOD has been purified

Organism	Subunit structure	Moles Mn per mole enzyme
Higher organisms		
Maize	Tetramer	2
Bovine adrenal cells	Tetramer	2
Luminous fungus (*Pleurotus olearius*)	Tetramer	2
Pea (*Pisum sativum*)	Tetramer	1
Chicken liver	Tetramer	2
Rat liver	Tetramer	4
Human liver	Tetramer	4
Saccharomyces cerevisiae	Tetramer	4
Bullfrog (*Rana catesbeiana*)	Tetramer	4
Bacteria		
Halobacterium halobium	Dimer	1–2
Rhodopseudomonas spheroides	Dimer	1
E. coli	Dimer	1
Bacillus stearothermophilus	Dimer	1
Mycobacterium phlei	Tetramer	2
Mycobacterium lepraemurium	Dimer	1
Thermus thermophilus	Tetramer	2
Paracoccus denitrificans	Dimer	1–2
Streptococcus faecilis	Dimer	1
Streptococcus mutans	Dimer	1–2
Propionibacterium shermanii	Dimer	3
Bacillus subtilis	Dimer	1
Serratia marcescens	Dimer	1–2
Gluconobacter cerinus	Dimer	1
Acholeplasma laidlawii	Dimer	1

enzyme. The iron in the resting state is Fe(III) and it probably oscillates between the Fe(III) and Fe(II) states during the catalytic cycle, i.e.

$$Fe^{3+}\text{-enzyme} + O_2^- \rightarrow Fe^{2+}\text{-enzyme} + O_2$$
$$Fe^{2+}\text{-enzyme} + O_2^- + 2H^+ \rightarrow Fe^{3+}\text{-enzyme} + H_2O_2$$

net reaction: $O_2^- + O_2^- + 2H^+ \rightarrow H_2O_2 + O_2$

although this may be an oversimplification of the mechanism in some cases. Like MnSODs, FeSODs show decreased catalytic activity at high pH values

Table 3.6. Some organisms from which iron SOD has been purified

Organism	Subunit structure	Moles Fe/per mole enzyme
Bacteria		
Streptococcus mutans	Dimer	1–2
E. coli	Dimer	1–1.8
Desulphovibrio desulphuricans	Dimer	1–2
Thiobacillus denitrificans	Dimer	1
Chromatium vinosum	Dimer	2
Photobacterium leiognathi	Dimer	1 (plus some 'non-specifically bound' iron)
Pseudomonas ovalis	Dimer	1–2
Methanobacterium bryantii	Tetramer	2–3
Thermoplasma acidophilum (this enzyme has very low activity and the question has been raised as to whether or not it is truly a FeSOD)	Tetramer	2
Azotobacter vinelandii	Dimer	2
Bacillus megaterium	Dimer	1
Mycobacterium tuberculosis	Tetramer	4
Propionibacterium shermanii	Dimer	2
Other organisms		
Tomato (*Lycopersicon esculentum*)	Dimer	1–2
Mustard (*Brassica campestris*)	Dimer	1–2
Water-lily (*Nuphar luteum*)	Dimer	1
Porphyridium cruentum (red alga)	Dimer	1
Spirulina platensis (blue-green alga)	Dimer	1
Plectonema boryanum (blue-green alga)	Dimer	1
Anacystis nidulans (blue-green alga)	Dimer	1
Euglena gracilis (alga)	Not reported	1
Crithidia fasciculata (trypanosome)	Dimer	2–3
Ginkgo biloba	Dimer	1

(compared to pH 7) and are not inhibited by CN^-. The rate constants for reaction with O_2^- are slightly lower for FeSODs than for the other types of SOD.

The amino-acid sequences of FeSODs are extremely similar to those of MnSODs from all sources, and very different from the sequences of CuZnSODs. The tertiary structure of FeSODs from *Pseudomonas ovalis* and

E. coli have been determined by X-ray crystallography; they are very different from that of CuZnSOD (see above).

Some bacteria contain both FeSOD and MnSOD, such as *E. coli*, whereas others contain only one enzyme. For example, *Bacillus cereus* contains only FeSOD, and *Streptococcus sanguis*, only MnSOD. However, *Propionibacterium shermanii*, which normally contains a FeSOD, has been reported to produce an MnSOD if it is grown on iron-deficient media and *Streptococcus mutans* has been claimed to use the same apoenzyme to produce a FeSOD or a MnSOD, depending on the metal provided in the growth medium. *Photobacterium leiognathi* contains a FeSOD in addition to its CuZnSOD, discussed previously, whereas the non-symbiotic, free-living, strain *Photobacterium sepia* contains a FeSOD but no CuZnSOD. The aerobic bacterium *Nocardia asteroides* has been reported to produce a SOD that contains *both* manganese and iron ions.

No animal tissues have been found to contain FeSOD, but a few higher plant tissues do. Of 43 plant families investigated by Salin's group in the USA, FeSOD was found in only 3, and it has been purified from water-lily, *Ginkgo biloba*, tomato, and mustard leaves (Table 3.6). Mitochondria from mustard leaves apparently contain CuZnSOD in the intermembrane space, and MnSOD in the matrix, but the FeSOD appears to be located in the chloroplasts of the plant. There is a great deal of interest in the information about the process of evolution that may be obtained by investigating the SOD enzymes present in 'intermediate' organisms. For example, the bacterium *Paracoccus denitrificans* shares many structural and biochemical features with mitochondria, and it has been proposed that both it and mitochondria might have evolved from a common ancestral bacterium, i.e. *P. denitrificans* resembles the symbiotic bacterium that 'fused' with the primitive eukaryote. Consistent with this, *P. denitrificans* contains a MnSOD. A Cu-protein with some CN^--sensitive SOD activity has also been isolated from *P. denitrificans* but it should not necessarily be assumed that it is related to a CuZnSOD since a number of other copper proteins react with O_2^- radical (Section 3.5.2)

Assays of superoxide dismutase activity
In investigating the biological importance of SOD, it is obviously necessary to determine the activity of the enzyme in the organism under investigation, and to determine which type of enzyme is present without having to go to the trouble of purification and metal determination. Immunological methods for detecting the amount of CuZnSOD and MnSOD proteins in animal tissues have been developed. Since these two enzymes are very different the antibodies do not cross-react. Table 3.7 shows some results for the amount of CuZnSOD protein in different human tissues. Because of the limited availability of human tissues for assay, these results should be taken as

Table 3.7.   CuZnSOD protein in human tissues

Tissue	CuZnSOD μg per mg protein
Cerebral grey-matter	3.7
Liver	4.71
Erythrocytes	0.52*
Renal cortex	1.93
Renal medulla	1.31
Thyroid	0.38
Testis	2.16
Cardiac muscle	1.82
Gastric mucosa	0.94
Pituitary	0.99
Pancreas	0.39
Lung	0.47

An immunological method was used which measures the enzyme protein rather than enzyme activity. Results were obtained from patients who died after accidents. Data abstracted from Hartz *et al.* (1973). *Clin. Chim. Acta*, **36**, 125–32. (*Erythrocyte value per mg of haemoglobin.)

guidelines only, but they do show an especial concentration of SOD in liver, which is broadly consistent with the more extensive data from animal studies (Table 3.8).

Direct determination of SOD activity can be carried out by pulse radiolysis (Fig. 3.8), which has been especially useful in investigations of the mechanism of enzyme action. Similarly, the loss of the ultraviolet absorbance of O_2^- when KO_2 is added to an aqueous solution can be observed in a spectrophotometer, although this method can only be used at alkaline pH values when the rate of non-enzymic O_2^- dismutation is low. Any assay carried out at alkaline pH values will underestimate the amount of FeSOD or MnSOD activity in relation to that of CuZnSOD, as explained previously, and appropriate corrections must be introduced. Italian scientists have developed two other methods of measuring SOD activity. One employs an electrolytic cell in which oxygen is reduced to O_2^- at a 'coated dropping mercury electrode'. Very simply, the current flowing in the system depends on the concentration of oxygen available at the electrode. Addition of SOD, by dismutating O_2^- more rapidly, increases oxygen production and hence the flow of current. This method is very useful at alkaline pH values for determination of catalytic rate constants, and is cheaper than pulse radiolysis apparatus. Their other method involves the fact that halide ions will bind to

Table 3.8. Superoxide dismutase activities in animal tissues

Animal used	Assay	Tissue	Total SOD activity (units per mg protein)
Mice (data from *Biochem. J.* [1981] **199**, 393–8.)	Disproportionation of KO_2 in alkaline solution (one unit of SOD causes O_2^- to decay at the rate of $0.1\,s^{-1}$ in a 3 ml reaction volume).	Pancreatic islets	331
		Liver	660
		Kidney	582
		Erythrocytes	52
		Heart	390
		Brain	408
		Skeletal muscle	282
Rat (data from *Biochem. J.* [1975] **150**, 31–9.)	Riboflavin–light–NBT system (one unit of SOD inhibits NBT reduction by 50%).	Liver	22
		Adrenal	20
		Kidney	13
		Erythrocytes	4
		Spleen	5
		Heart	9
		Pancreas (whole)	1.5
		Brain	3
		Lung	3
		Stomach	7
		Intestine	3
		Ovary	2
		Thymus	1
Rat (data from *FEBS Lett.* **114**, 42–43.)	Xanthine–xanthine oxidase–cytochrome c method (see text)	Adipose tissue	11

the active site of Mn and CuZnSOD enzymes. Observation of the nuclear magnetic resonance (NMR) behaviour of fluoride ($^{19}F^-$) ions so bound can be used to assess the amount of enzyme present, although the effect depends on the oxidation state of the metal and the assay is not highly sensitive.

However, most laboratories still use the so-called *indirect assay methods* for SOD activity. In these, O_2^- is generated by some mechanism and allowed to react with a detector molecule. SOD, by removing the O_2^-, inhibits the reaction with the detector. In their original work in 1968 on the SOD activity of the erythrocyte enzyme, McCord and Fridovich used an assay of this type. O_2^- was generated by a mixture of the enzyme xanthine oxidase and its substrate xanthine, and was allowed to reduce cytochrome c, this reduction being accompanied by a change in absorbance at 550 nm. SOD, by removing O_2^-, will inhibit the absorbance change. One unit of SOD activity was defined as that amount that would inhibit cytochrome c reduction by 50 per cent under their assay conditions, and so the units of SOD activity quoted in the literature bear no relation whatsoever to quoted units for other enzymes (1 enzyme unit is normally defined as that amount catalysing transformation of 1 μmole of substrate per minute). Figure 3.11 summarizes the principles of this assay. It may be seen that a decrease in the cytochrome c reduction, looking like a SOD activity, could also be produced by a reagent that inhibits O_2^- generation, i.e. by inhibiting the xanthine oxidase enzyme. Fortunately, this can be checked for by measuring the production of uric acid in the system as an index of xanthine oxidase activity; and such a control should always be done when the assay is being applied to a crude tissue extract. Many such extracts contain cytochrome oxidase, a mitochondrial enzyme complex that re-oxidizes reduced cytochrome c and so interferes with the assay. Chemical modification of the cytochrome c by attachment of acetyl

$$ \begin{matrix} O \\ \| \\ (CH_3-C-) \end{matrix} $$

(CH$_3$—C—) groups to some of its amino-acid side-chains prevents it from being a substrate for cytochrome oxidase but still allows it to react with O_2^-. Use of acetylated cytochrome c permits this assay to be used in extracts containing cytochrome oxidase. H_2O_2 produced by xanthine oxidase can slowly re-oxidize reduced cytochrome c. This is not usually a major problem in assays of SOD by the xanthine–xanthine oxidase method, but if necessary some catalase can be added to the reaction mixture. It is essential to ensure that any commercial catalase used is not itself contaminated with SOD; commercial cytochrome c also often contains traces of SOD.

Detector molecules for O_2^- other than cytochrome c can also be used, e.g. nitro-blue tetrazolium (NBT) which is reduced by O_2^- to a deep-blue coloured formazan (Chapter 2), or adrenalin, which is oxidized by O_2^- to form a pink product known as adrenochrome. The ability of O_2^- to oxidize NADH in the presence of lactate dehydrogenase (Chapter 2), accompanied by a fall in absorbance at 340 nm, has also been used. Table 3.9 summarizes these and some other methods (see also Fig. 3.11). The authors' laboratory routinely uses inhibition of NBT reduction in a xanthine–xanthine oxidase system to assay SOD in tissue extracts. Although this avoids problems with cytochrome oxidase, it must be remembered that the reaction of NBT with O_2^- is very complex (Chapter 2). Although formazan is only sparingly

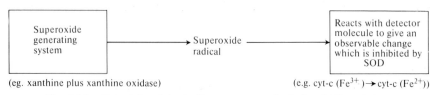

Superoxide generating system | Superoxide radical | Reacts with detector molecule to give an observable change which is inhibited by SOD

(eg. xanthine plus xanthine oxidase)   (e.g. cyt-c (Fe^{3+}) → cyt-c (Fe^{2+}))

Fig. 3.11. Principle of the indirect assay methods for SOD activity.

Table 3.9. Indirect methods that have been used to measure SOD activity

Source of superoxide	Detector of superoxide	Reaction measured[1]
Xanthine–xanthine oxidase	Cytochrome c	Reduction, ΔA
	Nitro-blue tetrazolium[2]	Reduction, ΔA
	Luminol	Light-emission
	Adrenalin	Oxidation, ΔA
	NADH + lactate dehydrogenase	Oxidation, ΔA
	Hydroxylamine[3]	Nitrite (NO_2^-) formation (colorimetric method)
	2-Ethyl-1-hydroxy-2,5,5-trimethyl-3-oxazolidine (hydroxylamine derivative)	Oxidation to nitroxide, detected by ESR (Chapter 2)
Autoxidation reactions	Adrenalin	Oxidation, ΔA
	Sulphite	O_2 uptake
	Pyrogallol	O_2 uptake, or ΔA
	6-Hydroxydopamine	Oxidation, ΔA
Directly added $K^+O_2^-$	—	Loss of O_2^-, ΔA in UV
	Nitro-blue tetrazolium	Reduction, ΔA
	Cytochrome c	Reduction, ΔA
	Tetranitromethane	Reduction, ΔA
Illuminated flavins	Nitro-blue tetrazolium	Reduction, ΔA
		O_2 uptake (SOD accelerates)
	Dianisidine	Oxidation, ΔA ('*positive*' assay)
NADH + phenazine methosulphate[4]	Nitro-blue tetrazolium	Reduction, ΔA

[1] ΔA: reaction results in an absorbance change that can be measured using a spectrophotometer.
[2] High O_2 concentrations can decrease NBT reduction by O_2^- (Chapter 2, Fig. 2.10).
[3] Hydroxylamine itself reacts slowly, if at all, with O_2^-, and nitrite formation may require hydroxyl radicals. Thus the chemistry of this assay is very complex.
[4] Not recommended; see text and also *J. Am. Chem. Soc.* (1982). **104**, 1666.

soluble in water, its precipitation can be avoided by keeping absorbance changes fairly low. The compound *luminol* emits light when exposed to O_2^- (again the mechanism is very complex) and has been used as a detector molecule (Table 3.9).

Xanthine oxidase
Since xanthine oxidase is so often used as a source of O_2^- in the laboratory, it is worthwhile saying a little about it. The commercially-available enzyme is

usually obtained from cream, and the purification process employed by some manufacturers involves the use of proteolytic enzymes to free the oxidase from the milk fat globule membranes. Sometimes these proteases are still present in the final preparation and this must be carefully checked for. One report of the damaging effects of O_2^- from a xanthine–xanthine oxidase system upon chloroplast membranes turned out to be an effect of the trypsin contamination of the enzyme preparation. Phospholipases may also contaminate commercial xanthine oxidase preparations, and chelating agents such as EDTA are often present.

Xanthine oxidase catalyses oxidation of both hypoxanthine and xanthine as shown below, but it will also act on a number of other substrates such as acetaldehyde (ethanal, CH_3CHO),

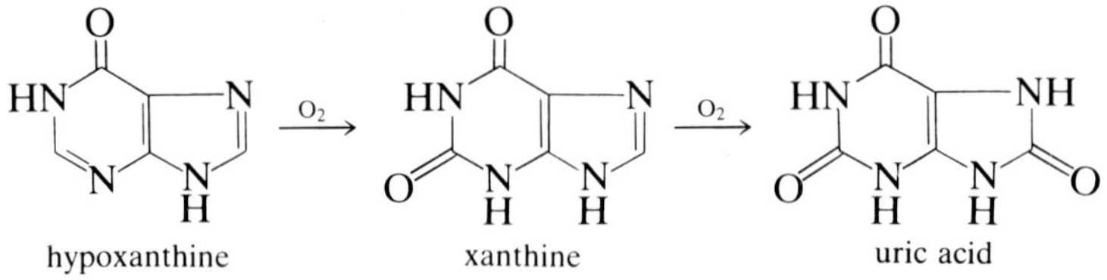

hypoxanthine   xanthine   uric acid

Both hypoxanthine and ethanal can be used instead of xanthine in O_2^--generating systems. A powerful inhibitor of xanthine oxidase is the structurally-related compound allopurinol (Fig. 3.12). Allopurinol is oxidized by the enzyme to give oxypurinol, which binds tightly to the active site and causes the inhibition. Hence allopurinol has been called a 'suicide substrate' of xanthine oxidase. Allopurinol is widely used in clinical medicine to inhibit uric acid accumulation in conditions such as gout, and oxypurinol is a major metabolite of allopurinol in the human body.

Commercial preparations of the water-soluble vitamin folic acid often inhibit xanthine oxidase. However, this inhibition is mostly caused by a contaminant of the folic acid, pterinaldehyde (2-amino-4-hydroxypteridine-6-aldehyde).

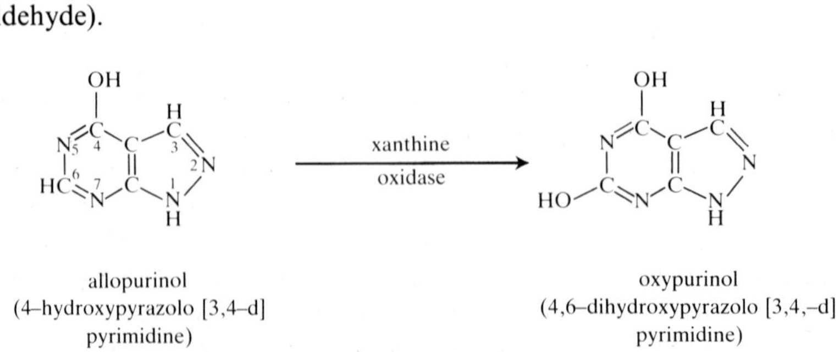

allopurinol
(4–hydroxypyrazolo [3,4–d]
pyrimidine)

oxypurinol
(4,6–dihydroxypyrazolo [3,4,–d]
pyrimidine)

Fig. 3.12. Structures of allopurinol and oxypurinol.

Other sources of superoxide

Sources of O_2^- other than xanthine oxidase have been used in SOD assays. A mixture of NADH and phenazine methosulphate produces some O_2^-, but we do not recommend it for use because it also directly generates hydroxyl radicals and creates many artefacts (Table 3.9). Illumination of a riboflavin solution in the presence of either EDTA or of the amino-acid methionine causes a reduction of the flavin. It then re-oxidizes and simultaneously reduces oxygen to O_2^-, which is allowed to react with a detector molecule such as NBT. SOD will inhibit the formazan production. Flavin photo-chemistry is extremely complicated, however, and singlet oxygen is also produced (see Chapter 2). In an interesting variation on this assay, an oxygen electrode is used to measure the rate of oxygen consumption during photochemical generation of O_2^- in the presence of NBT. Reduction of the dye by O_2^- is accompanied by stoichiometric oxygen-production, i.e.

$$NBT + O_2^- \rightarrow NBT^{\cdot} \text{ radical} + O_2$$

On addition of SOD, two O_2^- molecules are required to make one oxygen molecule, and the rate of oxygen uptake increases.

A number of compounds have been shown to oxidize in solution with simultaneous production of O_2^-; these include 6-hydroxydopamine, pyrogallol, the sulphite ion (SO_3^{2-}), and adrenalin (at alkaline pH values). O_2^-, once formed, participates in the oxidation of further molecules, so that addition of SOD greatly slows down the observed rates of oxidation of these compounds. This can be used as an assay for SOD, the rate of oxidation being measured either by an absorbance change (as with adrenalin oxidation to adrenochrome) or by oxygen uptake using an oxygen-electrode (as with pyrogallol or sulphite). The rates of these oxidations are often greatly accelerated by the presence of transition-metal ions, however, and this can cause problems in the assay of crude extracts containing traces of such ions.

Fridovich's laboratory in the USA has developed a 'positive' assay for SOD activity. A solution containing riboflavin and the detector molecule *ortho*-dianisidine is illuminated, whereupon the detector is slowly oxidized, accompanied by an absorbance change at 460 nm. Addition of SOD greatly increases the rate of dianisidine oxidation because it removes O_2^-, which interacts with an intermediate dianisidine radical and thereby decreases the net rate of oxidation. The assay is called 'positive' because addition of the SOD actually causes a reaction to accelerate, instead of inhibiting it. The reaction mechanism is very complicated, however. At alkaline pH values, SOD accelerates the oxidation of *haematoxylin*, a dye used by histologists as a 'stain'. This reaction has also been proposed as a positive assay for SOD.

Assays of superoxide dismutase: some general cautions
Whatever assay is used, it should first be calibrated with pure SOD enzyme, and a known amount of SOD enzyme, added to the crude tissue extract being examined, should be quantitatively detected on subsequent assay. The scientists performing the assay should also think carefully about possible artefacts, including interference with O_2^- generation. To take one example, the compound pamoic acid appeared to inhibit SOD activity in a number of indirect assays, but careful analysis showed that it was actually interfering with the assay mechanism. No one assay will be suitable for all systems, e.g. assays using NBT reduction cannot be applied to eye tissues because of the presence of enzyme systems that reduce NBT directly. Table 3.9 summarizes some of the methods that have been used by different workers; and Table 3.8 shows the activities detected in various animal tissues using some of these assay methods. The units quoted are different for each assay, but comparison of Tables 3.7 and 3.8 shows that the relative amount of SOD activity in different body organs is broadly similar in different mammals.

Distinction between the different SOD types in tissue extracts
We have seen that CuZnSOD is inhibited by cyanide ion (CN^-) whereas FeSOD or MnSOD is not. Inhibition by CN^- can therefore be used to identify CuZnSOD activity in assays of tissue homogenates or on polyacrylamide gels (e.g. Fig. 3.10).

Both CuZnSOD and FeSOD are inactivated on prolonged exposure to hydrogen peroxide, whereas MnSOD is not. The rate of inactivation of the CuZnSOD is faster at higher pH values. It may be due to a reduction of Cu(II) to Cu(I) at the active site by hydrogen peroxide, followed by a reaction of Cu(I) with hydrogen peroxide to generate OH·, or another oxidizing species (Chapter 2) which then destroys one of the histidine residues essential for the catalytic mechanism. Thus incubation of, say, a bacterial extract with hydrogen peroxide will inactivate FeSOD but not MnSOD and can be used to distinguish the two. Inactivation may be physiologically relevant under certain circumstances (Section 3.5.3).

Another method for distinguishing between different SOD types employs the fact that FeSODs are more sensitive to inhibition by azide. For example, at pH 7.8 azide at a concentration of 10 mmoles per litre inhibits CuZn, Mn and Fe enzymes by about 10 per cent, 30 per cent and 70 per cent, respectively. There is some variation however, e.g. the *Methanobacterium bryantii* FeSOD is less sensitive to azide than are other FeSODs and CuZnSOD from tomato leaves seems more sensitive to inhibition by azide than other CuZnSODs. A third approach has been to remove the metals from SOD proteins in bacterial extracts, and then to add either Fe(II) or Mn(II) back to the extract. If a particular band of enzyme activity observed on electrophoresis before metal removal re-appears on addition of, say, Fe^{2+}, then it most likely represented a FeSOD. Despite the close structural similarities between FeSODs and MnSODs, most of them will only work with the correct metal at the active site. There are exceptions to this rule, however. For example, SOD apoenzymes from *Bacteroides fragilis* and *Bacteroides thetaiotamicron* can be reconstituted to active enzyme with either iron or manganese ions.

3.2 Protection by small molecules

3.2.1 *Ascorbic acid (vitamin C)*

Pure ascorbic acid is a white crystalline solid, very soluble in water. Plants and most animals can synthesize it from glucose, but humans, other primates, guinea-pigs, and fruit-bats lost one of the necessary enzymes during their evolution, and so require ascorbic acid to be present in the diet, as vitamin C. We rely on the fact that plants can still make it. Ascorbic acid is required *in vivo* as a cofactor for several enzymes, of which the best known are proline hydroxylase and lysine hydroxylase, involved in the biosynthesis of collagen. Both these enzymes contain iron at their active sites. Ascorbate is also required for the action of the copper enzyme dopamine-β-hydroxylase which converts dopamine into noradrenalin. Deficiency of ascorbate from the human diet causes *scurvy*. Collagen synthesized in the absence of ascorbic acid is insufficiently hydroxylated and does not form fibres properly, giving rise to poor wound-healing and fragility of blood vessels.

The most striking chemical property of ascorbate is its ability to act as a reducing agent (electron donor). We have already seen (Chapter 2) that its ability to reduce Fe(III) to Fe(II) is important in promoting the uptake of iron in the gut. The observation that dietary ascorbate inhibits the carcinogenic action of several nitroso-compounds fed to animals (Chapter 8) can be attributed to its ability to reduce them to inactive forms. Ascorbate may help to detoxify various organic radicals *in vivo* (e.g. those formed by ionizing radiation—Section 3.4) by a similar reduction process. Indeed, ascorbate is probably required by the above hydroxylase enzymes in order to keep the iron or copper at the active site in the reduced form necessary for hydroxylation to occur.

Donation of one electron by ascorbate gives the semidehydroascorbate radical (Fig. 3.13), which can be further oxidized to give dehydroascorbate. The semidehydroascorbate radical is not particularly reactive and mainly undergoes a disproportionation reaction,

$$2 \text{ semidehydroascorbate} \rightleftarrows \text{ascorbate} + \text{dehydroascorbate}.$$

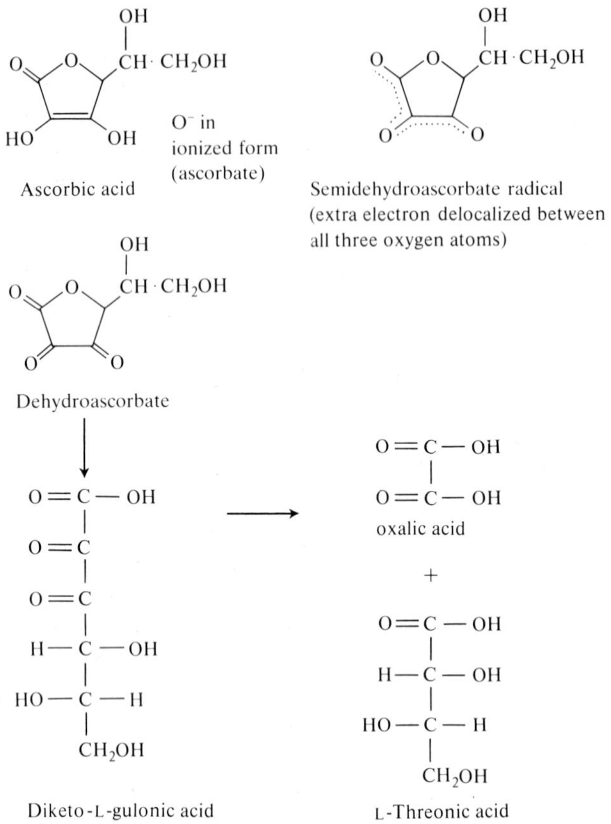

Fig. 3.13.   Structure of ascorbic acid and its oxidized forms.

Dehydroascorbate is unstable and breaks down rapidly in a very complex way, eventually producing oxalic and L-threonic acids. Aqueous solutions of ascorbic acid are stable unless transition-metal ions are present, which catalyse their rapid oxidation at the expense of molecular oxygen. Copper salts are the best catalysts—if you want plenty of vitamin C from your fruit and vegetables, don't cook them in copper pans! Copper-induced oxidation of ascorbate produces hydrogen peroxide and hydroxyl radicals. The reported ability of ascorbate to degrade DNA and damage various animal cells in culture, including cancer cells, can probably be attributed to the formation of these species in the presence of traces of copper ions in the solution. Ascorbic acid–Cu^{2+} mixtures inactivate many proteins, probably by formation of hydroxyl radicals and/or Cu(III) species (Chapter 1).

In Chapter 2 we saw that ascorbate reacts rapidly with O_2^- and $HO_2^{\cdot}$ and even more rapidly with $OH^{\cdot}$ to give semidehydroascorbate. It also scavenges singlet oxygen, reduces thiyl radicals (Chapter 1), and combines quickly with hypochlorous acid, a powerful oxidant generated at sites of inflammation (Chapter 7). The function of ascorbate peroxidase in removing hydrogen peroxide has already been discussed. Hence ascorbate may well help to protect against oxygen-derived species *in vivo*. For example, the lens of the human eye is low in SOD activity but rich in ascorbate, whereas rat lens has more SOD but less ascorbate. In agreement with such a protective role, exposure of animals to elevated oxygen concentrations, or to ozone, causes a decrease in the ascorbic acid content of the lungs; and administration of ascorbic acid to animals has sometimes been reported to lessen pulmonary damage caused either by ozone or by high-pressure oxygen. In the lung, ascorbate appears to accumulate in the fluid lining the air spaces and may thus act as an extracellular antioxidant, complementing the intracellular SOD and H_2O_2-removing enzymes. Intracellular ascorbate concentrations in several lung cells are in the millimolar range. Treatment with ascorbic acid greatly decreased the incidence of growth abnormalities induced by exposure of a strain of tobacco seedlings to pure oxygen.

Injection of dehydroascorbate into animals affects insulin secretion by the pancreas and induces diabetes. Its decomposition product, oxalic acid, is not very pleasant either. Hence, both animal and plant tissues have evolved mechanisms for converting the oxidized form of ascorbate back to the reduced form. A *dehydroascorbate reductase* enzyme was originally purified from many plant tissues, but later found in several animal and human tissues. It catalyses the overall reaction:

$$\text{dehydroascorbate} + 2\text{GSH} \rightarrow \text{GSSG} + \text{ascorbate}$$

but its true substrate may be the semidehydroascorbate radical, i.e.

$$2 \text{ semidehydroascorbate} + 2\text{GSH} \rightarrow \text{GSSG} + 2 \text{ ascorbate.}$$

Animals also contain a *NADH-semidehydroascorbate reductase* enzyme, which reduces the semidehydroascorbate radical back to ascorbate whilst oxidizing NADH to NAD^+ (Table 3.10). This enzyme has been reported in a few fungi and plant tissues.

Not everything is wonderful about vitamin C, however, whatever the health food shops say. Like O_2^-, ascorbate can reduce Fe(III) ions to Fe(II) and, in the presence of hydrogen peroxide, can stimulate $OH^{\cdot}$ formation by the Fenton reaction. Its overall effect will depend on the concentration of ascorbate present, since it also scavenges $OH^{\cdot}$. Administration of vitamin C to patients with iron-overload has sometimes provoked severe reactions, perhaps due to increased $OH^{\cdot}$ formation *in vivo*. Ascorbate can stimulate iron-dependent peroxidation of membrane lipids (Chapter 4) under certain circumstances.

Table 3.10.   Semidehydroascorbate reductase activity in rat
tissues

Tissue	Enzyme activity (mean ± SEM)
Adrenal cortex	49.6 ± 2.4
Brain	9.1 ± 0.6
Heart	0
Ileum	3.3 ± 0.3
Kidney	49.3 ± 4.9
Liver	30.9 ± 1.0
Lung	8.9 ± 1.8
Pancreas	16.3 ± 1.1
Skeletal muscle	0
Spleen	6.3 ± 0.3
Testis	11.4 ± 0.3
Thyroid gland	5.8 ± 0.3

Semidehydroascorbate reductase activity was assayed in homogenates
of several rat tissues. The enzyme reduces semidehydroascorbate
(SDA) to ascorbate at the expense of NADH. Data from Diliberto *et
al.*, *J. Neurochem.* **39**, 563–8 (1982). Enzyme activity is quoted as
nanomoles of NADH oxidized per minute per mg of protein. The high
activity in adrenal gland cortex may be due to the fact that SDA is
formed from ascorbate during the dopamine-β-hydroxylase reaction
(see text).

3.2.2 *Glutathione*

The role of GSH as a substrate for the H_2O_2-removing enzyme glutathione
peroxidase and for dehydroascorbate reductase has already been discussed.
In addition glutathione is a scavenger of hydroxyl radicals and singlet
oxygen. Since it is present at high concentrations in many cells (Table 3.2) it
may help to protect against these species. GSH can reactivate some enzymes
that have been inhibited by exposure to high oxygen concentrations.
Presumably the oxygen causes oxidation of essential —SH groups on the
enzyme, which are regenerated on incubation with GSH. Glutathione is not
essential for aerobic life since several strains of aerobic bacteria are known
that do not contain it, although they might contain other low-molecular-
mass thiol compounds serving a similar purpose. Mutants of *E. coli* unable to
synthesize GSH grow normally under air, as do mutants defective in
glutathione reductase, although their ability to tolerate elevated oxygen
concentrations has not been reported. GSH-deficient *E. coli* mutants seem to
leak K^+ ions across their plasma membranes, and cannot grow in media low
in K^+ ions. Deficiencies of GSH synthesis in animal cells have serious

consequences, however, such as haemolysis. Feeding diets deficient in sulphur-containing amino acids to rats potentiates the toxic effects of elevated oxygen concentrations.

GSH is a cofactor for several enzymes in widely different metabolic pathways, such as glyoxylase, maleylacetoacetate isomerase, protaglandin endoperoxide isomerase (Chapter 8), and DDT dehydrochlorinase and it may be involved in the synthesis of thyroid hormones. It plays a role in the degradation of insulin in animals and also in the metabolism of herbicides, pesticides, and 'foreign' compounds generally in both animal and plant tissues. For example, corn leaves contain an enzyme which detoxifies the herbicide atrazine by combining it with GSH. The higher the activity of this enzyme, the greater is the resistance of the plant to the herbicide. Many 'foreign compounds' supplied to animals are metabolized in the liver to yield *mercapturic acids* that are excreted. The first stage in this process is conjugation of the compound with GSH by *glutathione-S-transferase* enzymes, as shown in Fig. 3.14. Several enzymes of this type are present in liver, and also in many other animal tissues. Glutathione conjugates are usually excreted into bile, using the same transport mechanism that ejects GSSG when the liver is subjected to oxidative stress. Among compounds converted to mercapturic acids in the rat are chloroform, bromobenzene, naphthalene, and paracetamol. The presence of these compounds has the effect of decreasing hepatic GSH concentrations, which reduces the ability of the liver to cope with hydrogen peroxide and other oxygen radicals. Supplying metabolic precursors of glutathione can raise the GSH content of some tissues and protect against these effects. The compound 2-oxothiazolidine-4-carboxylate, a precursor of cysteine *in vivo*, seems especially effective. Injection of methyl esters of GSH, which can cross membranes easily, might also help to raise tissue GSH concentrations. Professor A. Meister in the USA has proposed a function for GSH in the transport of amino acids across the plasma membrane of animal cells.

If a tissue is exposed to a large flux of hydrogen peroxide and/or hydroxyl radicals, a point might be reached at which the GSH/GSSG ratio cannot be maintained at its normal high value (Table 3.2) and GSSG will accumulate. Unfortunately, GSSG can inactivate a number of enzymes, probably by forming mixed disulphides with them. Mixed disulphides with proteins and with such molecules as coenzyme A accumulate in tissues subjected to 'oxygen radical stress'.

$$\text{enzyme—SH} + \text{GSSG} \rightleftarrows \text{enzyme—S—S—G} + \text{GSH}.$$
$$\text{(active)} \qquad\qquad\qquad \text{(inactive)}$$

GSSG has been shown to inhibit protein synthesis in animal cells, and accumulation of GSSG has been suggested to contribute to the inhibition of

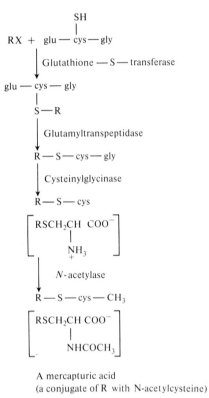

A mercapturic acid
(a conjugate of R with N-acetylcysteine)

Fig. 3.14. Mercapturic acid formation. RX represents the 'foreign' compound. Mercapturic acid formation is usually a detoxification pathway, but sometimes glutathione S— conjugates are themselves toxic, or react to form toxic products.

protein synthesis seen in dehydrated plant tissues, e.g. wheat germ and several mosses. GSSG also inhibits several enzymes, including adenylate cyclase, chicken liver fatty-acid synthetase, rabbit muscle phosphofructokinase, and phosphorylase phosphatase. This action of GSSG is presumably why cells maintain high GSH/GSSG ratios under normal conditions, and why organs such as the liver (Fig. 3.6) and to a lesser extent, the heart, release GSSG when they are under oxidant stress. It also shows how a protective mechanism can be turned into a damaging one at excessive rates of oxidant generation.

Other aspects of the biochemistry of GSH can cause problems. Thus the fast reaction of GSH with OH' ($k_2 > 10^9 \, \text{M}^{-1} \, \text{s}^{-1}$) gives thiyl radicals:

$$GSH + OH' \rightarrow GS' + H_2O$$

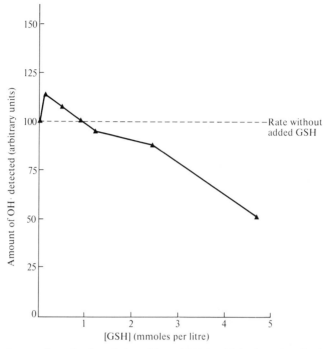

Fig. 3.15. Interaction of reduced glutathione (GSH) with hydroxyl radicals. GSH at various concentrations was added to a system generating hydroxyl radicals (a mixture of hypoxanthine and xanthine oxidase in the presence of an iron salt). At the concentrations present in most body tissues (Table 3.2) GSH decreased the amount of OH˙ detected because it can scavenge this radical. At low concentrations, however, GSH stimulated OH˙ formation because it can interact with Fe^{2+} salts and hydrogen peroxide (produced by the xanthine oxidase) to itself produce OH˙. Data from Rowley and Halliwell, *FEBS Lett.* **138**, 33–6 (1982). (Other thiol compounds showed similar effects.)

which can also be formed when GSH is oxidized by peroxidases (section 3.1.1), or by O_2 in the presence of transition metal ions such as Cu^{2+} and Fe^{2+} (Chapter 1). Thiyl radicals, although less reactive than OH˙, might cause some biological problems. Mixtures of Fe^{2+} salts, H_2O_2 and GSH can also *produce* OH˙ radicals, although this can only be demonstrated at low GSH concentrations (Fig. 3.15) and is thus unlikely to occur intracellularly. Wefers and Sies in West Germany reported that the slow reaction of GSH with superoxide radical ($k_2 < 7.7 \times 10^5\,\mathrm{M^{-1}\,s^{-1}}$) can lead to singlet O_2 formation: a *possible* series of reactions is shown below:

$$GSH + O_2^- + H^+ \rightarrow GS^{\cdot} + H_2O_2$$
$$(\text{or } GSH + HO_2^{\cdot} \rightarrow GS^{\cdot} + H_2O_2$$
$$GS^{\cdot} + O_2 \rightarrow GSO_2^{\cdot}$$
$$GSO_2^{\cdot} + O_2^- + H^+ \rightarrow {}^1O_2 + GSOOH$$
$$(\text{singlet } O_2)$$
$$GSO_2^{\cdot} + GSO_2^{\cdot} \rightarrow GSSG + 2{}^1O_2$$
$$GSO_2^{\cdot} + GS^{\cdot} \rightarrow GSSG + {}^1O_2$$

Several reactive radicals, such as $GSO_2^{\cdot}$ and $GSO^{\cdot}$, appear to form when $GS^{\cdot}$ radicals are exposed to O_2.

However, the reaction mixtures used by Wefers and Sies were saturated with pure O_2, and so whether singlet O_2 could be formed by this mechanism at more physiological O_2 concentrations is uncertain. Of course, thiols can also *scavenge* singlet oxygen.

3.2.3 *Uric acid*

Human tissues do not contain the enzyme urate oxidase, so that uric acid accumulates as an end-product of purine metabolism and it is present in human blood plasma at concentrations in the range 0.25–0.45 millimolar (mM). In 1981, Ames *et al.* in the USA pointed to the fact that uric acid is a powerful scavenger of singlet O_2, peroxyl ($RO_2^{\cdot}$) radicals, and $OH^{\cdot}$ radicals, and suggested that uric acid functions as an antioxidant *in vivo*, which means that loss of the urate oxidase enzyme during human evolution might have been beneficial. Japanese scientists had earlier reported the ability of uric acid to inhibit lipid peroxidation (Chapter 4). Uric acid is also a powerful scavenger of ozone and of hypochlorous acid, an oxidant produced by the enzyme myeloperoxidase (Chapter 7). This proposal of Ames *et al.* is supported by the observation, in the authors' laboratories, that allantoin, an oxidation product of uric acid, accumulates under some conditions in which 'oxidative stress' occurs. Glucose, present in the human bloodstream at 4.5 mM, or even higher concentrations shortly after a carbohydrate-rich meal, can also scavenge $OH^{\cdot}$ (Table 2.1).

Reaction of $OH^{\cdot}$ with uric acid produces a range of carbon-centred radicals, that mostly react with O_2 to give urate peroxyl radicals (Chapter 1):

$$R{>}C{-}H + OH^{\cdot} \longrightarrow R{>}C^{\cdot} + H_2O$$

$$R{>}C^{\cdot} + O_2 \longrightarrow R{>}CO_2^{\cdot}$$
$$\text{peroxyl radical}$$

These urate-derived radicals, although much less reactive than $OH^{\cdot}$, are not completely harmless, e.g. they can inactivate the enzyme alcohol dehydrogenase (from yeast) and human α-antiproteinase. Thus uric acid, like GSH or ascorbate, is not always a 'perfect' antioxidant.

3.3 Protection by sequestration of metal ions

Ions of such transition metals as iron and copper are involved in many free-radical reactions, and often they lead to generation of very-reactive species from less reactive ones; Table 3.11 lists some examples of this. In the case of iron, we have already seen that iron bound to proteins is not normally available to stimulate radical reactions, unless iron becomes released from proteins under the assay conditions being used. Hence the evolution of iron storage and transport proteins (ferritin and transferrin) provides not only a convenient way of moving iron around the body (Section 2.4.3), but may also be regarded as an *antioxidant defence*. For example, in normal human plasma there is an excess of transferrin, and the concentration of non-transferrin-bound iron ions is essentially zero. Indeed, this 'sequestration' of metals in forms unreactive in radical reactions provides an important part of extracellular antioxidant defence (Chapter 4). Uric acid (Section 3.2.3) can interfere with radical reactions not only by scavenging oxidants, but also by binding iron and copper ions in forms that do not participate in radical reactions. For example, uric acid is a very powerful inhibitor of the Cu^+-dependent formation of $OH^.$ (or Cu(III)) from H_2O_2, apparently by binding the copper ions. The absence of low-molecular-mass iron complexes from plasma also accounts for much of the bacteriostatic effect of this fluid; bacteria need iron in order to grow and, unless they can somehow obtain it from transferrin (e.g. by secretion of a suitable siderophore that removes iron from this protein), they cannot grow in plasma.

However, cells do contain a small 'low molecular mass' iron pool, used for the synthesis of iron proteins (Section 2.4.3). This pool could conceivably provide iron ions to accelerate radical reactions; it is difficult to make a firmer statement than this until our knowledge of the chemical nature and subcellular location of the pool improves. The existence of an intracellular iron pool may explain why cells rely heavily on enzymes that remove O_2^- (superoxide dismutase), H_2O_2 (catalase and peroxidases) and lipid peroxides (Chapter 4) for their antioxidant defence; it is important to remove these species before they come into contact with iron ions and generate more-damaging agents (Table 3.11).

Most plasma copper is attached to the protein caeruloplasmin, which has antioxidant properties. These are partly due to the ability of caeruloplasmin to oxidize Fe^{2+} to Fe^{3+} (Section 2.4.3), which decreases Fe^{2+}-dependent production of $OH^.$ from H_2O_2 and Fe^{2+}-dependent lipid peroxidation. The small amount of plasma copper not attached to caeruloplasmin is apparently

Table 3.11. Role of transition metal ions in converting less reactive into more reactive species

Starting agent	More reactive species	Metal involved	Comment
H_2O_2 ($\pm O_2$)	$OH^{\cdot}$ (and possibly other species)	Fe/Cu	Iron-dependent and copper-dependent conversion of H_2O_2 to $OH^{\cdot}$ (see Chapter 1 and Section 3.5)
Lipid peroxides	Peroxy radicals, alkoxy radicals, cytotoxic aldehydes	Fe/Cu	See Chapter 4
Thiols (R—SH)	O_2^{-}, H_2O_2, $RS^{\cdot}$, $OH^{\cdot}$	Fe/Cu	Oxidation of thiols such as GSH produces thiyl radicals and oxygen radicals (Chapter 1 and Section 3.2.2)
Ascorbic acid	$OH^{\cdot}$, possibly O_2^{-}, semidehydroascorbate radical	Fe, especially Cu	Oxidation of ascorbate produces cytotoxic species (Section 3.2.1)
Alloxan, adrenalin, dihydroxyfumarate, 6-hydroxydopamine, other 'autoxidizable' compounds	$OH^{\cdot}$, O_2^{-}, carbon-centred or other radicals derived from the toxin	Fe, Cu, Mn, often other metals	Most 'autoxidations' are dependent on the presence of traces of transition metal ions. Several examples are discussed in Chapter 6 and Section 3.5.

bound to histidine, small peptides, and albumin. None of these forms of copper can apparently generate reactive oxidants in free solution, and OH˙ or Cu(III) formed by reaction of bound copper ions with H_2O_2 appears to attack the binding molecule (Section 2.4.3). Indeed, the ability of albumin to bind copper ions may prevent them from binding somewhere more important, and thus albumin is acting as an antioxidant in this context.

Hence organisms appear to have evolved to minimize, as far as possible, the amount of reactive metal ions that are allowed to exist *in vivo*. This 'sequestration' of metal ions helps decrease damaging radical reactions; the consequences of its failure can be seen in such disorders as iron overload (Section 2.4.3).

3.3.1   *Metallothioneins*

Metallothioneins are low-molecular-mass (about 6500) proteins found in the cytosol of eukaryotic cells, especially in liver, kidney, and gut. They may also be able to enter the nucleus. Metallothioneins are very rich in sulphur (23–33 per cent cysteine) and possess the ability to bind ions of such metals as zinc (Zn^{2+}), copper (Cu^+), cadmium (Cd^{2+}) and mercury (Hg^{2+}). Binding is achieved by association of cysteine —SH groups with the metal ion, e.g. Cd^{2+} and Zn^{2+} are linked to four cysteine thiolate ligands $(cys—S^-)$ in a tetrahedral arrangement. The metallothionein content of liver, kidney, and gut can be increased by injection or oral administration of Cd, Cu, or Zn salts, and synthesis of these proteins is also increased by several hormones, including glucocorticoids, glucagon, and adrenalin, and by interleukin I produced during inflammation (Chapter 8). Proposed functions of metallothioneins include storage of heavy metals in a non-toxic form, and the regulation of both cellular copper and cellular zinc metabolism, and of the absorption of these metals from the gut. For example, cultured mammalian cells that have lost the ability to make metallothioneins are exceptionally sensitive to injury by Cd^{2+} ions, whereas cells overproducing these proteins are more resistant to Cd^{2+}. Thornalley and Vasak in England have suggested that metallothioneins might also have antioxidant properties. Sequestration of Cu^+ will diminish radical generation promoted by this metal (see above), Zn^{2+} released from zinc-metallothionein might inhibit lipid peroxidation and the high content of —SH groups in metallothioneins makes them excellent scavengers of singlet O_2 and OH˙ radicals (although the resulting sulphur-centred radicals must not be ignored). Metallothionein-enriched Chinese hamster cells are significantly more resistant to damage by H_2O_2 than are the parent cells, consistent with an antioxidant role for metallothionein.

3.4 Repair systems

Despite the plethora of defence systems, it is likely that some radical damage still occurs *in vivo*; indeed, it might contribute to the ageing process (Chapter 8). Hence systems have evolved to 'repair' radical damage. Thus peroxidized lipids can be cleaved out of membranes (Chapter 4), and several other repair systems exist.

3.4.1 *Methionine sulphoxide reductase*

The action of singlet oxygen, hypochlorous acid (HOCl; see Chapter 7), or hydroxyl radicals upon the amino acid methionine can cause its oxidation to methionine sulphoxide (Chapter 2). The proteins within the lenses of patients suffering from cataracts contain a significant amount of methionine sulphoxide and, as we saw in Chapter 2, there is considerable interest in the role of singlet oxygen produced by photosensitization reactions in damaging lens proteins. *E. coli*, yeast, rat tissues, rabbit tissues, the alga *Euglena gracilis*, human and bovine lens, human lung and neutrophils, spinach leaves, and the protozoan *Tetrahymena pyriformis* have all been found to contain an enzyme that reduces methionine sulphoxide back to methionine and hence can reactivate proteins damaged by previous oxidation of their methionine residues. The source of reducing power used by the enzyme is not clear, although there is evidence that NADPH reduces a thiol-containing protein (thioredoxin) which then, in the presence of the enzyme, reduces the sulphoxide. The actual importance of this reductase in repairing radical-induced damage to cells cannot yet be evaluated.

3.4.2 *DNA repair*

Most non-replicating nuclear DNA interacts with basic proteins called histones, which might help to protect the DNA against radicals such as OH˙. Despite this, Ames has calculated (Section 2.5.3) that there could be, on average, about 10^3 oxidant-mediated damaging events upon the DNA of each cell in the human body every day. This figure, combined with the significant rates of 'spontaneous' loss of purines from DNA and deamination of cytosine to uracil, as well as any errors made during DNA replication, emphasizes the importance of DNA repair mechanisms.

There are several different DNA repair systems. Those involved in correcting oxygen/hydroxyl radical damaged DNA generally act by an 'excision repair' mechanism. A *DNA glycosylase* enzyme removes a damaged base by cleaving the base–sugar bond to leave an apurinic/apyrimidinic (AP) site. The AP site is recognized by a *DNA AP endonuclease*, which may be a separate enzyme, or another enzyme activity present in a DNA glycosylase molecule. The endonuclease nicks the strand at the AP site. For example:

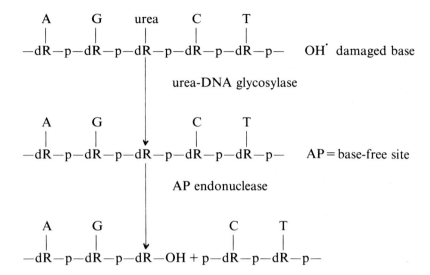

The damaged part of the strand is removed, new DNA synthesis fills the gap and finally a *DNA ligase* enzyme joins the newly synthesized DNA to the rest of the strand.

Prokaryotic and eukaryotic DNA glycosylases have been described for both purines and pyrimidines damaged by OH˙ attack. In *E. coli*, one multifunctional enzyme excises thymine glycol, 5-hydroxy-5-methyl-hydantoin, urea, and methyltartronylurea (i.e. ring saturated, contracted, and fragmented thymine), and also acts as an AP endonuclease. This enzyme has been known variously as DNA endonuclease III, urea-DNA glycosylase, and thymine glycol-DNA glycosylase; the latter name may be the most appropriate. An analogous enzyme exists in mammalian, including human, cells. 5-Hydroxymethyluracil-DNA glycosylase is unusual in that it is only found in differentiated mammalian cells.

Purines whose imidazole ring has been opened are also excised by a DNA glycosylase. The mechanism of repair of the miscoding lesion 8-hydroxyguanine (Section 2.5.3) is not known.

DNA strand breakage additionally results in activation of poly (ADP-ribose) synthetase within cells (Section 2.4.1), which may also play a part in the repair mechanism. GSH may also sometimes 'repair' radicals resulting from the attack of OH˙ on DNA bases (Section 3.6).

The importance of DNA-repair systems in man is revealed by inborn errors of metabolism. The best explained is *Bloom's syndrome*, in which an inborn deficiency in DNA ligase I leads to a failure to rejoin DNA. This results in genetic instability characterized by increased chromosome aberrations and sister chromatid exchanges, impaired growth, immunodeficiency, and greatly increased cancer risk. Another inborn defect in humans is

Xeroderma pigmentosum, in which a deficiency in the ability to cleave modified thymine bases from DNA results in a hypersensitivity to sunlight and a greatly-increased incidence of skin cancer.

3.4.3 *Degradation of abnormal proteins*

Oxidized methionine residues in proteins can be repaired (Section 3.4.1) but proteins in which other amino acids have been attacked by OH˙ seem to be irreversibly damaged and must be removed from the cell. The groups of R. T. Dean in England, and of R. A. Floyd and K. J. A. Davies in the USA, have studied in detail the action of OH˙ radicals upon several proteins. Superoxide (O_2^-) radical has little, if any, protein-damaging effects, as expected from its relatively-poor reactivity. However, O_2^- can sometimes exacerbate protein damage produced by OH˙ in the presence of oxygen, apparently because O_2^- reacts with some of the products of OH˙ attack.

Proteins modified by oxidants are degraded at an accelerated rate by proteolytic enzymes present within *E. coli*, human erythrocytes and other mammalian cells. For example, Stadtman, Levine, and their colleagues in the USA showed that inactivation of the *E. coli* enzyme glutamine synthetase by exposure to oxidants, which modifies an essential histidine residue, led to rapid degradation of the protein within the bacterium. Intact mammalian and plant cells have also been shown to degrade proteins faster when the cells are put under a limited degree of 'oxidative stress'. The ability of cells to degrade abnormal proteins may decrease with age, since there have been several reports of the accumulation of non-functional proteins during ageing (Chapter 8). Proteolytic systems that recognize oxidatively-modified proteins are present in both the cytosol and in the mitochondria of mammalian cells.

3.5 The superoxide theory of oxygen toxicity

The discovery of SOD in aerobic cells led directly to the proposal, from Fridovich's laboratory in the USA, that the superoxide radical is a major factor in oxygen toxicity and that the SOD enzymes are an essential defence against it. Even if it turned out to be untrue, this *superoxide theory of O_2 toxicity* would still have been extremely valuable because of the great amount of experimental work that it provoked, which has led to many important discoveries. Indeed, in the past few years more papers have been published on SOD than on any other type of enzyme. There is much truth in the theory, however; although, like all good theories, it has required modification in the light of new experimental facts. Let us evaluate the current status of the theory.

3.5.1 *Is superoxide formed* in vivo?

Most oxygen taken up by higher aerobes is reduced to form water by the operation of the mitochondrial electron transport chain. A few enzymes, such as glycollate oxidase, reduce oxygen directly to hydrogen peroxide as described in Chapter 2. However, a number of enzymes have been discovered that actually reduce oxygen to O_2^-*, which is released into free solution. Table 3.12 describes some of these. Xanthine oxidase and the oxidase action of peroxidase have already been discussed. However, work by Italian scientists has shown that 'xanthine oxidase' from liver, and probably in most other animal tissues and in plants, is in fact a xanthine dehydrogenase enzyme *in vivo*, which transfers electrons from the substrate onto NAD^+ rather than onto oxygen. Xanthine dehydrogenase becomes converted to xanthine oxidase during its purification, either because of attack by proteolytic enzymes or by oxidation of some thiol ($-SH$) groups. Thus the metabolism of xanthine or hypoxanthine by this enzyme, which is located in the cytosol of animal and plant tissues, would not normally produce O_2^- *in vivo*. However, there is evidence for some xanthine oxidase activity *in vivo* in thyroid gland and intestinal cells. For example, addition of xanthine to whole intestinal cells increases indoleamine dioxygenase activity *in vivo*, an effect suppressed by the xanthine-oxidase inhibitor, allopurinol (see Table 3.12). Conversion of xanthine dehydrogenase into xanthine oxidase, and formation of O_2^- by the latter, is important in mediating reoxygenation damage to tissues after ischaemia. This is discussed in detail in Chapter 8.

Autoxidation reactions
Several biologically-important molecules oxidize in the presence of oxygen to yield O_2^-; these include glyceraldehyde, the reduced forms of riboflavin and its derivatives FMN and FAD, adrenalin, tetrahydropteridines, and thiol compounds such as cysteine. Tetrahydropteridines act as cofactors for several oxygenase enzymes, including those that catalyse hydroxylation of the aromatic amino acids, phenylalanine and tyrosine. Oxidation of adrenalin (a very complex reaction) and of photochemically-reduced flavins has been employed as a source of O_2^- in SOD assays, as discussed in Section 3.1.2. The reduced forms of several biological flavoproteins and flavoenzymes have also been shown to release O_2^- in the presence of oxygen. We have seen (Table 3.11) that oxidation of adrenalin and thiol compounds (cysteine, GSH) *in vitro* is greatly accelerated by the presence of ions of transition metals such as iron and copper. Indeed, Swedish scientists have pointed out that several of the components present in the growth media used to culture anaerobic bacteria oxidize to produce O_2^- and hydrogen peroxide

---

* Reasons for writing superoxide as O_2^- instead of O_2^{--} are discussed in Chapter 1.

Table 3.12. Enzymes that generate the superoxide radical

Enzyme	Location	Comments
Peroxidases (nonspecific)	Plants and bacteria	O_2^- produced during the oxidase reaction (see Section 3.1.1).
Cellobiose oxidase	White-rot fungus *Sporotrichum pulverulentum*	Contains FAD and a b-type cytochrome. Oxidizes a range of disaccharides, O_2^- is primary reduced oxygen product.
Xanthine oxidase	Intestine, ischaemic tissues (Chapter 8)	See Section 3.5.1. Forms both O_2^- and H_2O_2
Nitropropane dioxygenase	*Hansenula mrakii* (a yeast)	Catalyses oxidation of 2-nitropropane into acetone. O_2^- produced and involved in the catalytic mechanism. SOD inhibits the oxidation.
Indoleamine dioxygenase	Most animal tissues, especially small intestine, not liver. Activity of enzyme in lung increases during virus infection or after injection of bacterial endotoxin (30–100 fold) but is not increased by exposure to elevated oxygen concentrations.	Cleaves the indole ring of tryptophan and several related compounds such as serotonin

		O_2^- produced and involved in the catalytic mechanism, SOD inhibits the oxidation. Inhibition of the SOD in isolated rabbit intestine cells by diethyldithiocarbamate increased tryptophan degradation, as did addition of xanthine.
Tryptophan dioxygenase	Liver	Same reaction as above but specific for tryptophan.
Galactose oxidase	Fungi	Copper enzyme. O_2^- produced and involved in the catalytic mechanism. Oxidizes a $-CH_2OH$ group of the sugar galactose to $-CHO$.
Aldehyde oxidase	Liver	Contains molybdenum, iron. Produces free O_2^-. Broad substrate specificity.

in the presence of oxygen and could thus contribute to the toxic effects seen when the bacterial cultures are exposed to oxygen.

In general it is difficult to assess the contribution of these oxidation reactions to O_2^- formation *in vivo*, but there is one such reaction that is of great significance to the cells in which it occurs—that of oxyhaemoglobin. The haemoglobin molecule has four protein subunits, two a amino-acid chains and two β chains. Each chain carries a haem group to which the oxygen reversibly attaches. The iron in the haem ring of deoxyhaemoglobin is in the Fe(II) state, but when oxygen attaches to it, an intermediate structure results, in which an electron is delocalized between the iron ion and the oxygen (see Appendix I), i.e.

$$Fe^{2+}-O_2 \leftrightarrow Fe^{3+}-O_2^-$$

The bonding is intermediate between Fe(II) bonded to O_2, and Fe(III) bonded to the superoxide radical. Every so often a molecule of oxy-haemoglobin undergoes decomposition and releases O_2^- (this is a gross oversimplification of the actual mechanism by which O_2^- is released but it is sufficient for our purpose):

$$haem-Fe^{2+}-O_2 \rightarrow O_2^- + haem-Fe^{3+}$$

The product that has Fe (III), present in the haem ring, is unable to bind oxygen and is thus biologically inactive; it is known as *methaemoglobin*. It has been estimated that about 3 per cent of the haemoglobin present in human erythrocytes undergoes such oxidation every day, and so these cells are exposed to a constant flux of O_2^-. Mature mammalian erythrocytes contain no nuclei or mitochondria, nor can they synthesize proteins or membrane lipids. Since they have to survive for an average of 120 days in the circulation of humans without membrane or protein renewal, they must carefully protect themselves against O_2^- and H_2O_2 using CuZnSOD, cata-lase, glutathione peroxidase, and pentose phosphate pathway enzymes, as described previously and summarized in Fig. 3.16. Erythrocytes can, how-ever, degrade oxidatively-modified proteins (Section 3.4.3). A *methaemoglo-bin reductase* enzyme is additionally present to reduce Fe(III) protein to Fe(II) protein and so to reactivate the haemoglobin. Haemoglobin oxidation is speeded up by the presence of transition metal ions, especially those of copper, and of nitrite ion (NO_2^-). The presence of large amounts of nitrate (NO_3^-) in the water supply of some rural areas, due to excessive use of inorganic fertilizers, can cause problems in young bottle-fed babies: the NO_3^- in the water used to make up feeds is reduced by gut bacteria to NO_2^-, which is then absorbed and causes sufficient methaemoglobin formation to interfere with oxygenation of the body tissues. Several mutant haemoglobins

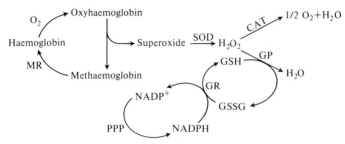

Fig. 3.16. Protection of erythrocytes against damage resulting from the oxidation of haemoglobin. MR, Methaemoglobin reductase; SOD, Cu–Zn superoxide dismutase; CAT catalase; GP, glutathione peroxidase; GR, glutathione reductase, PPP, pentose phosphate pathway (first enzyme: glucose 6-phosphate dehydrogenase). Haemoglobin is degraded by H_2O_2 with loss of iron ions (Section 2.4.3), so it is very important that erythrocytes dispose of H_2O_2 rapidly.

oxidize much more readily than normal, as do the isolated α or β chains that accumulate in thalassaemia. Oxymyoglobin, and the oxygenated form of the leghaemoglobin present in the root nodules of leguminous plants (Chapter 1), can also release O_2^- ions.

Superoxide formation has also been reported from negative air ionizers (supposed to improve your health and prevent allergies), during exposure of the amino acid tryptophan (Table 3.12) to near-UV light (which may be of relevance in the eye, since lens proteins contain tryptophan) and from light-exposed melanins. Melanins are pigments formed by oxidation and polymerization of the aromatic amino acid tyrosine,

$$CH_2 \cdot CH \cdot \overset{+}{N}H_3$$
$$COO^-$$
$$OH$$

and they contain high concentrations of *o*-quinone (oxidizing) and *o*-hydroquinone (reducing) groups as well as semiquinones. The brown or black melanins (*eumelanins*) of human skin afford protection against the ultraviolet light in sunlight. The melanin polymer contains many unpaired electrons left over from the polymerization process and these can be detected by ESR. Hence one can regard melanins as large free radicals. The movement of unpaired electrons between different energy levels helps to absorb the UV radiation. Illumination of eumelanins generates O_2^- within the molecule, but

142 *Protection against oxidants in biological systems*

this is very quickly scavenged; O_2^- can reduce melanin quinones to semiquinones, and oxidize hydroquinones, also to semiquinones (Table 2.7). Hence, overall, eumelanins are radical scavengers. The red-brown or yellow pigment found in the skin and hair of fair-skinned, red-headed humans is *pheomelanin*, which is less good as a radical scavenger. On exposure of pheomelanin to strong light it is degraded, and a net formation of O_2^- can be measured. Finally, photochemical degradation of organic compounds in sea-water also generates O_2^- and H_2O_2, and trace quantities of H_2O_2 are present in most natural water sources (Chapter 1).

Mitochondrial electron transport chains

Probably the most important sources of O_2^- *in vivo* in most aerobic cells are the electron transport chains of mitochondria and endoplasmic reticulum. The chloroplast electron transport chain is discussed in Chapter 5. The most important function of animal mitochondria is the oxidation of NADH and of $FADH_2$ produced during the Krebs cycle, β-oxidation of fatty acids, and other metabolic pathways. Oxidation is achieved by an electron transport chain located in the inner mitochondrial membrane and described in Fig. 3.17. The energy released by these oxidations is used to drive ATP synthesis. For every four electrons fed into the cytochrome oxidase complex (Fig. 3.17) a molecule of oxygen is reduced to two molecules of water. Cytochrome oxidase releases no dectable oxygen radicals into free solution, i.e. it has evolved so as to keep all the radical intermediates of oxygen reduction firmly bound to the proteins. The exact mechanism of the reduction is not yet known, but scientists in Sweden and the USA have identified a number of partially-reduced oxygen intermediates tightly bound to the cytochrome oxidase complex. Unlike cytochrome oxidase, some other components of the electron transport chain do 'leak' a few electrons onto oxygen whilst passing the great bulk of them on to the next component in the chain. As would be expected from the chemistry of oxygen (Chapter 1), this leakage produces a univalent reduction to give O_2^-. The main sites of leakage seem to be the NADH-coenzyme Q reductase complex, and the reduced forms of coenzyme Q itself. Mitochondria isolated from several animal tissues have been shown

---

Fig. 3.17. The electron transport chain of animal mitochondria. NADH is oxidized to NAD^+ by a multienzyme complex known as 'NADH-coenzyme Q reductase' and the two electrons are eventually passed onto coenzyme Q. The reductase complex contains flavoproteins (FMN at active site) and non-haem-iron proteins. Coenzyme Q accepts the electrons to form both semiquinone and fully reduced forms. It can also accept electrons from various reduced flavoproteins generated by the Krebs cycle and β-oxidation of fatty acids. From Q the electrons pass through another multienzyme complex ('coenzyme Q cytochrome c reductase', which contains cytochromes b) onto cytochrome c. Cytochromes are haem proteins which accept electrons by allowing Fe(III) at the centre of the haem ring to be reduced to Fe(II), i.e. they can accept one

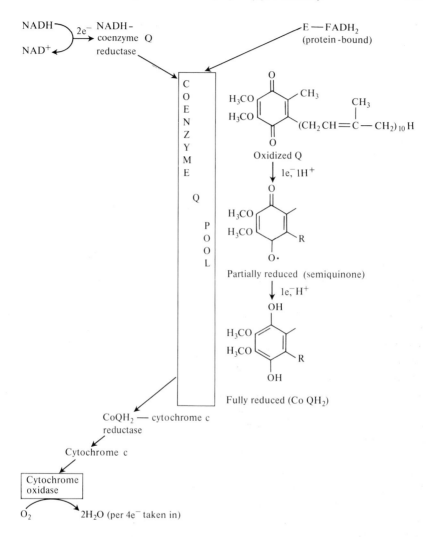

electron at a time per molecule. Different cytochromes, designated by small letters, contain different proteins and different haem groups. Finally, reduced cytochrome c is re-oxidized by the multienzyme complex, 'cytochrome c oxidase' which contains cytochrome a, cytochrome a_3 and copper. For every four electrons taken in by this complex, one oxygen molecule is fully reduced to two molecules of water. The enzyme dihydro-orotic acid dehydrogenase, which catalyses a step in pyrimidine synthesis, feeds electrons into the electron transport chain at several points in the region of CoQ. The ubisemiquinone radical can be detected in respiring mitochondria by observing its ESR signal.

to produce hydrogen peroxide *in vitro*, most, if not all, of which arises as a result of dismutation of O_2^- by mitochondrial SOD activity. Mitochondria possess MnSOD in the matrix and might have a little CuZnSOD in the space between the inner and outer membranes, as explained previously. The rate of O_2^-, and hence H_2O_2, production by mitochondria is increased at elevated oxygen concentrations. For example, in slices of rat lung exposed to air, about 9 per cent of total oxygen uptake could be attributed to O_2^- formation, the rest being due to cytochrome oxidase activity. In an atmosphere containing 85 per cent oxygen, however, O_2^- formation accounted for 18 per cent of the total oxygen uptake.

Mitochondria from some trypanosomes and from plant tissues (e.g. spinach leaves, mung beans, potato tubers, and Jerusalem artichokes) have also been shown to produce hydrogen peroxide, probably via O_2^- produced from CoQ and from NADH-coenzyme Q reductase, as in animal mitochondria. The electron transport chains located in the plasma membranes of several aerobic bacteria, e.g. *E. coli* and *Paracoccus denitrificans* have been shown to produce O_2^-, as have rabbit spermatozoa (although its exact origin is unknown).

Endoplasmic reticulum
The endoplasmic reticulum of many animal and some plant tissues contains cytochromes known collectively as cytochrome P-450. The name was given because the reduced forms of the cytochromes complex with carbon monoxide to produce a species that absorbs light strongly at 450 nm. Cytochrome P-450 is involved in the oxidation of a wide range of substrates at the expense of molecular oxygen. One atom of the oxygen enters the substrate and the other forms water, such a reaction being known as a *mono-oxygenase* or *mixed-function oxidase* reaction. The functioning of cytochrome P-450 requires a reducing agent (RH_2), and the overall reaction catalysed can be represented by the following equation, in which AH is the substrate:

$$AH + O_2 + RH_2 \rightarrow A{\cdot}OH + R + H_2O$$

Liver endoplasmic reticulum is especially rich in P-450, which metabolizes a large number of chemicals. Some of these compounds can increase synthesis of one or more forms of the cytochrome when fed to animals. One such inducer is the barbiturate phenobarbital, hydroxylation of which increases its solubility and aids its excretion from the body. Excessive intake of ethanol by mammals increases the ability of liver microsomal fractions to oxidize this substance, because of increased synthesis of a specific form of cytochrome P-450. This 'ethanol-inducible cytochrome P-450', is effective in oxidizing ethanol, other alcohols and acetone, and it contributes to the microsomal ethanol-oxidizing system in ethanol-treated animals (Section 2.4.5).

Substrates for the cytochrome P-450 system include insecticides such as heptachlor and aldrin, hydrocarbons such as benzpyrene, and drugs such as phenacetin, amphetamine, and paracetamol. Usually the product of reaction with P-450 is less toxic than the starting material, but this is not always the case: there is evidence that it is the hydroxylated products of paracetamol (Chapter 6) and of carcinogenic hydrocarbons such as benzpyrene (Chapter 8) that actually do the cellular damage that these compounds cause.

Often the initial product undergoes further reactions, e.g. *deamination* (removal of an amino group, $-NH_2$) in the case of amphetamine. Two hydroxylated intermediates may exist, i.e.

amphetamine   phenylpropanone

N-demethylation (removal of a methyl group, CH_3-, produces the aldehyde formaldehyde, HCHO. Metabolism of the carcinogen *dimethylnitrosamine* in this way produces methyl ions ($CH_3{}^+$) which can attack DNA:

dimethylnitrosamine   formaldehyde

$$N_2 + HO^- + H_3C^+$$

Removal of hydrocarbon groups attached to oxygen atoms (*O-dealkylation*) can also occur, e.g.

$$R-O-CH_3 \xrightarrow[O_2]{P_{450}} [R-O-CH_2-OH] \longrightarrow ROH + HCHO$$

In the liver the electrons required by the P-450 system are donated by NADPH via a flavoprotein enzyme *NADPH-cytochrome P-450 reductase*.

Figure 3.18 shows a possible mechanism for substrate hydroxylation by P-450 but the nature of the actual hydroxylating species at the active site of this protein is not yet clear (see figure legend). Adrenal cortex mitochondria contain a cytochrome P-450 which is involved in the hydroxylation of cholesterol to give the adrenal steroid hormones (e.g. aldosterone, hydrocortisone, and corticosterone) but the electrons it requires are donated by a non-haem-iron protein known as *adrenodoxin*. A flavoprotein enzyme transfers electrons from NADPH to adrenodoxin. Cytochrome P-450 is found in some bacteria, e.g. in *Pseudomonas putida* it serves to hydroxylate camphor. Here electrons are supplied by the non-haem-iron protein *putidaredoxin*, which is kept reduced at the expense of NADH by a flavoprotein enzyme. The hydroxylated camphor can then be metabolized by the cells to provide energy.

The reduced forms of adrenodoxin, putidaredoxin, and NADPH-cytochrome P-450 reductase from liver can 'leak' electrons onto O_2, reducing it to give O_2^-. In addition, there is some evidence that the oxygenated intermediates of cytochrome P-450 itself (Fig. 3.18) can decompose in a minor side-

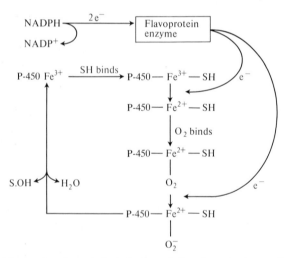

Fig. 3.18. Mechanism for substrate hydroxylation by cytochrome P-450 in liver. SH represents the substrate molecule. The mechanism of P-450 hydroxylation appears similar in other tissues or bacteria, but the source of reducing power used is different. The one shown is for liver endoplasmic reticulum. The mechanism by which the P-450—Fe(II)—O_2^- complex produces a hydroxylating species is not known. It is possible that there is a loss of water to give P-450—$[FeO]^{3+}$, in which the iron has a nominal valency of 5. It may be that the actual valency state of the iron is $+4$ (i.e. a ferryl species, FeO^{2+}) and the extra oxidizing capacity is located elsewhere on the molecule, as in compound I of horseradish peroxidase.

reaction and release O_2^-. Indeed, isolated microsomal fractions from various tissues have been shown to produce hydrogen peroxide rapidly *in vitro* in the presence of NADPH, presumably via dismutation of O_2^-. The rate of hydrogen peroxide production is increased at elevated oxygen concentrations. Increasing the amount of P-450 and its reductase by pretreatment of animals with phenobarbital increases the rates of H_2O_2 production by liver microsomes later isolated from the animals.

In perfused rat liver, however, the basal rate of GSSG release as an index of H_2O_2-production (Section 3.1.1) is smaller than expected from the rates observed with microsomes *in vitro*, and it does not increase on pretreatment of the animals with phenobarbital. This suggests that H_2O_2-formation and, by implication, O_2^- formation by the endoplasmic reticulum *in vivo* does not occur as rapidly as would be expected from experiments on microsomes. Perhaps during the fragmentation and membrane vesicle formation that occur on cell disruption to produce microsomes (Chapter 2) the arrangement of the components of the P-450 system within the membrane is altered so that electrons escape more easily to oxygen. This should be borne in mind in experiments with isolated microsomal fractions. The oxygen concentration adjacent to the endoplasmic reticulum *in vivo* must also be much lower than that seen by microsomes incubated *in vitro*, which will lead to less generation of O_2^- *in vivo*.

Liver endoplasmic reticulum also contains an enzyme system, *desaturase*, that introduces carbon–carbon double bonds into fatty acids. The system requires oxygen, NADH, or NADPH, and a special cytochrome known as cytochrome b_5. Electrons from NAD(P)H are transferred to b_5 by a flavoprotein enzyme, and reduced b_5 then donates electrons to the desaturase enzyme. Both cytochrome b_5 and the flavoprotein can leak electrons onto oxygen to make O_2^-, and this might be an additional source of O_2^- *in vivo*.

Nucleus

The membrane surrounding the cell nucleus also contains an electron transport chain, of unknown function, that can 'leak' electrons to give O_2^-, at a rate increasing with oxygen concentration, in the presence of NADH or NADPH. It resembles the microsomal electron transport system and may be of especial importance *in vivo* because the oxygen radicals that it generates are close to the cell's DNA.

Conclusion

Contemplation of all the information given above shows that production of O_2^- would be very likely to occur within all aerobic cells, but it is very difficult to assess the actual amount produced. In extracts of the bacterium *Streptococcus faecalis* incubated with NADH, 17 per cent of the oxygen consumed was used to form O_2^-, although this figure obviously need not apply to intact cells. It is probably an overestimate, since disruption of cellular structures probably allows electrons to escape to oxygen more easily, as may be the case for liver microsomes. Oxygen uptake associated with formation of O_2^- and hydrogen peroxide can be easily demonstrated in whole leaves or illuminated chloroplasts (Chapter 5) and the study of low-level chemiluminescence (Chapter 2) gives evidence for oxygen radical reactions in bacteria, animal organs both isolated and *in situ*, whole cells of the soil amoeba *Acanthamoeba castellanii*, and phagocytic cells (Chapter 7). Pretreatment of guinea-pigs with diethyldithiocarbamate to inactivate CuZnSOD (Section 3.1.3) decreases the rate at which liver catalase activity is inhibited by aminotriazole in these animals. Inhibition of catalase by aminotriazole requires hydrogen peroxide, so it may be concluded that inhibition of CuZnSOD had decreased H_2O_2-production and therefore that this enzyme must have been acting on O_2^- *in vivo*.

Thus our answer to the question posed at the head of this section would be 'Yes, but we don't know how much'.

3.5.2   *Is the true function of superoxide dismutase that of removing superoxide radicals?*

The specificity of SOD for reaction with O_2^- has frequently been used to probe for the involvement of this radical in biochemical systems. Yet there have been several claims that SOD is not specific for reactions with O_2^-. An early suggestion that SOD could catalyse the conversion of singlet oxygen back to the ground state has not been substantiated. The SOD *proteins* can, of course, react with singlet oxygen because they often contain histidine, tryptophan, and methionine residues which react with this species (Chapter 2). For the same reason the SOD proteins react with hydroxyl radicals. The same is true of any other protein, however, including heat-denatured SOD, the apoenzyme (the protein with metals removed from its active sites), catalase, and serum albumin. Thus if a large quantity of SOD or other protein is added to a system producing OH· or singlet oxygen then there can be scavenging. To prove an involvement of O_2^- one must keep the molar concentrations of SOD very low and carry out a control using heat-denatured enzymes, apoenzymes, or another protein, such as albumin, at the same concentration. A similar point applies to the use of catalase inhibition as a means of detecting the involvement of hydrogen peroxide in a reaction.

It should perhaps be mentioned here that inhibition by SOD must be interpreted with caution in systems containing quinones and semiquinones. As we saw in Chapter 2, many semiquinones react reversibly with oxygen:

$$semiquinone + O_2 \rightleftharpoons quinone + O_2^-$$

The equilibrium will tend to move over to the right because of the non-enzymic dismutation of O_2^-. Addition of SOD, by removing O_2^- much more rapidly, can further decrease the concentration of semiquinone present. As Sutton and Winterbourn in New Zealand have shown, a reaction that is actually caused by the semiquinone might be mistakenly attributed to O_2^- as a result of the inhibition on addition of SOD.

Some diphenols can bind to the active site of CuZnSOD and partially reduce the copper to the Cu(I) form, becoming themselves simultaneously oxidized to the semiquinone form. However, this is a reaction seen with many copper proteins. SOD does not *catalyse* the oxidation of diphenols but merely reacts stoichiometrically with them (i.e. only one mole of diphenol reacts per active site). Wardman and Fielden in England have studied the interaction of CuZnSOD with a range of radical species, including semiquinones and radicals derived from FMN, NAD, aromatic nitro compounds, and organic peroxyl ($RO_2^{\cdot}$) radicals, but no catalysis of radical removal was evident in any case. On the basis of present evidence we are able to conclude that the SOD enzymes are specific for O_2^- as substrates and therefore that their use as 'probes' for O_2^- is valid if suitable controls are employed.

A number of simple metal complexes are able to react with O_2^- *in vitro* and in some cases to catalyse its dismutation. Table 3.13 describes some of these. It must be pointed out here that the use of 'indirect' assay systems for SOD activity (Table 3.9) to study these low-molecular-mass SOD 'mimics' should be avoided in favour of pulse radiolysis or stopped-flow methods using $K^+O_2^-$. We recommend this because many of the complexes interfere with O_2^--generating systems (e.g. Cu^{2+} ions inhibit xanthine oxidase and stimulate the oxidation of adrenalin, sulphite, or pyrogallol) or with the detector molecules (e.g. Fe^{2+}-EDTA reduces cytochrome c directly). The high dismutase activity of Cu^{2+} complexes *in vitro* has led a few scientists to speculate that the biological role of SOD is not the dismutation of O_2^- but something else, perhaps just a metal storage protein, and that the dismutase activity of the enzyme is merely a consequence of the fact that it happens to contain copper. If this were so, however, then one might expect some other copper proteins to dismutate O_2^-. At physiological pH, the rate constant for O_2^- dismutation by each active site of CuZnSOD is about $1.6 \times 10^9 \, \text{M}^{-1} \text{s}^{-1}$. Some other copper proteins do react with O_2^- but the rate constants are much lower; some approximate values are given below

cytochrome c oxidase, $k = 2 \times 10^7 \, \text{M}^{-1} \text{s}^{-1}$
caeruloplasmin, $k = 7 \times 10^5 \, \text{M}^{-1} \text{s}^{-1}$
galactose oxidase, $k = 3 \times 10^6 \, \text{M}^{-1} \text{s}^{-1}$

These reactions with O_2^- are usually non-catalytic under physiological conditions; the O_2^- is merely reducing Cu^{2+} at the active site to Cu^+ and the reaction then stops. The ability of Cu^{2+} ions to react with O_2^- is decreased by the presence of proteins, such as albumin, whereas that of CuZnSOD is not. Indeed, CuZnSOD shows an arrangement of charged residues that seems designed to guide O_2^- into the active site (Section 3.1.3) and evidence for a similar 'electrostatic facilitation' of O_2^--binding has been presented for FeSOD and MnSOD enzymes. Finally, it must be noted that the latter two enzymes are far more effective in catalysing O_2^- dismutation than are equivalent amounts of free or complexed Mn or Fe ions under any conditions (Table 3.13).

Table 3.13.   Simple compounds that react with superoxide

Compound	Comments
Fe^{3+}—EDTA	Both Fe^{3+}—EDTA and Fe^{2+}—EDTA react with O_2^-, probably by the reactions: $$Fe^{3+}-EDTA + O_2^- \rightarrow O_2 + Fe^{2+}-EDTA$$ $$Fe^{2+}-EDTA + O_2^- \rightarrow [O_2^- - Fe^{2+}-EDTA]$$ complex. The reaction mechanism is complicated by the generation of OH· from reaction of Fe^{2+}—EDTA with H_2O_2. Probably overall not a significant catalyst of O_2^- dismutation. 'Unchelated' iron salts react more slowly with O_2^- than do iron—EDTA chelates.
Tetrakis-(4-N-methyl) pyridyl porphine-Fe^{3+}	Catalyses O_2^- dismutation with about 2–3% of the rate of the SOD enzyme. The manganese and cobalt complexes are also effective, but the copper complex is not.

Table 3.13.—*continued*

Compound	Comments
Desferal-manganese IV	A green complex formed by reaction of desferrioxamine B methanesulphonate (Table 2.2) with manganese dioxide, MnO_2. 1 μM equivalent to 1 unit of SOD in the cytochrome c assay.
Mn^{2+}-polyphosphate Mn^{2+}-lactate Mn^{2+}-succinate Mn^{2+}-malate	Slowly catalyse O_2^- dismutation, but Mn(II) in the absence of a complexing agent is a poor catalyst. EDTA decreases catalysis by Mn(II) (compare iron, above). Mn complexes may be an important catalyst of O_2^- dismutation in some *Lactobacillaceae* (see text). The most active complex, Mn-lactate, has about one-sixty-fifth the rate constant for O_2^- that MnSOD has.
Copper complexes	Catalyse O_2^- dismutation more rapidly at low pH than equimolar amounts of CuZnSOD. Much less effective when the Cu is bound to proteins such as albumin. Some Cu—amino-acid complexes (e.g. with lysine or with histidine) are still quite good catalysts at pH 7.4, as is a complex of Cu with the anti-inflammatory drug *indomethacin*. Many simple copper complexes are unstable, however, and problems can be caused due to the O_2^--scavenging ability of free Cu(II) ions present in the solution. Thus a Cu–penicillamine complex is a very poor catalyst of O_2^- dismutation (if a catalyst at all) under conditions where Cu^{2+} does not dissociate from it. EDTA decreases or abolishes reaction of O_2^- with Cu salts but has no effect on CuZnSOD.

3.5.3 *Evidence bearing on the superoxide theory of oxygen toxicity*
Mutation experiments

According to the superoxide theory of oxygen toxicity as originally formulated, the damaging effects of elevated O_2 are due to an increased formation of O_2^-, and SOD is essential in allowing organisms to survive in the presence of any oxygen at all. Indeed, SOD has been found to be present in almost all aerobes examined, even in species such as *Bacillus popilliae* or *Acholeplasma laidlawii* that contain no catalase or peroxidases.

The proposal that SOD is essential for aerobic life could be tested by obtaining a mutant strain of an aerobe that lacked SOD, and showing that it could not grow in the presence of O_2. Modern techniques of genetic engineering have been used to tackle such experiments. *E. coli* has two main SOD enzymes, FeSOD and MnSOD (Section 3.1.2). Touati *et al.* in France inactivated the genes encoding MnSOD and FeSOD in *E. coli*, such that the bacterium had no SOD activity. This mutant would not grow aerobically on a minimal glucose medium. Growth could be restored by removing O_2, by restoring SOD production to the cells as a result of introducing DNA bearing a gene coding for SOD (even for CuZnSOD), or by enriching the growth medium of the SOD-negative bacteria by providing 20 different amino acids. Even with this supplementation, the SOD-deficient mutant grew only half as fast as normal *E. coli*. It was also much more sensitive than normal *E. coli* to increased O_2 concentrations and to H_2O_2. The SOD-deficient *E. coli* mutants further showed greatly enhanced mutation rates during aerobic (but not anaerobic) growth on rich media. These studies show that failure to remove O_2^- within cells leads to DNA damage, a hypersensitivity to H_2O_2, and impairment of amino acid biosynthesis. Indeed, Brown *et al.* (Fig. 1.2, Chapter 1) had previously shown that certain amino acids could diminish the growth-inhibiting effects of hyperbaric O_2 upon *E. coli*. Schatz and others in Switzerland found that inactivating the gene for MnSOD in *Saccharomyces cerevisiae* caused this yeast to become hypersensitive to the presence of O_2. Similarly, yeast lacking CuZnSOD is hypersensitive to oxygen.

Another approach has been to isolate O_2-intolerant mutants, i.e. select strains for their inability to grow aerobically, and see what enzymes have been lost. Several oxygen-intolerant mutants of *E. coli* K12 have been isolated and examined. In one type of mutant, peroxidase and catalase were deficient, whereas, in a second type, peroxidase, catalase and MnSOD were all decreased. Some mutant strains eventually regained the capacity for growth in the presence of O_2; these were of two types. In one group, some or all of the missing enzyme activities had returned. In the second group the enzyme defects were still present, but the cells no longer respired. As they did not take up O_2, they presumably no longer produced O_2^- or H_2O_2 and thus could survive without protective enzymes.

Induction experiments

The experiments with mutants described above show the great importance of SOD in allowing rapid aerobic growth, without genetic damage, of *E. coli* and of yeast. Further evidence supporting this conclusion in these and other organisms comes from *induction experiments*. Exposure of organisms to elevated O_2 should, according to the superoxide theory of O_2 toxicity, cause them to form more $O_2{}^-$ *in vivo* and this might lead to synthesis of more SOD if insufficient is present to cope with the increased $O_2{}^-$ generation. Much data supports this interpretation. For example, elevated oxygen concentrations increase the total SOD activity in *E. coli* B cells due to increased synthesis of the MnSOD. *E. coli* B cells grown under an atmosphere of 100 per cent oxygen were much more resistant to the toxic effects of high-pressure oxygen than were cells grown under air. By contrast, exposure of *Bacillus subtilis* to elevated oxygen concentrations increases its catalase activity but not that of SOD, and this organism is equally sensitive to high-pressure oxygen, whether grown previously under air or under 100 per cent oxygen. Exposure of anaerobically grown *E. coli* K12 cells to oxygen causes induction of catalase and peroxidase activities, but much greater oxygen concentrations are required than those necessary to increase synthesis of MnSOD. This organism has two forms of catalase, one of which is always present and the other which is only synthesized in the presence of oxygen. Catalase induction may be due to the formation of hydrogen peroxide *in vivo*, since the addition of H_2O_2 can induce catalase synthesis in *E. coli* cultures under either aerobic or anaerobic conditions.

It is possible to increase the SOD activity of *E. coli* cells by supplying them with a compound that increases intracellular $O_2{}^-$ generation; such compounds include the antibiotic streptonigrin, the herbicide paraquat, juglone, menadione, pyocyanine, methylene blue, and phenazine methosulphate. Strains of *E. coli* with elevated SOD activity are resistant to the toxic effects of these compounds and, *vice versa*, strains with elevated SOD due to treatment with these compounds are less sensitive to toxic oxygen effects. Many of these drugs induce H_2O_2-degrading enzymes as well, perhaps as a result of H_2O_2-generation by the increased SOD activity. The above compounds appear to act by being reduced within the cell, followed by a reaction of the reduced forms with oxygen to make $O_2{}^-$ ('redox cycling'). Further discussion of the toxicity of paraquat and of quinones can be found in Chapter 6.

The SOD activity present within an aerobe should depend not only on the oxygen concentration, but also on what percentage of this oxygen is used to make $O_2{}^-$. For example, when *E. coli* K12 is grown in continuous culture both its capacity for respiration and its SOD activity increase in parallel with growth rate, whereas peroxidase and catalase activities do not. A sudden rise in the availability of carbon source for the bacteria does not immediately result in an increased growth rate: there is a lag period before the rate increases, which correlates with the time taken for the cells to raise their SOD activity to a concentration characteristic of the greater rate of growth. Here the SOD concentration is probably responding to the respiratory activity of cells i.e. as they grow more quickly, more oxygen is used and more $O_2{}^-$ is formed *in vivo*. Consistent with this interpretation is the observation that when *E. coli* is grown in an aerated, rich growth medium containing glucose, the cells take up only small amounts of oxygen as they first use up the glucose, and release lactate and other organic acids into the medium. When the glucose is exhausted, the cells increase their respiratory capacity and oxidize other components of the medium, this increase being accompanied by a rise in SOD concentrations.

Induction experiments in which SOD activity was correlated with oxygen exposure have also been carried out in *Streptococcus faecalis*, *Vibrio* spp, the microaerophile *Campylobacter sputorum*, the nitrogen-fixing bacterium *Rhizobium japonicum*, and the yeast *Saccharomyces cerevisiae*. A mutant strain of the alga *Chlorella* with increased SOD activity was much more resistant to elevated oxygen concentrations and to streptonigrin than was the wild-type; and oxygen has also been observed to increase SOD activity in potato tubers and in the blue-green alga *Anabaena cylindrica*. The oxygen produced by the photosynthetic activity of symbiotic algae has been observed to increase SOD activity in their host, the sea anemone *Anthopleura elegantissima*. Anemones containing symbiotic algae have much greater SOD and catalase activities than those which do not contain them.

Studies on animals

If adult rats are placed in pure oxygen gas they rapidly develop symptoms of lung damage and usually die after 60–72 hours. If, however, they are exposed to gradually increasing oxygen concentrations they can adapt to survive in 100 per cent oxygen and the gain of the ability to survive is strongly correlated with an increased content of SOD, as measured in lung homogenates. Catalase, glutathione reductase, and glutathione peroxidase activities are also increased in the lungs, as is the amount of glutathione (GSH) present. If rats are pretreated with diethyldithiocarbamate, which inhibits CuZnSOD both *in vitro* and *in vivo*, the toxic effects of high O_2 tensions are enhanced; although care must be exercised in interpreting this observation since diethyldithiocarbamate inhibits several other enzymes as well. The decrease in SOD activity is followed by a loss of glutathione peroxidase activity in the lung. Such changes also occur in lung homogenates treated with diethyldithiocarbamate *in vitro*, but the loss of glutathione peroxidase can be prevented by adding SOD. This indicates that *in vitro*, and presumably *in vivo* as well, the glutathione peroxidase is inactivated by O_2^- (or by other oxidants derived from O_2^-) as a consequence of the inhibition of SOD.

It must be pointed out, however, that lung is a very complex tissue containing many different cell types. Damage caused by exposure to oxygen may well destroy some cell types and encourage others to proliferate, so that changes in enzyme activities as assayed in homogenates of whole lungs may simply be due to changes in cell populations, or might well underestimate very large enzyme changes taking place in only a few cell types. For example, sublethal oxygen exposure causes a proliferation in rat lungs of the cells known as 'granular pneumocytes' or 'type II cells'. Cultures of such cells *in vitro* showed increased SOD activity if they were isolated from rats previously exposed to elevated oxygen concentrations, but activities of glutathione peroxidase or glucose -6-phosphate dehydrogenase were not altered.

Newborn rats are more resistant to oxygen toxicity than are adults. This may occur because the SOD activity of their lungs is increased and maintained more effectively under high O_2 than it is in adults. The same phenomenon is seen in newborn rabbits. Catalase and glutathione peroxidase activities increase as well. If induction of these three enzymes is prevented (e.g. by injection of protein synthesis inhibitors), newborn rats become highly sensitive to oxygen. Newborn rats kept in pathogen-free environments have lower SOD activities in lung than normal, and they are more sensitive to oxygen-poisoning. Treatment of adult rats with bacterial endotoxin enhances their resistance to oxygen and induces an increase in SOD, catalase, and glutathione peroxidase activities in the lungs. This might be because bacteria infecting mammals are dealt with by polymorphonuclear leucocytes, which produce O_2^- and other oxidants after engulfment of the bacteria (Chapter 7). The presence of bacterial endotoxin could be a 'signal' to the organism to produce more SOD in order to help protect the host tissues against O_2^- from the leucocytes. Simultaneous administration of diethyldithiocarbamate to the animals abolishes the protective effect of the endotoxin, even though it does not prevent the increase in catalase and glutathione peroxidase activities. Unlike adult rats, adult mice cannot be 'adapted' to survive in 100 per cent oxygen by pre-exposure to 60 or 80 per cent oxygen, and there is no increase in their lung SOD activity under such conditions. Similarly, injection of endotoxin does not increase SOD activity or confer protection against 100 per cent oxygen. The susceptibility of lung to oxidant damage is considered further in Chapter 8.

Other evidence for induction of SOD in animal tissues has been obtained. Petkau *et al.* in Canada studied breast cancers produced by treating female rats with a carcinogen. SOD activity at the centre of the tumour, which had a restricted oxygen supply, was less than at the outer edges. If the rats were exposed to 85 per cent oxygen for 5 days, SOD activities in both tumour regions increased, but the activity ratio of centre to margin remained about the same.

Several other observations are consistent with an important role of SOD in animal tissues. For example, the SOD activity of erythrocytes from a wide range of animals is virtually constant when expressed per unit haemoglobin, whereas catalase and glutathione peroxidase activities vary widely. As we saw in Chapter 1, the swim-bladders of certain fish have to tolerate high oxygen concentrations. The SOD activity of the swim-bladder of the toadfish *Opsanus tau* was found to be higher than that of any other body tissue examined, but its catalase and glutathione peroxidase activities are fairly low.

The above results on the correlation between SOD activity and resistance to oxygen toxicity are fully consistent with the idea that SOD plays an important protective role, supporting and extending the genetic experiments reviewed earlier. Since catalase and peroxidase activities often increase as well, they also must be important. Indeed it has been observed that both CuZnSOD and FeSOD are inactivated on prolonged exposure to hydrogen peroxide *in vitro*, although MnSOD is not. This has been used (Section 3.1) as a means of distinguishing between the different enzymes. Inactivation can happen in biological systems as well. For example, a decrease of SOD activity in the presence of H_2O_2-generating drugs has been observed in erythrocyte suspensions treated with aminotriazole to inhibit catalase. Feeding amino-triazole to rabbits inhibits catalase activity in the lens of the eye and causes loss of CuZnSOD activity. When wild-type strains of blue-green algae are brightly illuminated under 100 per cent oxygen (as can sometimes happen in their normal environment, e.g. in Israeli ponds) they are quickly killed. Strains of *Plectonema boryanum* sensitive to this 'photo-oxidative' death contain FeSOD as the major cellular SOD activity, but a resistant strain of this blue-green algae contains mainly MnSOD. Under photo-oxidative conditions, FeSOD activity is rapidly lost, presumably due to the action of hydrogen peroxide produced during photosynthesis, whereas MnSOD is not.

We have already seen that inhibition of SOD in rat lung can lead to inactivation of glutathione peroxidase, which further illustrates the interdependence of these enzymes. The superoxide radical can also partially inhibit catalase (Section 3.5.4) by reducing the Fe(III) iron in the haem ring of the active enzyme to an Fe(II)—O_2 form that is inefficient at degrading hydrogen peroxide. The survival time of rats exposed to pure oxygen was increased by about 70 per cent when liposomes containing both catalase and SOD were injected intravenously before and during O_2 exposure; liposomes containing either SOD or catalase alone had much less effect. Exposure of rats to high-pressure O_2 produces convulsions (Chapter 1). Rats injected with liposomes containing both SOD and catalase and then exposed to pure O_2 at 6 atmospheres pressure had their time before the onset of convulsions extended approximately threefold. Liposomes containing either SOD or catalase were again much less effective.

Conclusion

SOD is not the *sole mechanism* that operates against oxygen toxicity; rather it is a very important mechanism that operates in conjunction with a wide range of others. Which is the most important will depend on the organism, the nature of the oxidative challenge, the tissue in question, and even the time of day which can affect, among other things, GSH concentrations. For example, rabbit bone marrow contains little SOD, although it has substantial catalase activity. When cotton plants were exposed to elevated O_2, there was a significant rise in the glutathione reductase activity of leaves, but no change in SOD. Further, SOD may not be the only mechanism by which O_2^- is removed in aerobes. Ascorbate reacts rapidly with O_2^- and it may play an important role in this respect in isolated chloroplasts (Chapter 5) and in the lens of the human eye. The human lens is poor in SOD but rich in ascorbate, whereas the rat lens has more SOD but less ascorbate. As a final example of the complexity of higher organisms, however, it must be pointed out that exposure of rats to oxygen concentrations *lower* than normal (10 per cent oxygen) for several days also increases the SOD activity of the lung; and rats so pretreated can then survive prolonged exposure to 100 per cent oxygen. It is also possible for cells to increase their tolerance to O_2 by mechanisms that do not involve changes in SOD or H_2O_2-degrading enzymes. Thus Joenje *et al.* in Holland selected a strain of HeLa cells (a human cancer cell line) capable of growing under 80 per cent O_2, an atmospheric O_2 concentration that quickly damages 'normal' HeLa cells. However, the O_2-tolerant cells showed no increase in SOD, catalase or glutathione peroxidase activities. They suggested that the cells might have an increased *resistance* to radical attack by altering normal intracellular targets of such attack.

Heat-shock proteins

When many animal, plant, and bacterial cells are exposed to elevated, but non-lethal, temperatures they increase their synthesis of several proteins, which are known collectively as *heat-shock proteins*. Increased synthesis is due to the activation of a small number of specific genes, previously either non-expressed or expressed only at low levels. There are some five classes of heat-shock proteins, the most prominent being the classes with average relative molecular masses of around 70 000 and 84 000. Synthesis of heat-shock proteins appears to confer some thermotolerance upon the organism.

The major heat-shock proteins do not appear to be antioxidants, but some of the minor ones may be. Thus heat shock has been reported to increase SOD activity in both *E. coli* and in some mammalian cells. Treatment of *Salmonella typhimurium* with non-lethal concentrations of H_2O_2 induces synthesis of thirty proteins, five of which are also induced by heat shock. H_2O_2 treatment causes activation of some heat-shock protein genes in cultured hamster fibroblasts: the induced cells are not only more resistant to H_2O_2, but also slightly more thermotolerant. It seems likely that heat shock induces oxidant stress within cells, leading to increased free-radical reactions, but that free-radical reactions are not responsible for the majority of the heat-shock response. One protein that can be induced by H_2O_2 in human fibroblasts is the enzyme *haem oxygenase*, which catalyses oxidation of haem to bile pigments. The ability of H_2O_2 to react with haem rings to release 'catalytic' iron ions (Chapter 2) suggests that induction of a haem-removal system would be a valuable protective mechanism.

3.5.4   *Is superoxide a damaging species?*

The evidence reviewed in Section 3.5.3 clearly indicates that SOD is an important antioxidant enzyme. Since SOD seems specific (Section 3.5.2) for O_2^- as a substrate, it follows that O_2^- is a species that is well worth removing *in vivo*. One reason for this could be the toxicity of O_2^- itself. However, in chemical terms, O_2^- is a poorly-reactive radical in aqueous solution (Chapter 2). The protonated form of O_2^-, $HO_2^{\cdot}$, is somewhat more reactive than O_2^- itself. For example, $HO_2^{\cdot}$ can initiate peroxidation of fatty acids (Chapter 4), and $HO_2^{\cdot}$ might be responsible for the reported reaction of superoxide with GSH (Section 3.2.2), since a small amount of $HO_2^{\cdot}$ exists in equilibrium with O_2^- even at physiological pH (Chapter 2). The pH very close to a membrane surface may also be much more acidic than the pH in bulk solution, so that $HO_2^{\cdot}$ formation would be favoured. The pH beneath activated macrophages adhering to a surface has been reported to be 5 or less (Chapter 8), and so a considerable amount of any superoxide that they generate will exist as $HO_2^{\cdot}$. $HO_2^{\cdot}$ should be able to cross membranes as easily as H_2O_2 (Section 2.10). Much of the O_2^- generated within cells comes from membrane-bound systems (e.g. the electron transport chains of mitochondria and endoplasmic reticulum), and so $HO_2^{\cdot}$ formed close to the membrane could conceivably produce damage. Any O_2^- that was produced in the hydrophobic membrane interior could be very damaging, since O_2^- is highly reactive in organic media (Chapter 2). Thus O_2^- in such media *in vitro* can nucleophilically attack the carbonyl groups of the ester bonds that link fatty acids to the glycerol 'backbone' of membrane phospholipids.

However, it remains to be convincingly shown that O_2^- or $HO_2^{\cdot}$ mediate membrane damage *in vivo*. The fact that superoxide dismutase in animal tissues is located in both the cytosol and mitochondrial matrix, not apparently within membranes, perhaps suggests that membranes are not a major target of direct attack by O_2^- or $HO_2^{\cdot}$.

Several biologically-important molecules that can react with superoxide have been identified, although it is often difficult to decide if O_2^- itself, or the small amount of $HO_2^{\cdot}$ in equilibrium with it, is responsible for the reaction. Thus catalase is *partially* inhibited by O_2^-, mainly because the reaction,

$$\text{enzyme}-Fe^{3+} + O_2^- \rightleftarrows \text{enzyme}-Fe^{2+}-O_2$$

generates a 'ferroxycatalase' that does not rapidly decompose H_2O_2. Formation of ferroxycatalase is similar to generation of peroxidase compound III, which is also poorly reactive (Section 3.1.1). Almagor *et al.* in Israel have suggested that *Mycoplasma pneumoniae*, a pathogen that can cause infections of the human respiratory tract, decreases the catalase activity of its target cells by producing O_2^-, so that the cells are further damaged by H_2O_2

resulting from dismutation of the O_2^-. Superoxide-generating systems can also inactivate glutathione peroxidase in the absence of GSH. Since most cells are rich in GSH (Section 3.1.1), this reaction might seem unlikely to be generally important *in vivo*. However, it could explain why glutathione peroxidase is inactivated in the lungs of rats treated with diethyldithiocarbamate to decrease CuZnSOD activity (Section 3.5.3). Superoxide (or $HO_2^{\cdot}$) can also inactivate mammalian creatine kinase, and the *E. coli* enzyme dihydroxyacid dehydratase, which catalyses an essential step in the biosynthesis of branched-chain amino acids (Section 1.4). The ability of amino acids to overcome some of the effects of hyperbaric O_2 toxicity in *E. coli* (Chapter 1) and to permit SOD-negative *E. coli* to grow slowly in the presence of O_2 (Section 3.5.3) suggests that dihydroxyacid dehydratase may be an important target of attack by O_2^-, at least in *E. coli*. Superoxide (or $HO_2^{\cdot}$) can initiate a few radical chain reactions, such as the oxidation of adrenalin, nitropropane, dihydroxyacetone, bilirubin, and biliverdin. H_2O_2 produced by the dismutation of O_2^- can also directly inactivate some enzymes, such as fructose bisphosphatase and glyceraldehyde 3-phosphate dehydrogenase (Section 2.10).

Thus O_2 and $HO_2^{\cdot}$ are certainly not harmless, but they do not appear to be highly cytotoxic themselves (it is, of course, possible that the important cellular targets subject to attack by O_2^- or $HO_2^{\cdot}$ remain to be discovered). Despite the poor chemical reactivity of O_2^- in aqueous solutions, O_2^--generating systems (enzymic, chemical or phagocytic) can do considerable biological damage; Table 3.14 summarizes some of the huge literature on this point. The idea has therefore grown that O_2^-, like H_2O_2, exerts many or most of its damaging effects by leading to generation of more-reactive species. Let us examine some potential candidates.

Singlet oxygen
Dismutation of O_2^- catalysed by SOD produces H_2O_2 and ground-state O_2. One explanation that has been offered to explain the toxicity of O_2^--generating systems is that they produce singlet O_2 $^1\Delta g$, either in the non-enzymic dismutation of O_2^- (Chapter 2) or in the metal-catalysed Haber–Weiss reaction (see below). Addition of limited amounts of water to suspensions of potassium superoxide ($K^+O_2^-$) in chlorinated hydrocarbon solvents has been shown to produce $^1\Delta gO_2$, but O_2^- can react with chlorinated hydrocarbons (Chapter 2), which might confuse the results of such experiments. Thus the following reactions might account for singlet oxygen ($^1O_2$) production:

$$O_2^- + CCl_4 \rightarrow CCl_3O_2^{\cdot} + Cl^-$$
$$CCl_3O_2^{\cdot} + CCl_3O_2^{\cdot} \rightarrow 2^1O_2 + CCl_3CCl_3$$
$$O_2^- + CCl_3O_2^{\cdot} \rightarrow CCl_3O_2^- + {}^1O_2.$$

There is, as yet, no convincing evidence for O_2^--dependent production of $^1\Delta gO_2$ under physiological conditions. However, reaction of O_2^- (or $HO_2^{\cdot}$) with GSH has been suggested to produce singlet O_2 (Section 3.2.2). It must be noted that superoxide itself is a *quencher* of singlet O_2 by electron transfer, i.e.

$$O_2^- + {}^1O_2 \rightarrow O_2 + O_2^-.$$

Hydroxyl radical

In many of the systems in which O_2^--generation has been shown to do damage, there is protection by catalase and scavengers of $OH^{\cdot}$ (such as mannitol, formate, thiourea, and dimethylsulphoxide: see Chapter 2) as well as by SOD (Table 3.14). It was therefore suggested by Fridovich in the USA that O_2^- and H_2O_2 can react to form highly-reactive $OH^{\cdot}$, according to the overall equation:

$$H_2O_2 + O_2^- \rightarrow O_2 + OH^{\cdot} + OH^-.$$

Formation of $OH^{\cdot}$ in a wide range of systems generating O_2^- has been detected by the ability of $OH^{\cdot}$ to hydroxylate aromatic compounds, give a characteristic ESR signal with spin-traps, degrade tryptophan, decarboxylate benzoic acid, oxidize methional into ethene, dimethylsulphoxide into formaldehyde and methane, and deoxyribose into TBA-reactive material. Thus there is no doubt that $OH^{\cdot}$ is often formed in O_2^--generating systems.

The above reaction was first postulated by F. Haber and J. Weiss in 1934 and has become known as the *Haber–Weiss reaction*. Unfortunately, the second-order rate constant for the reaction in aqueous solution has since been shown to be virtually zero, and it certainly could not occur at the low steady-state concentrations of O_2^- and H_2O_2 that exist *in vivo*. Several scientists, including J. M. McCord in the USA and the authors, have shown that the $OH^{\cdot}$ formation can be accounted for if the Haber–Weiss reaction is catalysed by traces of transition metal ions, i.e.

oxidized metal + $O_2^- \rightarrow$ reduced metal + O_2
complex   complex

reduced metal + $H_2O_2 \rightarrow OH^{\cdot} + OH^- +$ oxidized metal
complex   complex

Net: $O_2^- + H_2O_2 \xrightarrow[\text{catalyst}]{\text{metal}} O_2 + OH^{\cdot} + OH^-.$

Transition metal ions, especially iron ions, contaminate most biochemical reagents. Several transition metal complexes (such as complexes of chro-

Table 3.14. Some deleterious effects of superoxide-generating systems

Source of O_2^-	System studied	Damage	Comments
Heart muscle sub-mitochondrial particles	NADH oxidase activity[a]	Activity lost	Damage prevented by SOD, not catalase: the O_2^- generated by the complex inactivates it unless SOD is present.
Xanthine + xanthine oxidase[a]	Rat heart mitochondria	Lowered P/O ratios and respiratory control	Both SOD and catalase protect.
Photosensitizing dyes (thiazines, xanthines, acridines, phenazines)	E. coli	Loss of viability	SOD, catalase, and OH· scavengers protect.
Xanthine + xanthine oxidase	Rat heart or liver mitochondria	Impaired Ca^{2+} uptake	Protection by SOD or mannitol (OH· scavenger).
Hypoxanthine + xanthine oxidase	Rat brain mitochondria	Inhibition of respiration	Catalase and OH· scavengers protect, not SOD: iron salts necessary.
Acetaldehyde + xanthine oxidase	Arachidonic acid	Oxidation	Both SOD and catalase protect (but inhibitory effects of catalase on lipid peroxidation can be ambiguous; see Chapter 4).
Hypoxanthine + xanthine oxidase	Cultured bovine articular cartilage	Inhibition of proteoglycan synthesis	Catalase protects. H_2O_2 may penetrate to form OH· inside the tissue (Chapter 8).
Tetramethylammonium superoxide	Erythrocyte ghosts	Disruption of membrane organization (detected by ESR spin-label)	SOD protects, not catalase.
Xanthine + xanthine oxidase	Microvasculature of hamster cheek pouch	Increased fluid leakage	SOD protects.

Table 3.14. — *continued*

Source of O_2^-	System studied	Damage	Comments
Redox cycling of naphthoquinones	Rat liver plasma membranes	Increased protein phosphorylation on tyrosine	Inhibited by SOD, catalase or desferrioxamine.
Hypoxanthine + xanthine oxidase	Injected into hind feet of rats	Oedema	Some protection by SOD and catalase, mannitol protects.
PMA-stimulated human monocytes	Erythrocytes	Haemolysis	Both catalase and SOD protect, but not OH· scavengers.
Streptonigrin	*E. coli*	Loss of viability	Cells protected by increased intracellular SOD; iron salts needed for toxicity.
Pulse radiolysis	Papain	Inactivation	Effect due to O_2^- only.
Pulse radiolysis	Acetylcholinesterase	[—]	OH· inactivates but O_2^- does not.
Illuminated FMN	*Photobacterium leiognathi*	Loss of viability	Protection by SOD.
Illuminated FMN	Ribonuclease	Inactivation	SOD protects partially.
Xanthine + xanthine oxidase	Bacteriophage R17	Inactivation	SOD protects partially.
Doxorubicin/iron salt/thiol mixture	Erythrocyte ghosts	Membrane lysis	SOD, catalase, mannitol protect.
Systemic activation of complement by cobra venom factor	Rat lung *in vivo*	Oedema, cell injury, and death	Protected by SOD, catalase, desferrioxamine, apolactoferrin; made worse by injected Fe(III) salts.
Activated neutrophils	Isolated rat lung cells	Membrane damage	Protection by catalase, some protection by SOD.
Hypoxanthine + xanthine	Isolated perfused rat lung	Oedema, cell injury, lipid peroxidation	SOD protects.

Xanthine oxidase + purine	Isolated perfused rabbit lung	Oedema	Catalase protects, not SOD.
Xanthine + xanthine oxidase	Instilled into rat lung *in vivo*	Acute injury, oedema	SOD protects, not catalase.
Instillation of PMA (phorbol myristate acetate) into the airway	Rats	Acute lung injury, oedema	Catalase protects, not SOD.
Cigarette smoke or activated phagocytes	α-1-antiprotease activity	Partial inactivation	Catalase and SOD protect, also $OH^{\cdot}$ scavengers.
Human neutrophils stimulated by PMA	Erythrocytes	Haemolysis, haemoglobin oxidation	SOD protects against haemolysis, not catalase or $OH^{\cdot}$ scavengers.
Xanthine + xanthine oxidase	Porcine thoracic aorta endothelial cells	Increased membrane fluidity (detected by ESR spin-label)	SOD protects, not catalase.
NADH/phenazine methosulphate[b]	*E. coli*	Loss of viability	Both SOD and catalase protect.
Acetaldehyde + xanthine oxidase	*Staphylococcus aureus*	Loss of viability	SOD, $OH^{\cdot}$ scavengers, and catalase protect: iron chelates needed for killing.
Xanthine + xanthine oxidase	*Neisseria gonorrhoeae*	Loss of viability	Catalase protective, SOD not.
Hypoxanthine + xanthine oxidase, or $K^{+}O_{2}^{-}$	DNA	Degradation, single strand breaks, deoxyribose degraded, base modification	SOD, $OH^{\cdot}$ scavengers, and catalase protect: metal ions required.
Hypoxanthine + xanthine oxidase or activated neutrophils	Mammalian cells in culture	Chromosome damage	May involve $OH^{\cdot}$ damage to DNA (Chapter 2) by a site-specific mechanism.

Table 3.14.—*continued*

Source of O_2^-	System studied	Damage	Comments
Xanthine + xanthine oxidase	Lymphocyte cultures	Decreased ability to bind sheep red blood cells and to form caps after concanavalin A treatment	Catalase protects, not SOD.
Xanthine + xanthine oxidase	Ehrlich ascites cells	Cell lysis	Catalase protects, not SOD.
Xanthine + xanthine oxidase	Rat heart ornithine decarboxylase	Inactivation	SOD and mannitol protect, not catalase.
Autoxidation of dihydroxyfumarate	Rat thymocytes	Inhibition of Na^+-dependent amino acid uptake	SOD protects, but not catalase.
Illuminated FMN	Calf myoblast cells	Growth abnormality, some cell death	SOD protects, partially.
Hypoxanthine + xanthine oxidase	Rat brain membrane Na^+, K^+ ATPase	Inactivation	SOD protects, partially.
Alloxan	Isolated mouse or rat pancreatic islet cells	Damage to membranes, cell death	SOD, catalase, OH scavengers, and iron chelators protect; SOD, mannitol, and DETAPAC also protect *in vivo*.
Xanthine + xanthine oxidase	Rabbit lung surfactant	Destruction of functional properties	—
Hypoxanthine or xanthine + xanthine oxidase, activated neutrophils	Hyaluronic acid	Depolymerization, loss of viscosity	Protection by SOD, catalase, OH scavengers, metal chelators (Chapter 8).
Hypoxanthine + xanthine oxidase	Collagen	Degradation, failure of gelation	SOD protects.

ª Be careful when using commercial samples of xanthine oxidase; they may be heavily contaminated with proteases such as trypsin, or with phospholipases, and with stabilizers such as salicylate and the chelating agent EDTA.
ᵇ Not recommended as a source of O_2^-; and it produces OH radicals directly.

mium [Cr^{2+}], cobalt [Co^{2+}], titanium [Ti^{3+}], and vanadium [Vanadyl]) can participate in OH$^{\cdot}$ formation in test-tube systems, but the best candidate for a role in generating OH$^{\cdot}$ by the above reactions *in vivo* seems to be iron, i.e.

$$Fe^{3+}-complex + O_2^- \rightarrow Fe^{2+}-complex + O_2$$
$$Fe^{2+}-complex + H_2O_2 \rightarrow OH^{\cdot} + OH^- + Fe^{3+}-complex$$

Net: $O_2^- + H_2O_2 \xrightarrow[\text{catalyst}]{\text{metal}} O_2 + OH^{\cdot} + OH^-$ (*the iron-catalyzed*

Haber–Weiss reaction).

Iron complexes able to catalyse OH$^{\cdot}$ formation by this mechanism *in vivo* seem to be provided by the low molecular mass intracellular iron pool, and by the ability of O_2^- and H_2O_2 to release catalytically-active iron ions from iron proteins. For example, O_2^- can release iron ions from ferritin, whereas H_2O_2 releases iron from haemoglobin (this is discussed in detail in Section 2.4.3). Thus increased generation of O_2^- and H_2O_2 can *create* the conditions that lead to OH$^{\cdot}$ formation. Hydroxyl radicals, once generated, react with the molecules in their immediate surroundings. The much lower reactivity of O_2^- and H_2O_2 means that they can diffuse away from their sites of formation, leading to OH$^{\cdot}$ generation in different parts of the cell whenever they meet a 'spare' iron ion. Hence the toxicity of O_2^- and H_2O_2 to cells is influenced by the intracellular distribution of metal ions, and sequestration of metal ions in 'safe' forms contributes to antioxidant defence mechanisms (Section 3.3).

Copper ions might also participate in OH$^{\cdot}$ generation, although $Cu^+-H_2O_2$ mixtures may generate a Cu(III) complex (e.g. $Cu(OH)_2^+$) as well as or instead of OH$^{\cdot}$ (Chapter 1):

$$Cu^{2+} + O_2^- \rightarrow Cu^+ + O_2$$
$$Cu^+ + H_2O_2 \rightarrow Cu^{2+} + OH^{\cdot} + OH^-$$

Net: $O_2^- + H_2O_2 \rightarrow OH^{\cdot} + OH^- + O_2$ (*copper-catalyzed Haber–Weiss reaction*).

Most copper complexes present *in vivo* are not able to generate OH$^{\cdot}$ in free solution (Section 2.4.3), but copper ions bound to specific targets (such as DNA or enzymes) might cause damage by generating OH$^{\cdot}$ that reacts immediately with the binding molecule, i.e. site-specific damage (Section 2.4.3).

Production of OH$^{\cdot}$ *in vitro* by O_2^--generating systems containing iron or copper ions (added or present as contaminants in the reagents) is inhibited by catalase, which removes the necessary H_2O_2. It is also inhibited by SOD

(Table 3.14). The only role apparent for O_2^- in the above equations is to reduce metal ions (Fe^{3+} or Cu^{2+}) that then react with H_2O_2 to form $OH^.$. Hence the iron-catalysed Haber–Weiss reaction can equally well be described as a *superoxide-assisted Fenton reaction*. Several scientists have argued that cells are full of reducing agents (GSH, NADH, NADPH, cysteine, ascorbic acid) that could reduce Fe^{3+} and Cu^{2+}, so the role of O_2^- as a reductant seems an unlikely one. However, the authors investigated the effect of adding other reducing agents to iron-dependent systems generating $OH^.$ from O_2^- and H_2O_2. They concluded that NADH, NADPH, or thiol compounds such as GSH and cysteine would be unlikely to prevent O_2^--dependent formation of $OH^.$ radicals *in vivo*. Indeed, NADH, NADPH and thiol compounds (Fig. 3.15) can interact with metal ions and H_2O_2 to increase $OH^.$ formation under certain circumstances, and the $OH^.$ generation is inhibited by SOD, i.e. it is superoxide-dependent.

Another important biological reducing agent is ascorbic acid (Section 3.2.1) and this can certainly replace O_2^- as an apparent reducing agent in several of the *in-vitro* systems that have been used to demonstrate O_2^--dependent formation of $OH^.$ radicals in the presence of iron or copper salts. When ascorbate is the reducing agent, SOD does not prevent the $OH^.$ production, but it is still inhibited by catalase. If both O_2^- and ascorbate are available, the relative contributions of each depend on their concentrations. Ascorbate at the concentrations normally present in human extracellular fluids can at best partially replace O_2^- in reducing Fe^{3+} (Fig. 3.19) but ascorbate is rapidly oxidized by direct reaction with O_2^- and with $OH^.$, so that $OH^.$ production eventually becomes completely O_2^--dependent (Fig. 3.19). On the other hand, at the high (millimolar) ascorbate concentrations present in some mammalian tissues such as the eye, spinal cord, or lung cells, it is difficult to imagine O_2^- competing as a reducing agent unless the ascorbate and O_2^- are present in different subcellular compartments or unless large amounts of O_2^- are produced at a localized site. Indeed, at such high concentrations the ability of ascorbate to scavenge $OH^.$ becomes significant.

An example of differential location is that ascorbate seems to be excluded from the phagocytic vacuole of human neutrophils (Chapter 7), even though these cells are rich in ascorbate. Hence any $OH^.$ radical production within the phagocytic vacuole (Chapter 8) might depend on O_2^- and H_2O_2. Localized production of large amounts of O_2^- and H_2O_2 can also occur; examples being in the inflamed rheumatoid joint and in some cases of the adult respiratory distress syndrome (Chapter 8). In both cases, large numbers of neutrophils accumulate (in the joint or in the lung, respectively) and become activated to produce O_2^- and H_2O_2. Rapid depletion of ascorbic acid by excessive O_2^- generation might decrease ascorbate concentration to the point at which it no longer scavenges $OH^.$, but assists its production (by reducing metal ions). It

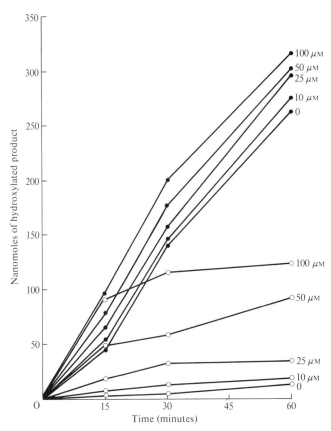

Fig. 3.19. Effect of ascorbic acid on hydroxyl radical production by a mixture of hypoxanthine and xanthine oxidase. OH˙ production by a mixture of hypoxanthine and xanthine oxidase is completely inhibited by SOD, catalase, or the iron chelator desferrioxamine. Addition of ascorbate to a hypoxanthine–xanthine oxidase system at concentrations below 50 μmol l^{-1} increases OH˙ production slightly but it is still largely inhibited by SOD, indicating that O_2^- is still the major reductant of Fe(III). At 50 μmol l^{-1} ascorbate, SOD only inhibited OH˙ formation after 15 minutes had elapsed. Presumably during this period ascorbate was reducing Fe(III) itself until its concentration had fallen so low that O_2^- once more became the major reductant. At 100 μmol l^{-1} ascorbate, SOD only inhibited after 30 minutes. Ascorbate concentrations in body fluids are usually 50 μmol l^{-1} or less (very much less in many diseases). On the other hand, lung cells contain millimolar concentrations of internal ascorbate (Section 3.2.1) and so the action of O_2^- in reducing Fe(III) would be difficult to envisage unless it happens at a specific site in the cell. Closed circles, no SOD added; open circles, SOD present. Data from Rowley and Halliwell, *Clin. Sci.* **64**, 649 (1983).

is interesting to note that patients with iron overload, and healthy Bantu (Section 2.4.3), often have abnormally-low contents of ascorbic acid in blood and tissues. Feeding them with ascorbate in the absence of desferrioxamine has produced deleterious, and sometimes lethal, consequences, possibly because of increased lipid peroxidation and generation of $OH^{\cdot}$ radical by ascorbate/iron salt mixtures. Simultaneous administration of desferrioxamine prevents these effects since it inhibits iron-dependent $OH^{\cdot}$ radical generation (see below). Thus it is safe to give ascorbic acid to iron-overloaded patients being treated with desferrioxamine.

Evidence that hydroxyl radical often contributes to the toxicity of $O_2^{\cdot-}$-generating systems
Detection of $OH^{\cdot}$ radicals, and inhibition of damage by SOD and catalase, all suggest that the metal-catalysed Haber–Weiss reactions account for many of the toxic effects of $O_2^{\cdot-}$-generating systems (Table 3.14). SOD does not usually protect mammalian cells against the toxicity of H_2O_2, either because $O_2^{\cdot-}$ is not involved in that toxicity or because SOD does not penetrate the plasma membrane. That the latter interpretation is correct is suggested by experiments of Farber *et al.* in the USA. They report that SOD protects isolated hepatocytes from H_2O_2 only in conditions under which the enzyme enters the cells.

We have already seen (Section 3.5.3) that SOD-negative *E. coli* is hypersensitive to H_2O_2; killing of *E. coli* K12 by low (2–3 mM) concentrations of H_2O_2 appears to involve DNA damage and it is inhibited by the metal chelator *o*-phenanthroline. These observations suggest that H_2O_2-dependent DNA damage is mediated by some form of metal-catalysed Haber–Weiss reaction within the cell, i.e. H_2O_2 penetrates the plasma membrane and interacts with $O_2^{\cdot-}$ generated intracellularly to form $OH^{\cdot}$, using transition metal ions bound to the cell's DNA (Chapter 2).

Further evidence that $OH^{\cdot}$ is involved in damage is often provided by the use of 'scavengers' of this radical. Most commonly used are formate, mannitol, ethanol, benzoate, thiourea, dimethylthiourea, dimethylsulphoxide, butan-1-ol, and glucose, all of which react quickly with $OH^{\cdot}$. Several commonly-used buffers (Tris, Hepes, Mes, Mops) also scavenge $OH^{\cdot}$, producing buffer-derived radicals that can sometimes affect the system under study. Urea, which reacts much more slowly with $OH^{\cdot}$, is often used as a 'negative control' (Chapter 2). The use of scavengers to implicate $OH^{\cdot}$ in causing damage is a valid approach if four points are borne in mind.

1. An inhibition by a single scavenger proves nothing, especially if that scavenger is thiourea or dimethylthiourea, which react with H_2O_2, hypochlorous acid, and alkoxyl ($RO^{\cdot}$) radicals and may also chelate metal ions necessary for $OH^{\cdot}$ production. Ethanol also reacts with alkoxyl radicals,

although mannitol and formate do not. A range of scavengers should be used and the degree of inhibition they produce should be correlated with the published rate constants for reaction of the scavengers with OH' (Chapter 2).

2. The scavenger and the molecule being used to detect OH' should show competition kinetics, i.e. they should be competing for the same species.

3. Reaction of OH' radical with a scavenger produces a secondary radical that might itself do damage in certain systems, e.g. radicals produced by attack of OH' on formate and ethanol can attack serum albumin, and azide radical (formed by reaction of OH' with azide anion, N_3^-) attacks tryptophan and tyrosine. Production of a scavenger-derived radical that can also damage the target may sometimes account for failure to 'protect' by one scavenger. For example, uric acid did not protect yeast alcohol dehydrogenase against inactivation by OH', because the urate-derived radicals produced could also inactivate the enzyme (Section 3.2.3). Reaction of some scavengers, eg mannitol, with OH' can lead to increased O_2^- production (Section 3.6).

4. Much production of OH' *in vivo* is probably *site-specific*, i.e. O_2^- and H_2O_2 interact with a metal ion bound to a cellular target and the OH' produced immediately attacks that target (Chapter 2). Site-specific damage is very difficult to protect against by OH' scavengers, since they cannot usually reach the localized site of OH' generation in sufficient concentration to protect the target. Detailed work in the authors' laboratories strongly suggests that protection is only achieved when the scavenger has sufficient metal-binding capacity to withdraw the metal from the target. Thiourea has significant metal-binding capacity, and both mannitol and deoxyribose can weakly bind iron ions at physiological pH.

Other evidence for the importance of metal ions in promoting damage by O_2^--generating systems via OH' formation comes from the use of metal-chelating agents. Table 2.2 shows the structures of some of these molecules. Iron ions bound to EDTA are still active in the iron-catalysed Haber–Weiss reaction, since Fe^{3+}–EDTA is reduced by O_2^- and Fe^{2+}–EDTA is oxidized by H_2O_2, leading to OH' production. Hence EDTA does not usually inhibit iron-dependent OH' formation or biological damage by O_2^--generating systems (Table 3.14). It can *sometimes* protect by preventing site-specific reactions: if iron ions are bound to a specific critical target that is destroyed by OH', then EDTA can protect by withdrawing iron ions from that site. The Fe–EDTA complex will still generate OH' in 'free solution', but less OH' will reach the target. Copper ions bound to EDTA seem poorly-reactive in producing OH', however.

The first chelating agent reported (in 1978) to decrease the rate of the iron-catalysed Haber–Weiss reaction was DETAPAC (diethylenetriaminepentaacetic acid: Table 2.2) It slows OH' generation because an Fe^{3+}–DETAPAC complex is reduced inefficiently by O_2^-. An Fe^{2+}–DETAPAC complex still reacts with H_2O_2 to form OH·, however, and more powerful reducing agents than O_2^- (such as paraquat radical: Chapter 6) *are* able to reduce Fe^{3+}–DETAPAC. Hence DETAPAC is not a *general* inhibitor of iron-dependent OH· generation, only a partial inhibitor of the O_2^--dependent OH· generation.

Complexes of iron salts with phytic acid, *o*-phenanthroline, bathophenanthroline sulphonate, and desferrioxamine also show diminished reactivity in formation of OH· from O_2^- and H_2O_2. The substance *o*-phenanthroline, which penetrates easily into cells, has been shown to diminish H_2O_2-dependent DNA damage in mammalian cells, perhaps by binding iron ions from the vicinity of the DNA and preventing DNA fragmentation by site-specific OH· generation (Chapter 2). Desferrioxamine (Table 2.2) is now the most widely-used chelator. It is highly (but not absolutely) specific for iron, which it binds as Fe^{3+}. The Fe^{3+}–desferrioxamine complex (ferrioxamine) is very difficult to reduce, not only by O_2^-, but also by more powerful reductants. Table 3.15 summarizes the properties of desferrioxamine that are relevant to its use in radical-generating systems.

Desferrioxamine was originally developed for the treatment of iron-overload disease (Chapter 2). Its limited toxicity allows it to be used *in vivo* to investigate the role of iron ions in animal models of some human diseases in which the involvement of radical reactions is suspected. Table 3.16 summarizes some of the results that have been obtained. These are *consistent* with a role for iron and OH· generation in mediating oxidant damage *in vivo*. Unfortunately, desferrioxamine is known to be unsuitable for prolonged therapeutic use in humans for diseases other than iron overload.

Reactive species other than OH· in the Fenton reaction
The proposal of OH· as the reactive oxidizing radical produced in the Fenton reaction was at first fiercely contested, and the argument has not yet ceased. However, most scientists now accept that OH· is formed. Formation of a *ferryl radical*, in which the iron has a valency of + 4, has been suggested to occur instead of, or as well as, OH· production

$$Fe^{2+} + H_2O_2 \rightarrow FeOH^{3+} \text{ (or } FeO^{2+}\text{)} + OH^-$$

Indeed, it is possible that the reaction of Fe^{2+} and H_2O_2 forms ferryl, which then decomposes to give OH·. Ferryl radical may well exist at the active site of peroxidase compounds I and II (Section 3.1.1) and in cytochrome P-450 (legend to Figure 3.18), and it clearly must have extensive substrate-oxidizing and hydroxylating properties. However, if the reactive species produced in O_2^--generating systems is not OH·, then it must still attack spin-traps such as DMPO to produce the correct 'OH'' signal (Chapter 2), react with sca-

Table 3.15. Properties of desferrioxamine[a] in relation to its action in radical-generating systems

Powerful chelator of Fe^{3+} (stability constant[b] $\sim 10^{31}$).

Chelates several other metal ions with affinity constants several orders of magnitude lower than for Fe^{3+} (e.g. $Al^{3+} \sim 10^{25}$, $Cu^{2+} \sim 10^{14}$).

Reacts *slowly* with O_2^- or $HO_2^{\cdot}$ ($k_2 \sim 10^3 \ M^{-1} s^{-1}$) to form a relatively-stable nitroxide radical that can inactivate yeast alcohol dehydrogenase.

Reacts *quickly* with $OH^{\cdot}$ ($k_2 \sim 10^{10} \ M^{-1} s^{-1}$). Hence desferrioxamine is an excellent $OH^{\cdot}$ scavenger. Ferrioxamine (Fe^{3+}—complex) also scavenges $OH^{\cdot}$ with the same rate constant. Nitroxide radical again produced.

Inhibits iron-dependent lipid peroxidation in most systems (Chapter 4).

Inhibits iron-dependent $OH^{\cdot}$ generation from H_2O_2 in most systems. Ferrioxamine does not inhibit, so a control with this substance can distinguish protection by iron binding (and inhibition of $OH^{\cdot}$ formation) from protection by $OH^{\cdot}$ scavenging.

Accelerates the oxidation of Fe^{2+} solutions, by binding the resulting Fe^{3+} more tightly than it does Fe^{2+}.

Penetrates only slowly into most animal cells.

Moderately-good scavenger of peroxyl ($RO_2^{\cdot}$) radicals in aqueous solution.

---

[a] Most studies with desferrioxamine are carried out with the commercially-available *desferal*, desferrioxamine B methanesulphonate.

[b] The stability constant is the value of the expression:

$$\frac{[\text{complex}]}{[\text{free metal ion}] \ [\text{free chelator}]}.$$ The higher its value, the further the equilibrium

$$\text{chelator} + \text{metal ion} \rightleftarrows \text{complex}.$$

lies to the right, and the 'stronger' is the chelating agent:

vengers of $OH^{\cdot}$ with similar rate constants to $OH^{\cdot}$ itself, attack aromatic compounds to give the same end-products as $OH^{\cdot}$, and require both O_2^- and H_2O_2 for its formation. Ferryl radicals have not yet been demonstrated to do any of these things. The failure of some $OH^{\cdot}$ scavengers to inhibit damage by a 'reactive species' in a system producing O_2^- and H_2O_2 has sometimes been presented as evidence that the species is not $OH^{\cdot}$. While this may be true, the concept of site-specificity of $OH^{\cdot}$ radical formation (see above) can also account for the observations made.

Other oxidants produced by interaction of metal ions and O_2^-

Reaction of Fe^{3+} with O_2^- produces *perferryl* as an intermediate complex:

$$Fe^{3+} + O_2^- \rightleftarrows [Fe^{3+}-O_2^- \leftrightarrow Fe^{2+}-O_2] \rightleftarrows Fe^{2+} + O_2$$

Perferryl appears to be present in peroxidase compound III (Section 3.1), in oxyferrocatalase and in oxyhaemoglobin (see above). It has weak oxidizing

Table 3.16. Demonstrated effects of desferrioxamine on animal models of human
disease or toxicology

Effect reported	Comments
Decreases neutrophil-mediated acute lung injury in rats after complement activation.	A model for some forms of the adult respiratory distress syndrome, in which O_2^- and H_2O_2 produced by activated neutrophils in the lung cause damage. Protection by desferal suggests involvement of OH˙ (Chapter 8).
Anti-inflammatory in several acute and acute-to-chronic models of inflammation.	See Chapter 8.
Inhibits the toxic action of alloxan and paraquat to animals, which are thought to be mediated by increased generation of O_2^- and H_2O_2 *in vivo*.	Results are contradictory: see Chapter 6.
Decreases liver damage by CCl_4 in animals.	Iron-dependent decomposition of lipid peroxides may be inhibited (Chapter 4).
Protective against reoxygenation injury after ischaemia in heart, kidney, gut, and skin.	Reperfusion injury involves O_2^- and H_2O_2 (Chapter 8). The action of desferrioxamine suggests that OH˙ is involved.
Inhibits the progress of an autoimmune disease model in rats.	Mechanism of action unclear (see Chapter 8).
Decreases antigen–antibody-induced kidney damage in rabbits.	Damage may be due to O_2^- and H_2O_2 produced by neutrophils. The action of desferrioxamine suggests that OH˙ is involved.

properties, but it is not a very reactive species, and seems an unlikely candidate to explain the cytotoxicity of O_2^--generating systems.

Oxidation of manganese (Mn^{2+}) ions by O_2^- can produce Mn^{3+} ions and/or manganese–oxygen complexes that are more oxidizing than is O_2^-. For example, O_2^- reacts very slowly with NADH, but addition of Mn^{2+} speeds up the reaction. O_2^- oxidizes Mn^{2+} to Mn^{3+}, which then oxidizes the NADH. However, it has been argued that accumulation of Mn^{2+} ions protects some *Lactobacillaceae* against the toxicity of O_2^- (Section 3.5.5 below), and so it seems very unlikely that products of reaction of O_2^- with manganese ions usually exacerbate O_2^- toxicity *in vivo*. Vanadate (VO_3^-) ions also stimulate the oxidation of NADH by O_2^-, perhaps by formation of an 'oxidizing complex' of vanadium ions with O_2^-. Another possibility is that O_2^- can not only reduce Cu^{2+} to Cu^+, but also oxidize it to the reactive oxidizing species Cu(III):

$$Cu^{2+} + O_2^- + 2H^+ \rightarrow Cu(III) + H_2O_2.$$

Thus it must not be assumed that OH˙ is the *only* reactive species that can be formed in systems containing O_2^- and metal ions. If formation of the above oxidizing species requires only O_2^- and metal ion, then it should not be inhibited by catalase. However, catalase usually does inhibit that damage done by O_2^--generating systems that is metal-ion-dependent (Table 3.14).

Conclusion
The wide range of evidence reviewed above (e.g. Table 3.14) shows clearly that O_2^--generating systems are very damaging to biological material. This may be direct damage by O_2^-, $HO_2˙$, or H_2O_2 formed by dismutation of O_2^-. A further illustration of the fact that O_2^- toxicity is not always iron-dependent comes from studies upon the bacterium *Streptococcus sanguis*, whose growth appears to be independent of the availability of iron. It contains no haem compounds, and lacks peroxidase or catalase activities. *S. sanguis* is damaged by exposure to O_2^--generating systems, but damage is not prevented by OH˙ scavengers.

Other reactive species can form by interaction of O_2^- with metal ions as an oxidant (perferryl, Mn^{3+}, possibly Cu(III)) or as a reductant followed by reoxidation of the reduced metal ion with H_2O_2 (OH˙). Singlet O_2 might result from some reactions of O_2^- or $HO_2˙$, such as that with GSH. Thus the data on cytotoxicity of O_2^--generating systems fits with the important biological role of SOD as an antioxidant. Although O_2^- itself was reported not to damage proteins, O_2^- does exacerbate the damage produced by OH˙ (Section 3.4.3).

3.5.5 *Challenges to the superoxide theory*

The studies reviewed above have shown that SOD is not strictly *essential* for aerobic life (but it is needed for rapid growth without genetic damage), and that HeLa cells can become resistant to O_2 without inducing SOD. There have also been some more direct challenges to the superoxide theory.

Anaerobes

The first question about the validity of the superoxide theory came from studies upon anaerobes, it being suggested that organisms living without O_2 would not make O_2^- and thus would not need a SOD. Indeed, many anaerobes completely lack SOD activity. However, some anaerobes do contain SOD, the first report of this coming from Morris *et al.* in Wales. Table 3.17 summarizes some of the reports of the presence of SOD in anaerobes. It may be seen that, when SOD is present, it is usually the FeSOD in small amounts (compare SOD activities with those of aerobically-grown *E. coli* K12 in Table 3.15).

Does the presence of SOD in some anaerobes mean that SOD does not function to remove O_2^-? Not necessarily, since the word 'anaerobe' covers a wide spectrum of oxygen tolerance (Chapter 1). Several anaerobes can survive brief exposure to O_2, albeit with growth inhibition. It seems reasonable to propose that SOD is present within them in order to aid survival during such exposures. Consistent with this interpretation, growth of *E. coli* under strictly anaerobic conditions for several generations causes loss of MnSOD, but FeSOD remains. On re-exposure to O_2, the MnSOD is promptly resynthesized. The FeSOD presumably aids survival when O_2 is restored until MnSOD can be made.

Table 3.17. Anaerobic bacteria that contain SOD activity

Anaerobe studied	Type of SOD	Activity (units per mg protein)
Chlorobium thiosulphatophilum	FeSOD	14
Chromatium	FeSOD	0.6
Desulphovibrio desulphuricans	FeSOD	0.6
Clostridium perfringens	FeSOD	15.6
Bacteroides distasonis	FeSOD	0.1–0.4 (most strains)
	FeSOD	3.2–3.9 (strain ATCC8503)
Actinomyces naeslundii	MnSOD	—
Propionibacterium shermanii	FeSOD (but produces MnSOD if grown under Fe-restricted conditions)	
Bacteroides fragilis	FeSOD	—
Bacteroides thetaiotamicron	FeSOD	—

SOD activity can vary with the growth medium and with position in the cell cycle, so do not take the values too literally. For comparison, aerobically grown *E. coli* contained 44 units SOD per mg protein under the same assay conditions. Most anaerobes examined contain no SOD activity at all.

Some scientists have found a correlation between O_2 tolerance and the SOD content of anaerobes, but others have not. This is perhaps not surprising, since the bacterial contents of catalase, peroxidase, and NADH oxidase enzymes will also be important (Section 3.1.1). Even if SOD is absent, an NADH oxidase enzyme can reduce O_2 to water and remove it from the immediate environment of the bacteria. Some anaerobes, such as *Bacteroides fragilis*, synthesize more SOD if oxygen is present.

Aerobes without superoxide dismutase activity
The biological role of SOD has also been questioned as a result of the discovery of aerobic organisms that contain none of this enzyme. A few *Leptospira* strains and three virulent strains of the aerobic gonococcus *Neisseria gonorrhoeae* were found to contain no SOD activity, although they are exceptionally rich in catalase and non-specific peroxidase. *Mycoplasma pneumoniae* contains O_2^--generating systems, but neither SOD- nor catalase-activities. However, if one accepts that the metal-ion-catalysed Haber–Weiss reaction is a major explanation of the toxicity of O_2^--generating systems (Section 3.5.4), it follows that protection against $OH\cdot$ formation could be achieved either by the efficient removal of O_2^- or by the efficient removal of hydrogen peroxide. Both are not necessarily required. Hence one can find organisms with SOD but no catalase or peroxidase activities, such as *Bacillus popilliae*. The gonococcus, by contrast, has invested in exceptionally high catalase activities and can do without SOD (the inhibition of catalase by O_2^- is only partial). Consistent with this argument, catalase-negative strains of *Listeria monocytogenes* have increased SOD compared with catalase-positive strains. Perhaps *Mycoplasma pneumoniae*, a pathogen of the human respiratory tract, contains no metal complexes that catalyse $OH\cdot$ formation. Addition of *M. pneumoniae* to isolated human cells causes membrane damage, an effect prevented by simultaneous addition of SOD and perhaps related to the ability of this organism to generate sufficient O_2^- to decrease catalase activity in the infected cells (see Section 3.5.4). Cells taken from trisomy 21 patients, which contain elevated SOD concentrations (Chapter 8), show much less damage by *M. pneumoniae*. Given that no protection mechanism is 100 per cent efficient *in vivo* and that FeSOD and CuZnSOD are inactivated by hydrogen peroxide, most organisms have probably evolved to contain both SOD and H_2O_2-removing systems to make doubly sure that $OH\cdot$ is not generated to an unacceptable extent under normal conditions.

Several aerotolerant strains of *Lactobacillaceae* contain SOD activity, but a few do not, such as *Lactobacillus plantarum*. This organism accumulates manganese ions from its growth medium to an internal concentration of 25 mmoles per litre or more. If accumulation is prevented by removing the metal ions from the medium, the organism will not grow in the presence of oxygen. Since simple complexes of manganese salts with some biological components slowly catalyse O_2^- dismutation (Table 3.11) Archibald and Fridovich have suggested that these complexes function to remove O_2^- *in vivo*. Strains without high internal Mn or SOD are unable to survive in air, consistent with this argument. *L. plantarum* also possesses an H_2O_2-degrading enzyme that contains manganese ions (Section 3.1).

The existence of a few aerobic organisms without SOD clearly means that this enzyme is not *essential* to aerobic life, a conclusion already reached as a result of genetic experiments (Section 3.5.3). The massive accumulation of evidence reviewed in Section 3.5, however, shows that SOD is a major protector against O_2 toxicity.

Pro-oxidant effects of superoxide dismutase
Dismutation of O_2^- by SOD produces H_2O_2. If an organism is more sensitive to H_2O_2 than it is to O_2^-, increasing its content of SOD can have a pro-oxidant effect, i.e. 'oxidative damage' is increased. Thus two American groups have shown that increasing the SOD content of *E. coli* can sometimes make it more susceptible to the toxic effects of paraquat, a herbicide whose action is thought to be mediated by increased intracellular formation of O_2^- and H_2O_2 (Chapter 6). In these experiments H_2O_2-degrading systems were not increased. These studies are in no way inconsistent with the conclusion (Section 3.5) that SOD is a major antioxidant defence that usually works in co-operation with several other defence mechanisms, including H_2O_2-removing enzymes.

3.5.6　Regulation of bacterial SOD synthesis by metal ions

In view of the roles played by metal ions in potentiating the toxicity of O_2^- and H_2O_2, it is interesting to note that metal ions are also involved in the *regulation* of SOD activity in several bacteria. This has been most studied in *E. coli* B. Some metal-chelating agents that can bind iron (such as nalidixic acid, dipyridyl and 1,10-phenanthroline) increase the synthesis of MnSOD by *E. coli* B, and even cause the appearance of MnSOD when *E. coli* B is grown under anaerobic conditions. On the other hand, increasing the Fe^{2+} content of the growth medium accelerated synthesis of FeSOD, but depressed that of MnSOD. How these events are brought about remains to be established. Hassan in the USA has suggested that synthesis of MnSOD is controlled by a repressor protein containing Fe^{2+}. Removal of iron from the protein, or oxidation of that iron to Fe^{3+} (e.g. in the presence of oxygen) inactivates the repressor and allows the gene for MnSOD to be transcribed.

3.6   Oxygen radicals and radiation damage

It has been known for a long time that the damaging effects of ionizing radiation on cells are aggravated by the presence of oxygen (Chapter 1). Much of the initial damage done is due to formation of the hydroxyl radical, $OH^{.}$, which can react with other cellular components to produce organic radicals (Chapter 2).

Often these organic radicals can be 'repaired' by reaction with ascorbic acid or glutathione. If $R^{.}$ is used to denote them, the 'repair' can be represented by the equations:

$$R^{.} + GSH \rightarrow RH + GS^{.}$$
$$2GS^{.} \rightarrow GSSG$$
$$R^{.} + \text{ascorbate} \rightarrow RH + \text{semidehydroascorbate}.$$

We have already seen that mechanisms exist *in vivo* for removing GSSG and semidehydroascorbate. However, the presence of oxygen may 'fix' the damage by forming other radicals that cannot be repaired, e.g.

$$R^{.} + O_2 \rightarrow RO_2^{.} \text{ (peroxyl radical)}.$$

Mammalian, and some bacterial, cells that cannot synthesize GSH do not usually show the same extent of protection against ionizing radiation seen in normal cells when oxygen is removed. This observation is consistent with the 'repair' role of thiols proposed above. Similarly, treatment of human lymphoid cells with *buthionine sulphoximine*, an inhibitor of GSH synthesis, increases their sensitivity to irradiation. Indeed, GSH, its precursors (Section 3.2.2) and other thiol compounds such as cysteamine have been used as radioprotectors. However, it seems unlikely that a single mechanism accounts for the radioprotective action of thiols, and the possible biological actions of sulphur-centred radicals produced by loss of $H^{.}$ from thiols (Chapter 1) must not be ignored. GSH may be radioprotective not only because of 'repair', but also because it provides a substrate for glutathione peroxidase in animal cells and possibly because it can scavenge $OH^{.}$ radicals directly, although this last mechanism is probably far less important in radioprotection that is the 'repair' function of GSH.

In the presence of O_2, the hydrated electrons formed by ionizing radiation can produce O_2^- (Chapter 2). In addition (although quantitatively less important), some peroxyl radicals formed as a result of attack by $OH^{.}$ on organic molecules can decompose to give superoxide, e.g. *a*-hydroxyalkyl-peroxyl radicals:

$$R - \underset{\underset{OH}{|}}{\overset{\overset{R^1}{|}}{C}} - O_2^- \rightarrow R - \overset{\overset{O}{\|}}{C} - R + H^+ + O_2^- .$$

Hydroxyalkylperoxyl radicals formed from alcohols such as ethanol, methanol, and mannitol, or from some carbohydrates (e.g. glucose) can decompose in this way. If metal ions are present, O_2^- can give more $OH\cdot$ by a catalysed Haber–Weiss reaction, and thus increase the damage. Consistent with this, the presence of Fe(III) salts increased the damaging effects of ionizing radiation upon macrophages. Addition of SOD to the growth medium could partially protect *Acholeplasma laidlawii* or *E. coli* cells against damage by ionizing radiation, an effect also seen with several animal cell lines in culture. Treatment of Chinese hamster cells or human lymphocytes with diethyldithiocarbamate (DIECA) increased their sensitivity to radiation, consistent with a protective role of CuZnSOD *in vivo*. In such experiments, it is essential to remove the inhibitor completely before irradiating because, being a thiol compound, DIECA is an efficient radioprotector. Several radioresistant bacterial strains such as *Arthrobacter radiotolerans* and *Micrococcus radiodurans* contain exceptionally high activities of SOD. Canadian scientists have shown that injection of CuZnSOD into the bloodstream of mice reduces the lethal effects of X-irradiation, whereas injection of inactivated enzyme does not. The effects depend very much on the time and dose of SOD administered, and on the intensity of radiation used; but SOD appears to be particularly protective towards the bone marrow. The mechanism of this radioprotective effect of SOD remains to be established.

Hence the importance (if any) of O_2^- formation in accounting for the oxygen enhancement of radiation damage, in comparison to damage 'fixation' by oxygen (see above), is impossible to evaluate from present data. A mutant of *E. coli* lacking both FeSOD and MnSOD showed the normal oxygen enhancement of radiation damage, suggesting that O_2^- is not a significant component of the oxygen enhancement of radiation damage in this organism.

3.6.1   *Food irradiation*

The treatment of foodstuffs with ionizing radiation for the purposes of sterilization, or prevention of germination and ripening, is slowly becoming an accepted method of food processing. Strict laws govern the types of food that may be irradiated, and the dose of radiation used. However, enforcement of these laws requires a suitable method to detect and measure the usage of radiation. Developing such methods is an interesting problem in free-radical biology. For example, most irradiated spices and some other foodstuffs give substantially more chemiluminescence from alkaline luminol solutions than do un-irradiated control samples. Bone and calcified cuticle (e.g. from chicken, crustacea, and some fish) develop long-lived ESR signals upon irradiation. Determination of end-products arising from attack of free radicals upon lipids (Chapter 4), DNA, or aromatic amino acids (Chapter 2) might be an additional approach. Thus thymine glycol, 8-hydroxyguanine, or hydroxylated phenylalanine, can be assayed. As yet, however, there is no single method applicable to all food substances.

3.7 Further reading

Ackermann, A. D. *et al.* (1988). Partial monosomy 21, diminished activity of superoxide dismutase, and pulmonary oxygen toxicity. *New Engl. J. Med.* **318**, 1666.

Almagor, M., Kahane, S., and Yatziv, S. (1984). Role of superoxide anion in host cell injury induced by *Mycoplasma pneumoniae* infection. A study in normal and trisomy 21 cells. *J. Clin. Invest.* **73**, 842.

Ames, B. N. *et al.* (1981). Uric acid provides an antioxidant defense in humans against oxidant- and radical-caused ageing and cancer: a hypothesis. *Proc. Natl. Acad. Sci. USA* **78**, 6858.

Anonymous (1988). A termination codon specifies selenocysteine in glutathione peroxidase. *Nutr. Rev.* **46**, 202.

Archibald, F. S. (1985). Manganese: its acquisition by and function in the lactic acid bacteria. *CRC Crit. Rev. Microbiol.* **13**, 63.

Archibald, F. S. and Duong, M. N. (1986). Superoxide dismutase and oxygen toxicity defenses in the genus *Neisseria*. *Infect. Immunity.* **51**, 631.

Arthur, J. R. *et al.* (1987). Stimulation of peroxidation in rat liver microsomes by (copper,zinc)-metallothioneins. *Free Radical Res. Commun.* **4**, 15.

Autor, A. P. (Ed.) (1982). *Pathology of oxygen*. Academic Press, New York.

Bannister, J. V. and Calabrese, L. (1987). Assays for superoxide dismutase. *Meth. Biochem. Anal.* **32**, 279.

Bendich, A. *et al.* (1986). The antioxidant role of vitamin C. *Adv. Free Radical Biol. Med.* **2**, 419.

Beyer, R. E., Nordenbrand, K., and Ernster, L. (1987). The function of coenzyme Q in free radical production and as an antioxidant: a review. *Chemica Scripta* **27**, 145.

Beyer, W. F. Jr. (1987). Examination of the role of arginine-143 in the human copper and zinc superoxide dismutases by site-specific mutagenesis. *J. Biol. Chem.* **262**, 11182.

Beyer, W. F. Jr. and Fridovich, I. (1987). Assaying for superoxide dismutase: some large consequences of minor changes in conditions. *Analyt. Biochem.* **161**, 559.

Biaglow, J. E. *et al.* (1983). The role of thiols in cellular response to radiation and drugs. *Radiat. Res.* **95**, 437.

Blake, D. R. *et al.* (1985). Cerebral and ocular toxicity induced by desferrioxamine. *Quart. J. Med.* **219**, 345.

Boyce, N. W. and Holdsworth, S. R. (1986). Hydroxyl radical mediation of immune renal injury by desferrioxamine. *Kidney Intl.* **30**, 813.

Breimer, L. H. (1988). Ionizing radiation induced mutagenesis. *Br. J. Cancer*, **57**, 6–18.

Brigelius, R. *et al.* (1983). Identification and quantitation of glutathione in hepatic protein mixed disulphides and its relationship to glutathione disulphide. *Biochem. Pharmacol.* **32**, 2529.

Britton, L. and Fridovich, I. (1977). Intracellular location of the superoxide dismutases of *E. coli*—a re-evaluation. *J. Bacteriol.* **131**, 815.

Brot, N. and Weissbach, H. (1983). Biochemistry and physiological role of methionine sulphoxide residues in proteins. *Arch. Biochem. Biophys.* **223**, 271.

Burdon, R. H., Gill, V. M., and Rice-Evans, C. (1987). Oxidative stress and heat shock protein induction in human cells. *Free Radical Res. Commun.* **3**, 129.

Butler, J., Hoey, B. M., and Swallow, A. J. (1986). Radiation Chemistry. *Ann. Rep. Prog. Chem. Sect. C*, p. 129.

Cabelli, D. E. and Bielski, B. H. J. (1984). Pulse radiolysis study of the kinetics and mechanisms of the reactions between manganese (II) complexes and HO_2/O_2^- radicals. 1. Sulfate, formate and pyrophosphate complexes. *J. Phys. Chem.* **88**, 3111.

Cameron, J. S. and Simmonds, H. A. (1987). Use and abuse of allopurinol *Br. Med. J.* **294**, 1504.

Carlioz, A. and Touati, D. (1986). Isolation of superoxide dismutase mutants in *Escherichia coli*: is superoxide dismutase essential for aerobic life? *EMBO J.* **5**, 623.

Carlsson, J. (1987). Salivary peroxidase: an important part of our defense against oxygen toxicity. *J. Oral Pathol.* **16**, 412.

Carter, M. H. and Josephy, P. D. (1986). Mutagenicity of thionitrites in the Ames test. The biological activity of thiyl free radicals. *Biochem. Pharmacol.* **35**, 3847.

Chan, T. M. *et al.* (1986). Stimulation of tyrosine-specific protein phosphorylation in the rat liver plasma membrane by oxygen radicals. *Biochem. Biophys. Res. Commun.* **139**, 439.

Chance, B., Sies, H., and Boveris, A. (1979). Hydroperoxide metabolism in mammalian organs. *Physiol. Rev.* **59**, 527.

Chedekel, M. R. *et al.* (1978). Photodestruction of pheomelanin: role of O_2. *Proc. Natl. Acad. Sci. USA*, **75**, 5395.

Clare, D. A., Blum, J., and Fridovich, I. (1984). A hybrid superoxide dismutase containing both functional iron and manganese. *J. Biol. Chem.* **259**, 5932.

Corey, E. J., Mehrota, M. M., and Khan, A. U. (1987). Water induced dismutation of superoxide anion generates singlet molecular oxygen. *Biochem. Biophys. Res. Commun.* **145**, 842.

Cudd, A. and Fridovich, I. (1982). Electrostatic interactions in the reaction mechanism of bovine erythrocyte superoxide dismutase. *J. Biol. Chem.* **257**, 11443.

Darr, D. and Fridovich, I. (1986). Irreversible inactivation of catalase by 3-amino-1,2,4-triazole. *Biochem. Pharmacol.* **35**, 3642.

Davies, K. J. A. (1986). Intracellular proteolytic systems may function as secondary antioxidant defenses: a hypothesis. *J. Free Radical Biol. Med.* **2**, 155.

Davies, M. J. *et al.* (1987). Desferrioxamine (Desferal) and superoxide free radicals. *Biochem. J.* **246**, 725.

De Boer, E. *et al.* (1986). Bromoperoxidase from *Ascophyllum nodosum*: a novel class of enzymes containing vanadium as a prosthetic group? *Biochim. Biophys. Acta*, **869**, 48.

De Croo, S. *et al.* (1988). Isoelectric focusing of superoxide dismutase: report on the unique SOD A*2 allele in a US white population. *Human Heredity*, **38**, 1.

De Mello, M. P. *et al.* (1980). Excited indole-3-aldehyde from the peroxidase-catalysed aerobic oxidation of indole-3-acetic acid. Reaction with an energy transfer to *t*RNA. *Biochemistry*, **19**, 5270.

Deuschle, U. and Weser, U. (1985). Decomposition and superoxide dismutase activity of the copper complex of D-penicillamine $(Cu(II)_6Cu(I)_8$ (D-penicillamine$)_{12}Cl)^{5-}$. *Inorg. Chim. Acta*, **107**, 275.

Dhindsa, R. S. (1987). Glutathione status and protein synthesis during drought and subsequent rehydration in *Tortula ruralis*. *Plant Physiol.* **83**, 816.

Dice, J. F. (1987). Molecular determinants of protein half-lives in eukaryotic cells. *FASEB J.* **1**, 349.

Diplock, A. T. (1981). Metabolic and functional defects in selenium deficiency. *Phil. Trans. Roy. Soc. Lond.* **B294**, 105.

Dunn, M. A., Blalock, T. L., and Cousins, R. J. (1987). Metallothionein. *Proc. Soc. Exp. Biol. Med.* **185**, 107.

Dykens, J. A. and Shick, J. M. (1984). Photobiology of the symbiotic sea anemone, *Anthopleura elegantissima*: defences against photodynamic effects, and seasonal photoacclimatization. *Biol. Bull.* **167**, 683.

Ekström, G., Cronholm, T., and Ingelman-Sundberg, M. (1986). Hydroxyl radical production and ethanol oxidation by liver microsomes isolated from ethanol-treated rats. *Biochem. J.* **233**, 755.

England, S. and Seifter, S. (1986). The biochemical functions of ascorbic acid. *Ann. Rev. Nutr.* **6**, 365.

Escribano, J. *et al.* (1988). A kinetic study of hypoxanthine oxidation by milk xanthine oxidase. *Biochem. J.* **254**, 829.

Ewing, D. and Jones, S. R. (1987). Superoxide removal and radiation protection in bacteria. *Arch. Biochem. Biophys.* **254**, 53.

Farr, S. B., D'Ari, R., and Touati, D. (1986). Oxygen-dependent mutagenesis in *Escherichia coli* lacking superoxide dismutase. *Proc. Natl. Acad. Sci. USA* **83**, 8268.

Frank, L. (1985). Effects of oxygen on the newborn. *Fed. Proc.* **44**, 2328.

Freeman, B. A., Rosen, G. M., and Barber, M. J. (1986). Superoxide perturbation of the organization of vascular endothelial cell membranes. *J. Biol. Chem.* **261**, 6590.

Fridovich, I. (1983). Superoxide radical: an endogenous toxicant. *Ann. Rev. Pharmacol. Toxicol.* **23**, 239.

Fridovich, I. (1986). Superoxide dismutases. *Meth. Enzymol.* **58**, 61.

Fridovich, I. (1986). Biological effects of the superoxide radical. *Arch. Biochem. Biophys.* **247**, 1.

Gonder, J. C., Proctor, R. A., and Will, J. A. (1985). Genetic differences in oxygen toxicity are correlated with cytochrome P450 inducibility. *Proc. Natl. Acad. Sci. USA* **82**, 6315.

Graf, E., Empson, K. L., and Eaton, J. W. (1987). Phytic acid. A natural antioxidant. *J. Biol. Chem.* **262**, 11647.

Grootveld, M. and Halliwell, B. (1987). Measurement of allantoin and uric acid in human body fluids. *Biochem. J.* **243**, 803.

Guengerich, F. P. *et al.* (1986). Human-liver cytochromes P-450 involved in polymorphisms of drug oxidation. *Xenobiotica*, **16**, 367.

Gutteridge, J. M. C. (1987). Ferrous-salt-promoted damage to deoxyribose and benzoate. The increased effectiveness of hydroxyl-radical scavengers in the presence of EDTA. *Biochem. J.* **243**, 709.

Gutteridge, J. M. C. and Wilkins, S. (1983). Copper salt-dependent hydroxyl radical formation. Damage to proteins acting as antioxidants. *Biochim. Biophys. Acta* **59**, 38.

Halliwell, B. (1982). Ascorbic acid, iron overload and desferrioxamine. *Br. Med. J.* **285**, 296.

Halliwell, B. (1988). Albumin—an important extracellular antioxidant? *Biochem. Pharmacol.* **37**, 569.

Halliwell, B. and Gutteridge, J. M. C. (1986). Iron and free radical reactions. Two aspects of antioxidant protection. *Trends Biochem. Sci.* **11**, 372.

Halliwell, B., Grootveld, M., and Gutteridge, J. M. C. (1988). Methods for the measurement of hydroxyl radical in biochemical systems: deoxyribose degradation and aromatic hydroxylation. *Meth. Biochem. Anal.* **33**, 59.

Halliwell, B., Wasil, M., and Grootveld, M. (1987). Biologically-significant scavenging of the myeloperoxidase-derived oxidant hypochlorous acid by ascorbic acid. *FEBS Lett.* **213**, 15.

Harman, L. S. *et al.* (1986). One- and two-electron oxidation of reduced glutathione by peroxidases. *J. Biol. Chem.* **261**, 1642.

Harrison, R. *et al.* (1987). Fulminant hepatic failure after occupational exposure to 2-nitropropane. *Ann. Intl. Med.* **107**, 466.

Hass, M. A. and Massaro, D. (1987). Differences in CuZn superoxide dismutase induction in lungs of neonatal and adult rats. *Am. J. Physiol.* **253**, C66.

Hassan, H. M., Dougherty, H., and Fridovich, I. (1980). Inhibitors of superoxide dismutases: a cautionary tale. *Arch. Biochem. Biophys.* **199**, 349.

Henderson, G. B. *et al.* (1988). Subversive substrates for the enzyme trypanothione disulfide reductase: Alternative approach to chemotherapy of Chagas disease. *Proc. Natl. Acad. Sci. USA* **85**, 5374.

Igwe, O. J. (1986). Biologically active intermediates generated by the reduced glutathione conjugation pathway. *Biochem. Pharmacol.* **35**, 2987.

Imlay, J. A. and Linn, S. (1988). DNA damage and oxygen radical toxicity. *Science*, **240**, 1302.

Joenje, H. *et al.* (1985). Some characteristics of hyperoxia-adapted HeLa cells. A tissue culture model for cellular oxygen tolerance. *Lab. Invest.* **52**, 420.

Kanofsky, J. R. (1986). Singlet oxygen production in superoxide ion-halocarbon systems. *J. Am. Chem. Soc.* **108**, 2977.

Keeping, H. S. and Lyttle, C. R. (1984). Monoclonal antibody to rat uterine peroxidase and its use in identification of the peroxidase as being of eosinophil origin. *Biochim. Biophys. Acta* **802**, 399.

Keyse, S.M. and Tyrrell, R.M. (1989) Heme oxygenase is the major 32-KDa stress protein induced in human skin fibroblasts by UVA radiation, hydrogen peroxide and sodium arsenite. *Proc. Natl. Acad. Sci USA* **86**, 99.

Kirkman, H. N., Galiano, S., and Gaetani, G. F. (1987). The function of catalase-bound NADPH. *J. Biol. Chem.* **262**, 660.

Kittridge, K. J. and Willson, R. L. (1984). Uric acid substantially enhances the free radical-induced inactivation of alcohol dehydrogenase. *FEBS Lett.* **170**, 162.

Korytowski, W. *et al.* (1986). Reaction of superoxide anions with melanins: electron spin resonance and spin trapping studies. *Biochim. Biophys. Acta* **882**, 145.

Kosower, N. S. and Kosower, E. M. (1978). The glutathione status of cells. *Intl. Rev. Cytol.* **54**, 109.

Kozlov, Y. N. and Berdnikov, V. M. (1973). Photodecomposition of hydrogen peroxide in the presence of copper ions. IV. Determination of rate constants of elementary reactions. *Russ. J. Phys. Chem.* **47**, 338.

Kubota, S. and Yang, J. T. (1984). *Bis*[cyclo(histidylhistidine)]copper(II) complex that mimicks the active center of superoxide dismutase has its catalytic activity. *Proc. Natl. Acad. Sci. USA* **81**, 3283.

Kuo, C. F., Mashino, T. and Fridovich, I. (1987). α, β-dihydroxyisovalerate dehydratase. A superoxide-sensitive enzyme. *J. Biol. Chem.* **262**, 4724.

Kyle, M. E. *et al.* (1988). Endocytosis of superoxide dismutase is required in order for the enzyme to protect hepatocytes from the cytotoxicity of hydrogen peroxide. *J. Biol. Chem.* **263**, 3784.

Liochev, S. and Fridovich, I. (1985). Further studies of the mechanism of the enhancement of NADH oxidation by vanadate. *J. Free Radical Biol. Med.* **1**, 287.

Marklund, S. L. (1987). Caeruloplasmin, extracellular superoxide dismutase, and scavenging of superoxide anion radicals. *J. Free Radical Biol. Med.* **2**, 255.

Marklund, S. L., Beckman, G., and Stigbrand, T. (1976). A comparison between the common type and rare genetic variant of human cupro-zinc superoxide dismutase. *Eur. J. Biochem.* **65**, 415.

Martin, J. P. and Logsdon, N. (1987). The role of oxygen radicals in dye-mediated photodynamic events in *Escherichia coli* B. *J. Biol. Chem.* **262**, 7213.

Martin, M. E. *et al.* (1986). A *Streptococcus mutans* superoxide dismutase that is active with either manganese or iron as a cofactor. *J. Biol. Chem.* **261**, 9361.

Meister, A. (1983). Selective modification of glutathione metabolism. *Science* **220**, 472.

Mello Filho, A. C. de and Meneghini, R. (1985). Protection of mammalian cells by *o*-phenanthroline from lethal and DNA-damaging effects produced by active oxygen species. *Biochim. Biophys. Acta* **847**, 82.

Mello Filho, A. C., Chubatso, L. S. and Meneghini, R. (1988). V79 Chinese-hamster cells rendered resistant to high cadmium concentration also become resistant to oxidative stres. *Biochem. J.* **256**, 475.

Moreno, S. N. J. *et al.* (1988). Oxidation of cyanide to the cyanyl radical by peroxidase/H_2O_2 systems as determined by spin trapping. *Arch. Biochem. Biophys.* **265**, 267.

Morgan, R. W. *et al.* (1986). Hydrogen-peroxide inducible proteins in *Salmonella typhimurium* overlap with heat shock and other stress proteins. *Proc. Natl. Acad. Sci. USA* **83**, 8059.

Morpeth, F. F. (1985). Some properties of cellobiose oxidase from the white-rot fungus *Sporotrichum pulverulentum*. *Biochem. J.* **228**, 557.

Morris, S. M. and Albright, J. T. (1984). Catalase, glutathione peroxidase, and superoxide dismutase in the rete mirabile and gas gland epithelium of six species of marine fishes. *J. Exp. Zool.* **232**, 29.

Morse, M. L., Touati, D., and Smith, D. S. (1988). An oxygen enhancement ratio in an *Escherichia coli* strain lacking both the iron and manganese superoxide dismutases. *Biochem. Biophys. Res. Commun.* **150**, 866.

Nagy, I.Zs. and Floyd, R. A. (1984). Hydroxyl free radical reactions with amino acids and proteins studied by electron spin resonance spectroscopy and spin-trapping. *Biochim. Biophys. Acta* **790**, 238.

Nakajima, R. and Yamazaki, I. (1987). The mechanism of oxyperoxidase formation from ferryl peroxidase and hydrogen peroxide. *J. Biol. Chem.* **262**, 2576.

Naqui, A., Chance, B., and Cadenas, E. (1986). Reactive oxygen intermediates in biochemistry. *Ann. Rev. Biochem.* **55**, 137.

Natvig, D. O. *et al.* (1987). Human copper–zinc superoxide dismutase complements superoxide-dismutase deficient *Escherichia coli* mutants. *J. Biol. Chem.* **262**, 14697.

Oberley, L. W. (Ed.) (1982). *Superoxide dismutase*, vols. I, II. CRC Press, Florida; vol. III (1985).

Okuyama, S. *et al.* (1987). Copper complexes of non-steroidal anti-inflammatory agents: analgesic activity and possible opioid receptor activation. *Agents Actions*, **21**, 130.

Ortiz de Montellano, P. R. (1987). Control of the catalytic activity of prosthetic heme by the structure of hemoproteins. *Acc. Chem. Res.* **20**, 289.

Osman, R. and Basch, H. (1984). On the mechanism of action of superoxide dismutase: a theoretical study. *J. Am. Chem. Soc.* **106**, 5710.

Parker, M. W. and Blake, C. C. F. (1988). Iron- and manganese-containing superoxide dismutases can be distinguished by analysis of their primary structures. *FEBS Lett.* **229**, 377.

Pasternack, R. F. *et al.* (1981). Catalysis of the dismutation of O_2^- by metalloporphyrins. *J. Inorg. Biochem.* **15**, 261.

Patole, M. S., Gullapalli, S., and Ramasarma, T. (1988). Vandate stimulated NADH oxidation requires polymeric vanadate, phosphate and superoxide. *Free Radical Res. Commun.* **4**, 201.

Pennington, C. D. and Gregory, E. M. (1986). Isolation and reconstitution of iron- and manganese-containing superoxide dismutases from *Bacteroides thetaiotamicron*. *J. Bacteriol.* **166**, 528.

Porter, D. J. T. and Bright, H. J. (1983). The mechanism of oxidation of nitroalkanes by horseradish peroxidase. *J. Biol. Chem.* **258**, 9913.

Privalle, C. T. and Fridovich, I. (1987). Induction of superoxide dismutase in *Escherichia coli* by heat shock. *Proc. Natl. Acad. Sci. USA* **84**, 2723.

Pugh, S. Y. R. and Fridovich, I. (1985). Induction of superoxide dismutase in *Escherichia coli* B by metal chelators. *J. Bacteriol.* **162**, 196.

Puntarulo, S. and Cederbaum, A. I. (1988). Effect of oxygen concentration on microsomal oxidation of ethanol and generation of oxygen radicals. *Biochem. J.* **251**, 787.

Rabinowitch, H. D., Privalle, C. T., and Fridovich, I. (1987). Effects of paraquat on the green alga *Dunaliella salina*: protection by the mimic of superoxide dismutase, desferal-Mn(IV). *J. Free Radical Biol. Med.* **3**, 125.

Rao, K. K. and Cammack, R. (1981). Evolution of ferredoxin and superoxide dismutases in micro-organisms. In *Molecular and cellular aspects of microbial evolution* (Carlile *et al.*, Eds) p. 175. Cambridge University Press, England.

Reed, D. J. (1986). Regulation of reductive processes by glutathione. *Biochem. Pharmacol.* **35**, 7.

Rigo, A. *et al.* (1979). Nuclear magnetic relaxation of ¹⁹F as a novel assay method of superoxide dismutase. *J. Biol. Chem.* **254**, 1759.

Rowley, D. A. and Halliwell, B. (1983). Superoxide-dependent and ascorbate-dependent formation of hydroxyl radicals in the presence of copper salts: a physiologically significant reaction? *Arch. Biochem. Biophys.* **225**, 279.

Rush, J. D. and Koppenol, W. H. (1986). Oxidizing intermediates in the reaction of ferrous EDTA with hydrogen peroxide. Reactions with organic molecules and ferrocytochrome c. *J. Biol. Chem.* **261**, 6730.

Sarna, T. *et al.* (1986). Interaction of radicals from water radiolysis with melanin. *Biochim. Biophys. Acta* **883**, 162.

Schupp, T., Waldmeier, U., and Divers, M. (1987). Biosynthesis of desferrioxamine B in *Streptomyces pilosus*: evidence for the involvement of lysine decarboxylase. *FEMS Microbiol. Lett.* **42**, 135.

Scott, M. D., Meshnick, S. R., and Eaton, J. W. (1987). Superoxide dismutase-rich bacteria. Paradoxical increase in oxidant toxicity. *J. Biol. Chem.* **262**, 3640.

Seib, P. A. and Tolbert, B. M. (Eds) (1982). *Ascorbic acid: chemistry, metabolism and uses*. Advances in Chemistry, series 200. American Chemical Society, Washington DC, USA.

Shiavone, J. R. and Hassan, H. M. (1988). The role of redox in the regulation of manganese-containing superoxide dismutase biosynthesis in *Escherichia coli*. *J. Biol. Chem.* **263**, 4269.

Sies, H. *et al.* (1978). Glutathione efflux from perfused rat liver after phenobarbital treatment, during drug oxidations, and in selenium deficiency. *Eur. J. Biochem.* **89**, 113.

Sies, H. (Ed.) (1985). *Oxidative stress*. Academic Press, London and New York.

Sinet, P. M., Heikkila, R. A., and Cohen, G. (1980). Hydrogen peroxide production by rat brain *in vivo*. *J. Neurochem.* **34**, 1421.

Singh, A. and Singh, H. (1982). Time scale and nature of radiation biological damage. Approaches to radiation protection and post-irradiation therapy. *Prog. Biophys. Mol. Biol.* **39**, 69.

Slot, J. W. *et al.* (1986). Intracellular localization of the copper–zinc and manganese superoxide dismutases in rat liver parenchymal cells. *Lab. Invest.* **35**, 363.

Smith, A. M., Morrison, W. L., and Milham, P. J. (1982). Oxidation of indole-3-acetic acid by peroxidase: involvement of reduced peroxidase and compound III with superoxide as a product. *Biochemistry* **21**, 4414.

Smith, M. W. and Neidhardt, F. C. (1983). Proteins induced by aerobiosis in *Escherichia coli*. *J. Bacteriol.* **154**, 344.

Spector, T., Hall, W. W., and Krenitsky, T. A. (1986). Human and bovine xanthine oxidases. Inhibition studies with allopurinol. *Biochem. Pharmacol.* **35**, 3109.

Spitz, D. R., Dewey, W. C., and Li, G. C. (1987). Hydrogen peroxide or heat shock induces resistance to hydrogen peroxide in Chinese hamster fibroblasts. *J. Cell. Physiol.* **131**, 364.

Stadtman, E. R. (1986). Oxidation of proteins by mixed-function oxidation systems:

implication in protein turnover, ageing and neutrophil function. *Trends Biochem. Sci.* **11**, 11.

Stallings, W. C. *et al.* (1984). Manganese and iron superoxide dismutases are structural homologs. *J. Biol. Chem.* **259**, 10695.

Steinman, H. M. (1982). Copper–zinc superoxide dismutase from *Caulobacter crescentus* CB15. *J. Biol. Chem.* **257**, 10283.

Sutton, H. C. and Sangster, D. F. (1982). Reactivity of semiquinone radicals and its relation to the biochemical role of superoxide. *J. Chem. Soc. Faraday Trans. 1*, **78**, 695.

Tainer, J. A. *et al.* (1982). Determination and analysis of the 2A° structure of copper–zinc superoxide dismutase. *J. Mol. Biol.* **160**, 181.

Téoule, R. (1987). Radiation-induced DNA damage and its repair. *Intl. J. Radiat. Biol.* **51**, 573.

Thomas, J. P., Bachowski, G. J., and Girotti, A. W. (1986). Inhibition of cell membrane lipid peroxidation by cadmium- and zinc-metallothioneins. *Biochim. Biophys. Acta* **884**, 448.

Touati, D. (1988). Molecular genetics of superoxide dismutases. *Free Radical Biol. Med.* **5**, 393.

Turrens, J. F., Crapo, J. D., and Freeman, B. A. (1984). Protection against oxygen toxicity by intravenous injection of liposome-entrapped catalase and superoxide dismutase. *J. Clin. Invest.* **73**, 87.

Turrens, J. F. and McCord, J. M. (1988). How relevant is the reoxidation of ferrocytochrome c by hydrogen peroxide when determining superoxide anion production? *FEBS Lett.* **227**, 43.

Van der Valk, P. *et al.* (1988). Characterization of oxygen-tolerant Chinese hamster ovary cells. II Energy metabolism and antioxidant status. *Free Radical Biol. Med.* **4**, 345.

Van Loon, A. P. G. M., Pesold-Hurt, B., and Schatz, G. (1986). A yeast mutant lacking mitochondrial manganese-superoxide dismutase is hypersensitive to oxygen. *Proc. Natl. Acad. Sci. USA* **83**, 3820.

Von Sonntag, C. (1987). *The chemical basis of radiation biology.* Taylor and Francis, London.

Watkins, J. A., Kawanishi, S., and Caughey, W. S. (1985). Autoxidation reactions of hemoglobin A free from other red cell components: a minimal mechanism. *Biochem. Biophys. Res. Commun.* **132**, 742.

Wefers, H. and Sies, H. (1983). Oxidation of glutathione by the superoxide radical to the disulfide and the sulfonate yielding singlet oxygen. *Eur. J. Biochem.* **137**, 29.

Weinraub, D., Levy, P., and Faraggi, M. (1986). Chemical properties of water-soluble porphyrins·5. Reaction of some manganese(III) porphyrins with the superoxide and other reducing radicals. *Int. J. Radiat. Biol.* **50**, 649.

Winston, G. W. *et al.* (1983). The generation of hydroxyl and alkoxyl radicals from the interaction of ferrous bipyridyl with peroxides. Differential oxidation of typical hydroxyl-radical scavengers. *Biochem. J.* **216**, 415.

Winterbourn, C. C. and Sutton, H. C. (1986). Iron and xanthine oxidase catalyse formation of an oxidant species distinguishable from OH˙: comparisons with the Haber–Weiss reaction. *Arch. Biochem. Biophys.* **244**, 27.

Wislocki, P. G., Miwa, G. T., and Lu, A. Y. H. (1980). Reactions catalysed by the cytochrome P_{450} system. In *Enzymatic Basis of Detoxication*, vol. I. Academic Press, New York.

Wolff, S. P., Garner, A., and Dean, R. T. (1986). Free radicals, lipids and protein degradation. *Trends Biochem. Sci.* **11**, 27.

Yamazaki, I. and Yokota, K. (1973). Oxidation states of peroxidase. *Mol. Cell. Biochem.* **2**, 39.

Zafiriou, O. C. (1987). Is sea water a radical solution? *Nature* **325**, 481.

4

Lipid peroxidation: a radical chain reaction

Lipid peroxidation has been broadly defined by A. L. Tappel in the USA as the 'oxidative deterioration of polyunsaturated lipids', i.e. lipids that contain more than two carbon–carbon double covalent bonds ($\diagup C{=}C\diagdown$). Oxygen-dependent deterioration, leading to *rancidity*, has been recognized since antiquity as a problem in the storage of fats and oils and is even more relevant today with the popularity of 'polyunsaturated' margarines and cooking oils, and the importance of paints, plastics, lacquers, and rubber, all of which can undergo oxidative damage. The first attempts to study this problem began in 1820 when de Saussure, using a simple mercury manometer, observed that a layer of walnut oil on water exposed to air absorbed three times its own volume of air in the course of 8 months. This initial lengthy period was followed by a second phase of rapid air-absorption, the oil taking up sixty times its own volume of air in 10 days. During the following 3 months, the rate of air uptake gradually diminished, so that the oil had eventually taken up 145 times its own volume of oxygen. Parallel with these changes, the oil became viscous and evil-smelling. Commenting on these experiments a few years later, the famous chemist Berzelius suggested that oxygen absorption might account for not only the autoxidation of oil exposed to air but also a host of similar phenomena. In particular, it was thought that autoxidation might be involved in the spontaneous ignition of wool after its lubrication with linseed oil, a common cause of disaster in textile mills at that time. Interestingly, Berzelius also discovered the element selenium, of vital importance as a protective mechanism as discussed later in this chapter.

The sequence of reactions which is now recognized as the basis of lipid peroxidation was worked out in detail by Farmer and others at the British Rubber Producers Association research laboratories in the 1940s. The relevance of these reactions to biological systems was not appreciated until later, however.

The membranes surrounding cells and cell organelles contain large amounts of polyunsaturated fatty-acid side-chains. What therefore stops us from going rancid ourselves?

4.1 Membrane structure

The major constituents of biological membranes are lipid and protein, the amount of protein increasing with the number of functions the membrane performs. In the nerve myelin sheath, which serves merely as an insulator of the nerve axon, only 20 per cent of the dry weight of the membrane is protein, but most membranes have 50 per cent or more protein, and the highly complicated inner mitochondrial membrane (Chapter 3) and chloroplast thylakoid membrane (Chapter 5) have 80 per cent protein. Some proteins are loosely attached to the surface of membranes (*extrinsic proteins*), but most are tightly attached (*intrinsic proteins*), being partially embedded in the membrane or, in some cases, located in the membrane interior or completely traversing the membrane. Lipid peroxidation can damage membrane proteins as well as the lipids.

Membrane lipids are generally *amphipathic* molecules, i.e. they contain hydrocarbon regions that like to stay together and have little affinity for water, together with polar parts that like to associate with water. In animal cell membranes the dominant lipids are *phospholipids* based on glycerol (Fig. 4.1) but some membranes, particularly plasma membranes, contain significant proportions of sphingolipids and the hydrophobic molecule *cholesterol* (Fig. 4.1). The commonest phospholipid in animal cell membranes is *lecithin*

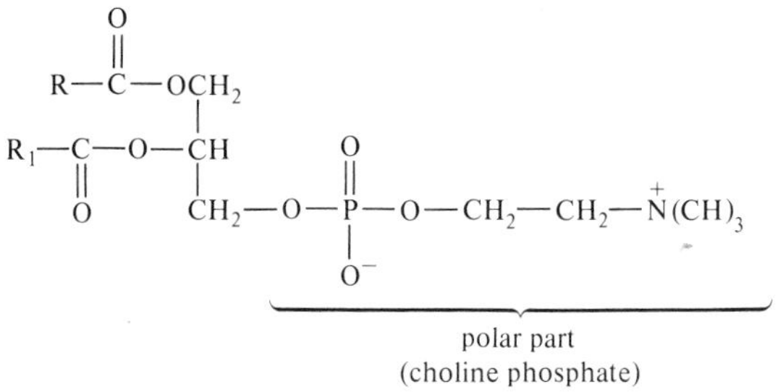

Phosphatidylcholine
(lecithin)

polar part
(choline phosphate)

Fig. 4.1. Lipid molecules found in membranes, R_1, R_2, etc. represent long, hydrophobic, fatty-acid side-chains (for structures, see Fig. 4.2). In most lipids these are joined by *ester* bonds to the alcohol glycerol (Fig. 4.2), i.e.

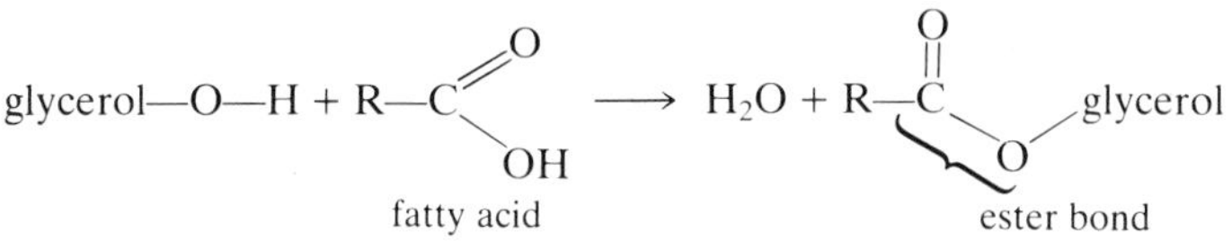

glycerol—O—H + R—C(O)(OH) fatty acid $\longrightarrow$ H_2O + R—C(O)—O—glycerol ester bond

In sphingomyelins, however, the fatty acids are attached to the $-NH_2$ group of sphingosine (Fig. 4.2). All these lipid molecules contain a polar (hydrophilic) part that can interact with water, but in cholesterol this is very small (only an $-OH$ group) so that, overall, cholesterol is a very hydrophobic molecule.

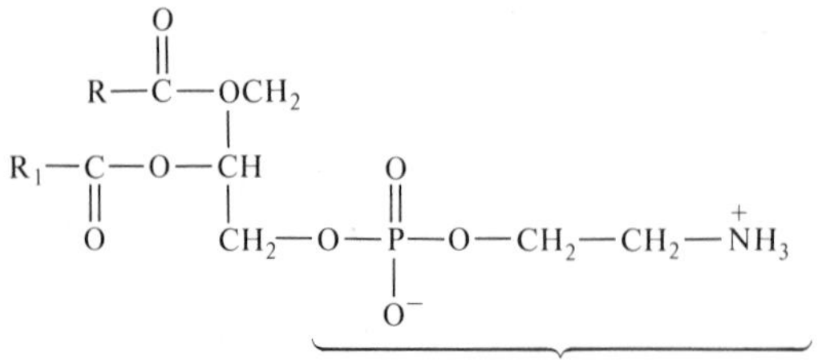

$$R-\overset{\overset{\displaystyle O}{\|}}{C}-OCH_2$$

$$R_1-\overset{\overset{\displaystyle }{}}{\underset{\underset{\displaystyle O}{\|}}{C}}-O-\overset{}{\underset{}{CH}}$$

$$CH_2-O-\overset{\overset{\displaystyle O}{\|}}{\underset{\underset{\displaystyle O^-}{|}}{P}}-O-CH_2-CH_2-\overset{+}{N}H_3$$

Phosphatidylethanolamine

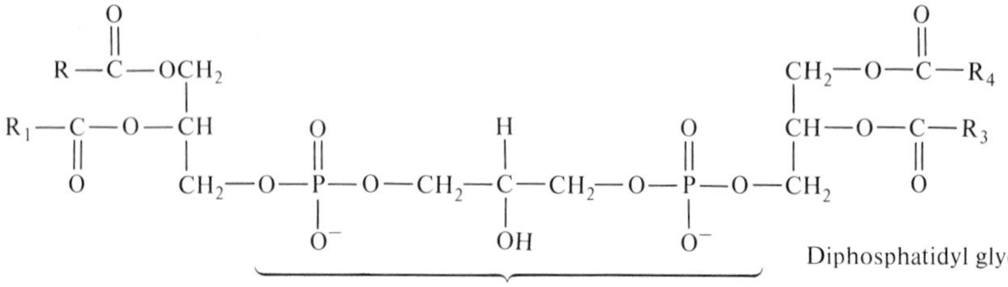

polar part
(ethanolamine phosphate)

Diphosphatidyl glycerol

polar part

Sphingomyelin

$$H_3C-(CH_2)_{12}-CH=CH-\overset{\overset{\displaystyle H}{|}}{\underset{\underset{\displaystyle OH}{|}}{C}}-\overset{\overset{\displaystyle H}{|}}{\underset{\underset{\displaystyle NH}{|}}{C}}-CH_2-O-\overset{\overset{\displaystyle O}{\|}}{\underset{\underset{\displaystyle O^-}{|}}{P}}-O-CH_2-CH_2-\overset{+}{N}(CH_3)_3$$

O=C
|
R₁

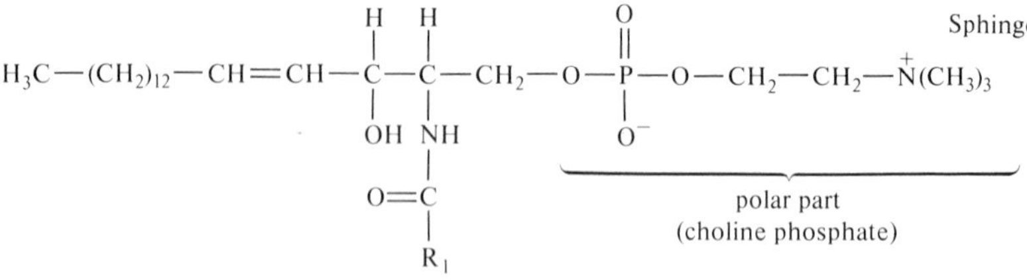

polar part
(choline phosphate)

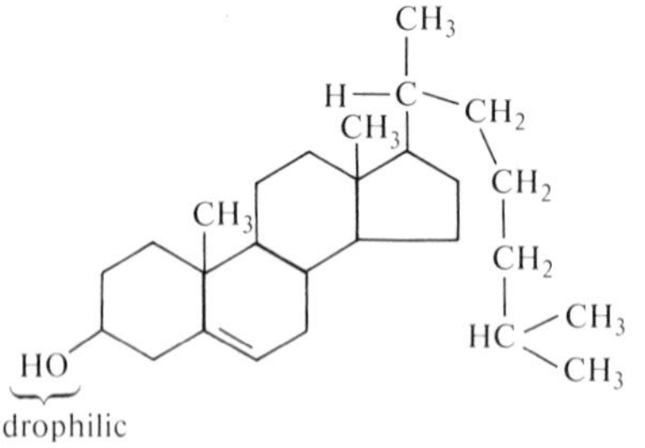

Cholesterol

HO

hydrophilic
part

CH₂OCOR₂
|
CHOCOR₁
|
O-P-OCH₂
|
O⁻

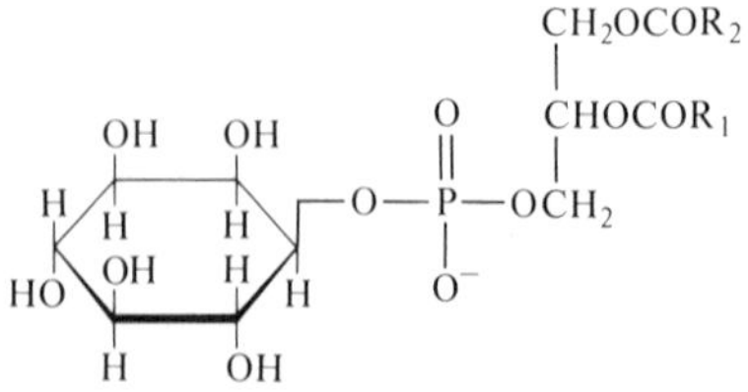

Phosphatidylinositol
(polar part is the sugar inositol)

(phosphatidylcholine). By contrast, the membranes of subcellular organelles such as mitochondria or nuclei rarely contain much sphingolipid or cholesterol. The fatty-acid side-chains of membrane lipids in animal cells have unbranched carbon chains and contain even numbers of carbon atoms, mostly in the range 14–24, and the double bonds are of the *cis*-configuration. A double bond in a carbon chain prevents rotation of the groups attached to the carbon atoms forming it, so that they are forced to stay on one side of the double bond or the other. For example if the two hydrogen atoms of the acid $CH_3(CH_2)_6CH{=}CH(CH_2)_6COOH$ are on the same side of the double bond the *cis*-configuration results, i.e.

$$
\begin{array}{ccc}
H & & H \\
\diagdown & & \diagup \\
& C{=}C & \\
\diagup & & \diagdown \\
(CH_2)_6 & (CH_2)_6 & \text{(esterified to} \\
\diagup & \diagdown & \text{lipid normally)} \\
CH_3 & COOH &
\end{array}
$$

rather than the *trans*-arrangement

$$
\begin{array}{ccc}
 & & \text{(esterified to} \\
 & & \text{lipid normally)} \\
H & (CH_2)_6COOH & \\
\diagdown & \diagup & \\
& C{=}C & \\
\diagup & \diagdown & \\
(CH_2)_6 & H & \\
\diagup & & \\
CH_3 & &
\end{array}
$$

Thus the membrane fatty-acid side-chains have 'kinks' in them whenever a *cis* double bond occurs. Phospholipids contain a number of unsaturated and polyunsaturated fatty-acid side-chains, the structures of some of which are summarized in Fig. 4.2.

The lipid composition of bacterial membranes depends very much on the species and even on the culture conditions and stage in the growth cycle. Membrane fractions usually contain 10–30 per cent lipid. In Gram-positive bacteria (i.e. those that take up *Gram's stain*, used by microscopists) phosphatidylglycerol is present, but phosphatidylethanolamine (Fig. 4.1) is more common in Gram-negative species. *Mycoplasma* contain a lot of cholesterol in their membranes.

As the number of double bonds in a fatty-acid molecule increases, its melting point drops. For example, stearic acid (Fig. 4.2) is solid at room temperature whereas linoleic acid is a liquid. Fatty acids with zero, one, or two double bonds are more resistant to oxidative attack than are the polyunsaturated fatty acids; yet polyunsaturated fatty-acid side-chains are present in many membrane phospholipid molecules.

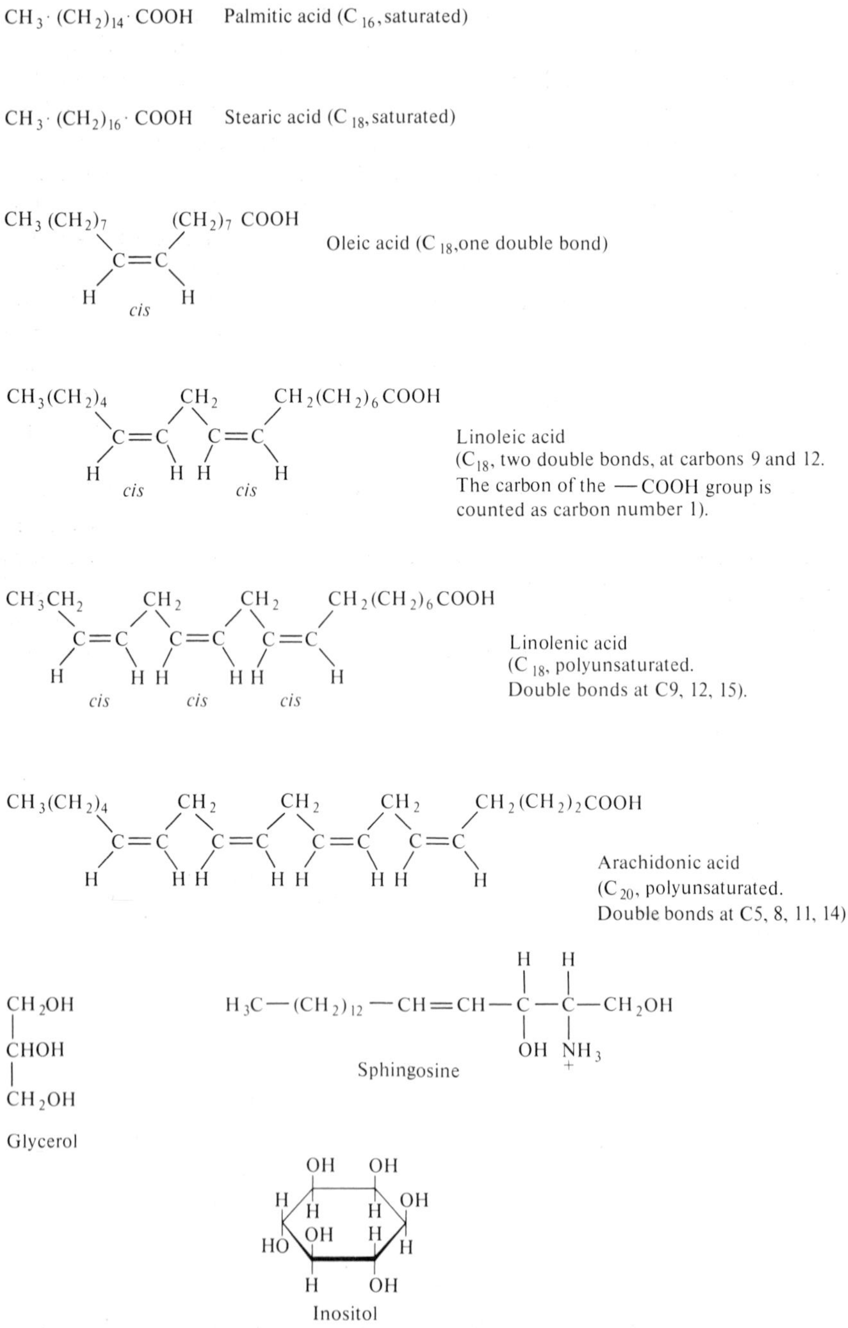

$CH_3 \cdot (CH_2)_{14} \cdot COOH$ Palmitic acid (C_{16}, saturated)

$CH_3 \cdot (CH_2)_{16} \cdot COOH$ Stearic acid (C_{18}, saturated)

$CH_3 (CH_2)_7$ $(CH_2)_7 COOH$

$C=C$ Oleic acid (C_{18}, one double bond)

H   H
cis

$CH_3(CH_2)_4$ CH_2 $CH_2(CH_2)_6 COOH$

$C=C$ $C=C$

H   H H   H   Linoleic acid
cis   *cis*   (C_{18}, two double bonds, at carbons 9 and 12. The carbon of the —COOH group is counted as carbon number 1).

CH_3CH_2 CH_2 CH_2 $CH_2(CH_2)_6 COOH$

$C=C$ $C=C$ $C=C$

H   H H   H H   H   Linolenic acid
cis   *cis*   *cis*   (C_{18}, polyunsaturated. Double bonds at C9, 12, 15).

$CH_3(CH_2)_4$ CH_2 CH_2 CH_2 $CH_2(CH_2)_2COOH$

$C=C$ $C=C$ $C=C$ $C=C$

H   H H   H H   H H   H   Arachidonic acid
(C_{20}, polyunsaturated. Double bonds at C5, 8, 11, 14)

CH_2OH

$CHOH$

CH_2OH

Glycerol

$H_3C-(CH_2)_{12}-CH=CH-\overset{\overset{\displaystyle H}{|}}{C}-\overset{\overset{\displaystyle H}{|}}{C}-CH_2OH$
                         $\underset{OH}{|}$ $\underset{\overset{+}{NH_3}}{|}$

Sphingosine

Inositol

Fig. 4.2. Fatty acids and other 'building blocks' of the membrane lipids.

Since membrane lipids are amphipathic molecules, on exposure to water they will tend to aggregate with their hydrophobic regions clustered together away from the water and their hydrophilic regions in contact with it so that, for example, ions (see the structures in Fig. 4.1) can be stabilized by hydration (Appendix, Section A.2.3). How this arrangement is achieved depends on the relative amounts of lipid and water present. When phospholipids are shaken or sonicated in aqueous solution they form *micelles*, but as more phospholipids are added *liposomes* result, bags of aqueous solution bounded by a *lipid bilayer*. Figure 4.3 shows the structures involved. Liposomes can be surrounded by a single lipid bilayer (*unilamellar*) or several bilayers, as shown in the electron micrograph of a liposome preparation in Fig. 4.4. The interior of liposomes contains a portion of the aqueous solution in which they were made, so they have sometimes been used to study membrane permeability, and as 'parcels' for transporting drugs to target tissues.

There is considerable evidence that the lipid bilayer (Fig. 4.3) is the basic structure of all cell and organelle membranes, proteins being inserted in different parts of the bilayer (as explained previously and summarized diagrammatically in Fig. 4.5). In each half of the lipid bilayer, protein and lipid molecules can diffuse extremely quickly—indeed a lipid molecule can get from one end of a bacterial cell to the other in one or two seconds.

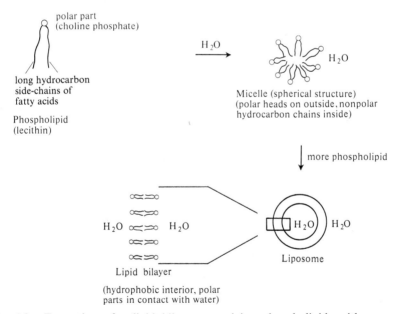

Fig. 4.3. Formation of a lipid bilayer on mixing phospholipids with aqueous solutions.

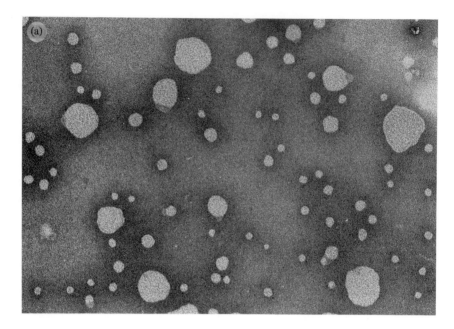

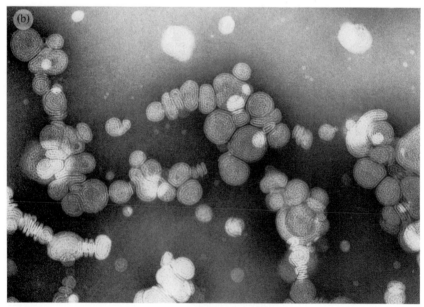

Fig. 4.4. Electron micrograph of a liposome preparation. (a) Unilamellar; (b) multilamellar liposomes.

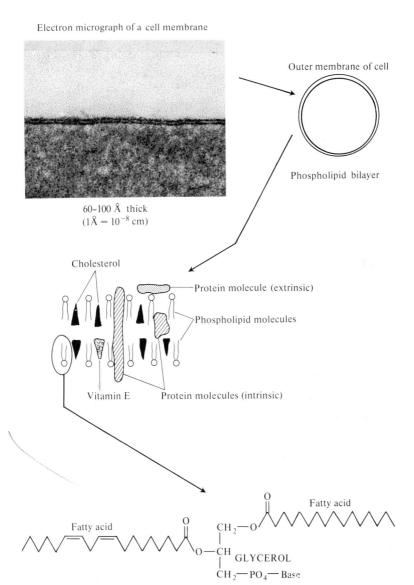

Fig. 4.5. The lipid bilayer as the basic structure of a cell membrane. Cholesterol is found in cell plasma membranes but not usually in organelle membranes. The membrane shown is that of a sheep red blood cell as seen under the electron microscope.

However, exchange of lipid molecules between the two halves of the bilayer is rare. This *membrane fluidity* is due to the presence of unsaturated and polyunsaturated fatty-acid side-chains in many membrane lipids, which lower the melting point of the membrane interior so that it effectively gains the chemical nature and viscosity of a 'light oil'. Damage to polyunsaturated fatty acids tends to reduce membrane fluidity, which is known to be essential for the proper functioning of biological membranes.

4.2 The peroxidation process

4.2.1 Initiation and propagation

When discussing lipid peroxidation, it is essential to use clear terminology for the sequence of events involved; imprecise use of such terms as 'initiation' has caused considerable confusion in the literature. In a completely peroxide-free lipid system, *first-chain initiation* of a peroxidation sequence in a membrane or polyunsaturated fatty acid refers to the attack of any species that has sufficient reactivity to abstract a hydrogen atom from a methylene $(-CH_2-)$ group. Hydroxyl radical can certainly do this:

$$-CH_2- + OH^{\cdot} \rightarrow -\dot{C}H- + H_2O.$$

Hence radiolysis of aqueous solutions, which produces $OH^{\cdot}$, is well-known to stimulate peroxidation of any lipids present; this has been shown not only for biological membranes and fatty acids but also for food lipids. The peroxidation is inhibited by scavengers of $OH^{\cdot}$, such as mannitol and formate. Indeed, stimulation of lipid peroxidation is a problem in the use of ionizing radiation to preserve foods (Chapter 3). The rate constant for reaction of $OH^{\cdot}$ with artificial lecithin bilayers is about $5 \times 10^8 \, M^{-1} s^{-1}$.

By contrast, O_2^- is insufficiently reactive to abstract H from lipids; in any case, it would not be expected to enter the hydrophobic interior of membranes because of its charged nature. In agreement with this, O_2^- does not readily cross biological membranes, the only exception to this rule known to date being the erythrocyte membrane. Here O_2^- can travel via the 'anion channel', through which Cl^- and bicarbonate (HCO_3^-) ions normally pass.

The protonated form of O_2^-, $HO_2^{\cdot}$, is more reactive (Chapter 2) and appears to be capable of abstracting H from some fatty acids, such as linoleic acid.

$$-CH_2- + HO_2^{\cdot} \rightarrow -\dot{C}H- + H_2O_2$$

$HO_2^{\cdot}$, being uncharged, should enter membranes fairly easily (in the same way that H_2O_2 crosses membranes readily). However, $HO_2 \cdot$ has not yet been proved to initiate peroxidation in cell membranes, although it has been suggested to with isolated low-density-lipoprotein fractions. Various iron–oxygen complexes (Section 4.2.2) may also be capable of abstracting H and initiating peroxidation.

Since a hydrogen atom has only one electron (see Appendix I), abstraction of H from a $-CH_2-$ group leaves behind an unpaired electron on the carbon ($-\dot{C}H-$). The presence of a double bond in the fatty acid weakens the C$-$H bonds on the carbon atom adjacent to the double bond and so makes H removal easier. The carbon radical tends to be stabilized by a molecular rearrangement to form a *conjugated diene* (Fig. 4.6). These can undergo various reactions, e.g. if two of them came into contact within a membrane they could cross-link the fatty acid molecules:

$$R-\underset{|}{\dot{C}}H + R-\underset{|}{\dot{C}}H \rightarrow R-\underset{|}{C}H-\underset{|}{C}H-R.$$

However, by far the most likely fate of conjugated dienes under aerobic conditions is to combine with O_2, especially as O_2 is a hydrophobic molecule that concentrates into the interior of membranes. Reaction with oxygen gives a *peroxyl radical*, ROO˙ (or $RO_2 \cdot$). The name is often shortened to *peroxy radical*:

$$\diagup\!\!\!\diagdown \text{CH}^{\cdot} + O_2 \longrightarrow \diagup\!\!\!\diagdown \text{CHO}_2^{\cdot}$$

Of course, very low O_2 concentrations might favour self-reaction of carbon-centred radicals, or perhaps their reaction with other membrane components such as proteins. Hence the O_2 concentration in a biological system might to some extent alter the pathway of peroxidation.

Formation of peroxy radicals has been demonstrated during peroxidation of rat liver microsomes, using the spin-trapping method. Their formation is important because peroxy radicals are capable of abstracting H from another lipid molecule, i.e. an adjacent fatty-acid side-chain:

$$\diagup\!\!\!\diagdown \text{CHO}_2^{\cdot} + \diagup\!\!\!\diagdown \text{CH}_2 \longrightarrow \diagup\!\!\!\diagdown \text{CHO}_2\text{H} + \diagup\!\!\!\diagdown \text{CH}^{\cdot}$$

This is the *propagation stage* of lipid peroxidation. The carbon radical formed can react with O_2 to form another peroxy radical and so the *chain reaction* of lipid peroxidation can continue (Fig. 4.6). The peroxy radical combines with the hydrogen atom that it abstracts to give a *lipid hydroperoxide*. This is sometimes shortened to *lipid peroxide*, although the latter term includes cyclic peroxides as well as L$-$OOH species. A probable alternative fate of peroxy radicals is to form cyclic peroxides (Figs. 4.6 and 4.7). Of course, the initial H abstraction from a polyunsaturated fatty acid can occur

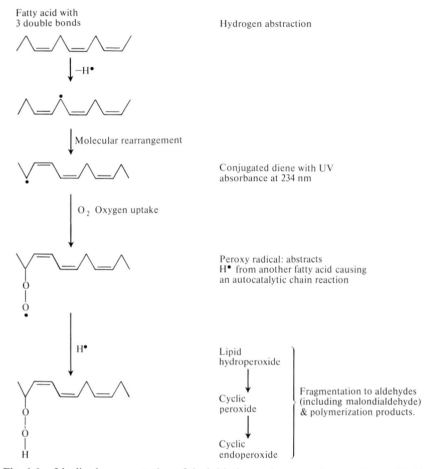

Fig. 4.6. Idealized representation of the initiation and propagation reactions of lipid peroxidation. The peroxidation of a fatty acid with three double bonds is shown.

at different points on the carbon chain (see legend to Fig. 4.7). Thus peroxidation of arachidonic acid gives at least six lipid hydroperoxides as well as cyclic peroxides and other products (Fig. 4.7).

We have defined *first-chain initiation* as the initial H abstraction in a completely peroxide-free lipid system. Abstraction of H atoms by peroxy radicals during the propagation stage of lipid peroxidation is also a form of

Fig. 4.7. Proposed mechanism for formation of lipid hydroperoxides and cyclic peroxides from arachidonic acid. MDA, malondialdehyde (malonaldehyde). Initial abstraction of H at C-13 is shown. H can also be abstracted at C-10 or C-7, giving several other peroxide end-products.

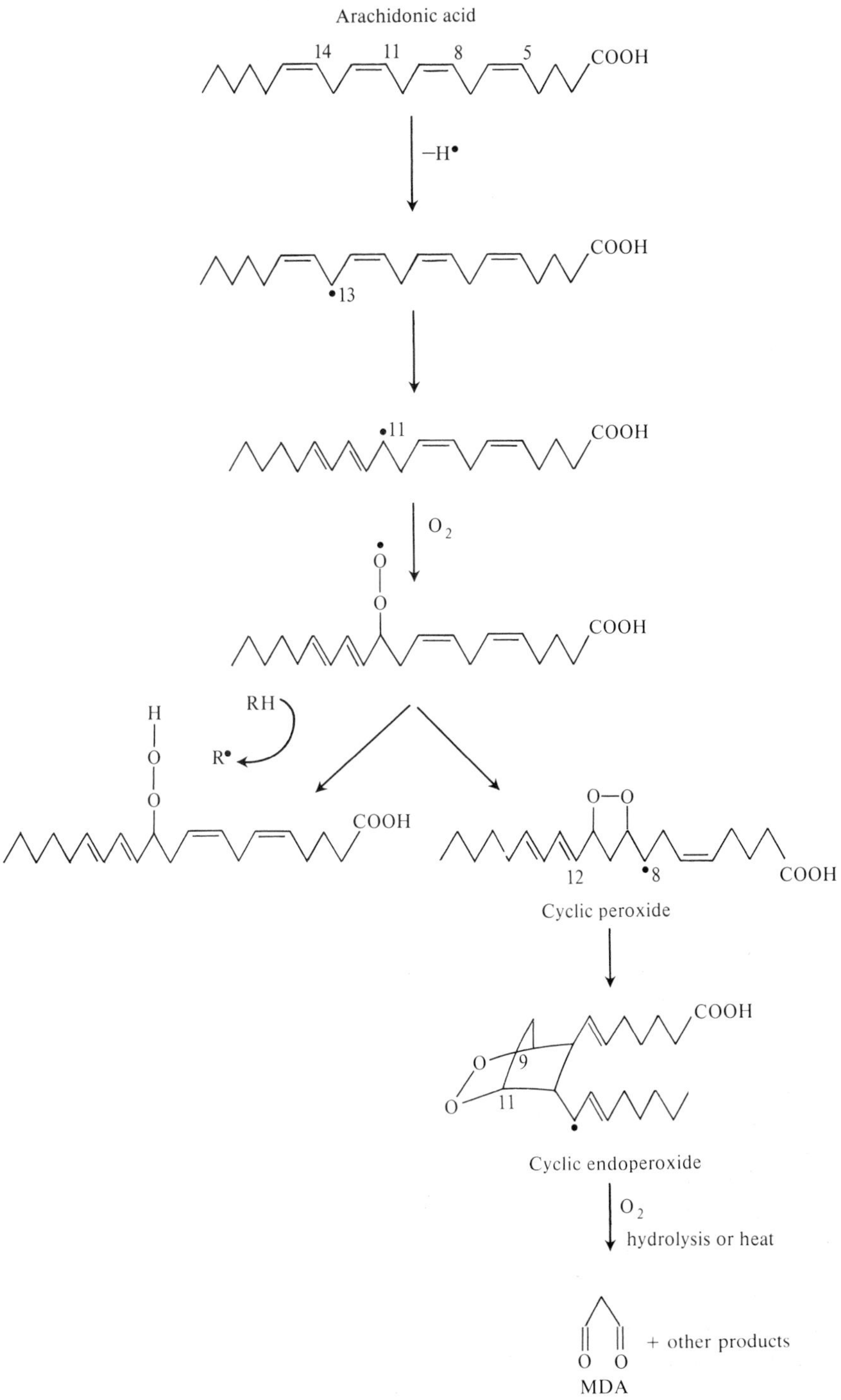

Arachidonic acid
14
11
8
5
COOH
−H•
•13
•11
O_2
RH
R•
Cyclic peroxide
12
•8
9
11
Cyclic endoperoxide
O_2
hydrolysis or heat
+ other products
MDA

'initiation', but, in order to avoid confusion with first-chain initiation, the authors prefer to use the term 'stimulation' of lipid peroxidation to describe continued H abstraction by peroxy radicals.

4.2.2 The importance of iron in lipid peroxidation

The exact role played by iron ions in accelerating lipid peroxidation is at present an area of great confusion. As we have discussed (Chapter 1), iron ions are themselves free radicals and iron(II) ions (Fe^{2+}) can take part in electron transfer reactions with molecular oxygen:

$$Fe^{2+} + O_2 \rightleftarrows (Fe^{2+} - O_2 \leftrightarrow Fe^{3+} - O_2^-) \rightleftarrows Fe^{3+} + O_2^-$$

Superoxide can dismute to form hydrogen peroxide, giving all the essential ingredients for Fenton chemistry and the formation of $OH^{\cdot}$ radicals:

$$2O_2^- + 2H^+ \rightarrow H_2O_2 + O_2$$

$$Fe^{2+} + H_2O_2 \rightarrow Fe^{3+} + OH^- + OH^{\cdot}$$

Thus the addition of an iron(II) salt to a peroxide-free unsaturated lipid should bring about a first-chain initiation of lipid peroxidation (H abstraction by $OH^{\cdot}$). The resulting peroxidation should be inhibitable by H_2O_2-removing enzymes (catalase or selenium glutathione peroxidase), scavengers of $OH^{\cdot}$ and chelating agents that bind iron and prevent its participation in free-radical reactions. Reactions such as these have sometimes been demonstrated in fatty-acid systems solubilized by detergents. However, most scientists, including the authors, find that when catalase or scavengers of $OH^{\cdot}$ are added to isolated cellular *membrane* fractions (e.g. microsomes) or to liposomes undergoing peroxidation in the presence of Fe(II) salts or Fe(III) salts and a reducing agent (such as ascorbate), peroxidation is not inhibited. However, $OH^{\cdot}$ radicals can usually be detected in the reaction mixtures by such techniques as aromatic hydroxylation, spin-trapping, and the deoxyribose method (Chapter 2). Formation of these radicals *is* inhibited by H_2O_2-scavenging enzymes.

The effect of chelating agents on membrane lipid peroxidation is very variable. Desferrioxamine is a powerful inhibitor of iron-stimulated lipid peroxidation in biological buffer systems. EDTA and DETAPAC can either stimulate or inhibit iron-mediated lipid peroxidation, depending on the ratio of the concentration of chelator to that of iron salt (chelator/iron concentration ratios > 1 usually inhibit, lower ratios usually stimulate). Iron ions complexed to EDTA are still capable of forming $OH^{\cdot}$ from O_2^- and H_2O_2, whereas Fe(III) ions complexed with DETAPAC react slowly, if at all, with O_2^- (Chapter 3).

The inability of catalase or OH˙ scavengers to inhibit iron-stimulated peroxidation in most membrane systems, even though OH˙ radicals are formed in such systems, suggests strongly that OH˙ formation is not required for the peroxidation to take place. The lack of action of OH˙ scavengers might be interpreted to mean that the required OH˙ formation is 'site-specific', involving iron ions bound to the membrane, so that any OH˙ formed reacts immediately with the membrane components and is not amenable to scavenging (Chapter 3). However, a source of H_2O_2 would still be required and catalase should still inhibit. The fact that it does not has led several scientists to suggest that first-chain initiation of lipid peroxidation in membrane systems incubated with iron salts in the presence of O_2 is achieved by reactive species other than OH˙. Ferryl (Chapter 3) is a possibility, although ferryl formation by reaction of Fe(II) with H_2O_2 would still require a source of H_2O_2 and inhibition by peroxide-removing enzymes would be expected. Perferryl ($Fe^{2+} - O_2 \leftrightarrow Fe^{3+} - O_2^-$) could also be involved, as first suggested by Hochstein in the USA. However, what is known of the chemistry of perferryl complexes (Chapter 3, Section 3.5.4) suggests that they would be insufficiently reactive to abstract H or to directly insert oxygen into fatty-acid side-chains. Studies on the kinetics of microsomal or liposomal lipid peroxidation in the presence of Fe(II) and/or Fe(III) salts have led Aust *et al.* in the USA to propose that first-chain initiation requires an iron(II)/ iron(III)/oxygen complex, or at least some specific critical ratio of Fe(II) to Fe(III). This proposal explains a number of experimental observations, although attempts to isolate and characterize this complex have been unsuccessful to date. Recent work by the authors has shown that lead (Pb^{2+}) ions can replace Fe(III) in stimulating Fe(II)-dependent peroxidation in such experiments, which argues against a requirement for a specific Fe(II)-Fe(III) complex.

However, iron plays a second important role in lipid peroxidation. Pure lipid peroxides are fairly stable at physiological temperatures, but, in the presence of transition metal complexes, including iron and copper salts, their decomposition is greatly accelerated. Thus a reduced iron complex can react with lipid peroxide in a similar way to its reaction with H_2O_2: it causes fission of an O—O bond to form an *alkoxyl* radical (often shortened to *alkoxy* radical).

$$R-OOH + Fe^{2+}\text{-complex} \rightarrow Fe^{3+}\text{-complex} + OH^- + R-O˙$$
lipid hydroperoxide   alkoxy radical

An iron(III) complex can form both peroxy and alkoxy radicals, according to the overall equation:

$$R-OOH + Fe^{3+}\text{-complex} \rightarrow RO_2˙ + H^+ + Fe^{2+}\text{-complex}$$
peroxy radical   (further reaction to give alkoxy radical).

The reactions of Fe^{2+} ions with lipid hydroperoxides are several orders of magnitude faster than their reactions with H_2O_2 (k_2 for $Fe^{2+} + H_2O_2$ is about $76 \, M^{-1} s^{-1}$; for $R—OOH + Fe^{2+}$, about $1.5 \times 10^3 \, M^{-1} s^{-1}$); reactions of Fe^{3+} with hydroperoxides are much slower than those of Fe^{2+}. The variable effects of chelating agents on lipid peroxidation can, at least in part, be explained by their ability to influence these different reactions. For example, EDTA appears to stimulate the reaction of iron ions with H_2O_2, but slows their reaction with lipid peroxides.

Commercially-available lipids are all contaminated with lipid peroxides, so that liposomes or micelles made from them will already contain traces of lipid peroxides. When cells are injured, lipid peroxidation is favoured (see Section 4.9) and traces of lipid peroxides are formed enzymically in tissues by cyclo-oxygenase and lipoxygenase enzymes (Chapter 7). Thus membrane fractions isolated from disrupted cells also contain lipid peroxides. When iron salts are added to isolated membrane fractions, the lipid peroxides present will be decomposed by the above two reactions to generate peroxy and alkoxy radicals. Both alkoxy and peroxy radicals can abstract $H^{\cdot}$ and stimulate lipid peroxidation. Thus there is no need for $OH^{\cdot}$ formation and first-chain initiation; the added metal ions are doing no more than 'stimulating' further lipid peroxidation. It is certainly possible that the putative abilities of ferryl, perferryl, and $Fe(II)/Fe(III)/O_2$ complexes to initiate lipid peroxidation are explicable by the abilities of these complexes to efficiently degrade traces of lipid peroxides in the membrane systems that were being studied. Experiments on peroxide-free lipid systems (which are very difficult to obtain) are urgently required to establish firmly whether or not various iron–oxygen complexes are really capable of abstracting hydrogen.

We saw in Chapters 2 and 3 that simple iron complexes can react with H_2O_2 to form $OH^{\cdot}$, but that reaction of iron proteins with H_2O_2 and/or O_2^- does not result in formation of $OH^{\cdot}$ detectable outside the protein, unless iron is released from the protein under the reaction conditions being used. The range of iron complexes that can stimulate lipid peroxidation is wider (Table 4.1). Thus not only Fe^{2+} salts and simple complexes (e.g. Fe^{2+}-ADP) are effective, but also free haem, met- and oxy-haemoglobin, metmyoglobin, cytochromes (including cytochromes c and P-450), horseradish peroxidase, and lactoperoxidase (under certain circumstances). Ferritin stimulates lipid peroxidation to an extent proportional to the amount of iron it contains, whereas haemosiderin stimulates much less strongly (on a unit iron basis). Sometimes the stimulation of lipid peroxidation in these systems is due to the release of iron from the protein under the conditions of the experiment. For example, stimulation of lipid peroxidation in liposomes by ferritin or haemosiderin is almost completely inhibitable by desferrioxamine, suggesting that it is mediated by released iron ions. Stimulation of peroxidation by myoglobin and haemoglobin involves both iron release from the protein and

Table 4.1. Biological iron complexes and their possible participation in oxygen radical reactions

Type of iron complex	Decomposition of lipid peroxides to form alkoxy and peroxy radicals	Hydroxyl radical formation by Fenton chemistry
Loosely-bound iron		
Iron ions attached to:		
(a) Phosphate esters (e.g. ATP)	Yes	Yes
(b) Carbohydrates and organic acids (e.g. citrate, picolinic acid, deoxyribose)	Yes	Yes
(c) DNA	Yes	Yes
(d) Membrane lipids	Yes	Yes
Iron tightly bound to proteins:		
(a) Non-haem iron		
Ferritin (4500 mol Fe/mol protein)	Yes	Yes (when iron is released)
Haemosiderin	Weakly	Weakly
Lactoferrin (iron saturated, 2 mol Fe^{3+}/mol protein)	Weakly? not if < 2 mol Fe^{3+}/mol	No (only if iron released)
Transferrin (iron saturated, 2 mol Fe^{3+}/mol protein)	Weakly? not if < 2 mol Fe^{3+}/mol	No (only if iron released)
(b) Haem iron		
Haemoglobin	Yes	Yes (when iron is released)
Leghaemoglobin	Yes	Yes (when iron is released)
Myoglobin	Yes	Yes (when iron is released)
Cytochrome c	Yes	Yes (when iron is released)
Catalase	Weakly	No

also reactions brought about by the intact protein itself, e.g. decomposition of lipid peroxides into various radicals by interaction with the haem ring. It has been suggested that metmyoglobin–H_2O_2 or methaemoglobin–H_2O_2 mixtures can initiate lipid peroxidation by forming a ferryl species that abstracts H, but an alternative explanation is that decomposition of pre-formed peroxides in the lipid fractions being studied is responsible for the stimulation of lipid peroxidation.

Some other metal ions, e.g. cobalt(II), can also decompose lipid peroxides. However, neither Zn^{2+} nor Mn^{2+} is effective, and both zinc(II) and manganese(II) salts have been reported to inhibit lipid peroxidation in some systems.

4.2.3 Products of the reaction of metal complexes with lipid peroxides

The reaction of iron and copper complexes with lipid peroxides generates a huge range of products, including epoxides, compounds containing the carbonyl (C=O) group, and hydrocarbon gases. For example, if Fe(II) reacts with a hydroperoxide on the fifth carbon from the methyl end of the fatty acid, pentane gas can be produced. This can happen with linoleic acid and arachidonic acid:

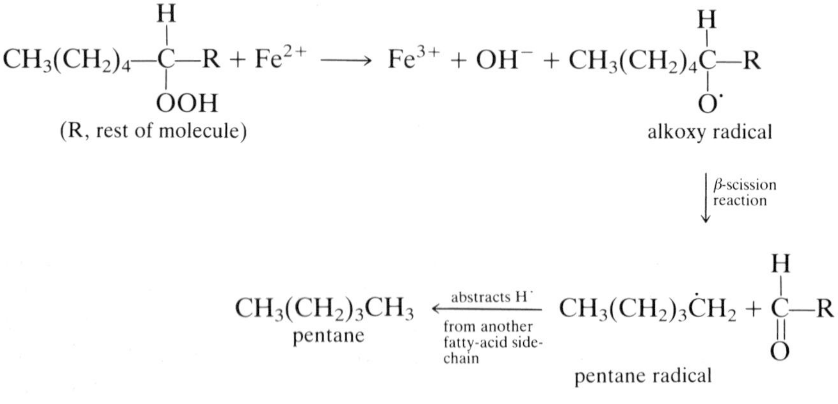

$$CH_3(CH_2)_4-\overset{\overset{\displaystyle H}{|}}{\underset{\underset{\displaystyle OOH}{|}}{C}}-R + Fe^{2+} \longrightarrow Fe^{3+} + OH^- + CH_3(CH_2)_4\overset{\overset{\displaystyle H}{|}}{\underset{\underset{\displaystyle O^\cdot}{|}}{C}}-R$$

(R, rest of molecule)   alkoxy radical

$\downarrow$ β-scission reaction

$$CH_3(CH_2)_3CH_3 \xleftarrow[\substack{\text{from another}\\\text{fatty-acid side-}\\\text{chain}}]{\text{abstracts } H^\cdot} CH_3(CH_2)_3\dot{C}H_2 + \overset{\overset{\displaystyle H}{|}}{\underset{\underset{\displaystyle O}{||}}{C}}-R$$

pentane   pentane radical

Ethane (C_2H_6) and ethylene (ethene, H_2C=CH_2) gases are produced in similar reactions from linolenic acid. *β-scission*, shown above, is a well-known reaction of radicals, especially alkoxy radicals.

Many of the carbonyl compounds produced by metal ion-dependent decomposition of lipid peroxides are aldehydes (RCHO), as shown in Table 4.2. One aldehyde, isolated by Austrian scientists from peroxidizing membranes, is the unsaturated aldehyde 4-hydroxy-2,3,*trans*-nonenal:

Table 4.2. Formation of carbonyl compounds from fatty acids, microsomes and whole cells during lipid peroxidation stimulated by addition of iron salts

System	Iron added	MDA formed (nmol/mg fatty acid or protein)	Other carbonyl compounds formed (nmol/mg fatty acid or protein)	percentage of carbonyls as MDA
Fatty acids				
—Oleic	none	—	—	—
	20 μM FeSO$_4$	0	62	0
—Linoleic	none	0	98	0
	20 μM FeSO$_4$	12	317	4
—Linolenic	none	—	—	—
	20 μM FeSO$_4$	13	562	2
—Arachidonic	none	19	588	3
	20 μM FeSO$_4$	45	1728	3
Whole hepatocytes	none	0.17	4.37	4
	ADP-Fe^{2+}	1.71	8.43	20
Liver microsomes	none	5.6	21.9	26
	ADP-Fe^{2+}	76.8	144.2	53

Dates were abstracted from the articles by Esterbauer *et al.* in McBrien and Slater (1982) *Free Radicals, Lipid Peroxidation and Cancer*, Academic Press, London and in (1987) *First Vienna Shock Forum, part A*. A. R. Liss, New York.
Note the high percentage of MDA formed by rat liver microsomes (Section 4.3).

$$CH_3-(CH_2)_4-\underset{\underset{H}{|}}{\overset{\overset{OH}{|}}{CH}}-C=C-CHO$$

Malonaldehyde (sometimes called *malondialdehyde*) is also formed in small amounts in most tissues. Larger amounts of malonaldehyde are formed during the peroxidation of liver microsomal fractions than during the peroxidation of other cell membrane fractions. For example, Table 4.2 shows the amounts of malonaldehyde or other carbonyl compounds produced when peroxidation was stimulated by adding iron salts to fatty acids, to hepatocytes or to liver microsomal fractions. The structure of malonaldehyde is:

$$O=\underset{H}{\overset{}{C}}\diagdown^{CH_2}\diagdown\underset{H}{\overset{}{C}}=O$$

In most membranes, the propagation reactions of lipid peroxidation will not proceed very far before they meet a protein molecule, which can then be attacked and damaged. Proteins can be attacked by alkoxy radicals, peroxy radicals (at a slower rate), and possibly by carbon-centred radicals. For example, alkoxy radicals can attack tryptophan and cysteine residues. Aldehydes can react with —SH groups on proteins, e.g. for hydroxynonenal:

$$CH_3(CH_2)_4\!-\!\overset{\overset{\displaystyle OH}{|}}{CH}\!-\!\overset{}{C}\!=\!\overset{\overset{}{\underset{\displaystyle H}{|}}}{C}\!-\!CHO \xrightarrow{\text{protein—SH}}$$

$$CH_3(CH_2)_4\!-\!CH\!-\!\underset{\underset{\displaystyle S\!-\!protein}{|}}{CH}\!-\!CH_2\!-\!CHOH$$
(with an O bridge between the first CH and the terminal CHOH)

In addition, dialdehydes such as MDA can attack amino groups on protein molecules to form both intramolecular cross links and also cross-links between different protein molecules, e.g. for malonaldehyde:

$$OHC\!\cdot\!CH_2\!\cdot\!CHO + protein\!\!\begin{array}{c}NH_2\\ \\NH_2\end{array} \longrightarrow protein\!\!\begin{array}{c}NH\!-\!CH\\ \searrow CH\\N\!=\!CH\end{array}$$

intramolecular cross-link

$$OHC\!\cdot\!CH_2CHO + 2\ protein\!-\!NH_2 \longrightarrow$$

$$protein\!-\!NHCH\!=\!CH\!-\!CH\!=\!N\!-\!protein$$

intermolecular cross-link

Treatment of membrane with malonaldehyde *in vitro* has frequently been observed to cross-link and aggregate membrane proteins. Enzymes that require —NH$_2$ or —SH groups for their activity are usually inhibited during lipid peroxidation, e.g. the glucose 6-phosphatase enzyme found in liver microsomal fractions is inhibited as its —SH groups are attacked. Hydroxy-*trans*-nonenal and other low-molecular-mass products of lipid peroxidation have been shown to inhibit protein synthesis and to interfere with the growth of bacteria and animal cells in culture (Fig. 4.8). Indeed, hydroxy-*trans*-nonenal and several similar unsaturated hydroxy-aldehydes (e.g. 4,5-dihydroxydecenal) are far more toxic than is malonaldehyde. Also produced during peroxide decomposition are unsaturated aldehydes such as hexenal and nonenal, saturated aldehydes such as propanal, butanal, and hexanal, and a range of ketones such as butanone, pentanones, and octanones. Some aldehydes (e.g. 4-hydroxyoctenal) may exert chemotactic actions upon

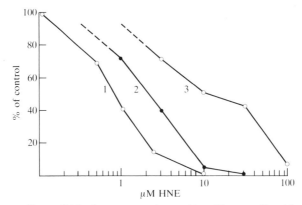

Fig. 4.8. Some effects of 4-hydroxynonenal on cultured human fibroblasts, showing the high cytotoxicity of this compound. Figure reproduced, by courtesy of Professor H. Esterbauer, from *First Vienna Shock Forum*, part A, A. R. Liss Inc. (1987), p. 245. *Line 1*: inhibition of proliferation. *Line 2*: inhibition of colony forming efficiency. *Line 3*: depletion of cellular thiol content. Line 1 used skin fibroblasts, Lines 2 and 3 used bronchial fibroblasts.

neutrophils in biological fluids, and several aldehydes are mutagenic in bacterial test systems.

These complex reactions of radicals, aldehydes, and other products of lipid peroxidation can cause severe damage to membrane proteins. Peroxidation of liver or erythrocyte membranes is known to cause formation of high-molecular-mass protein aggregates within the membrane. The surface receptor molecules that allow cells to respond to hormones can be inactivated during lipid peroxidation (Fig. 4.9), as are enzymes such as glucose 6-phosphatase or the $Na^+K^+ATPase$ involved in maintenance of correct ion balance within cells. In general, the overall effects of lipid peroxidation are to decrease membrane fluidity, increase the 'leakiness' of the membrane to substances that do not normally cross it (such as Ca^{2+} ions), and inactivate membrane-bound enzymes. Continued fragmentation of fatty-acid side-chains to produce aldehydes and hydrocarbons such as pentane will eventually lead to complete loss of membrane integrity. Rupture of, say, the membranes of lysosomes in this way can spill hydrolytic enzymes into the rest of the cell to cause amplification of damage. Lipid peroxidation of erythrocyte membranes causes them to lose their ability to change shape and squeeze through the smallest capillaries ('deformability') and it will eventually lead to haemolysis. It has been suggested that the loss of viability of mammalian sperm on prolonged incubation at 37 °C is due to the accumulation of products of lipid peroxidation, and the loss of the germinating ability of soybean seeds, stored under warm, damp conditions, has been attributed to

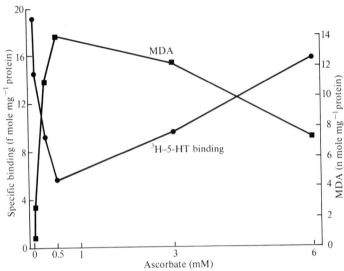

Fig. 4.9. Effect of ascorbate on lipid peroxidation. Cortex membranes were pre-pared from rat brains by centrifugation—brain fractions usually contain significant amounts of endogenous iron. The effect of ascorbic acid on the rate of lipid peroxidation, measured by the TBA test and expressed as the amount of malonalde-hyde produced is shown. The binding of tritium [^{3}H]-labelled 5-hydroxytryptamine to receptors on the membrane was studied—loss of binding means that the receptors have been damaged. Data from Muakkassah-Kelly *et al.*, *Biochem. Biophys. Res. Commun.* **104**, 1003–10 (1982), with permission.

the same reason. In some bacteria the DNA is close to or attached to the cell membrane and it can be damaged during peroxidation. Studies on a mutant of *E. coli* that could not synthesize fatty acids, and thus incorporated into its membrane lipids whatever acids it was given in the grown medium, showed that the toxicity of hyperbaric oxygen became greater as the percentage of polyunsaturated fatty acids in the membrane was increased.

Injection of peroxidized lipids into experimental animals produces del-eterious effects, e.g. damage to the heart in rats and 'fatty liver' in both rats and rabbits. Exposure of animals to elevated oxygen concentrations causes increased peroxidation of their cell membranes. Oral administration of large doses of linoleic acid peroxides and their decomposition products to rats leads to liver injury and necrosis. In mice, damage to lymphoid tissues has been observed after feeding them with an ester of linoleate hydroperoxide.

It is quite likely that cholesterol in membranes can also become oxidized during lipid peroxidation, since cholesterol included in liposomes subjected to peroxidation becomes oxidized to a mixture of products that include cholesterol-7-hydroperoxides, epoxides and cholestanediols. Fig. 4.10 shows

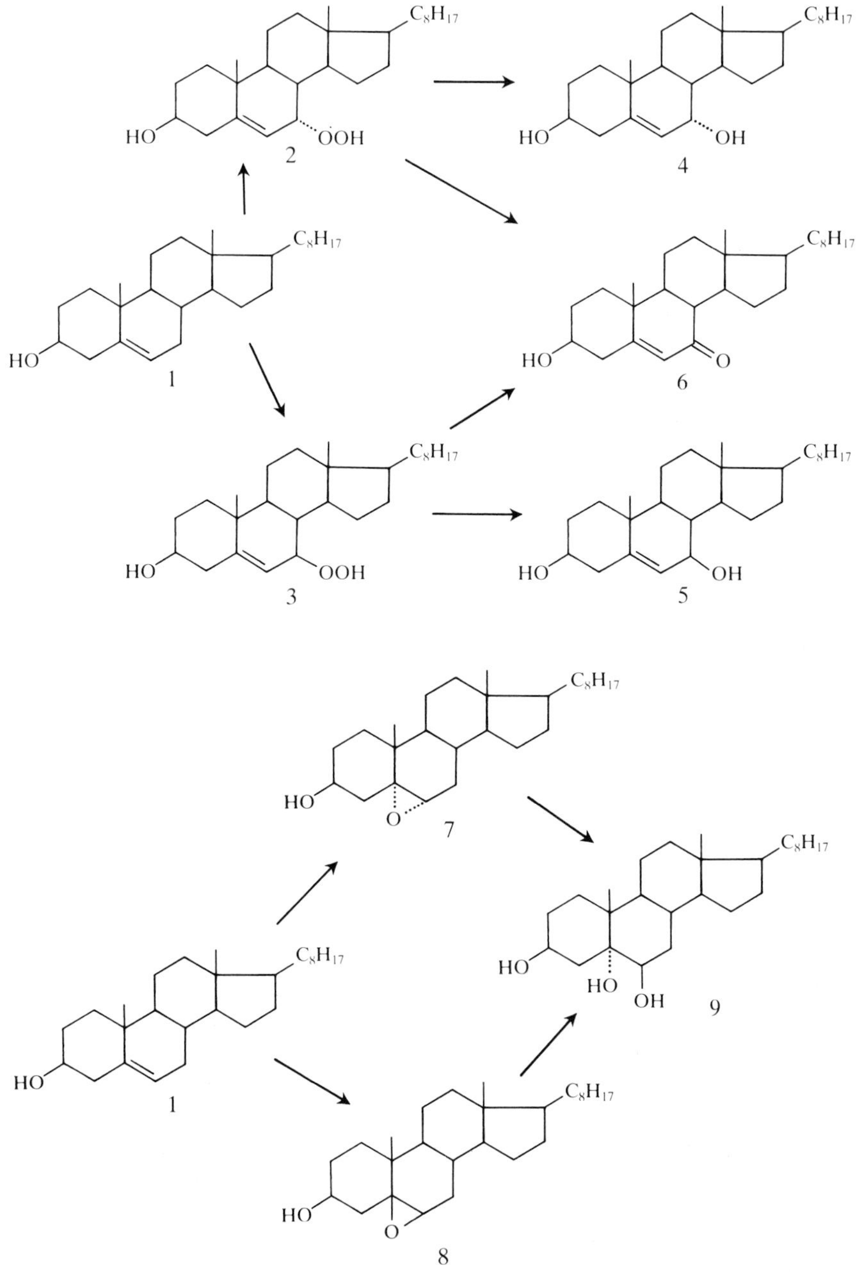

Fig. 4.10. Some products of cholesterol oxidation. 1 cholesterol; 2 and 3 cholesterol 7-hydroperoxides; 4 and 5 cholest-5-ene-3-β,7-diols; 6 3β-hydroxycholest-5-en-7-one; 7 and 8 5,6β-epoxy-5β- and 5,6α-epoxy-5α-cholestan-3-β-ols; 9 5α-cholestane-3β,5,6,β-triol. Diagrams modified by courtesy of Professor L. L. Smith (see *Chem. Phys. Lipids*, **44,** 87 [1987]). Reaction of singlet O_2 with cholesterol yields primarily the 5-α-hydroperoxide (Chapter 2) whereas free-radical oxidation gives 7-β and 7-α-hydroperoxides and the other products shown.

the structure of some of these. However, there are conflicting reports on the cytotoxicity of cholesterol oxidation products, possibly due to the fact that most studies have used complex mixtures rather than purified single products. For example, oxidation products of cholesterol have variously been claimed to stimulate or to suppress atherogenesis (Section 4.9).

4.3 Enzymic and non-enzymic lipid peroxidation

Scientists often loosely refer to iron salts and other iron complexes as 'initiating' peroxidation, but in most cases what is really happening is that they are causing the decomposition of preformed lipid peroxides, as explained above, to generate alkoxy or peroxy radicals. Copper salts also stimulate peroxidation, probably by a similar mechanism. (Indeed, it has been suggested that the copper wire in one form of the intrauterine contraceptive device stimulates lipid peroxidation in the cervical mucus, hence damaging spermatozoa.)

Iron(II) and its complexes stimulate membrane peroxidation more than does iron(III). This may be explained by the greater solubility of iron(II) salts in solution, the faster rate at which lipid peroxides are decomposed by iron(II) and the high reactivity of the alkoxy radicals so produced. Hence the rate of peroxidation of purified membrane lipids or microsomal fractions in the presence of added Fe(III) complexes can be greatly increased by addition of ascorbic acid, a biological reducing agent (Chapter 3). For a given concentration of iron salt, low concentrations of ascorbate stimulate peroxidation, probably by reducing Fe(III) to Fe(II), whereas high concentrations inhibit the peroxidation (Fig. 4.9). It has been suggested that high concentrations of ascorbate may reduce some of the lipid peroxy radicals directly to hydroperoxides, and thus interfere with the chain reaction. Ascorbate can also regenerate the chain-breaking antioxidant a-tocopherol in biological membranes (Section 4.7.3).

Mixtures of GSH and iron salts have been reported to stimulate lipid peroxidation, but the general view is that GSH is protective *in vivo* because it is a substrate for glutathione peroxidase (Section 4.1.5), as well as being able to react directly with various aldehydes produced during peroxidation and thus protect the —SH groups of membrane proteins. It should be noted, however, that although the adducts formed by reaction of such cytotoxic aldehydes as 4-hydroxynonenal with GSH are far less reactive than the aldehydes themselves, they are not completely harmless. Consistent with the view that GSH is generally protective against lipid peroxidation *in vivo*, treatment of rats or mice with reagents that decrease liver GSH concentrations seems to increase their susceptibility to peroxidation.

Peroxidation of a cell or organelle membrane stimulated by Fe(II), or Fe(III) plus ascorbate, *in vitro* does not require the activity of any enzymes.

Another way of stimulating lipid peroxidation *in vitro* is to add lipid hydroperoxides or artificial organic hydroperoxides such as *tert*-butyl hydroperoxide or cumene hydroperoxide (as used in Fig. 4.11). The decomposition of these to alkoxy or peroxy radicals accelerates the chain reaction of lipid peroxidation. Decomposition is again aided by metal ions and their complexes, e.g. by methaemoglobin or cytochrome P-450. Indeed, cytochrome P-450 will catalyse a hydroperoxide-dependent hydroxylation of certain substrates, such as cyclohexane or toluene. This activity is sometimes known as the 'peroxygenase' action of cytochrome P-450. It probably has no physiological significance, but studies of it have given interesting information about the mechanism of hydroxylation by P-450. If SH is the hydroxylatable substrate and $-OOH$ the peroxide, peroxygenase activity can be represented by the equation:

$$SH + XOOH \rightarrow S{-}OH + X{-}OH$$

peroxide   hydroxylated   alcohol
              substrate

The decomposition of cumene and *tert*-butyl hydroperoxides in the presence of a metal ion such as Fe^{2+} can be written:

$$Fe^{2+} + H_3C{-}\overset{\displaystyle H_3C}{\underset{\displaystyle H_3C}{C}}{-}OOH \longrightarrow H_3C{-}\overset{\displaystyle H_3C}{\underset{\displaystyle H_3C}{C}}{-}O^{\cdot} + OH^- + Fe^{3+}$$

tert-butyl hydroperoxide   alkoxy radical

cumene hydroperoxide   alkoxy radical

Azo initiators

Yet another way of generating peroxy radicals to stimulate lipid peroxidation is the use of 'azo initiators', which decompose at a temperature-controlled rate to give a known flux of radicals. For example, 2,2′-azo*bis*(2-amidinopropane) dihydrochloride (AAPH) is a water soluble radical generator. On heating, it forms carbon-centred radicals that can react swiftly with O_2 to yield peroxy radicals capable of abstracting H from membrane lipids:

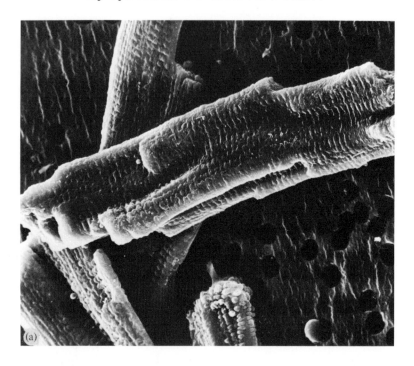

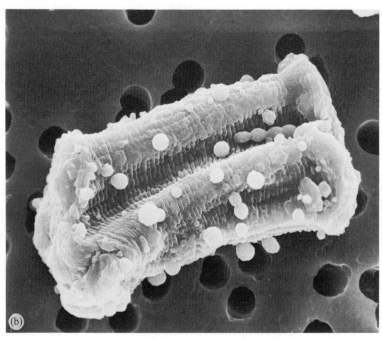

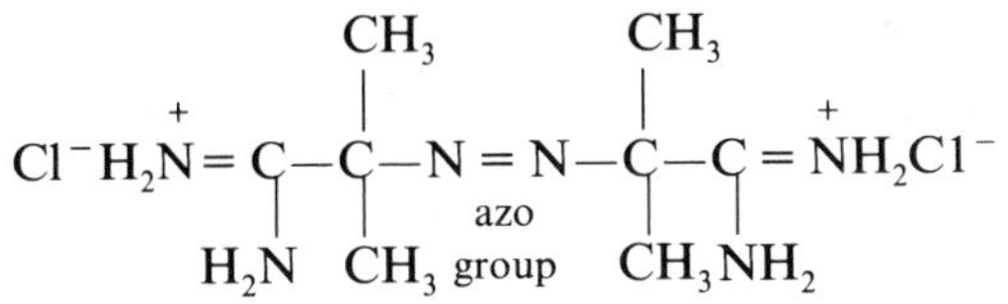

$$
\underset{\text{AAPH}}{\underset{\underset{H_2N \quad CH_3}{| \quad |}}{Cl^- \; \overset{+}{H_2N} = C - C} \underset{\text{azo}}{- N = N -} \underset{\underset{CH_3NH_2}{| \quad |}}{C - C = \overset{+}{N}H_2Cl^-}}
$$

with CH_3 groups above the carbons.

$$
A - N = N - A \xrightarrow{\text{heat}} N_2 + A^{\textbf{·}} + A^{\textbf{·}}
$$
carbon-centred radicals

$$
A^{\textbf{·}} + O_2 \longrightarrow AO_2^{\textbf{·}}
$$
peroxy radical
(abstracts H)

Lipid soluble azo initiators, containing more hydrophobic groups, also exist, e.g. 2,2′-azo*bis*(2,4-dimethylvaleronitrile),

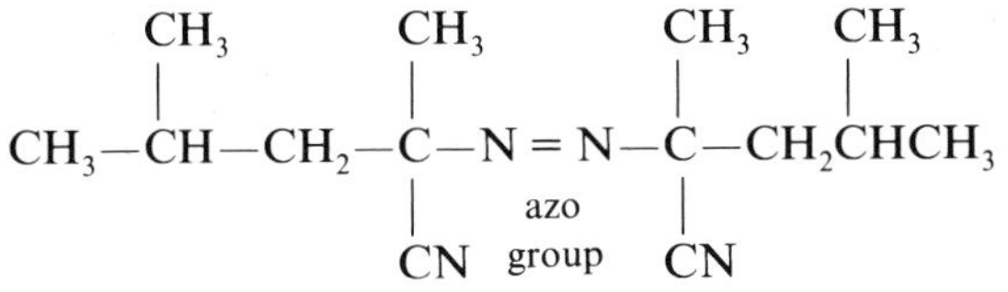

$$
\underset{\underset{CN}{|}}{CH_3 - CH - CH_2 - \overset{CH_3}{\overset{|}{C}}} \underset{\text{azo}}{- N = N -} \underset{\underset{CN}{|}}{\overset{CH_3}{\overset{|}{C}} - CH_2CHCH_3}
$$

Enzymic peroxidation

Peroxidation stimulated by adding Fe^{2+}, Fe^{3+}/ascorbate, azo initiators or synthetic hydroperoxides to lipids is often called 'non-enzymic' lipid peroxidation. The term 'enzymic' lipid peroxidation is also often found in the literature. The enzymes cyclo-oxygenase and lipoxygenase (Chapter 7) catalyse the controlled peroxidation of their fatty acid substrates to give hydroperoxides and endoperoxides that are *stereospecific* and have important biological functions. In the view of the authors, the term 'enzymic lipid peroxidation' should be reserved for these two types of enzyme. Unfortunately, the term has become more widely used in a way that has often led to confusion. For example, if some Fe(III) complexes are added to membrane

---

Fig. 4.11. Lipid peroxidation as a mechanism of injury to heart cells. Isolated beating cardiac myocytes were prepared from rat hearts. (a) Shows their normal appearance as seen by scanning electron microscopy. (b) The cells shown had been treated with cumene hydroperoxide for 30 minutes. Note the contraction and formation of 'blebs' of membrane on the cell surface. The cumene hydroperoxide stimulated peroxidation of the membrane lipids, as followed by diene conjugation. Photograph by courtesy of Dr A. A. Noronha-Dutra.

lipids, the rate of peroxidation can be increased by generating O_2^- in the reaction mixture (e.g. by xanthine plus xanthine oxidase). However, the enzyme-generated O_2^- is merely serving to reduce Fe(III) to Fe(II), which stimulates peroxidation. The function of the enzyme thus seems no different from that of, say, ascorbic acid in 'non-enzymic peroxidation' and the peroxidation products derived from each fatty-acid side-chain are complex and non-stereospecific.

The term 'enzymic' is also often used to describe microsomal lipid peroxidation. Microsomal fractions from several animal tissues (e.g. liver, kidney, and skin) undergo lipid peroxidation in the presence of NADPH and Fe(III) salts (often added as complexes with ADP, pyrophosphate, and EDTA). An antibody raised against the microsomal flavoprotein NADPH-cytochrome P-450 reductase (Chapter 3) inhibits this peroxidation by more than 90 per cent. During it, cytochromes b_5 and P-450 are attacked, the haem groups being degraded. Apparently the P-450 reductase enzyme, as well as reducing cytochrome P-450, can donate electrons to some Fe(III)-complexes and so generate Fe(II), which stimulates peroxidation.

Again, although an enzyme is involved, microsomal lipid peroxidation in the presence of iron complexes is in principle no different from non-enzymic peroxidation, in that the enzyme is serving only to reduce the iron complexes. It must also be remembered that microsomes are a heterogeneous mixture of vesicles from both endoplasmic reticulum and plasma membrane (Chapter 3). Their peroxidation in the presence of iron ions seems to yield more malonaldehyde than the peroxidation of liposomes or of whole cells (e.g. Table 4.2), for a reason that is not understood but presumably relates to overall composition of fatty-acid side-chains. Another problem in the use of microsomes is the variable amounts of endogenous antioxidants (e.g. vitamin E) that will be present in different preparations. Nuclear membranes contain an electron transport chain (Chapter 3) and undergo peroxidation on incubation with NADPH and Fe(III) salts *in vitro*; and a similar phenomenon has been shown with the inner mitochondrial membrane in the presence of NADH or NADPH. Here iron is reduced by the NADH-coenzyme Q reductase complex. During mitochondrial lipid peroxidation, coenzyme Q is destroyed.

4.4   Acceleration of lipid peroxidation by species other than oxygen radicals

4.4.1   Singlet oxygen $^1\Delta g$

Unlike ground-state oxygen, singlet $O_2{}^1\Delta g$ can react directly with carbon–carbon double bonds by an *ene* reaction to give peroxides (Chapter 2). Thus one possible reaction of singlet O_2 with the double bond between C12 and C13 of linoleic acid (Fig. 4.2) is to produce two hydroperoxides:

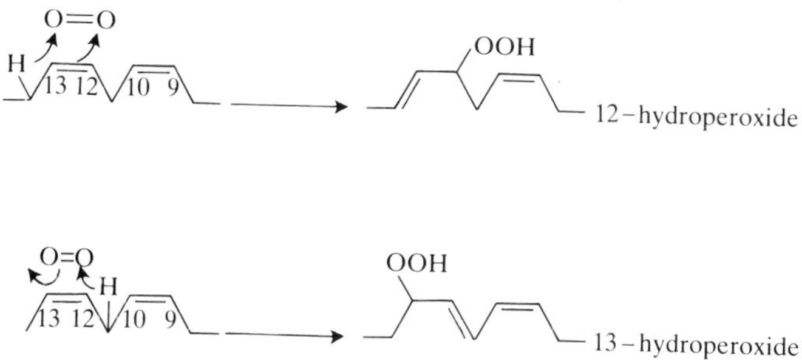

12-hydroperoxide

13-hydroperoxide

9- and 10-hydroperoxides are also produced by reaction of $^1O_2$ with the double bond between carbons 9 and 10. Thus four products result when linoleic acid is exposed to singlet oxygen. By contrast, free-radical peroxidation of linoleate gives mainly, but not exclusively, the 9- and 13-hydroperoxides.

This direct addition of singlet O_2 to membrane lipids can rapidly cause peroxidation. However, the authors think that it is best not to refer to singlet O_2 as *initiating* peroxidation: unless the peroxides are decomposed (e.g. by a metal complex) to give peroxy and alkoxy radicals, the chain reaction of lipid peroxidation will not begin as a result of the above reactions in the presence of singlet O_2. Hence we have reserved the term initiation to cover only H abstraction, leading to a chain reaction.

The lifetime of singlet oxygen in the hydrophobic interior of membranes is much greater than it is in aqueous solution. Illumination of unsaturated fatty acids in the presence of sensitizers of singlet-oxygen formation, such as chlorophyll, rose bengal, methylene blue, bilirubin, or porphyrins, induces rapid peroxide formation. Similar effects have been seen with intact membranes. For example, the rod outer segments in the retina of the frog are rich in polyunsaturated lipids and form peroxides upon illumination, probably due to singlet oxygen formation sensitized by retinal (this is further discussed in Chapter 5). Illumination of phospholipid liposomes containing the porphyrin haematin (Chapter 2) together with cholesterol not only peroxidizes the fatty-acid side-chains but also converts some of the cholesterol into its 5-α-hydroperoxide (Chapter 2), probably by reaction with singlet oxygen. Illumination of erythrocytes in the presence of porphyrins or bilirubin causes peroxidation, aggregation of membrane proteins, inactivation of enzymes and transport carriers, and eventually haemolysis. Fungi of the genus *Cercospora* produce a toxin known as *cercosporin* that can attack plant cells in the light, but not in the dark. It has been suggested that cercosporin sensitizes formation of singlet oxygen, which then generates lipid peroxides in the plant cell, producing disruption of membranes and leakage of ions.

It has been further suggested that singlet oxygen may be *formed* during the complex degradation reactions undergone during lipid peroxidation and might then cause more lipid peroxide formation. A suggested mechanism for this singlet O_2 formation is the reaction of two peroxy radicals together to form a cyclic intermediate which decomposes to give singlet O_2:

$$\diagdown\!\!CHO_2\cdot \; + \; \diagdown\!\!CHO_2\cdot \longrightarrow \; \diagdown\!\!C{=}O \; + \; \diagdown\!\!C{-}OH + {}^1O_2\cdot$$
$$\text{carbonyl}$$

Singlet O_2 formation might account for some of the chemiluminescence that accompanies lipid peroxidation (Chapter 2). Other sources are discussed in Section 4.6.6. Good evidence for singlet O_2 formation during the peroxidation of membrane lipids or during the reaction of artificial organic peroxides with iron complexes (e.g. *tert*-butylhydroperoxide with cytochrome c) has been obtained by showing that the spectrum of the emitted light is consistent with that expected from singlet O_2 production (see Chapter 2 for further details). The use of 'singlet O_2 scavengers' in investigating the exact role played by singlet O_2 in the peroxidation process has been less satisfactory because of their lack of specificity: it has been shown that DABCO, diphenylfuran and β-carotene can react with at least one organic peroxy radical (the trichloromethylperoxy radical, $CCl_3O_2\cdot$). Thus the contribution of singlet O_2 produced during lipid peroxidation to further peroxide formation is uncertain, but seems likely to be small; the self-reaction of peroxy radicals is unlikely to be a very significant reaction until they have accumulated within the membrane, i.e. until peroxidation has been underway for some time.

4.4.2 *Ozone*

Exposure of animals and humans to ozone (O_3) damages the lungs and the damage is associated with lipid peroxidation. Although ozone is not a radical itself (Chapter 1), it can react with a wide variety of organic molecules, including membrane lipids, to produce radical species, and it can thus stimulate lipid peroxidation. For example, exposure of erythrocytes to ozone inactivates membrane-bound enzymes and causes formation of protein aggregates. Reduction of fluidity has been observed in ozone-treated plant cell membranes.

The exact chemistry of ozone-mediated stimulation of lipid peroxidation is unclear. Thus ozone might add on across carbon–carbon double bonds:

$$\begin{array}{cc} H & H \\ | & | \\ R{-}C{=}C{-}R' + O_3 \longrightarrow \end{array} \begin{array}{cc} H & H \\ | & | \\ R{-}C{-}\!\!-C{-}R' \\ | \quad | \\ O{-}O{-}O \end{array}$$

This product could decompose to form radicals
$$\text{R}-\overset{\overset{\displaystyle \text{O}^{\cdot}}{|}}{\underset{\underset{\displaystyle \text{H}}{|}}{\text{C}}}-\overset{\overset{\displaystyle \text{OO}^{\cdot}}{|}}{\underset{\underset{\displaystyle \text{H}}{|}}{\text{C}}}-\text{R}^{1}$$

under suitable conditions, or an aldehyde could be formed:

$$\text{R}-\overset{\overset{\displaystyle \text{H}}{|}}{\underset{\underset{\displaystyle \text{O}-\text{O}-\text{O}}{|}}{\text{C}}}-\overset{\overset{\displaystyle \text{H}}{|}}{\text{C}}-\text{R}^{1} \rightarrow \text{R}-\underset{+}{\text{C}}\,\text{H}-\text{O}-\text{O}^{-} + \text{R}^{1}\text{CHO}$$

4.4.3 *Halogen radicals*

A mixture of the enzyme lactoperoxidase with H_2O_2 and iodide (I^-) ions has been reported to stimulate lipid peroxidation, perhaps by formation of an iodine radical (e.g. the iodine atom, which has one unpaired electron) that could abstract $H^{\cdot}$. Cl^- cannot replace I^- in this reaction. Singlet O_2 production from a lactoperoxidase/H_2O_2/halide system (Chapter 7) has also been reported and might conceivably contribute to the observed peroxidation.

4.5 Peroxidation of other molecules

Membrane lipids are not the only molecules, found in cells, that contain many $\overset{\diagdown}{\diagup}C = C\overset{\diagup}{\diagdown}$ bonds. Several others are present, and they can be peroxidized under appropriate conditions. For example, the aldehyde retinal is probably oxidized as it sensitizes singlet oxygen formation (Chapter 2). Retinal is obtained in the body by the oxidation of *retinol*, which has a $-CH_2OH$ group in place of the $-CHO$ in retinal. Retinol is essential in the human diet, where it is usually known as *vitamin A*, one of the fat-soluble vitamins. Like retinal, vitamin A can undergo peroxidation–exposure of it to Fe(II) salts *in vitro* causes rapid oxidation to a mixture of products. Any excess vitamin A absorbed from the gut is stored in the fat deposits of adipose tissue, but it must be protected against peroxidation. Similarly, several antifungal antibiotics such as candicidin, nystatin, and amphotericin contain conjugated double bonds and are easily oxidized, resulting in loss of the antifungal activity (see Chapter 6, Section 6.11, for further discussion of antibiotics).

The human bloodstream contains several *lipoproteins*—complexes of protein molecules with lipid. For example, the very low density lipoproteins carry fats made in the liver to the adipose tissue. Lipoproteins contain phospholipids, which are susceptible to peroxidation reactions that can, in turn, damage the proteins. Products of lipid peroxidation can be shown to accumulate in plasma stored at 4°C, and peroxidation of the lipids alters the biological properties of the lipoproteins. On storage of plasma at 4°C, the copper-containing protein caeruloplasmin (Chapter 2) appears to break down with release of copper ions, which can aid the peroxidation of lipoproteins. Peroxidation of *low density lipoproteins* has been implicated in the pathogenesis of atherosclerosis (Section 4.9).

4.6 Measurement of lipid peroxidation

A number of techniques are available for measuring the rate of peroxidation of membrane lipids or fatty acids. Each technique measures something different, however, and no one method by itself can be said to be an accurate measure of 'lipid peroxidation'.

4.6.1 *Uptake of oxygen*

Since peroxidation is accompanied by the uptake of oxygen in the formation of peroxy radicals (Fig. 4.6), and also in subsequent decomposition reactions, measurement of the rate of oxygen uptake in an oxygen-electrode is a useful overall index of the progress of peroxidation. Indeed, de Saussure used this method in his pioneering studies on walnut oil, although only a simple mercury manometer was available to him.

Most modern oxygen electrodes are Clark-type arrangements which measure dissolved oxygen. Calibration is made by exposing the electrode to known concentrations of dissolved oxygen, although it should be appreciated that, strictly speaking, the electrode measures the 'activity' of oxygen and not concentration. It is usually assumed that air-saturated pure water has $0.258\,\mu mol/ml$ of dissolved oxygen at 25°C and 1 atmosphere air pressure, although the presence of dissolved solutes will decrease this solubility slightly.

4.6.2 *Measurement of peroxides*

Iodine liberation
Lipid peroxides are capable of oxidizing iodide (I^-) ions into iodine (I_2), which may be estimated by titration with sodium thiosulphate:

$$ROOH + 2I^- + 2H^+ \rightarrow I_2 + ROH + H_2O$$

Although this method can be used in studies upon purified lipids, it is rarely of value in biological systems because of the presence of many other oxidizing agents that can cause iodine production from I^-, such as hydrogen peroxide. The amounts of peroxide present at a given time will obviously depend not only on the rate of initiation of peroxidation but also on how quickly they are decomposing to give other products.

Haem degradation of peroxides

As discussed in Section 4.2.2, isolated haem and the haem moiety of proteins can decompose lipid peroxides with the formation of reactive intermediates. This reaction has been made the basis of several techniques to quantitate peroxides. For example, *microperoxidase*, a haem-peptide formed by proteolytic degradation of cytochrome c, is particularly effective in decomposing peroxides. The radicals produced can be reacted with isoluminol under appropriate conditions to produce light. The sensitivity of the method is high, allowing detection of picomoles (10^{-12} moles) of peroxide. For analysis of biological fluids, HPLC can be used to separate the peroxides for assay; the identity of peroxides separated from biological fluids by HPLC must be confirmed chemically, and not just based on retention times. A simpler approach has been to link the haem-stimulated decomposition of peroxide to a reaction with a redox dye, giving a sensitivity of detection down to 7 nmol (10^{-9} moles) of hydroperoxide.

Glutathione peroxidase

Glutathione peroxidase reacts with H_2O_2 and fatty acid hydroperoxides, simultaneously oxidizing GSH to GSSG (Chapter 3 and Section 4.7.5). In the presence of excess glutathione peroxidase, GSH, and glutathione reductase (to convert GSSG back to GSH at the expense of oxidizing NADPH to $NADP^+$), the rate of NADPH consumption can be related to the peroxide content of the system. Prior addition of catalase can be used to remove H_2O_2. The sensitivity of the method is claimed to be 3 nmol of peroxide per ml. However, glutathione peroxidase will not act on peroxides within membranes (Section 4.7.5), so they must be cleaved out, e.g. by phospholipase treatment, before this assay can be applied.

Cyclo-oxygenase

Cyclo-oxygenase catalyses the first step in prostanoid synthesis and its rate of reaction with arachidonic acid can be increased by the presence of sub-micromolar amounts of lipid peroxides (Chapter 7). Lands *et al.* in the USA have applied this stimulation of cyclo-oxygenase activity to measure trace quantities of peroxides in biological fluids. Enzyme activity is measured by O_2 uptake and the method can detect picomoles of peroxide. One possible problem in these studies is that different lipid peroxides activate cyclo-oxygenase to different extents, so that values obtained for the 'total peroxide' content of a system will depend to some extent upon what peroxides are present. When applied to samples of human plasma, a 'total peroxide concentration' of about 0.5 μM was measured by this method.

4.6.3 Diene conjugation

The oxidation of unsaturated fatty acid side-chains is accompanied by the formation of conjugated diene structures (Fig. 4.6), which absorb ultraviolet light in the wavelength range 230–235 nm (Fig. 4.12). Measurement of this UV absorbance as an index of peroxidation is extremely useful in studies upon pure lipids and it measures an early stage in the peroxidation process, but it often cannot be used directly on biological materials because many of the other substances present, such as haem proteins, chlorophylls, purines, and pyrimidines, absorb strongly in the ultraviolet, and create such a high background that spectrophotometric measurements become grossly inaccurate. Another source of error is that the polyunsaturated fatty acids themselves absorb UV light at only slightly lower wavelengths than the conjugated dienes (Fig. 4.12). The breakdown of lipid peroxides produces several carbonyl compounds that absorb in the ultraviolet, and this must be borne in mind.

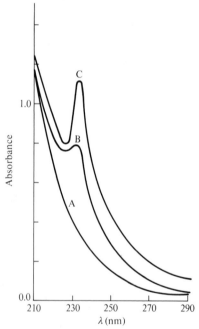

Fig. 4.12. Lipid peroxidation followed by the formation of conjugated dienes. Ethyl linoleate (an ester of linoleic acid and the alcohol ethanol) was purified, and its light absorption plotted at different wavelengths (plot A). Plot B shows the sample after oxidation by air at 30°C for 8 h; and plot C, a sample in which peroxidation had been speeded up by addition of nitrogen dioxide (see Chapter 6). In both cases the 'shoulder' of UV-absorbance due to conjugated diene formation is clearly visible. By courtesy of Prof. W. Pryor.

Corongiu *et al.* in Italy have achieved greater sensitivity for the diene conjugation method by applying second-derivative spectroscopy to lipid systems. A spectrum plots absorbance (A) against wavelength (λ). A first-derivative spectrum plots *rate of change of absorbance with wavelength* ($dA/d\lambda$) against wavelength. The second derivative spectrum plots the rate of change of this rate of change ($d^2A/d^2\lambda$). Figure 4.13 shows the appearance of the first and second derivatives of an absorption band. Fig. 4.14 shows the application of this technique to linolenic acid, and it may be seen that the change in the second-derivative spectrum upon oxidation of the fatty acid gives much more information than the change in the simple UV-absorption spectrum. Thus the 'hump' that appears in the UV-absorption spectrum as a result of peroxidation translates into a sharp minimum 'peak' at 233 nm in the second-derivative spectrum (Fig. 4.14). The height of this peak is a good measure of the conjugated dienes present. The increased resolution of this technique may allow discrimination between different conjugated diene structures present, by observing slight changes in the second-derivative spectrum.

Although conjugated diene methods have often been used to study peroxidation in animal body fluids and tissue extracts, their application to human body fluids has produced very serious problems. Dormandy *et al.* in

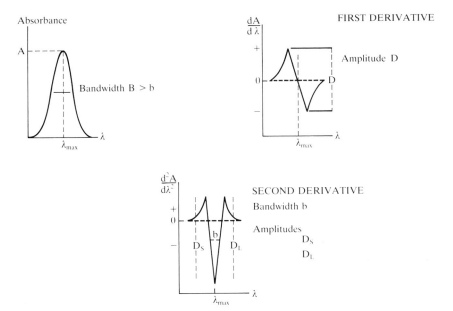

Fig. 4.13. Principal features of the first and second derivatives of an idealized absorption band. Diagram by courtesy of Philips Analytical.

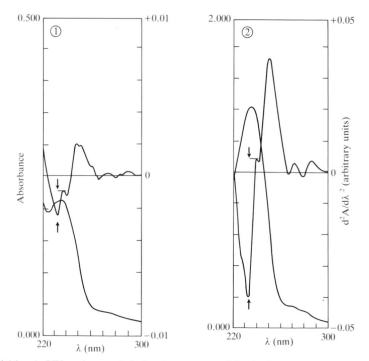

Fig. 4.14. 1. UV and second-derivative spectra of linolenic acid ('pure' but slightly oxidized). The arrows show in the second-derivative spectra the height of the minimum peak at $\lambda = 233$ nm that is expressed as $d^2A/d\lambda^2$ in absorbance arbitrary units. 2. UV and second-derivative spectra of linolenic acid after 24 hours of oxidation. The arrows show the height of the peak $d^2A/d\lambda^2$ at $\lambda = 233$ nm. (See also legend of 1.) Diagram reproduced by permission of Professor F. Corongiu and Elsevier (see *Chem-Biol. Interact.* **44**, 289).

England used HPLC techniques to separate the UV-absorbing 'diene conjugates' from human body fluids. They found that the UV-absorbing material consisted almost entirely of a non-oxygen-containing isomer of linoleic acid, octadeca-9(*cis*)11(*trans*)-dienoic acid. They proposed that this compound is produced by reaction of the carbon-centred radicals, obtained when H is abstracted from linoleic acid, with protein. Such a reaction is certainly possible, but examination of reaction rate constants strongly suggests that reaction with oxygen is the preferred fate of carbon-centred radicals (Fig. 4.6), except at very low O_2 concentrations. In any case, peroxidation of biological membranes produces carbon-centred radicals from several fatty acids, not only linoleic acid, and would not be expected to give only a 9(*cis*)11(*trans*) isomer. Also, this UV-absorbing product is not found in the plasma of animals subjected to oxidant stress (e.g. rats given the hepatotoxin bromotrichloromethane, a potent inducer of lipid peroxidation [Chapter 6]). Thus Thompson and Smith in the USA have argued that octadeca-9,11-dienoic acid is most unlikely to arise by lipid peroxidation and may be ingested in food and produced by the metabolism of gut bacteria.

It follows that application of any diene conjugation methods to human body fluids, or to extracts of them, is probably not measuring lipid peroxidation, although diene conjugation methods still seem applicable to animal body fluids. Despite this problem, detailed studies by Dormandy and others have shown that octadeca-9,11-dienoic acid *may* be a useful marker of disease activity.

4.6.4 *Measurement of hydrocarbon gases*

This technique, developed by Riely, Cohen, and Lieberman in the USA, is based on the formation of hydrocarbon gases such as pentane and ethane during lipid peroxidation (Section 4.2). These gases can be easily measured by the technique of gas–liquid chromatography. It must be emphasized that they are minor end-products of peroxidation, and their formation depends on the presence of transition-metal ions to decompose peroxides. Hence an increased rate of gas production might reflect increased availability of such metal ions rather than increased initiation of peroxidation. It has also been reported that, at least in microsomal systems, the formation of hydrocarbons is affected by O_2 concentrations, being favoured at low O_2 concentrations. This may be because the carbon-centred radicals that lead to hydrocarbon production (e.g. the pentane radical $CH_3(CH_2)_3\dot{C}H_2$; Section 4.2) can also react with O_2 to form peroxy radicals. Thus O_2 concentration at the site of peroxidation is another variable that must be considered. Some hydrocarbons produced by lipid peroxidation are metabolized in the liver, e.g. pentane is hydroxylated by cytochrome P-450. This can introduce a further source of error. For example, drugs that affect liver metabolism might alter pentane exhalation and so be thought to be altering the role of lipid peroxidation *in vivo*.

Bearing all these problems in mind, the value of the hydrocarbon gas exhalation technique is that formation of these gases can be measured in the expired breath of whole animals and of humans, i.e. this is potentially an assay for lipid peroxidation *in vivo*. The expired breath is passed through an adsorbent at low temperature to adsorb and concentrate the hydrocarbons, which are then desorbed and assayed in a gas chromatograph. The greatest care must be taken in experiments of this kind to control for hydrocarbon production from the bacteria always present on the skin (and fur) and in the gut. It is quite possible to ignite the flatulence from cows because of the rich variety of hydrocarbons present. 'Normal' air in large cities is contaminated with hydrocarbons from motor vehicles, creating a high 'background', so that the air breathed by animals or human subjects must first be purified. In fact, the major hydrocarbon in expired human breath is isoprene, which is probably formed during biosynthesis of cholesterol. Increased production of pentane from human volunteers during severe exercise has been reported, and injection of cumene hydroperoxide into animals increases their ethane production. *Tert*-butyl hydroperoxide has been shown to increase ethane production in the isolated rat liver.

4.6.5 *Loss of fatty acids*

Lipid peroxidation results in the ultimate destruction of unsaturated fatty-acid side-chains, so it is possible to measure the overall rate of the process by measuring the loss of each fatty acid. The membrane under study must be disrupted and the lipids hydrolysed to release the fatty acids, which can then be measured by HPLC or chemically converted into volatile products (e.g. by formation of esters with methanol) and separated by gas–liquid chromatography. Great care must be taken to avoid peroxidation of the fatty acids during the hydrolysis and extraction procedures, e.g. by carrying out the reactions under nitrogen gas. Additional information can be gained by separating the different classes of membrane lipids (Fig. 4.1) before hydrolysis to release the fatty acids. Table 4.3 shows the effect of lipid peroxidation on the erythrocyte membrane. The rapid loss of polyunsaturated fatty acids is obvious.

4.6.6 *Light emission*

As we saw in Chapter 3, east European and American scientists have pioneered the study of chemiluminescence as an index of radical reactions taking place in intact animal organs, either perfused or *in situ*.

Table 4.3. Loss of fatty acids during peroxidation of the red blood cell membrane

Number of carbon atoms in fatty acid	Number of $\diagup$C$=$C$\diagdown$ bonds in fatty acid	Percentage of total fatty acids in normal membrane	Percentage after lipid peroxidation
16	0	21	21
18	0	14	14
18	1	12	11
18	2	10	8
20	4	15	5
22	0	3	3
22	4	2	1

Stimulation of peroxidation in isolated membrane systems or in perfused animal organs (e.g. by infusing *tert*-butyl hydroperoxide) is accompanied by increased light emission. Hence one source of light *in vivo* is the peroxidation of membrane lipids, although basal light emission is not solely a measure of lipid peroxidation (Chapter 3).

Some of the light produced in peroxidizing lipid systems arises when two peroxy radicals meet:

$$\begin{array}{cccc}
\diagdown\!\!\!\diagup\text{CHO}_2\!\cdot + \diagdown\!\!\!\diagup\text{CHO}_2\!\cdot \longrightarrow & \diagdown\!\!\!\diagup\text{C}{=}\text{O} + & \diagup\!\!\!\diagdown\text{C}{-}\text{OH} + {}^1\text{O}_2\!\cdot \\
 & \text{carbonyl} &
\end{array}$$

This type of reaction is often called a *Russell-type mechanism*. Alternatively, the carbonyl compound produced might be in the excited state and emit light as it decays to the ground state (Chapter 3). Other possible sources of light could be self-reaction of alkoxy radicals to give excited-state ketones:

$$\diagdown\!\!\!\diagup\text{CHO}\cdot + \diagdown\!\!\!\diagup\text{CHO}\cdot \longrightarrow \diagdown\!\!\!\diagup\text{C}{=}\text{O}^* + \diagup\!\!\!\diagdown\text{C}{-}\text{OH}$$

and the formation of a dioxetane by reaction of singlet O_2 with unsaturated fatty-acid side-chains: dioxetanes can decompose to give excited-state carbonyl compounds (Chapter 2):

$$\diagdown\!\!\!\diagup\text{C}{=}\text{C}\diagup\!\!\!\diagdown + {}^1\text{O}_2 \longrightarrow -\overset{|}{\underset{|}{\text{C}}}-\overset{|}{\underset{|}{\text{C}}}- \longrightarrow \diagdown\!\!\!\diagup\text{C}{=}\text{O} + \diagdown\!\!\!\diagup\text{C}{=}\text{O}^*$$
$$\qquad\qquad\qquad\quad \text{O}{-}\text{O}$$

Measurements of light emission in isolated cells or membranes are reasonably correlated with results from other techniques of measuring lipid peroxidation, except that the peak light emission often occurs slightly later than the 'peak' recorded by other techniques. This is because maximum light production by the Russell-type mechanism will tend to occur during the later stages of lipid peroxidation, when sufficient peroxy radicals are present in the membrane to allow frequent collision between them.

4.6.7 *Measurement of fluorescence*

Fluorescent pigments partly derived from lipids are known to accumulate in some tissues as a function of age. These so-called *age-pigments* are complexes of lipid oxidation products with protein, and more will be said about them in Chapter 8.

The reaction of carbonyl compounds, such as malonaldehyde, with side-chain amino groups of proteins, free amino acids, phosphatidylethanolamine

(Fig. 4.1), the related lipid phosphatidylserine ($-\overset{\overset{\displaystyle COO^-}{|}}{CH}\overset{+}{N}H_3$ instead of

$-CH_2\overset{+}{N}H_3$), or with nucleic acid bases, produces products known as *Schiff bases.*

$$R_1-C\overset{\displaystyle H}{\underset{\displaystyle O}{<}} + R_2-NH_2 \xrightarrow[\text{reaction}]{\text{chemical}} R_1-C\overset{\displaystyle H}{\underset{\displaystyle N-R_2}{<}}$$

aldehyde   amino
            compound   Schiff base

Malonaldehyde, with two carbonyl groups, can cross-link two amino com-pounds to produce fluorescent molecules that have the general formula R_1- $N{=}CH-CH{=}CH-NH-R_2$ where R_1 and R_2 are the compounds to which the amino groups are attached. The products are called *aminoiminopropene Schiff bases* (Fig. 4.15) and they form readily at acidic pH values. However, their significance in the total formation of fluorescent products in peroxidiz-ing membranes is unclear, since several recent reports suggest that MDA is only a minor contributor to the formation of fluorescent products. Other aldehydes, such as 4-hydroxynonenal, may be far more important; 4-hydroxynonenal, of course, has only a single carbonyl group.

Further, Japanese scientists have questioned the role of Schiff bases in the formation of fluorescence accompanying lipid peroxidation at neutral pH values. They have proposed that dihydropyridines are the major fluorescent products formed. For example, in the case of MDA, a possible product is a 1,4 disubstituted 1,4 dihydropyridine 3,5 dicarbaldehyde.

$$\overset{\displaystyle H}{\underset{\displaystyle O}{>}}C-CH_2-C\overset{\displaystyle H}{\underset{\displaystyle O}{<}} \xrightarrow[\text{pH 7}]{\begin{array}{c}RNH_2\\ \text{(amino compound)}\end{array}}$$
MDA

4-methyl 1-substituted 1,4 dihydropyridine
3,5 dicarbaldehyde

If both MDA and an aldehyde with only one carbonyl group are present, the reaction may be written:

$$MDA + R_1CHO \xrightarrow[\text{pH7}]{RNH_2}$$

A common property of aldehydes is their *polymerization*—several molecules joining together to form a larger molecule. This can happen with malonaldehyde, and the polymeric products are fluorescent (Fig. 4.15). Autoxidizing polyunsaturated fatty acids develop fluorescence even though no compounds with—NH_2 groups are present, which may be due to polymers from malonaldehyde and other aldehydes.

Thus, as summarized in Fig. 4.15, mechanisms for the formation of fluorescent products are complex. Despite this, fluorescence is a very sensitive method for measuring peroxidation and the results obtained correlate reasonably well with those of other methods, although it is, of course, a measurement of a late stage in the peroxidation process. The results of fluorescence studies are usually expressed in RFI (relative fluorescence

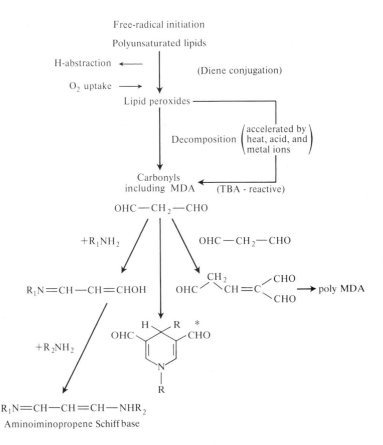

Fig. 4.15. Formation of fluorescent products during lipid peroxidation. The starred product (*) is a 1,4 disubstituted 1,4 dihydropyridine 3,5 dicarbaldehyde.

intensity) units by comparison with a stable standard solution, such as quinine sulphate in dilute sulphuric acid, or tetraphenylbutadiene. One problem that must be considered in applying the technique to biological membranes or cell extracts is that fluorescence can also develop in both protein and in DNA subjected to oxidant attack, e.g. by exposure to metal ions and H_2O_2. Thus it is not exclusively a lipid phenomenon.

4.6.8　*The thiobarbituric acid (TBA) test*

The TBA test is one of the oldest and most frequently used tests for measuring the peroxidation of fatty acids, membranes, and food products. It is the easiest method to use (the material under test is merely heated with thiobarbituric acid under acidic conditions, and the formation of a pink colour measured at about 532 nm) and it can be applied to crude biological systems. Unfortunately, the simplicity of the test has led many scientists into using it as an index of peroxidation without understanding exactly what it measures.

Small amounts of 'free' malonaldehyde are formed during the peroxidation of most membrane systems; somewhat greater amounts have been detected in peroxidizing rat liver microsomes (Table 4.2). MDA can react in the TBA test to generate a coloured product, probably by the reaction below.

In acid solution the product absorbs light at 532 nm and fluoresces at 553 nm, and it is readily extractable into organic solvents such as butan-1-ol. Construction of a calibration curve for the assay is complicated by the fact that malonaldehyde is unstable and must therefore be prepared immediately before use by hydrolysing its derivatives, 1,1,3,3-tetramethoxypropane or 1,1,3,3-tetraethoxypropane:

$$RO\diagdown CH—CH_2—CH \diagup OR$$
$$RO\diagup \qquad\qquad \diagdown OR$$

R = C$_2$H$_5$ (ethyl)
or CH$_3$ (methyl).

In studying the biological properties of malondialdehyde formed in this way it is essential to ensure complete hydrolysis of these derivatives and to bear in mind that the solution will contain four molecules of ethanol (or methanol) per molecule of malonaldehyde. For example, some reports that malonaldehyde is carcinogenic probably originated from the biological activities of partially hydrolysed derivatives of the above compounds.

Because the TBA test is calibrated with malonaldehyde (MDA), the results are often expressed in terms of the amount of MDA produced in a given time. This has sometimes given the impression that the TBA test detects only free MDA, and so measures the amount of free MDA in the peroxidizing lipid system. Thus the molar extinction coefficient of the MDA–TBA adduct (1.54×10^5 at 532 nm) is often used to calculate the amount of 'MDA formed'. However, the amount of free MDA produced in most peroxidizing lipid systems is extremely low and would be insufficient to give a substantial colour yield. Indeed, it was shown as long ago as 1958, in studies with peroxidizing fish oil, that 98 per cent of the MDA that reacts in the TBA test was not present in the sample assayed but formed by decomposition of lipid peroxides during the acid heating stage of the TBA assay.

Peroxidizing fatty acids with only one or two $>$C$=$C$<$ bonds give little pink colour in the normal TBA assay. Formation of MDA has been suggested by Pryor in the USA to be due to the formation of cyclic peroxides and endoperoxides that undergo fragmentation (Figs. 4.6 and 4.7). Fragmentation requires the presence of iron salts in the reagents used in the test (Table 4.4). Indeed, removal of such iron salts prevents TBA-reactivity from developing. This can lead to artefacts in studies of the action of metal chelating agents on lipid peroxidation; they will also affect colour development in the TBA test itself.

Several compounds other than MDA give products that absorb at, or close to, 532 nm on heating with TBA; Table 4.5 lists some of these. Simple measurement at 532 nm after a TBA test could therefore include contributions from these substances, although fluorescence measurements can distinguish the products they form from the 'real' TBA–MDA adduct. An alternative approach, to achieve greater specificity, is to separate the TBA–MDA adduct from the reaction mixture before measurement. This is best done by modern HPLC methods. Even so, it must be noted that exposure of several carbohydrates and amino acids to hydroxyl radicals, produced by ionizing radiation or metal-ion/H$_2$O$_2$ systems, yields products that do give a genuine TBA–MDA adduct on heating with TBA (Table 4.5). The exact amount of colour developed with these alternative compounds depends on

Table 4.4.   The importance of iron in the TBA test

	TBA reactivity at pH 3.5 of 0.19 mM linolenic acid peroxide	
	A_{532} nm	Percentage inhibition or stimulation
Reagents not Chelex-treated	0.250	
Reagents Chelex-treated	0.047	81% Inhibition
Chelex treatment + FeCl$_3$ (30 nM)	0.080	68% Inhibition
Chelex treatment + FeCl$_3$ (300 nM)	0.096	62% Inhibition
Chelex treatment + FeCl$_3$ (3 μM)	0.130	48% Inhibition
Chelex treatment + FeCl$_3$ (7 μM)	0.215	14% Inhibition
Chelex treatment + FeCl$_3$ (70 μM)	0.281	12% Stimulation
Iron concentration in the TBA reagents measured by the bleomycin–iron method	14 μM	

Linoleic acid hydroperoxide was heated with TBA at acid pH. Removal of iron salts from the reagents by Chelex treatment markedly decreased its reactivity in the TBA assay, presumably by preventing the peroxide from decomposing to MDA. See J. M. C. Gutteridge and G. J. Quinlan, *J. App. Biochem.* **5**, 293–9 (1983).

the type and strength of acid used in the TBA test, and on the time of heating. Application of the TBA assay to human body fluids will also measure endoperoxides produced enzymically by the prostaglandin synthesis pathway (Chapter 7). The lack of specificity of the TBA assay when applied to human plasma is shown by the work of Lands *et al.*; by their specific enzymic method (Section 4.6.2) they measured a mean peroxide content in human plasma of around 0.5 μM, whereas expression of the results from a TBA test in terms of 'peroxide equivalents' gave a mean value of 38 μM. For example, bile pigments in plasma react with TBA to produce coloured products.

Finally, it must be pointed out that any free MDA that is formed *in vivo* will be rapidly metabolized. For example, it seems to be a substrate for mitochondrial aldehyde dehydrogenase activity, being oxidized into malonic acid that is then decarboxylated to carbon dioxide and acetate. When radioactive (^{14}C) MDA is administered orally to rats, 60–70 per cent of it is excreted as ^{14}CO$_2$. However, 9–17 per cent of the radioactivity appears in urine, which has long been known to contain several substances that react to form chromogens on heating with TBA. Draper in Canada purified one metabolite of MDA from rat urine and identified it as arising by reaction of MDA with the amino acid lysine, eventually being excreted as N^{α}-acetyl-ϵ

Table 4.5. TBA reactivity of various molecules before and after radical damage

Compound	Damaging system (if any)	TBA chromogen (wavelength of maximum absorption, nm)
Compounds forming chromogens on heating with TBA		
Biliverdin	not required:	TBA-? A 532, 560
Acetaldehyde-sucrose	Direct reaction	TBA-? A 532
β-Formylpyruvic acid	with TBA	TBA-? A 510, 550
Glyoxal		TBA-? A 522, 550
Compounds forming yellow chromogens (430–450 nm) on heating with TBA		
Streptomycin	not required:	—
Aldehydes e.g. hydroxymethyl-furfuraldehyde	Direct reaction	—
Some carbohydrates	with TBA	—
Lipids		
PUFA	$Fe,Cu,OH^{\cdot},LO^{\cdot},LO_2^{\cdot}$	MDA–TBA A 532
Carbohydrates		
Deoxyribose	$Fe,Cu,OH^{\cdot}$	MDA–TBA A 532
Deoxyglucose	$Fe,Cu,OH^{\cdot}$	MDA–TBA A 532
Deoxygalactose	$Fe,Cu,OH^{\cdot}$	MDA–TBA A 532
Ascorbate	$Cu,OH^{\cdot}$	TBA-? A 550
Amino acids		
Glutamic acid	$Fe,OH^{\cdot}$	MDA–TBA A 532
Methionine	$Fe,OH^{\cdot}$	MDA–TBA A 532
Homocysteine	$Fe,OH^{\cdot}$	MDA–TBA A 532
Aminobutyric acid	$Fe,OH^{\cdot}$	MDA–TBA A 532
Proline	$Fe,OH^{\cdot}$	MDA–TBA A 532
Arginine	$Fe,OH^{\cdot}$	MDA–TBA A 532

Table 4.5.—*continued*

Compound	Damaging system (if any)	TBA chromogen (wavelength of maximum absorption, nm)
Nucleic acids and nucleosides		
d-Adenosine	Fe,OH$^{\cdot}$	MDA–TBA A 532
d-Inosine	Fe,OH$^{\cdot}$	MDA–TBA A 532
DNA	Fe-bleomycin, 'iron-oxygen species'	MDA–TBA A 532
DNA	Cu-phenanthroline, OH$^{\cdot}$	MDA–TBA A 532
DNA	Fe,OH$^{\cdot}$ (weak)	MDA–TBA A 532
DNA	Cu-rifamycin, OH$^{\cdot}$	MDA–TBA A 532
DNA	Cu-β-lactams, OH$^{\cdot}$	MDA–TBA A 532
Aromatic carboxylic acids		
Benzoate	Fe,OH$^{\cdot}$	MDA–TBA A 532
Picolinate	Fe,OH$^{\cdot}$	MDA–TBA A 532
Antibiotics		
Polyene antifungals	Fe,LO$^{\cdot}$,LO$_2^{\cdot}$	MDA–TBA A 532
Rifamycin	Cu,OH$^{\cdot}$	MDA–TBA A 532
Cephalosporin c		MDA–TBA A 532
Cefazolin	Cu,OH$^{\cdot}$	MDA–TBA A 532

*MDA–TBA, indicates that the absorption and fluorescence spectra were indistinguishable from authentic MDA–TBA adducts.
Data abstracted from J. M. C. Gutteridge, *Adv. Biosci.* **64**, 1–22 (1987); also see *Biochem. Pharmacol.* **36**, 3629 (1987) and *Free Radical Res. Commun.* **5**, 149–58 (1988).

(2-propenal) lysine. This product may be formed when MDA generated *in vivo* reacts with amino groups in proteins. It has also been detected in human urine. It is certainly possible that measurement of this product could serve as a crude index of lipid peroxidation *in vivo*, although dietary lipids might also contain protein-bound MDA.

4.6.9 *Measurement of aldehydes other than MDA*

There are many 'end products' of peroxidation in addition to hydrocarbon gases and MDA (Table 4.2) and these could be measured as well. Thin-layer chromatography has allowed separation and identification of a wide variety of carbonyl compounds and other products; and carbonyl compounds can also be separated and assayed by gas–liquid chromatography, sometimes after reaction with *2,4-dinitrophenylhydrazine* to form products known as *2,4-dinitrophenylhydrazones*. Thus Dratz *et al.* in the USA have developed a sensitive method for measuring hydroxyalkenals by converting them into volatile products that can be separated by gas–liquid chromatography and identified by mass spectrometry. Esterbauer's group in Austria have described a sensitive HPLC method for the measurement of 4-hydroxynonenal. Measurement of hydroxyalkenals is probably more meaningful than measurements of free MDA, since the 4-hydroxyalkenals are far more cytotoxic than MDA. Again, however, it must be pointed out that these aldehydes arise by peroxide decomposition, which can often be controlled by the availability of metal complexes to decompose peroxides rather than by the rate of peroxide formation. Further, these reactive aldehydes, like MDA, can be metabolized by cells, e.g. 4-hydroxynonenal is a substrate for some glutathione transferase enzymes (Section 4.7).

4.6.10 *Summary—which is the method of choice?*

The simple answer is 'none of them'. Each assay measures something different. Diene conjugation tells one about the early stages of peroxidation, as does direct measurement of lipid peroxides. In the absence of metal ions to decompose lipid peroxides there will be little formation of hydrocarbon gases, carbonyl compounds, or their fluorescent complexes, which does not necessarily mean therefore that nothing is happening. Whereas most scientists studying lipid peroxidation in isolated membrane systems add an excess (50–200 μM) of iron salt or iron complexes, the availability of metal ions to decompose lipid peroxides *in vivo* is usually very limited (Chapter 3 discusses results on the amount of redox-active 'bleomycin-detectable' iron in human body fluids, which always show concentrations less than 5 μM except in iron-overload conditions). Thus the concentration of metal ions may limit the rate of decomposition of lipid peroxides *in vivo*.

Even if peroxides do not decompose, the TBA test can still detect them because of decomposition of peroxides during the assay itself. Note that traces of iron salts must be present in the reagents to allow this, however. Analysis of hydrocarbon gases is a measure of a minor side-reaction. Changes in the mechanism of peroxide decomposition might conceivably alter the amounts produced without any change in the overall rate of peroxidation.

Whatever method is chosen, one should think clearly *what* is being measured and *how* it relates to the overall lipid peroxidation process. Whenever possible, two or more different assay methods should be used.

4.7   Protection against lipid peroxidation

4.7.1   *Protection by structural organization of the lipids*

Unsaturated fatty acids dispersed in organic solvents, or with detergents in aqueous solution, can be peroxidized easily. Similarly, phospholipids in simple micellar structures (Fig. 4.3) are peroxidized more rapidly than they are in lipid bilayers. Ingold in Canada has suggested that this may occur because lipid peroxy radicals are more polar than other hydrocarbon side-chains, and tend to move away from the membrane interior towards the surface of the bilayer, hence reducing the efficiency of initiation. The presence of cholesterol in cell surface membranes can influence their susceptibility to peroxidation, probably both by intercepting some of the radicals present and by affecting the internal structure of the membrane by interaction of its large hydrophobic ring structure with fatty-acid side-chains. Figure 4.16 shows this effect in phospholipid liposomes. As lipid peroxidation proceeds in any membrane, several of the products produced have a detergent-like activity, especially released fatty acids or phospholipids with one of their fatty-acid side-chains removed (*lysophospholipids*). This will contribute to increased membrane disruption and further peroxidation. It is especially true of chloroplast membranes, in which peroxidation activates enzymes that hydrolyse membrane lipids (Chapter 5).

The binding of positively charged species to a membrane (probably to the negatively-charged head-groups of phospholipids) can substantially alter the susceptibility of the membrane to oxidative damage. This can be seen as either an enhancement or an inhibition of the rate of lipid peroxidation. Several metal ions such as Ca^{2+}, Co^{2+}, Cd^{2+}, Al^{3+}, Hg^{2+}, and Pb^{2+} have been shown to alter the rate of peroxidation of liposomes, erythrocytes, and microsomal membranes, often stimulating peroxidation induced by addition of iron ions (e.g. Table 4.6 shows how Al^{3+} and Pb^{2+} can stimulate peroxidation of erythrocytes induced by adding 10 mM H_2O_2 in the presence

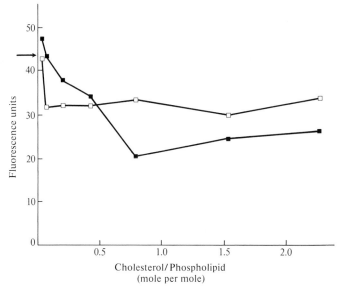

Fig. 4.16. Effect of cholesterol and cholesterol acetate on the rate of peroxidation of ox-brain phospholipid liposomes. Cholesterol (black square-points) or its ester with acetic acid (open squares) was incorporated into liposomes in various amounts during their formation, and peroxidation (induced by iron salt plus ascorbate) was measured by a fluorescence method. Small amounts of cholesterol promote peroxidation, but larger amounts depress it. Arrow indicates peroxidation control with no cholesterol added. Data from J. M. C. Gutteridge, *Res. Comm. Chem. Pathol. Pharmacol.* **22,** 563–71 (1978).

Table 4.6. Action of Pb^{2+}, Cd^{2+}, and Al^{3+} salts on peroxidation of human erythrocytes. Erythrocytes were incubated with H_2O_2 in the presence of azide to inhibit catalase and, where indicated, 400 μM (final concentrations) of Al^{3+}, Cd^{2+}, or Pb^{2+} were added. Peroxidation was measured by the TBA method. A, B, and C are three different healthy adult male subjects. Data from *Biochim. Biophys. Acta* (1988) **962,** 196–200. Metal ions in the absence of H_2O_2 did not cause any peroxidation.

Metal added to reaction mixture	Extent of peroxidation (A_{532})		
	Subject A	Subject B	Subject C
None (H_2O_2 only)	0.164	0.149	0.109
Al^{3+}	0.328	0.294	0.179
Pb^{2+}	0.219	0.367	0.159
Cd^{2+}	0.123	0.174	0.106

of the catalase inhibitor azide). Other charged molecules, such as Zn^{2+}, Mn^{2+} or polyamines may be able to displace iron ions from phospholipid binding sites and inhibit lipid peroxidation under certain circumstances.

The susceptibility of a membrane to peroxidation can be altered by changing the lipid composition of that membrane. Studies have shown that the ratio of phospholipids to cholesterol molecules (Fig. 4.16), the type of phospholipid present, and the fatty acids they contain all contribute to the structural integrity of the membrane. Any subtle changes in these variables, as occurs during tissue damage, renders the membrane more susceptible to oxidant stress—an important point to consider when interpreting the role of lipid peroxidation in disease processes (Section 4.9).

4.7.2   *Protection by 'antioxidants': general principles*

The term 'antioxidant' is frequently used in the biomedical literature, but it is rarely defined. Often, the term is implicitly restricted to chain-breaking antioxidants, such as vitamin E (see below). However, the authors take a much broader view and define an antioxidant as 'any substance that, when present at low concentrations compared to those of an oxidizable substrate, significantly delays or inhibits oxidation of that substrate'. Antioxidants can act at different levels in the oxidative sequence. As far as lipid peroxidation is concerned, they could act by:

1. *Decreasing localized O_2 concentrations* (e.g. sealing of foodstuffs under nitrogen).
2. *Preventing* first-chain initiation by scavenging initiating radicals such as $OH^{\cdot}$.
3. *Binding metal ions* in forms that will not generate such initiating species as $OH^{\cdot}$, ferryl, or $Fe^{2+}/Fe^{3+}/O_2$ and/or will not decompose lipid peroxides to peroxy or alkoxy radicals.
4. *Decomposing peroxides* by converting them to non-radical products, such as alcohols.
5. *Chain breaking*, i.e. scavenging intermediate radicals such as peroxy and alkoxy radicals to prevent continued hydrogen abstraction. Chain-breaking antioxidants are often phenols or aromatic amines.

Antioxidants acting by mechanisms 1, 2, and 3, can be called *preventative antioxidants*. Those acting by mechanism 3 are not usually consumed during the course of the reactions. Antioxidants of the fourth type are also preventative antioxidants, but they may or may not be consumed during the reaction, depending on their chemical behaviour (e.g. glutathione peroxidase acts by this mechanism and being an enzyme, is a catalyst and is not consumed). Chain-breaking antioxidants, acting by combining with the intermediate radicals, will be consumed, as will antioxidants of type 2 above.

It should be stressed that many antioxidants have multiple mechanisms of action, as we shall see. In addition, rapid repair of oxidative damage (e.g. cleavage of peroxidized fatty acids from membrane lipids) will tend to diminish the rate of lipid peroxidation; repair mechanisms for damage to DNA and proteins have already been described in Chapter 3. Thus repair reactions might be classified as antioxidant mechanisms.

4.7.3 *Vitamin E*

Vitamin E was first reported in 1922 by Evans and Bishop in the USA, who demonstrated that a fat-soluble nutritional factor is essential for normal reproduction in rats. Purification of this 'factor' revealed that it is a closely-related family of compounds, the *tocopherols* (Table 4.7). Four tocopherols are known, but *a-tocopherol* is the most biologically-important and the terms 'a-tocopherol' and 'vitamin E' are now used in the literature almost interchangeably. (This is, strictly speaking, incorrect because the other tocopherols do have some antioxidant activity.) Later work showed that vitamin E is essential in the diet of other animals: lack of it causes a wide variety of symptoms including sterility in male rats, dogs, cocks, rabbits, and monkeys, haemolysis in rats and chicks, muscular degeneration in rabbits, guinea pigs, monkeys, ducks, mice, and minks, 'white-muscle disease' in lambs and calves, and cerebellar degeneration in chicks. By contrast, short-term absence of vitamin E from the human diet does not cause any specific deficiency disease, although there is some evidence that lack of it in premature babies can predispose to haemolytic anaemia, probably due to increased fragility of the erythrocyte membrane (see below). There is no rigorous evidence that human muscular dystrophy or multiple sclerosis respond in any way to vitamin E administration.

Vitamin E is a fat-soluble molecule whose structure is shown in Fig. 4.17. Being hydrophobic, it will tend to concentrate in the interior of membranes. For example, the mitochondrial membranes contain about one molecule of a-tocopherol per 2100 molecules of phospholipid, but there is much more vitamin E than this in the chloroplast thylakoid membrane and in the outer segment membranes of the retinal rods. Both those membranes must be especially protected against peroxidation *in vivo*. a-Tocopherol is also concentrated in blood lipoproteins and in the adrenal glands, where both the cortex and medulla are susceptible to oxidative damage because of their high content of oxygenase enzymes, which can sometimes 'leak' radicals from the active site during the catalytic cycle.

Since the damaging effects caused by lack of vitamin E in animals are partially or completely alleviated by feeding synthetic antioxidants, it seems

Table 4.7. Antioxidant inhibitors of peroxidation

Name	Structure	Comments
Tocopherols	α-T = R_1, R_2, R_3, all CH_3 β-T = R_1, R_3, CH_3; R_2, H γ-T = R_1, R_2, CH_3; R_3, H δ-T = R_1, CH_3; R_2, R_3, H.	β-, γ-, δ-Tocopherols less good at reacting with peroxy radicals than α-tocopherol; less well absorbed in gut ($\alpha > \gamma > \beta \geqslant \delta$) (thus less important). (See *J. Am. chem. Soc.* (1981) **103**, 6472–7.) Conversion of the ring structure to improves the antioxidant capacity of α-tocopherol *in vivo* and *in vitro* (*FEBS Lett.* **205**, 117 [1986]).
Butylated hydroxyanisole (BHA)		Very often added to foodstuffs. Acts as an antioxidant by hydrogen donation, which is common to all the phenolic (and amine) antioxidants. For example, addition of **BHA** to fat (e.g. butter) increases its storage life from a few months to a few years.
Butylated hydroxytoluene (BHT)		Very often added to foodstuffs. Non-toxic, but some evidence for metabolism by liver cytochrome P-450 system. Very large doses cause liver damage in mice. Detected in the body fat of US citizens.

Trolox

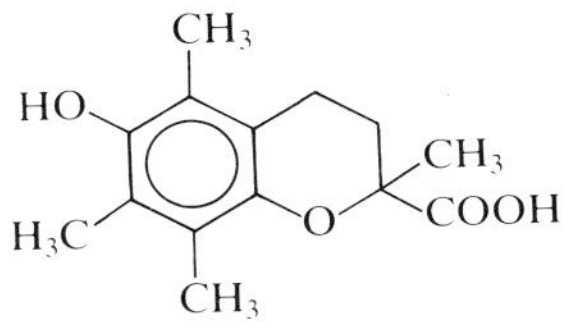

Water-soluble form of α-tocopherol; the hydrophobic side-chain is replaced by a hydrophilic —**COOH** group. Good scavenger of peroxy and alkoxy radicals, giving a Trolox radical that can be 'repaired' by ascorbate.

Propyl gallate

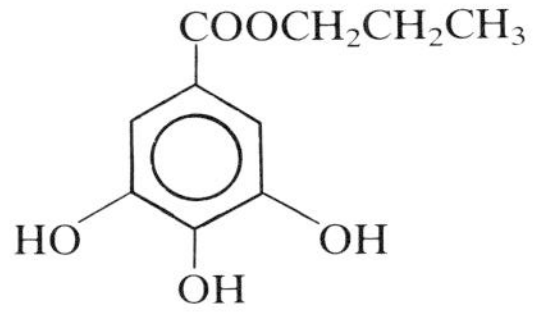

Fairly water-soluble. Good inhibitor of lipid peroxidation. Often added to foodstuffs. Binds iron ions.

Nordihydroguaiaretic acid (NDGA)

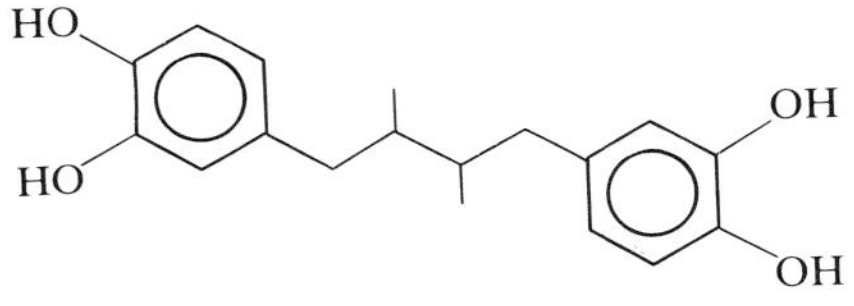

Occurs naturally in resinous exudate of *Larrea divaricata* (American creosote bush) and in some other plants. Often added to foodstuffs and several polymers (e.g. rubber, lubricants). Binds iron ions.

N,N'-Diphenyl-*p*-phenylene diamine (**DPPD**)

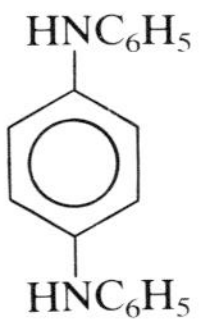

Popular antioxidant *in vitro*. Has been widely used for animal studies of *in vivo* lipid peroxidation.

Table 4.7. — *continued*

Name	Structure	Comments
6 Hydroxy-1,4-dimethylcarbazole (HDC)		Very powerful inhibitor of lipid peroxidation *in vitro*.
Promethazine		Inhibits lipid peroxidation *in vitro*. Prepared by the drug industry as an anti-histamine and sedative.
Chlorpromazine		Used in the drug industry as a tranquillizer. Antioxidant action of chlorpromazine in microsomes may partly depend on its enzymic conversion into hydroxylated products.
Ethoxyquin (Santoquin)		Frequently used in fruit canning. Powerful enzyme inducer *in vivo*. Widely used in longevity experiments in animals (Chapter 8).

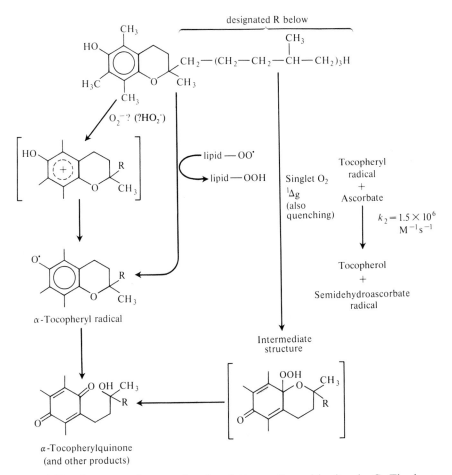

Fig. 4.17. Structure of α-tocopherol and its reaction with vitamin C. The long, hydrophobic side-chain makes the molecule lipid-soluble whereas the ring structure allows reaction with radicals. During peroxidation of cell membranes *in vitro* some α-tocopherol becomes converted into the quinone form.

to function *in vivo* as a protector against lipid peroxidation. How could it do this? Firstly, α-tocopherol both quenches and reacts with singlet oxygen and could therefore protect the membrane against this species (Chapters 2 and 5). α-Tocopherol is also oxidized by superoxide-generating systems (probably largely, if not completely, by reaction with $HO_2^{\cdot}$). Tocopherols, like most molecules, also react with $OH^{\cdot}$ at an almost diffusion-controlled rate. However, their major antioxidant action in biological membranes under most conditions is to react with lipid peroxy and alkoxy radicals, donating labile hydrogen to them (Fig. 4.17) and so terminating the chain reaction of peroxidation by scavenging chain-propagating radicals:

$$RO_2^{\cdot} + TH \rightarrow RO_2H + T^{\cdot}$$
$$RO^{\cdot} + TH \rightarrow ROH + T^{\cdot}$$

The tocopheryl radical T˙ (full structure shown in Fig. 4.17) is insufficiently reactive to abstract H˙ from membrane lipids because the unpaired electron on the oxygen atom can be delocalized into the aromatic ring structure (see Appendix I), so increasing its stability......

During its action as a chain-breaking antioxidant in membranes, α-tocopherol is consumed and converted to the radical form. However, mechanisms exist *in vivo* for reducing the radical back to α-tocopherol. A synergism between vitamin E and vitamin C in trapping free radicals was first suggested by Tappel in the USA in 1968. Pulse radiolysis studies by Slater's group in England confirmed that ascorbic acid can reduce the α-tocopheryl radical back to α-tocopherol and this reaction has since been shown to occur in intact membranes. Thus addition of ascorbate to membranes containing vitamin E can have antioxidant as well as pro-oxidant effects (see Section 4.2). Reduction presumably occurs close to the membrane surface since it is unlikely that the polar vitamin C molecule would enter the hydrophobic interior. Feeding guinea-pigs on a diet restricted in vitamin C has been reported to decrease the vitamin E content of liver and lung, even though dietary intake of this vitamin was normal, consistent with an interaction between the two vitamins *in vivo*.

Professor J. A. Lucy in England has suggested that vitamin E may also protect against peroxidation by modifying membrane structure (Section 4.7.1). If phospholipid liposomes are prepared, and peroxidation stimulated by adding iron salts and ascorbic acid, subsequent addition of α-tocopherol to the reaction mixture has little effect. However, if α-tocopherol is incorporated into the liposomes during their preparation it has a powerful protective effect as shown in Fig. 4.18. Esterification of the —OH group with acetic acid to give tocopherol acetate, which prevents reaction with lipid —OO˙ radicals (Fig. 4.17), greatly decreases the protective action at low concentrations, which suggests that any 'structural organization' effect makes only a small contribution to protection. The protective effect of tocopherol acetate at higher concentrations (Fig. 4.18) might be attributed to a structural effect, however, and there is evidence from experiments upon cell cultures that the presence of vitamin E can affect the types of fatty acids that become incorporated into membrane lipids.

There is considerable evidence that vitamin E functions to protect against lipid peroxidation *in vivo*, apart from the experiments with synthetic anti-

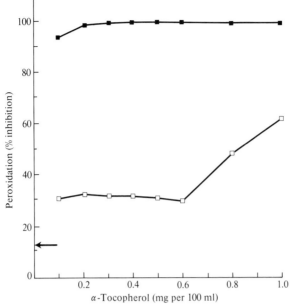

Fig. 4.18.   Effect of α-tocopherol and tocopherol acetate on peroxidation of ox-brain phospholipid liposomes. Liposomes were prepared incorporating tocopherol at the concentrations stated, and the effect on peroxidation stimulated by iron salts and ascorbate was measured. Black square-points, tocopherol; open squares, tocopherol acetate. The arrow shows the inhibition of peroxidation observed when tocopherol at 1.0 mg per 100 ml was added to preformed liposomes. For further details see J. M. C. Gutteridge, *Res. Comm. Chem. Pathol. Pharmacol.* **22**, 563–72 (1978). Tocopherol added to the outside of preformed liposomes has much smaller inhibitory effects against lipid peroxidation.

oxidants mentioned above. Feeding unsaturated fats to animals increases their requirement for vitamin E, e.g. for every 1 per cent of corn oil fed to young pigs above 4 per cent of the diet, 100 mg extra vitamin E is required. Chicks fed on lard, a mainly saturated fat, can remain healthy without vitamin E for weeks. Tissue samples taken from vitamin E-deficient animals show evidence of peroxidation (e.g. as TBA-reactive material), and tissue homogenates or subcellular fractions from such animals peroxidize much more rapidly than normal when incubated *in vitro*. Vitamin E-deficient mice are more sensitive to the toxic effects of pure oxygen than are controls. Vitamin E-deficient rats exhale more pentane and ethane gas, and accumulate fluorescent pigments much more rapidly than normal, especially if they are fed a diet rich in polyunsaturated fatty acids. This increased pigment accumulation is seen especially in the retinal pigment epithelium (Chapter 5).

Although a short-term lack of vitamin E in the diet of adult humans does not produce obvious acute signs of disease, probably because of the presence of other mechanisms of protection against peroxidation (see below) and because ascorbic acid can regenerate vitamin E in membranes, such a lack does increase susceptibility to peroxidation. For example, the amount of pentane in the expired air of some human volunteers during severe exercise could be decreased by feeding them extra vitamin E. Depletion of body vitamin E stores occurs in adult humans only after prolonged intravenous feeding, as a result of abnormalities of fat absorption by the gut, or as a result of some inborn error in vitamin E uptake.

Since tocopherols are highly-hydrophobic fat-soluble compounds, their efficient uptake is linked to that of fats, so that uptake requires adequate concentrations of bile salts in the intestinal lumen, and also chylomicron formation. Vitamin E is transported in plasma packaged into lipoproteins, particularly the low-density-lipoprotein (LDL) fraction, from which it is taken up into cells. Severe vitamin E deficiency occurs in pathological states that interfere with fat metabolism, e.g. in patients suffering the rare inborn error of lipid metabolism known as *abetalipoproteinaemia*. In this disease, dietary fat is ingested and absorbed, but not transported out of the intestinal mucosal cells, because of an inherited inability to synthesize apoprotein B, an essential component of chylomicrons. Patients with abetalipoproteinaemia have negligible plasma vitamin E concentrations and eventually show neuropathy, retinal degeneration, and abnormally-shaped erythrocytes (*acanthocytes*). Lloyd and Muller in England have shown that both the neuropathy and the retinopathy can be prevented by giving patients very large oral doses of vitamin E (sufficient to ensure some oral absorption). The brain is fairly poor in antioxidants and prone to undergo lipid peroxidation (Section 4.9) and so its vitamin E content may be an important part of its antioxidant mechanisms. Neurological and retinal disorders have also been observed in some patients with cystic fibrosis or with congenital defects in the biliary tree; both conditions result in impaired fat absorption. Figure 4.19 shows the concentrations of a-tocopherol in different regions of the brain and spinal cord of rats.

Newborn babies have low concentrations of vitamin E in plasma, especially if they are premature. Their erythrocytes are more susceptible to lipid peroxidation *in vitro*, although this does not normally cause a clinical problem. Sometimes haemolysis is seen *in vivo* in premature babies, and this *haemolytic syndrome of prematurity* responds to vitamin E therapy. Vitamin E also protects premature babies against retrolental fibroplasia (Chapter 5). The occasional haemolysis occurring in patients with thalassaemia or glucose 6-phosphate dehydrogenase deficiency (Chapters 2 and 3) can be decreased by administration of extra oral vitamin E, and a similar protective effect has been suggested to occur in sickle cell anaemia. In these diseases there is extra

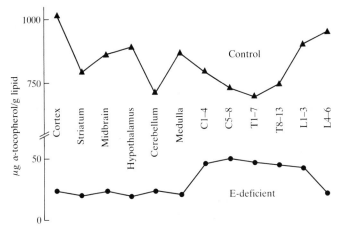

Fig. 4.19. Concentrations of α-tocopherol in regions of brain and spinal cord of control and deficient rats after 52 weeks from weaning. C = cervical, T = thoracic, L = lumbar regions of spinal cord. Diagram by kind permission of Dr D. P. R. Muller. See also *J. Neurol. Sci,* **87**, 25–35 (1988).

'oxidant stress' or a decrease in other protective mechanisms so that the effects of vitamin E are more readily seen.

Just as there have been dietary 'fads' for the consumption of increased amounts of vitamin C (Chapter 3), vitamin E has also been claimed to be a miracle-worker. Since adult humans can survive for long periods without it unless some other abnormality is present, this is unlikely to be true. If large doses are taken orally, most is not absorbed and is excreted in the faeces. Suggestions have been made that vitamin E may be of use in the treatment of degenerative arterial diseases. It must not be assumed that large doses are harmless, however, since they have been shown in adult human volunteers to decrease the ability of white blood cells to kill invading bacteria and to increase the dietary requirement for vitamin K, necessary for the blood coagulation system (Chapter 7).

Finally, it must not be assumed that the only biological role of vitamin E is to act as an antioxidant. There is evidence that it influences the metabolism of arachidonic acid in platelets and leucocytes (Chapter 7), that a 'tocopherol oxidase' enzyme is involved in the response of plant tissues to certain hormones and that vitamin E affects the metabolism of vitamin A.

4.7.4 *Synthetic antioxidants*

Over 100 years ago, Hoffman showed that the O_2 in the atmosphere was the primary cause of the slow progressive deterioration of rubber. This 'ageing' could be inhibited by adding small amounts of various compounds, usually aromatic amines or phenols. This observation led polymer chemists to investigate the mechanism of action of these 'antioxidants' and to develop new ones. Thus copper-chelating agents are used to prevent copper-catalysed oxidative degradation of polypropylene used to insulate copper cables, peroxide decomposers (e.g. dithiocarbamates and dithiophosphates) are used in several polymers, and 'UV absorbers' can be incorporated into plastics exposed to sunlight to prevent UV-induced free-radical degradation of the plastics. Food technologists also recognized the importance of antioxidants: it is necessary to protect food lipids against oxidative damage (and consequent *rancidity*) during storage, heat sterilization, or sterilization by ionizing radiation. Thus a huge range of synthetic antioxidants is available, although antioxidants developed for the polymer industry are not necessarily suitable for foodstuffs. Thus an antioxidant used in the rubber industry, *N*-isopropyl-*N'*-phenyl-*p*-phenylenediamine, is a haemolytic agent (Chapter 6).

Table 4.7 summarizes the structures of several antioxidants used in biology or food technology. *a*-Tocopherol itself is widely used in foodstuffs, as are BHA, BHT, and propyl gallate. A water-soluble analogue of vitamin E (*Trolox*: Table 4.7) is available for experimentation. Ingold *et al.* in Canada have shown that a slight chemical modification to the ring structure of vitamin E improves its antioxidant capacity both *in vivo* and *in vitro* (Table 4.7).

Most antioxidants shown in Table 4.7 are phenols or aromatic amines and usually act by a chain-breaking mechanism similar to that of vitamin E. They efficiently donate a hydrogen atom to a peroxy or alkoxy radical, so interfering with the propagation of lipid peroxidation, e.g.

$$A-OH + RO_2^{\cdot}(RO^{\cdot}) \rightarrow A-O^{\cdot} + RO_2H(ROH)$$
$$A-NH- + RO_2^{\cdot}(RO^{\cdot}) \rightarrow A-N^{\cdot}- + RO_2H(ROH)$$

The nitrogen- or oxygen-centred antioxidant radical so produced is insufficiently reactive to abstract hydrogen because of delocalization of the unpaired electron into an aromatic ring structure and, unless a mechanism exists for reducing it back to the antioxidant (e.g. ascorbic acid in the case of *a*-tocopheryl radical), the antioxidant radical can disappear by several mechanisms. These include self-reaction of radicals, e.g. if A· is any antioxidant radical:

$$
\left.
\begin{array}{l}
RO_2^{\cdot} + A^{\cdot} \\
RO^{\cdot} + A^{\cdot} \\
A^{\cdot} + A^{\cdot}
\end{array}
\right\} \rightarrow \text{non-radical products.}
$$

Many of the compounds shown in Table 4.7 have properties other than a chain-breaking action. For example, most of them have metal ion-complexing ability, especially those antioxidants with adjacent —OH groups such as propyl gallate or NDGA. Complexes of Fe(III) ions with these antioxidants can undergo a redox reaction to generate Fe(II) ions. Such reactions can alter the behaviour of antioxidants in iron-containing, radical-generating systems and can sometimes cause them to exacerbate free radical damage, e.g. damage to DNA by the antitumour antibiotic bleomycin (this is discussed further in Chapter 8). However, the chain-breaking action is predominant in peroxidizing lipid systems, causing phenolic antioxidants to be powerful inhibitors of the peroxidation process.

Several products of plant origin (some flavonoids, flavones, flavonols) have chain-breaking antioxidant activity; Table 4.8 shows some examples. Several of these compounds, such as quercetin and catechin, also have metal-binding capacity. Japanese scientists have been particularly interested in the antioxidant properties of compounds present in 'natural medicines' of plant origin.

4.7.5 *Glutathione peroxidase, a peroxide-decomposing antioxidant*

We saw in Chapter 3 that glutathione peroxidase, a selenium-containing enzyme found in the cytosol and mitochondria of animal tissues, helps to dispose of hydrogen peroxide by catalysing the reaction:

$$2GSH + H_2O_2 \rightarrow 2H_2O + GSSG$$

This enzyme is specific for GSH as a substrate but will act *in vivo* on a range of peroxides in addition to hydrogen peroxide, including cumene hydroperoxide, *tert*-butylhydroperoxide, progesterone 17a-hydroperoxide, several other steroid hydroperoxides (including cholesterol 7β-hydroperoxide at a low rate), and thymine hydroperoxide, as well as linoleic and linolenic acid hydroperoxides. In each case the peroxides (ROOH) are reduced to alcohols (ROH). Indeed, infusion of artificial organic hydroperoxides into animal organs *in situ* causes an increased formation and release of GSSG, much more so if the organs are taken from animals fed on diets deficient in vitamin E. Selenium deficiency in animals produces a variety of diseases that are strikingly similar to those induced by vitamin E deficiency, and to a considerable extent the effects of selenium deficiency can be overcome by giving excess vitamin E and vice versa. Exceptions to this generalization include the observation that selenium cannot protect female rats against foetal reabsorption caused by vitamin E lack, nor can vitamin E protect rats against the damage to the pancreas that occurs on selenium-deficient diets. A

Table 4.8.   Structure of some plant compounds with antioxidant activity in certain systems

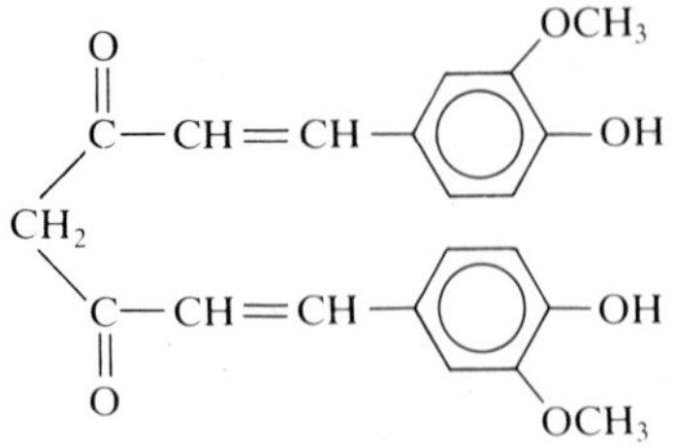

Cucurmin (two *trans* double bonds)

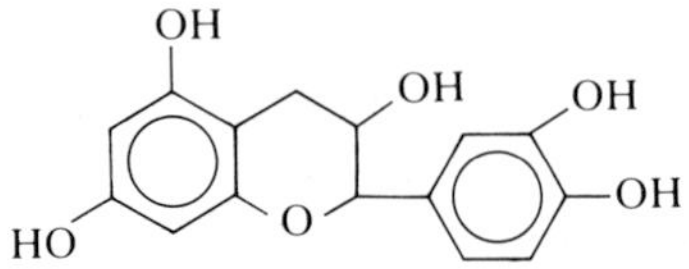

Catechin

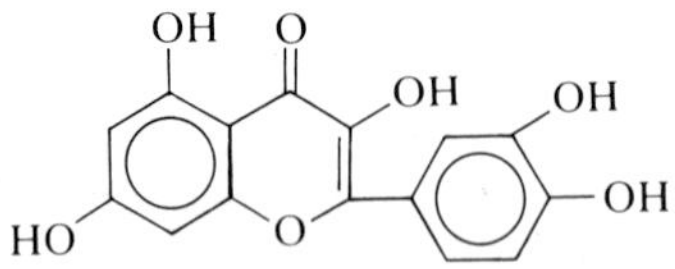

Quercetin

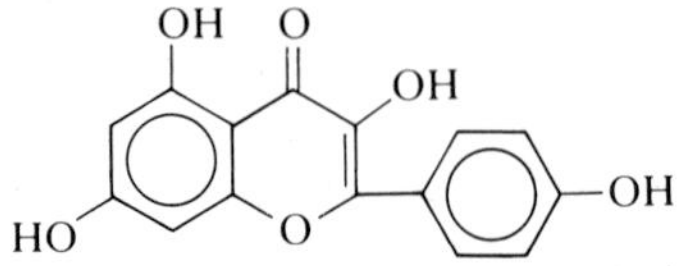

Kaempferol

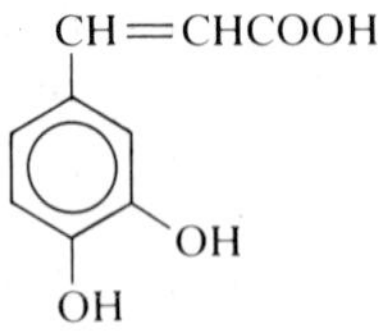

Caffeic acid

combined deficiency of selenium and vitamin E in the diet of animals is usually eventually fatal.

Feeding excess vitamin E to rats has been observed to decrease glutathione peroxidase activity in the tissues. Erythrocytes from selenium-deficient animals are more susceptible to haemolysis *in vitro* under conditions favouring lipid peroxidation than are normal erythrocytes. If rats on E-deficient diets are fed peroxidized methyl linoleate, the glutathione peroxidase activity of the small intestine increases, but not that of the liver. (Presumably peroxides in food lipids are dealt with by the intestinal enzyme.) Injection of peroxidized lipids does increase the activity of the liver enzyme, however. Injection of iron or copper salts into rats fed on diets deficient in both selenium and vitamin E causes a marked increase in the ethane content of the expired breath. The increase is smaller if either vitamin E or selenium is resupplied. Induction of lipid peroxidation by addition of Fe(III) salts chelated with ADP to isolated liver cells caused a decrease in their internal concentration of GSH. Administration of vitamin E to a child with an inborn deficiency of glutathione synthetase in white blood cells was observed to improve the functioning of these cells. Selenium-deficient rats are more susceptible to the toxic effects of elevated oxygen concentrations than normal.

This accumulation of evidence indicates that glutathione peroxidase plays an important role in protection against lipid peroxidation *in vivo*. As is usual, however, things are not so simple. Feeding a selenium-deficient diet to rats causes a drop in glutathione peroxidase activity in liver and erythrocytes, measured with hydrogen peroxide as substrate, to virtually zero within four weeks. However, if cumene hydroperoxide is used as substrate, substantial activity remains in liver but not in erythrocytes. Further study has revealed that several animal tissues contain a *non-selenium glutathione peroxidase activity* that acts on artificial organic hydroperoxides but not on hydrogen peroxide. This activity appears to be due to some of the glutathione-S-transferase enzymes involved in the conjugation of GSH with 'foreign compounds' (Chapter 3). Not all these transferases will act on hydroperoxides; for example, the transferases found in the bovine eye lens will not, nor will 'transferase c' in guinea-pig liver. The balance between selenium-dependent glutathione peroxidase activity and the non-selenium-dependent activity varies in different body tissues and animal species as summarized in Table 4.9. Perfused livers from rats fed a selenium-deficient diet release no GSSG when hydrogen peroxide is infused, but do so when *tert*-butylhydroperoxide is infused, indicating that these non-selenium enzymes can function in whole organs. Non-selenium peroxidase activity (like 'real' glutathione peroxidase) is mainly located in the cytosol of tissues with small amounts in the mitochondria.

Placing rats on a selenium-deficient diet has been reported to increase liver glutathione transferase activity. Several glutathione transferases can participate in metabolism of cytotoxic aldehydes produced during lipid peroxidation, such as 4-hydroxynonenal (Section 4.2.3). The transferases accelerate the reaction of these aldehydes with GSH to give less-toxic conjugates.

In view of their action upon hydroperoxides *in vitro*, one would expect both types of glutathione peroxidase to protect against lipid peroxidation in membranes by reducing lipid hydroperoxides (lipid—OOH) to stable hydroxylipids (lipid—OH), and thus preventing their decomposition to form

Table 4.9. Selenium and non-selenium glutathione peroxidase activities in different animals

Animal	Organ studied	Percentage of total glutathione peroxidase activity detected that was due to the non-selenium enzymes
Rat	adrenal	38
	spleen	0
	liver	35
	lung	0
	heart	0
	kidney	31
	testis	91
Hamster	liver	43
Sheep	liver	81
Pig	liver	67
Chicken	liver	70
Human	liver	84
Rabbit	liver	50 (approx)
Mouse	liver	30

Total activity was measured using cumene hydroperoxide as substrate. Results are mostly abstracted from H. Sies *et al.* (1982) *Proc. Third Int. Symp. Oxidases Relat. Redox Systems* (eds. T. E. King *et al.*), Pergamon Press, Oxford, p. 169. It should be noted that results can vary widely, depending on the strain and sex of animal selected, and on the dietary content of selenium.

alkoxy radicals. McCay *et al.* in the USA showed that peroxidation of rat liver microsomes in the presence of NADPH and ADP—Fe(III) was inhibited by addition of a 'soluble fraction' from rat liver in the presence of GSH. However, the inhibition of peroxidation was *not* accompanied by formation of hydroxylipids in the membrane. Separation of the proteins present in the soluble fraction showed that this GSH-dependent inhibition of peroxidation was due neither to selenium glutathione peroxidase nor to a glutathione transferase. The nature of this GSH-dependent cytosolic factor remains to be clarified. Another possibility is that GSH can, like ascorbate, participate in the regeneration of a-tocopherol from tocopheryl radicals in microsomal membranes.

Several scientists have shown that neither selenium glutathione peroxidase nor GSH transferases are capable of acting on peroxidized lipids in organized membrane structures, i.e. they reduce only unesterified fatty acid peroxides. Purdy and Tappel in the USA, and several other groups, have proposed that membrane-bound phospholipase A_2 is activated by peroxidation and cleaves out peroxidized fatty-acid side-chains from membrane lipids, so that they become substrates for glutathione peroxidases. Cleavage may be helped by the movement of the hydroperoxides, which are more polar than unperoxidized fatty-acid side-chains, away from the hydrophobic membrane interior (Fig. 4.20). Of course, the normal rapid 'turnover' of biological membrane lipids will tend to liberate peroxidized fatty acids in any case. Alternatively, Ursini *et al.* in Italy have purified, from rat, dog, and pig organs, a selenium-containing enzyme that does act upon phospholipid hydroperoxides in membrane structures and is not identical with cytosolic selenium glutathione peroxidase.

Obviously, the situation is confused and much more work needs to be done to investigate the mechanism by which selenium- and non-selenium-enzymes protect against peroxidation *in vivo*, as a great mass of evidence indicates that they do. One should not lose sight of the fact, however, that removal of hydrogen peroxide by the selenium enzyme *in vivo* will decrease the formation of hydroxyl radicals by the Fenton reaction and will thus prevent one route for the initiation of peroxidation.

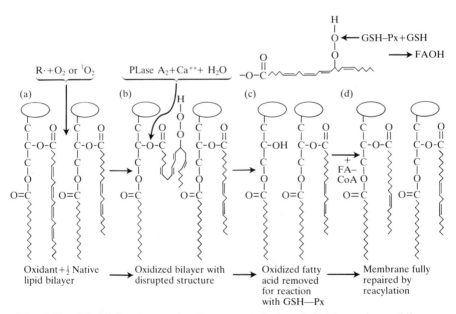

Fig. 4.20. Model for the repair of unsaturated phospholipid membrane bilayers damaged by oxidants. Formation of peroxides in the membrane causes rearrangement as the more-polar hydroperoxides move to the membrane surface. There they are preferentially cleaved by phospholipase A_2 (PLase A_2) in the presence of Ca^{2+} ions. Glutathione peroxidase (GSH—Px) in the surrounding fluid reduces the released fatty acid peroxides to alcohols (FAOH). Repair is completed by reacylation with a fatty-acyl-coenzyme A. Diagram adapted with permission from van Kuijk *et al.*, *Trends Biochem. Sci.* **7** 31 (1987).

Selenium deficiency

How much, if any, vitamin E is required in the normal human diet is debatable, but the situation is clearer for selenium. This follows the discovery in the People's Republic of China of a selenium-responsive degenerative heart disease known as *Keshan disease*. The name comes from an episode in 1935 in which 57 out of 286 inhabitants of a village in Keshan county of Heilongjiang province died of the disease. Careful epidemiological studies showed that the incidence of the disease was correlated with that of various animal degenerative diseases that are known to be related to selenium deficiency, such as white muscle disease. Both Keshan disease and the animal diseases could be prevented by administration of small doses of sodium selenite (Na_2SeO_3). In all affected areas, selenium concentrations in food-stuffs or in animals eating them were found to be extremely low. Thus a very low selenium intake is the major cause of Keshan disease, but it cannot be the only factor involved. For example, Keshan disease is not endemic in all low-selenium areas of China and seasonal variations in the disease incidence do not seem to be caused by changes in selenium intake. (Variations in vitamin E intake do not seem to be responsible for these observations.) Since gluta-thione peroxidase is the only known mammalian selenoprotein, it seems logical to attribute Keshan disease to a lack of active glutathione peroxidase enzyme. Of course, it is possible that selenium plays other biochemical roles, including perhaps an involvement in haem metabolism, and selenium has been observed to protect against the action of certain carcinogens (Chapter 8).

Selenium deficiency has also been implicated in *Kashin–Beck disease*, a disabling joint disease seen in children of age range 5–13 years in such areas as Northern China, North Korea and East Siberia. The name comes from the Russian scientists who first described the disease. Administration of sodium selenite and vitamin E during the early stages of the disease has been reported to have beneficial effects. The geographical distribution of Kashin–Beck disease resembles that of Keshan disease, but is not identical with it.

A study of blood selenium concentrations led Chinese scientists to conclude that a minimum adequate dietary intake for humans is about 30 μg. Normal daily intakes in advanced countries are in the range 60–200 μg per day. Larger amounts (> 350–600 μg/day) can produce toxic effects: an early sign of excess selenium intake is deformation and loss of fingernails, toenails, and sometimes hair. Indeed, selenium-accumulating plants such as *Astraga-lus* species that can poison cattle are a particular nuisance to farmers in certain parts of the world. Selenium deficiency has sometimes been claimed to accompany protein–calorie malnutrition or prolonged intravenous feed-ing, but often the reports are inconclusive. Residents of low-selenium areas in Finland or New Zealand appear to suffer no obvious ill-effects, although it has been suggested that low blood selenium might predispose to cardiovascu-lar disease or to complications of pregnancy.

Ebselen

The key role played by glutathione peroxidase in the metabolism of H_2O_2 (Chapter 3) and lipid peroxides has raised the possibility that this enzyme might be used as a therapeutic agent in diseases where oxidative damage is thought to make an important contribution to the disease pathology (Chapter 8). However pure proteins are expensive as well as potentially immunogenic, and glutathione peroxidase is an unstable enzyme, easily inactivated by oxidants. Thus there has been considerable interest in developing low-molecular-mass compounds with glutathione-peroxidase-like activity. One of these is Ebselen (Fig. 4.21). In the presence of GSH, the selenium bound to Ebselen catalyses the degradation of H_2O_2 and organic peroxides in a manner similar to that of selenium glutathione peroxidase. Ebselen can also inhibit phagocyte 5-lipoxygenase activity (Chapter 7). It has been shown to be anti-inflammatory in a number of animal models of inflammation, and to protect hepatocytes against the toxic effects of the herbicide diquat (Chapter 6). Ebselen is much less sensitive to inactivation by oxidants than is glutathione peroxidase.

4.7.6 *Antioxidants of extracellular fluids*

The extracellular fluids of the human body, such as blood plasma, tissue fluid, cerebrospinal fluid (CSF), synovial fluid, or seminal plasma contain no significant catalase activity and only low activities of superoxide dismutase and selenium glutathione peroxidase. There is also very little reduced glutathione (GSH)—about 20 μM in the plasma of normal rats and less in humans. GSH appears to originate from the liver. The selenium-containing glutathione peroxidase has been purified from human plasma, and shown to be different from the intracellular enzyme. Similarly, extracellular superoxide dismutase (EC-SOD) has been purified by Marklund in Sweden. Although EC-SODs are copper–zinc enzymes, they are very different from intracellular CuZnSOD (Chapter 3), in that EC-SODs have a much higher relative molecular mass (about 135 000), and have attached carbohydrate. The biological role of EC—SOD is unclear; such low activities are present in plasma and other extracellular fluids that bulk-scavenging of O_2^- in extracellular fluids seems an unlikely function. Marklund has shown that some EC-SODs bind to heparin and has proposed that, *in vivo*, they may be associated

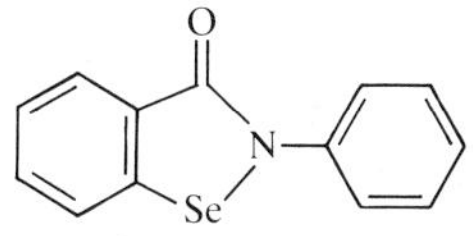

Fig. 4.21.   Structure of Ebselen (PZ51). The full chemical name is 2-phenyl-1,2-benzisoselanazol-3(2H)-one.

with endothelial cell surfaces as a protective antioxidant 'layer' over the cells. Trace amounts of EC-SODs are found attached to cells of various tissues, so the term 'extracellular' must not be taken to imply a location only within extracellular fluids.

It follows that enzymes contribute little to the antioxidant activity of extracellular fluids. Blood lipoproteins contain vitamin E as the major, if not the only, chain breaking antioxidant present. However, oxidants can still be generated in the aqueous phase of plasma, e.g. O_2^- and H_2O_2 can arise from enzymes, autoxidizing compounds, and activated phagocytic cells (Chapters 3 and 7).

In blood, O_2^- and H_2O_2 might diffuse into erythrocytes for metabolism, which would explain the existence of an anion channel in the erythrocyte membrane. However, for this to happen, it is essential that O_2^- and H_2O_2 be prevented from immediately reacting to form the highly-toxic hydroxyl radical, OH˙ (Chapter 3). The authors have thus argued that a major antioxidant defence in human plasma is to bind transition metal ions in forms that will not stimulate free-radical reactions, or to otherwise prevent the metal ions from participating in such reactions. Thus the *transferrin* present in normal human plasma is only 20–30 per cent loaded with iron, so that the content of free ionic iron in plasma is effectively nil. Iron bound to transferrin will not participate in OH˙ radical formation or lipid peroxidation. Hence if plasma from normal humans, or purified transferrin, is added to liposomes or microsomes undergoing iron-dependent peroxidation, the peroxidation is inhibited (provided that the iron present does not overwhelm the iron-binding capacity of the transferrin). The protein *lactoferrin*, secreted by phagocytic cells (Chapter 7), acts as an antioxidant in the same way.

We have seen that haemoglobin can stimulate lipid peroxidation, both by decomposing to release iron and by a direct interaction of peroxides with the protein. Isolated haem is also a powerful stimulator of peroxidation. Plasma contains the haemoglobin-binding proteins known as *haptoglobins*, as well as a haem-binding protein (*haemopexin*). Binding of haemoglobin to hapto-globin, or of haem to haemopexin, retards the effectiveness of these iron compounds in stimulating lipid peroxidation (Fig. 4.22). The haemoglobin–haptoglobin or haem–haemopexin complexes are rapidly cleared from the circulation.

The plasma copper-containing protein *caeruloplasmin* is thought to play an essential role in iron metabolism (Chapter 3), but it also has antioxidant properties. Firstly, as we have seen, caeruloplasmin has a ferroxidase activity—it oxidizes Fe(II) to Fe(III) while reducing oxygen to the level of water.

$$4Cu^+ \text{ (on protein)} + O_2 + 4H^+ \rightarrow 4Cu^{2+} \text{ (on protein)} + 2H_2O.$$

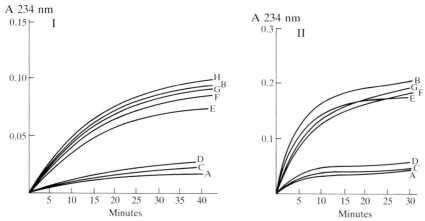

Fig. 4.22. Part I. Haemin-stimulated peroxidation of fatty-acid micelles measured as diene conjugation at 234 nm. A. Fatty-acid micelles at pH 7.4 incubated for the time indicated at 25°C. B. Fatty-acid micelles plus haemin (0.012 mM). C. as B, with BHT (0.07 mM). D. as B with apohaemopexin (0.013 mM). E. as B with albumin (0.012 mM). F. as B with haptoglobin (0.011 mM), G. as B with desferrioxamine (0.153 mM). H. as B with apotransferrin (0.010 mM). Final reaction concentrations are shown. Part II. Haemoglobin-stimulated lipid peroxidation measured as conjugated dienes. A. Fatty-acid micelles at pH 6.4 incubated for the time indicated at 25°C. B. Fatty-acid micelles pH 6.4, desferrioxamine (0.076 mM) and methaemoglobin (0.001 mM). C. as B with haptoglobin (0.001 mM). D. as B with BHT (0.03 mM). E. as B with apohaemopexin (0.001 mM). F. as B with apotransferrin (0.001 mM). G. as B with albumin (0.001 mM). Final reaction concentrations are shown. Data abstracted from *Biochem. J.* **256,** 861 (1988).

Thus the caeruloplasmin-catalysed oxidation of Fe^{2+} ions does not release any damaging oxygen radicals: they are kept on the active site of the protein. The ferroxidase action of caeruloplasmin allows it to inhibit iron-dependent lipid peroxidation or OH˙ formation from H_2O_2, and this is its major antioxidant activity. In addition, the protein reacts with H_2O_2 and with superoxide (O_2^-): both are stoichiometric reactions and so caeruloplasmin has no significant 'superoxide dismutase like' or 'catalase like' activity, i.e. it does not *catalyse* O_2^- dismutation or H_2O_2 decomposition at a significant rate (it should be noted that batches of commercial caeruloplasmin are often contaminated with EC-SOD, which has caused considerable confusion in some studies). Reaction of caeruloplasmin with H_2O_2 does not lead to the formation of 'free' OH˙, nor does it inactivate the protein. Caeruloplasmin also non-specifically binds copper ions and can inhibit copper-stimulated OH˙ formation and lipid peroxidation.

Albumin can also bind copper ions, and it inhibits copper-dependent lipid peroxidation and OH' radical formation. In fact, the reactions tend to continue on the protein surface and damage the protein, but the high plasma concentration of albumin and its rapid turnover means that such damage is probably biologically insignificant. Thus the binding of non-caeruloplasmin copper in plasma (Chapter 3) to albumin may lead to albumin damage under certain circumstances, but it prevents the copper ions from binding to more important targets. Albumin is also a powerful scavenger of hypochlorous acid (HOCl), an oxidant produced by the enzyme myeloperoxidase in activated phagocytic cells (Chapter 7). Again, the albumin acts by reacting with the HOCl and protecting more important targets, such as a_{-1}-*antipro-tease* (the major plasma inhibitor of proteolytic enzymes, such as elastase; Chapters 6 and 7). Albumin also transports free fatty acids in the blood, and the bile pigment bilirubin is bound to it. Stocker, Glazer, and Ames (1987) in the USA have suggested that albumin-bound bilirubin acts as an antioxidant in some lipid systems, although the prevention of copper-mediated or HOCl-mediated damage by albumin is a property of the protein itself.

　　Ascorbic acid, present in plasma from healthy humans at concentrations of 50–200 μM, is also a scavenger of HOCl and of O_2^- (Chapter 3) and would help to regenerate any vitamin E that became oxidized in plasma lipoproteins. *Uric acid* is also an inhibitor of lipid peroxidation (Chapter 3); it acts both by binding iron and copper ions in forms that do not stimulate radical reactions, and by direct scavenging of such oxidants as OH', peroxy radicals, singlet oxygen, and HOCl. Table 4.10 summarizes the antioxidant action of components of extracellular fluids. It should also be noted that several proteins with antioxidant action are *acute-phase proteins* (Table 4.11).

　　It thus seems that the major antioxidant action of plasma is to prevent transition metal ions from accelerating damaging free-radical reactions such as lipid peroxidation. However, the preventative antioxidant defences of other extracellular fluids may be much weaker. Thus human CSF contains little transferrin, albumin, or caeruloplasmin but has high concentrations of ascorbic acid (about ten-times those in plasma) and also contains uric acid. Synovial fluid has lower concentrations of albumin, transferrin, and caeruloplasmin than plasma, whereas the fluid that lines the alveoli of the lung also has a very low protein content but a high content of ascorbic acid.

　　Ingold *et al.* in Canada have attempted to define *quantitatively* the antioxidant capacity in the aqueous phase of extracellular fluids. The temperature-controlled decomposition of azo*bis*(2-amidinopropane) hydrochloride (Section 4.3) was used to produce water-soluble peroxy radicals at a defined rate. The TRAP value of plasma (the number of moles of peroxy radical that can be scavenged per dm^3 of plasma) was found to be around 8×10^{-4} (800 μmoles/dm^3) and is largely determined by the plasma content of ascorbic acid, uric acid, and plasma proteins such as albumin (whose

Table 4.10. Extracellular antioxidant defences in human plasma

Defence	Mode of action	Comments
Transferrin/lactoferrin	Bind iron and stop or slow its participation in lipid peroxidation and iron-catalysed OH⁻ radical formation.	Proteins not easily damaged by H_2O_2 or lipid peroxides; only release iron ions at acidic pH.
Caeruloplasmin	Catalytically oxidizes Fe(II) to Fe(III) *without release* of oxygen radical intermediates. Inhibits iron- and copper-dependent lipid peroxidation. It also reacts stoichiometrically with O_2^-, but this is probably much less significant. Reports of greater O_2^- scavenging activity may be due to contamination of the samples studied with EC—SOD, which closely resembles caeruloplasmin (S. L. Marklund, personal communication). Reacts with H_2O_2.	A significant part of the antioxidant activity of human plasma is due to caeruloplasmin. Acute-phase protein (Table 4.11).
Albumin	Binds copper tightly and iron weakly. Present at high concentrations (40–60 mg/ml). Copper bound to albumin may still participate in oxidant formation, but any OH⁻ would be formed on the albumin surface and scavenged by it, not allowed to escape into free solution. This is an example of a *biologically insignificant* site-specific reaction, i.e. the albumin is a *sacrificial antioxidant*. Also rapidly scavenges the myeloperoxidase-derived oxidant HOCl.	Liver synthesis and plasma concentration drop during liver injury, sometimes called 'a negative acute-phase response'.

Table 4.10. —*continued*

Defence	Mode of action	Comments
Haptoglobin/haemopexin	Bind free haemoglobin/haem. Haemoglobin and methaemoglobin release iron readily on exposure to H_2O_2 or peroxides and can accelerate lipid peroxidation; binding to haptoglobin/haemopexin hinders this.	Acute-phase proteins. See Fig. 4.22.
Urate (uric acid)	Inhibits lipid peroxidation and scavenges radicals	Ability to tightly bind iron and copper is an important part of its antioxidative role, as is direct scavenging of peroxy, and alkoxy radicals, and HOCl.
Vitamin E	Lipid-soluble antioxidant; breaks chains by trapping peroxy and alkoxy radicals.	Major, if not the only, *lipid-soluble*, chain-breaking antioxidant in human plasma. Important in protecting lipoproteins against oxidation.
Glucose	Scavenger of OH· radical; rate constant comparable to that of mannitol (Chapter 2).	Normal plasma concentration around 4.5 mM, greater soon after carbohydrate-rich meals.
Bilirubin	Bile pigment produced by haem catabolism; transported bound to albumin. May protect albumin-bound fatty acids from peroxidation.	Bilirubin is also a sensitizer of $^1O_2$ production (Chapter 2).

Table 4.11. Some of the acute-phase plasma proteins in humans

Protein	Typical increase in concentration	Biological function
Caeruloplasmin	50%	See text
C3	50%	Third component of complement reaction with antigens
Antiproteases (e.g. *a*-antiprotease)	2- to 4-fold	Inhibit proteolytic enzymes
Haptoglobin	2- to 4-fold	Binds haemoglobin
C-reactive protein (CRP)	Several-hundred-fold (very little present normally)	Function unknown (discovered by its ability to precipitate the 'C-polysaccharide' of *Pneumococci*)
Serum amyloid A protein (SAA)	Several-hundred-fold	See Chapter 8

—SH groups can scavenge peroxy radicals). Thus if the *preventative* antioxidant defences of plasma (binding metal ions in forms that will not stimulate free-radical reactions, or otherwise preventing them from stimulating such reactions) is overwhelmed, a second line of *chain-breaking* defence exists; scavenging peroxy radicals. This secondary defence may be of subsidiary importance in normal human plasma, but may be very significant in cerebrospinal fluid, which contains little transferrin, albumin, or caeruloplasmin, but does contain urate and high ascorbate concentrations.

Why should extracellular fluids possess mechanisms of antioxidant protection different from intracellular mechanisms; why do extracellular fluids not contain significant amounts of catalase, GSH, and selenium glutathione peroxidase? During the 'respiratory burst' of phagocytic cells, some O_2^- and H_2O_2 is released into the surrounding fluids (Chapter 7). Provided that these species can be stopped from forming destructive OH˙, they may act as *useful* signals between cells. Thus O_2^- may be involved in phagocyte chemotaxis (Chapter 8) and inactivation of endothelium-derived relaxing factor, EDRF (Chapter 6) whereas H_2O_2 may help the aggregation of platelets. The presence of too much SOD or H_2O_2-removing enzymes might interfere with these 'useful' processes.

The acute-phase response and interleukin-1
Several proteins with antioxidant activity are *acute-phase proteins* (Table 4.11). The human body responds to infections, inflammatory, or immunologically-mediated diseases with a range of biochemical and clinical alterations collectively called the *acute-phase response*. Clinical changes include fever; biochemical changes include increased synthesis of a group of proteins by the liver (acute-phase proteins; Table 4.11), decreased synthesis of

albumin by the liver, decreases in plasma iron and zinc, and negative nitrogen balance. 'Acute' signifies that the changes can be measured within hours of the noxious agent. The acute phase response also follows the inflammatory reaction to traumatic injury. Many features of the acute-phase response can be measured in chronic conditions, such as chronic 'low-grade' infections, rheumatoid arthritis (Chapter 8, Section 8.2) and many forms of cancer. The function of some acute-phase proteins (e.g. CRP, SAA; Table 4.11) is unknown, but it is tempting to speculate that the rises in caeruloplasmin and haptoglobin that occur are part of an increased 'antioxidant response' to cell injury.

Many, and perhaps all, aspects of the acute-phase response are mediated by the action of substances produced by the human body. The most important of these are interleukins-1, two small labile proteins (IL-1α, IL-Iβ), produced by several cells, especially mononuclear phagocytes (monocytes, macrophages: Chapter 7) exposed to micro-organisms or their toxins, pro-inflammatory agents, antigen–antibody complexes, and several other compounds, such as muramyl dipeptide (Table 4.12). Interleukin-1 'primes' cells to destroy foreign organisms (see Chapter 7) and the fever it induces can have beneficial effects by assisting the elimination of foreign organisms. However, continued production of IL-1 can give rise to deleterious changes, as in rheumatoid arthritis (Chapter 8, Section 2).

4.8   Erythrocyte lipid peroxidation

The erythrocyte is the most readily-available source of a pure membrane in the human body. This cell contains high concentrations of haemoglobin (and hence a ready supply of O_2) and its membrane is not only rich in polyunsatur-

Table 4.12.   Some biological activities attributed to interleukin-1

Fever
Falls in plasma iron and zinc
Neutrophilia
Increased hepatic synthesis of acute-phase proteins
Decreased hepatic synthesis of albumin
Increased immune responses (T cell, B cell, natural killer cells; see Appendix II)
Stimulation of neutrophil activation (degranulation, O_2^- production)
Chemoattractant for monocytes, neutrophils, lymphocytes
Proliferation of fibroblasts, glial cells, synovial fibroblasts (in arthritis)
Stimulation of prostaglandin E_2 synthesis (hypothalamus, cortex, muscle, fibroblasts, chondrocytes, endothelium [PGI$_2$], macrophages, monocytes)
Promotes phagocyte adhesion to endothelium

ated fatty acids, but also has an anion channel through which O_2^- can enter. The erythrocyte may also contain a 'low-molecular-mass iron pool', but data on this are conflicting. We have seen that haemoglobin can stimulate lipid peroxide decomposition, and that excessive exposure of it to peroxides can lead to release of iron ions in a form that promotes both lipid peroxidation and OH˙ generation. The normal formation of O_2^- from haemoglobin exposes the cell to a constant mild oxidative stress (Chapter 3). In addition, the lack of biosynthetic capacity of the erythrocyte means that membrane lipids or proteins damaged by peroxidation are not easily replaced.

Despite all the above problems, the normal erythrocyte is highly resistant to oxidative damage, because of its efficient protective mechanisms. The cell interior is rich in catalase, superoxide dismutase, GSH, glutathione reductase, and glutathione peroxidase (Chapter 3), and contains a proteolytic system that can hydrolyse oxidatively-modified proteins (Chapter 3). Indeed, erythrocytes might act as 'sinks' for H_2O_2 and O_2^- produced in plasma (Section 4.7). At physiological rates of H_2O_2 generation, the glutathione peroxidase/GSH/glutathione reductase system is much more important in catabolizing H_2O_2 than is catalase (Chapter 3). The erythrocyte membrane is rich in vitamin E (the major, if not the only, lipid-soluble chain-breaking antioxidant present) and its cholesterol content may also help to diminish peroxidation. Indeed, older human erythrocytes contain a higher *a*-tocopherol/arachidonic acid ratio than do younger ones due to the fact that lipid content of the membrane decreases with age whereas vitamin E content remains approximately constant.

Thus it is rather difficult to persuade erythrocytes from healthy subjects to peroxidize at all. Indeed, the standard test for the peroxidizability of erythrocytes is to incubate washed cells in a buffer containing azide (to inhibit catalase) but no glucose (no substrate for glucose 6-phosphate dehydrogenase, so no NADPH generation to maintain GSH/GSSG ratios) in the presence of 10 *millimolar* H_2O_2 (e.g. Table 4.6). The formation of peroxides can then be measured (e.g. by the TBA test) as can the generation of oxidatively-modified and cross-linked membrane proteins. Fig. 4.23 shows the *echinocytes* that can be produced under these conditions. The susceptibility of erythrocytes to peroxidation under these unphysiological conditions *in vitro* seems to be a reflection of their vitamin E content and membrane fatty-acid composition. Thus erythrocytes from animals or humans maintained on E-deficient diets for long periods are more 'peroxidizable' in this assay, although they do not contain any significant amount of lipid peroxides when freshly isolated. Susceptibility to peroxide stress is increased in erythrocytes from neonates (due partly to low vitamin E) and in some old people (reason unknown).

Peroxidation of erythrocytes can also be induced by adding organic hydroperoxides, such as *tert*-butyl hydroperoxide or linoleic acid hydroper-

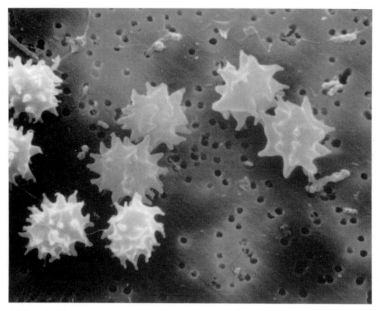

Fig. 4.23. Echinocytes produced by 'peroxide stress' of erythrocytes. Magnification × 1640. Electron micrograph by courtesy of Dr C. Rice-Evans.

oxide. Haemoglobin plays an important role in the 'peroxide stress test': prior conversion of oxyhaemoglobin to methaemoglobin by chemical treatment of the cells diminishes the peroxidation observed on subsequent addition of H_2O_2, or *tert*-butyl hydroperoxide. Perhaps the interaction of oxyhaemoglobin with H_2O_2, which produces a reactive species not identical with OH˙ (Chapter 3), is important in starting off H_2O_2-induced peroxidation. Reaction of an Fe^{2+} protein with an organic peroxide would also generate reactive alkoxy radicals.

Susceptibility to peroxide stress is increased in erythrocytes from patients with several conditions, including sickle-cell anaemia, glucose 6-phosphate dehydrogenase deficiency, thalassaemia, and autoimmune haemolytic disease. The relationship of the first two conditions to malaria has been discussed in Chapter 3. Unstable haemoglobins may readily tend to precipitate onto the erythrocyte membrane under peroxide stress, allowing site-specific free-radical reactions that are especially effective at promoting lipid peroxidation and damage to membrane proteins. Iron may also be released to bind to the membrane.

4.9   The biomedical significance of lipid peroxidation

Many of the assay techniques described in Section 4.6 have been used to show that lipid peroxidation in animal tissues increases in a number of disease states and in poisoning by several toxins (Chapter 6 examines the role of free radicals in toxicology). Behind some of these reports is the unspoken assumption that the disease or toxin causes increased lipid peroxidation, which is then responsible for the toxicity. In some cases, this assumption is probably valid (e.g. in poisoning by carbon tetrachloride; Chapter 6).

However, it was established many years ago by Barber in the USA and by Stocks, Gutteridge, and Dormandy in England that disrupted tissues often undergo lipid peroxidation more quickly than healthy ones, especially after mechanical disruption. For example, lipid peroxides accumulate in a brain homogenate much more quickly than they do in an isolated intact brain. Reasons for this increased peroxidizability of damaged tissues include inactivation or 'dilution out' of some antioxidants, and the release of metal ions (especially iron and copper) from intracellular storage sites, and from metalloproteins hydrolysed by proteolytic enzymes released from damaged lysosomes. Hence the sequence of events:

$$\text{disease or toxin} \rightarrow \begin{array}{c}\text{cell damage}\\\text{or death}\end{array} \rightarrow \begin{array}{c}\text{increased lipid}\\\text{peroxidation}\end{array}$$

can explain many of the reports of elevated lipid peroxidation in disease or toxicology. Often, the lipid peroxidation can be inhibited without preventing the cellular injury, as in the case of the lung damage produced by the herbicide paraquat (Chapter 7). The increased lipid peroxidation seen in the muscles of patients with muscular dystrophy or multiple sclerosis may simply be a consequence of the tissue degeneration; there is no evidence that increased lipid peroxidation is in any way causative for these diseases in humans. The detailed experiments needed to *prove* that lipid peroxidation is a major cause of tissue injury, and not merely an accompaniment to it, are best illustrated by studies on the toxicity of carbon tetrachloride (Chapter 6). It must be shown that the peroxidation precedes or accompanies the cell damage and that prevention of the peroxidation by antioxidants prevents the cell damage.

Of course, even if peroxidation is not the prime cause of damage, its occurrence as a *consequence* of damage may still be biologically important in worsening tissue injury, in view of the cytotoxicity of some end-products of the peroxidation process, such as hydroxynonenal. Therapy with chain-breaking antioxidants, or with chelating agents that can bind metal ions and prevent them from participating in free-radical reactions (e.g. desferrioxamine, Chapter 2) may be helpful in some diseases to the extent that further damage is done by the end-products of the lipid peroxidation secondary to the tissue injury, and to the extent that the antioxidants can reach the correct site of action. Antioxidants would certainly not be expected to cure most diseases. However, there are at least two human conditions in which lipid peroxidation *does* appear to play a major role in the pathology. One of these is atherosclerosis, discussed in detail in Chapter 8. The other is the degeneration of the brain and spinal cord that occurs *after* injury. Ischaemic brain damage (e.g. in 'stroke', where a cerebral blood vessel becomes blocked) is a major clinical problem. Indeed, ischaemic or traumatic injuries to the brain or spinal cord often result in more extensive tissue damage than do equivalent insults to other tissues. Free-radical reactions have often been implicated in such damage. The brain and nervous system may be especially prone to oxidant damage for a number of reasons. Firstly, the membrane lipids are especially enriched in polyunsaturated fatty-acid side-chains. Secondly, the brain is poor in catalase activity and has only moderate amounts of superoxide dismutase and glutathione peroxidase. Thirdly, some areas of the brain (e.g. globus pallidus, substantia nigra, circumventricular organs) are rich in iron. Brain iron has been poorly characterized, but it is likely that most of it is contained within ferritins or similar proteins. Iron seems to play an important role in brain function since moderately-severe anaemia in children produces behavioural modifications and diminished learning ability. Iron(II) salts are involved in the binding of serotonin to its receptors in rat brain, and there is evidence that serotonin- and dopamine-mediated behavioural responses induced by amphetamine or apomorphine are decreased in rats fed an iron-deficient diet. Co-administration of the sedative prochlorperazine (which is a weak chelator of iron salts and crosses the blood–brain barrier) and desferrioxamine produces prolonged unconsciousness in iron-deficient Wistar rats, and two human rheumatoid patients given prochlorperazine during experimental therapy with desferrioxamine also became unconscious. Hence brain iron seems to serve an important purpose.

When brain cells are injured, some of the iron they contain can be easily released, but cerebrospinal fluid (CSF) has no significant iron-binding capacity (Section 4.7.6). There is a high concentration of ascorbic acid in the grey and white matter of the brain. (The choroid plexus has a specific active transport system that raises ascorbate concentrations in the CSF to about ten times the plasma level, and neural tissue cells have a second transport system that concentrates intracellular ascorbate even more.) If the 'free' iron content of the system were raised by cell injury, ascorbate/iron salt mixtures would be expected to promote rapid lipid peroxidation and OH˙ radical formation. Indeed, injecting aqueous solutions of iron salts or haemoglobin into the cortex of rats causes transient focal epileptiform discharges, lipid peroxidation, and persistent behavioural and electrical abnormalities. Such rises in the iron content may occur *in vivo* as a result of bleeding after head injury. We have seen that haemoglobin accelerates lipid peroxidation, and the release of low-molecular-mass iron complexes both from injured cells and by the breakdown of haemoglobin from released blood would stimulate peroxidation and OH˙ formation, hence potentiating the damage. Homogenates of brain peroxidize extremely rapidly *in vitro*; indeed, the peroxidation of ox-brain homogenate has been used for a long time as an assay to measure 'antioxidant activity' and the peroxidation is largely prevented by metal chelating agents such as desferrioxamine. Cerebral ischaemia, followed by reperfusion, also stimulates lipid peroxidation, again probably because of metal release from injured cells. Since lipid peroxidation is known to generate cytotoxic products, such as hydroxynonenal, intense lipid peroxidation after traumatic injury or ischaemia in the brain would be expected to exacerbate the tissue damage. Indeed, pre-treatment of animals with *a*-tocopherol has been reported to lessen secondary damage in several models of ischaemic/traumatic injury to the central nervous system.

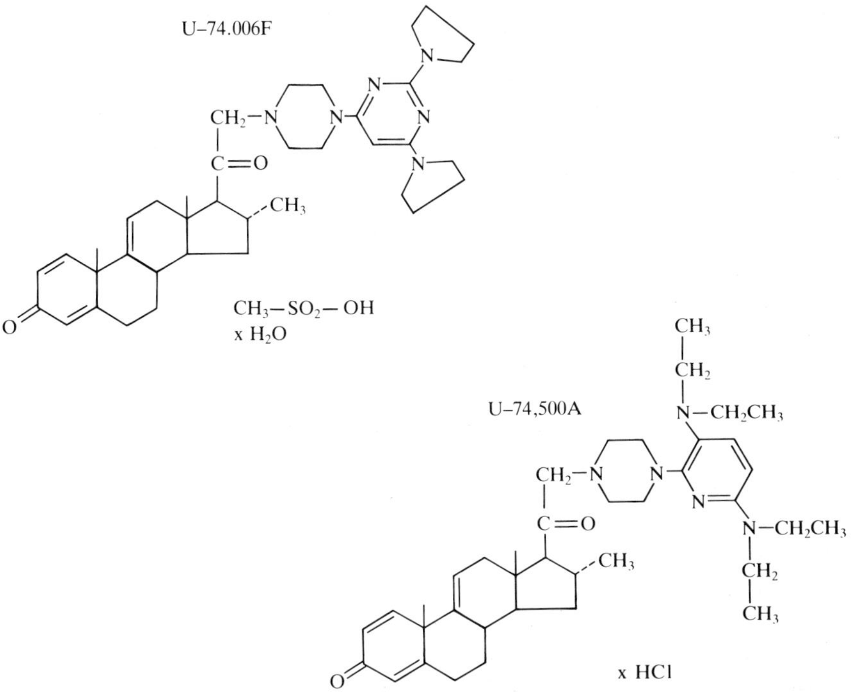

Fig. 4.24. Structures of two hydrophobic aminosteroid compounds (U74006F and U74500A) that are powerful inhibitors of iron-dependent lipid peroxidation. Reproduced by courtesy of Dr J. M. Braughler (see *J. Biol. Chem.* **262**, 10438 [1987]).

The likelihood that lipid peroxidation contributes significantly to further tissue injury in the brain and spinal cord after ischaemia or traumatic injury led Braughler, Hall *et al.* in the USA to develop a series of antioxidants (the *lazaroids*) in which the capacity to bind iron ions in forms incapable of catalysing free-radical reactions is present within a hydrophobic steroid-based ring structure. Fig. 4.24 shows the structure of two of these compounds. Injury to the spinal cord often results in a drop in blood flow in the injured segment, which may lead to spinal cord degeneration. Intravenous injection of U74006F into cats before contusive injury to the spinal cord was found to almost prevent this drop in blood flow. U74006F has also been claimed to enhance the early neurological recovery of mice after concussive head injury and to diminish the effects of ischaemia/reperfusion injury to the brains of cats or gerbils. These promising results clearly warrant further investigation.

Watch out for autoxidation!

Arachidonic Acid Warning

The Department of Health and Social Security has issued a Safety Information Bulletin, *SIB(7)10,* containing a warning about storage of opened vials of arachidonic acid and its salts. This followed an incident in which a 50 mg sample of the substance violently self-ignited on reaching room temperature after being frozen for several months. The DHSS is anxious to draw this incident to the attention of all laboratory workers.

4.10   Further reading

Abrams, B. A. *et al.* (1973). Vitamin E in neonatal hyperbilirubinaemia. *Arch. Dis. Child.* **48**, 271.

Allerheiligen, S. R. B., Ludden, T. M., and Burk, R. F. (1987). The pharmacokinetics of pentane, a by-product of lipid peroxidation. *Drug Metab. Disp.* **15**, 794.

Anonymous (1988). Vitamin E deficiency without fat malabsorption. *Nutr. Rev.* **46**, 189.

Aoshima, H., Anan, M., and Ishii, H. (1987). Effect of lipid hydroperoxide on *Xenopus* oocytes and on neurotransmitter receptors synthesized in *Xenopus* oocytes injected with exogenous mRNA. *Arch. Biochem. Biophys.* **258**, 324.

Arudi, R. L. *et al.* (1983). Purification of oleic acid and linoleic acid. *J. Lipid Res.* **24**, 485.

Aruoma, O. I. *et al.* (1989). The mechanism of initiation of lipid peroxidation. Evidence against a requirement for an iron(II)–iron(III) complex. *Biochem. J,* **258**, 617.

Ashida, H. *et al.* (1987). Effect of orally administered secondary oxidation products of linoleic acid on carbohydrate metabolism in rat liver. *Arch. Biochem. Biophys.* **259**, 114.

Aust, S. D., Morehouse, L. A., and Thomas, C. E. (1985). Role of metals in oxygen radical reactions. *J. Free Radical Biol. Med.* **1**, 3.

Barber, A. A. (1963). Addendum: mechanisms of lipid peroxide formation in rat tissue homogenates. *Radiat. Res.* suppl. **3**, 33.

Barber, D. J. W. and Thomas, J. K. (1978). Reaction of radicals with lecithin bilayers. *Radiat. Res.* **74**, 51.

Barclay, L. R. C. and Ingold, K. U. (1981). Autoxidation of biological molecules. 2. The autoxidation of a model membrane. A comparison of the autoxidation of egg lecithin phosphatidylcholine in water and in chlorobenzene. *J. Am. Chem. Soc.* **103**, 6478.

Benedetti, A. and Comporti, M. (1987). Formation, reactions and toxicity of aldehydes produced in the course of lipid peroxidation in cellular membranes. *Biolectrochem. Bioenerg.* **18**, 187.

Ben-Shachar, D. B., Finberg, J. P. M., and Youdim, M. B. H. (1985). Effect of iron chelators on dopamine D_2 receptors. *J. Neurochem.* **45**, 999.

Bereza, U. L., Brewer, G. J., and Hill, G. M. (1985). Effect of dietary cholesterol on erythrocyte peroxidant stress *in vitro* and *in vivo*. *Biochim. Biophys. Acta* **835**, 434.

Bieri, J. G., Corash, L., and Van Hubbard, S. (1983). Medical uses of vitamin E. *New Engl. J. Med.* **308**, 1063.

Bjørneboc, A. *et al.* (1986). Transport and distribution of a-tocopherol in lymph, serum and liver cells in rats. *Biochim. Biophys. Acta* **889**, 310.

Bond, A. M. *et al.* (1980). Malonaldehyde in cervical mucus associated with copper IUD. *Lancet* **i**, 1087.

Brajtburg, J. *et al.* (1985). Involvement of oxidative damage in erythrocyte lysis induced by amphotericin B. *Antimicrob. Agents Chemother.* **27**, 172.

Brash, A. R., Porter, A. T., and Maas, R. L. (1985). Investigation of the selectivity of hydrogen abstraction in the nonenzymatic formation of hydroxyeicosatetraenoic acids and leukotrienes by autoxidation. *J. Biol. Chem.* **260**, 4210.

Braughler, J. M., Duncan, L. A., and Chase, R. L. (1986). The involvement of iron in lipid peroxidation. *J. Biol. Chem.* **261**, 10282.

Brown, L. R. and Wüthrich, K. (1977). A spin label study of lipid oxidation catalyzed by heme proteins. *Biochim. Biophys. Acta* **464**, 356.

Brown, T. A. and Shrift, A. (1982). Selenium: toxicity and tolerance in higher plants. *Biol. Rev.* **57**, 59.

Buege, J. A. and Aust, S. D. (1976). Lactoperoxidase-catalyzed lipid peroxidation of microsomal and artificial membranes. *Biochim. Biophys. Acta* **444**, 192.

Burton, G. W. and Ingold, K. U. (1984). β-carotene: an unusual type of lipid antioxidant. *Science*, **62**, 569.

Burton, G. W. and Ingold, K. U. (1986). Vitamin E: application of the principles of physical organic chemistry to the explanation of its structure and function. *Acc. Chem. Res.* **19**, 194.

Burton, G. W. *et al.* (1986). Vitamin E in young and old human red blood cells. *Biochim. Biophys. Acta* **860**, 84.

Cadenas, E., Boveris, A., and Chance, B. (1980). Low-level chemiluminescence of hydroperoxide-supplemented cytochrome c. *Biochem. J.* **187**, 131.

Calabrese, L. and Carbonaro, M. (1986). An epr study of the non-equivalence of the copper sites of caeruloplasmin. *Biochem. J.* **238**, 291.

Cao, Y. Z. *et al.* (1987). Regulation by vitamin E of phosphatidylcholine metabolism in rat heart. *Biochem. J.* **247**, 135.

Chan, H. W. S. and Newby, U. K. (1980). Haemoprotein- and transition metal ion-catalysed oxidation of linoleic acid. Selectivity of the position of oxygenation. *Biochim. Biophys. Acta* **617**, 353.

Clemens, M. R. and Waller, H. D. (1987). Lipid peroxidation in erythrocytes. *Chem. Phys. Lipids* **45**, 251.

Clough, R. L., Yee, B. G., and Foote, C. S. (1979). Chemistry of singlet oxygen. The unstable primary product of tocopherol photooxidation. *J. Am. Chem. Soc.* **101**, 683.

Comporti, M. (1987). Glutathione depleting agents and lipid peroxidation. *Chem. Phys. Lipids* **45**, 143.

Coombs, G. F. *et al.* (1987). *Selenium in biology and medicine* Part A. Van Nostrand Reinhold, New York.

Corash, L. *et al.* (1980). Reduced chronic hemolysis during high-dose vitamin E administration in mediterranean-type glucose-6-phosphate dehydrogenase deficiency. *New Engl. J. Med.* **303**, 416.

Corongiu, F. P. *et al.* (1986). Lipid peroxidation and molecular damage to polyunsaturated fatty acids in rat liver. Recognition of two classes of hydroperoxides formed under conditions *in vivo*. *Chem. Biol. Interac.* **59**, 147.

Cotgreave, I. A. *et al.* (1987). N-acetylcysteine and glutathione-dependent protective effects of PZ51 (Ebselen) against diquat-induced cytotoxicity in isolated hepatocytes. *Biochem. Pharmacol.* **36**, 2899.

Coxon, D. T. *et al.* (1987). The occurrence of hydrogen peroxide in edible oils; chemical and microbiological consequences. *J. Sci. Food Agric.* **40**, 367.

Crompton, M., Costi, A., and Hayat, L. (1987). Evidence for the presence of a reversible Ca^{2+}-dependent pore activated by oxidative stress in heart mitochondria. *Biochem. J.* **245**, 915.

Curzio, M. *et al.* (1987). Possible role of aldehydic lipid peroxidation products as chemoattractants. *Intl. J. Tiss. React.* **9**, 295.

Daniel, J. W. (1986). Metabolic aspects of antioxidants and preservatives *Xenobiotica* **16**, 1073.

Daniels, V. (1984). The Russell effect—a review of its possible uses in conservation and the scientific examination of materials. *Studies in Conservation* **29**, 57.

Danielson, U. H., Esterbauer, H., and Mannervik, B. (1987). Structure–activity relationships of 4-hydroxyalkenals in the conjugation catalysed by mammalian glutathione transferases. *Biochem. J.* **247**, 707.

Daub, M. E. and Briggs, S. P. (1983). Changes in tobacco cell membrane composition and structure caused by cercosporin. *Plant Physiol.* **71**, 763.

Davies, M. J. (1987). Applications of electron spin resonance spectroscopy to the identification of radicals produced during lipid peroxidation. *Chem. Phys. Lipids* **44**, 149.

Dean, R. T., Thomas, S. M., and Garner, A. (1986). Free-radical-mediated fragmentation of monoamine oxidase in the mitochondrial membrane. *Biochem. J.* **240,** 489.

De Groot, H. and Noll, T. (1987). The role of physiological oxygen partial pressures in lipid peroxidation. Theoretical considerations and experimental evidence. *Chem. Phys. Lipids* **44,** 209.

Del Principe, D. *et al.* (1985). Hydrogen peroxide has a role in the aggregation of human platelets. *FEBS Lett.* **185,** 142.

Denko, C. W. (1979). Protective role of ceruloplasmin in inflammation *Agents Actions* **9,** 333.

Dillard, C. J., Gavino, V. C., and Tappel, A. L. (1983). Relative antioxidant effectiveness of a-tocopherol and γ-tocopherol in iron-loaded rats. *J. Nutr.* **113,** 2266.

Diplock, A. T. (1985). Vitamin E. In *Fat soluble vitamins* (A. T. Diplock, Ed.), p 154. Heinemann, London.

Dix, T. A. and Marnett, L. J. (1985). Conversion of linoleic acid hydroperoxide to hydroxy, keto, epoxyhydroxy, and trihydroxy fatty acids by hematin. *J. Biol. Chem.* **260,** 5351.

Dormandy, T. L. and Wickens, D. G. (1987). The experimental and clinical pathology of diene conjugation. *Chem. Phys. Lipids* **45,** 353.

Eluashvili, I. A. *et al.* (1977). Effect of chlorpromazine on enzymic oxidation of lipids. *Byull. Eksp. Biol. Med.* **84,** 323.

Erben-Russ, M., Bors, W., and Saran, M. (1987). Reactions of linoleic acid peroxyl radicals with phenolic antioxidants: a pulse radiolysis study. *Intl. J. Radiat. Biol.* **52,** 393.

Esterbauer, H. *et al.* (1986). Possible involvement of the lipid-peroxidation product 4-hydroxynonenal in the formation of fluorescent chromolipids. *Biochem. J.* **239,** 405.

Esterbauer, H. *et al.* (1987). Cytotoxic lipid peroxidation products. In *First Vienna shock forum*, Part A, p 245. A. R. Liss Inc, New York.

Fisher, D., Lichti, F. U., and Lucy, J. A. (1972). Environmental effects on the autoxidation of retinol. *Biochem. J.* **130,** 259.

Forni, L. G., Davies, M. J., and Willson, R. L. (1988). Vitamin E analogue Trolox C: ESR and pulse radiolysis studies on free radical reactions. *Biochem. J.* **255,** 513.

Fraga, C. G. *et al.* (1987). Effect of vitamin E- and selenium-deficiency on rat liver chemiluminescence. *Biochem. J.* **242,** 383.

Frankel, E. N. (1987). Biological significance of secondary lipid oxidation products. *Free Radical Res. Commun.* **3,** 213.

Frei, B. *et al.* (1988). Evaluation of an isoluminol chemiluminescence assay for the detection of hydroperoxides in human blood plasma. *Anal. Biochem.* **175,** 120.

Fukuzawa, K. and Gebicki, J. M. (1983). Oxidation of a-tocopherol in micelles and liposomes by the hydroxyl, perhydroxyl and superoxide free radicals. *Arch. Biochem. Biophys.* **226,** 242.

Gardner, H. W. (1989). Oxygen radical chemistry of polyunsaturated fatty acids. *Free Radical Biol. Med.* **7,** 65.

Gaunt, J. K., Matthews, G. M., and Plumpton, E. S. (1980). Control *in vitro* of tocopherol oxidase by light and by auxins, kinetins, gibberellic acid, abscisic acid and ethylene. *Biochem. Soc. Trans.* **8,** 186.

Gebicki, J. M. and Bielski, B. M. J. (1981). Comparison of the capacities of the perhydroxyl and the superoxide radicals to initiate chain oxidation of linoleic acid. *J. Am. Chem. Soc.* **103**, 7020.

Gelmont, D., Stein, R. A., and Mead, J. F. (1981). Isoprene—the main hydrocarbon in human breath. *Biochem. Biophys. Res. Commun.* **99**, 1456.

Giasuddin, A. S. M. and Diplock, A. T. (1981). The influence of vitamin E on membrane lipids of mouse fibroblasts. *Arch. Biochem. Biophys.* **210**, 348.

Gibson, D. G., Hawrylko, J., and McCay, P. B. (1985). GSH-dependent inhibition of lipid peroxidation: properties of a potent cytosolic system which protects cell membranes. *Lipids* **20**, 704.

Girotti, A. W., Bachowski, G. J., and Jordan, J. E. (1987). Lipid peroxidation in erythrocyte membranes: cholesterol product analysis in photosensitized and xanthine oxidase-catalyzed reactions. *Lipids* **22**, 401.

Gray, J. I. (1978). Measurement of lipid oxidation: a review *J. Am. Oil Chem. Soc.* **55**, 539.

Gutteridge, J. M. C. (1978). The membrane effects of vitamin E, cholesterol and their acetates on peroxidative susceptibility. *Res. Comm. Chem. Pathol. Pharmacol.* **22**, 563.

Gutteridge, J. M. C. (1983). Antioxidant properties of caeruloplasmin towards iron- and copper-dependent oxygen radical formation. *FEBS Lett.* **157**, 37.

Gutteridge, J. M. C. (1984). Ferrous ion-EDTA-stimulated phospholipid peroxidation. A reaction changing from alkoxyl-radical- to hydroxyl-radical-dependent initiation. *Biochem. J.* **224**, 697.

Gutteridge, J. M. C. (1986). Aspects to consider when detecting and measuring lipid peroxidation. *Free Radical Res. Commun.* **1**, 173.

Gutteridge, J. M. C. (1988). Lipid peroxidation: some problems and concepts. In *Proceedings of the Upjohn Symposium on Oxidants and Diseases* (B. Halliwell, Ed.). Lawrence Press, Kansas, USA.

Gutteridge, J. M. C. and Kerry, P. J. (1982). Detection by fluorescence of peroxides and carbonyls in samples of arachidonic acid. *Br. J. Pharmacol.* **76**, 459.

Gutteridge, J. M. C. and Quinlan, G. J. (1983). Malondialdehyde formation from lipid peroxides in the thiobarbituric acid test: the role of lipid radicals, iron salts and metal chelators. *J. Appl. Biochem.* **5**, 293.

Gutteridge, J. M. C., Lamport, P., and Dormandy, T. L. (1976). The antibacterial effect of water-soluble compounds from autoxidising linolenic acid. *J. Med. Microbiol.* **9**, 105.

Gutteridge, J. M. C., Thomas, A. H., and Cuthbert, A. (1983). Free radical damage to polyene antifungal antibiotics: changes in biological activity and thiobarbituric acid reactivity. *J. Appl. Biochem.* **5**, 53.

Gutteridge, J. M. C., Beard, A. P. C., and Quinlan, G. J. (1983). Superoxide-dependent lipid peroxidation. Problems with the use of catalase as a specific probe for Fenton-derived hydroxyl radicals. *Biochem. Biophys. Res. Commun.* **117**, 901.

Halliwell, B. (1988). Albumin—an important intracellular antioxidant? *Biochem. Pharmacol.* **37**, 569.

Halliwell, B. and Gutteridge, J. M. C. (1984). Lipid peroxidation, oxygen radicals, cell damage and antioxidant therapy. *Lancet* **i**, 1396.

Halliwell, B. and Gutteridge, J. M. C. (1985). Oxygen radicals and the nervous system. *Trends Neurosci.* **8**, 22.

Harding, A. E. (1987). Vitamin E and the nervous system. *CRC Crit. Rev. Neurobiol.* **3**, 89.

Hauptlorenz, S. *et al.* (1985). Effect of the lipid peroxidation product 4-hydroxynonenal and related aldehydes on proliferation and viability of cultured Ehrlich ascites tumor cells. *Biochem. Pharmacol.* **34**, 3803.

Hill, J. M. and Switzer, R. C. (1984). The regional distribution and cellular localization of iron in the rat brain. *Neuroscience* **11**, 595.

Hochstein, P. and Jain, S. K. (1981). Association of lipid peroxidation and polymerization of membrane proteins with erythrocyte aging. *Fed. Proc.* **40**, 183.

Hochstein, P., Sree Kumar, K., and Forman, S. J. (1980). Lipid peroxidation and the cytotoxicity of copper. *Ann. NY Acad. Sci.* **355**, 240.

Hoekstra, W. G. (1975). Biochemical function of selenium and its relation to vitamin E. *Fed. Proc.* **34**, 2083.

Hunter, M. I. S. *et al.* (1985). Lipid peroxidation products and antioxidant proteins in plasma and cerebrospinal fluid from multiple sclerosis patients. *Neurochem. Res.* **10**, 1645.

Igene, J. O. *et al.* (1985). Evaluation of 2-thiobarbituric acid reactive substances (TBRs) in relation to warmed-over flavor (WOF) development in cooked chicken *J. Agric. Food Chem.* **33**, 364.

Ilio, T. and Yoden, K. (1987). Fluorescence formation and heme degradation at different stages of lipid peroxidation. *Life Sci.* **40**, 2297.

Infante, J. P. (1986). Vitamin E and selenium participate in fatty acid desaturation. A proposal for an enzymatic function of these nutrients. *Mol. Cell. Biochem.* **69**, 93.

Ingold, K. U. *et al.* (1987). Vitamin E remains the major lipid-soluble, chain-breaking antioxidant in human plasma even in individuals suffering severe vitamin E deficiency. *Arch. Biochem. Biophys.* **259**, 224.

Jain, S. K. (1985). *In vivo* externalization of phosphatidylserine and phosphatidylethanolamine in the membrane bilayer and hypercoagulability by the lipid peroxidation of erythrocytes in rats. *J. Clin. Invest.* **76**, 281.

Jones, R., Mann, T., and Sherins, R. J. (1978). Adverse effects of peroxidized lipid on human spermatozoa *Proc. Roy. Soc. Lond. Ser. B* **201**, 413.

Kaneda, T. and Miyazawa, T. (1987). Lipid peroxides and nutrition. *Wld Rev. Nutr. Diet.* **50**, 186.

Kanner, J., German, J. B., and Kinsella, J. E. (1987). Initiation of lipid peroxidation in biological systems. *CRC Crit. Rev. Food Sci. Nutr.* **25**, 317.

Kanofsky, J. R. (1983). Singlet oxygen production by lactoperoxidase. *J. Biol. Chem.* **258**, 5991.

Karlsson, K. and Marklund, S. L. (1987). Heparin-induced release of extracellular superoxide dismutase to human blood plasma *Biochem. J.* **242**, 55.

Kaschnitz, R. M. and Hatefi, Y. (1975). Lipid oxidation in biological membranes. Electron transfer proteins as initiators of lipid autoxidation. *Arch. Biochem. Biophys.* **171**, 292.

Kikugawa, K. and Beppu, M. (1987). Involvement of lipid oxidation products in the formation of fluorescent and cross-linked proteins. *Chem. Phys. Lipids* **44**, 277.

Kostrucha, J. and Kappus, H. (1986). Inverse relationship of ethane or n-pentane and malondialdehyde formed during lipid peroxidation in rat liver microsomes with different oxygen concentrations. *Biochim. Biophys. Acta* **879**, 120.

Kosugi, H., Kato, T., and Kikugawa, K. (1987). Formation of yellow, orange, and red pigments in the reaction of alk-2-enals with 2-thiobarbituric acid. *Anal. Biochem.* **165**, 456.

Kramer, K. *et al.* (1986). Influence of lipid peroxidation on a-adrenoceptors. *FEBS Lett.* **198**, 80.

Kunert, K. J. and Tappel, A. L. (1983). The effect of vitamin C on *in vivo* lipid peroxidation in guinea pigs as measured by pentane and ethane production. *Lipids* **18**, 271.

Kuross, S. A. and Hebbel, R. P. (1988). Nonheme iron in sickle erythrocyte membranes: association with phospholipids and potential role in lipid peroxidation. *Blood* **72**, 1278.

Kushner, I. and Mackiewicz, A. (1987). Acute phase proteins as disease markers. *Dis. Markers* **5**, 1.

Lai, E. K., Fong, K. L., and McCay, P. B. (1978). Studies on the properties of the singlet oxygen-like factor produced during lipid peroxidation. *Biochim. Biophys. Acta* **528**, 497.

Landi, L. *et al.* (1987). Effect of oxygen free radicals on ubiquinone in aqueous solution and phospholipid vesicles. *Biochim. Biophys. Acta* **902**, 200.

Lang, J., Celotto, C., and Esterbauer, H. (1985). Quantitative determination of the lipid peroxidation product 4-hydroxynonenal by high-performance liquid chromatography. *Anal. Biochem.* **150**, 369.

Levander, O. A. (1987). A global view of human selenium nutrition. *Ann. Rev. Nutr.* **7**, 227.

Little, C. (1972). Steroid hydroperoxides as substrates for glutathione peroxidase. *Biochim. Biophys. Acta* **284**, 375.

Lynch, R. E. and Fridovich, I. (1978), Permeation of the erythrocyte stroma by superoxide radical. *J. Biol. Chem.* **253**, 4697.

Maddipati, K. R. and Marnett, L. J. (1987). Characterization of the major hydroperoxide-reducing activity of human plasma. *J. Biol. Chem.* **262**, 17398.

Mak, I. T., Misra, H. P., and Weglicki, W. B. (1983). Temporal relationship of free radical-induced lipid peroxidation and loss of latent enzyme activity in highly enriched hepatic lysosomes. *J. Biol. Chem.* **258**, 13737.

Mak, I. T., Kramer, J. H., and Weglicki, W. B. (1986). Potentiation of free radical-induced lipid peroxidative injury to sarcolemmal membranes by lipid amphiphiles. *J. Biol. Chem.* **261**, 1153.

Malvy, C. *et al.* (1980). Lipid peroxidation in liver: hydroxydimethyl carbazole a new potent inhibitor. *Biochem. Biophys. Res. Commun.* **95**, 734.

Marklund, S. L. (1984). Extracellular superoxide dismutase in human tissues and human cell lines. *J. Clin. Invest.* **74**, 1398.

Marklund, S. L. (1985). Product of extracellular-superoxide dismutase catalysis. *FEBS Lett.* **184**, 237.

Marklund, S. L. (1986). Ceruloplasmin, extracellular-superoxide dismutase, and scavenging of superoxide anion radicals. *J. Free Radical Biol. Med.* **2**, 255.

Marnett, L. J. *et al.* (1985). Naturally occurring carbonyl compounds are mutagens in *Salmonella* tester strain TA104. *Mut. Res.* **148**, 25.

Marnett, L. J. *et al.* (1986). Reaction of malondialdehyde with guanine nucleosides:

formation of adducts containing oxadiazabicyclononene residues in the base-pairing region. *J. Am. Chem. Soc.* **108**, 1348.

Marshall, P. J., Warso, M. A., and Lands, W. E. M. (1985). Selective microdetermination of lipid hydroperoxides. *Anal. Biochem.* **145**, 192.

Masini, A. *et al.* (1985). The effect of ferric iron complex on isolated rat liver mitochondria. 1. Respiratory and electrochemical responses. *Biochim. Biophys. Acta* **810**, 20.

McCay, P. B. (1985). Vitamin E: interactions with free radicals and ascorbate. *Ann. Rev. Nutr.* **5**, 323.

McGirr, L. G., Hadley, M., and Draper, H. H. (1985). Identification of N$^{\alpha}$-acetyl-ε-(2-propenal) lysine as a urinary metabolite of malondialdehyde. *J. Biol. Chem.* **260**, 15427.

Miccadei, S. *et al.* (1988). Toxic consequences of the abrupt depletion of glutathione in cultured rat hepatocytes. *Arch. Biochem. Biophys.* **265**, 311.

Minotti, G. and Aust, S. D. (1987). The role of iron in the initiation of lipid peroxidation. *Chem. Phys. Lipids* **44**, 191.

Moshage, H. J. *et al.* (1987). Studies of the molecular-mechanism of decreased liver synthesis of albumin in inflammation *J. Clin. Invest.* **79**, 1635.

Nakakimura, H. *et al.* (1980). Studies on lipid peroxidation in biological systems. I. Effects of various factors on lipid peroxide levels in blood. *Chem. Pharm. Bull.* **28**, 2101.

Negishi, H., Fujimoto, K., and Kaneda, T. (1980). Effect of autoxidized methyl linoleate on glutathione peroxidase. *J. Nutr. Sci. Vitaminol.* **26**, 309.

Noll, T., de Groot, H., and Sies, H. (1987). Distinct temporal relationship among oxygen uptake, malondialdehyde formation and low-level chemiluminescence during microsomal lipid peroxidation. *Arch. Biochem. Biophys.* **252**, 284.

Noronha-Dutra, A. A. and Steen, E. M. (1982). Lipid peroxidation as a mechanism of injury in cardiac myocytes. *Lab. Invest.* **47**, 346.

Oarada, M., Miyazawa, T., and Kaneda, T. (1986). Distribution of ^{14}C after oral administration of (U-^{14}C) labeled methyl linoleate hydroperoxides and their secondary oxidation products in rats. *Lipids* **21**, 150.

Oarada, M. *et al.* (1988). The effect of dietary lipid hydroperoxide on lymphoid tissues in mice. *Biochim. Biophys. Acta* **960**, 229.

O'Brien, P. J. (1969). Intracellular mechanisms for the decomposition of a lipid peroxide. I. Decomposition of a lipid peroxide by metal ions, heme compounds, and nucleophiles. *Can. J. Biochem.* **47**, 485.

O'Connell, M. J. and Garner, A. (1983). Radiation-induced generation and properties of lipid hydroperoxide in liposomes. *Intl. J. Radiat. Biol.* **44**, 615.

O'Connell, M. J. and Peters, T. J. (1987). Ferritin and haemosiderin in free radical generation, lipid peroxidation and protein damage. *Chem. Phys. Lipids* **45**, 241.

Organisciak, D. T. *et al.* (1987). Vitamin E in human neural retina and retinal pigment epithelium: effect of age. *Curr. Eye Res.* **6**, 1051.

Overbaugh, J. M. and Fall, R. (1985). Characterization of a selenium-independent glutathione peroxidase from *Euglena gracilis*. *Plant Physiol.* **77**, 437.

Pacht, E. R. and Davies, W. B. (1988). Role of transferrin and ceruloplasmin in antioxidant activity of lung epithelial lining fluid. *J. Appl. Physiol.* **64**, 2092.

Parnham, M. J. and Graf, E. (1987). Seleno-organic compounds and the therapy of hydroperoxide-linked pathological conditions. *Biochem. Pharmacol.* **36**, 3095.

Pauls, K. P. and Thompson, J. E. (1980). *In vitro* simulation of senescence-related membrane damage by ozone-induced lipid peroxidation. *Nature* **283**, 504.

Pietronigro, D. D. *et al.* (1977). Interaction of DNA and liposomes as a model for membrane-mediated DNA damage. *Nature* **267**, 78.

Reiter, R. and Wendel, A. (1983). Selenium and drug metabolism. *Biochem. Pharmacol.* **32**, 3063.

Rice-Evans, C. (1987). Oxidative modifications in erythrocytes induced by iron. In *Free radicals, oxidant stress and drug action*, p 307. Richelieu Press, London.

Rice-Evans, C. *et al.* (1985). *t*-Butyl hydroperoxide-induced perturbations of human erythrocytes as a model for oxidant stress. *Biochim. Biophys. Acta* **815**, 426.

Richter, C. (1987). Biophysical consequences of lipid peroxidation in membranes. *Chem. Phys. Lipids* **44**, 175.

Rosen, G. and Rauckman, E. J. (1981). Spin trapping of free radicals during hepatic microsomal lipid peroxidation. *Proc. Natl. Acad. Sci. USA* **78**, 7346.

Sandmann, G. and Böger, P. (1980). Copper deficiency and toxicity in *Scenedesmus*. *Z. Pflanzenphysiol.* **98**, 53.

Sandstrom, B. E. R., Carlsson, J., and Marklund, S. L. (1987). Variations among cultured cells in glutathione peroxidase activity in response to selenite supplementation. *Biochim. Biophys. Acta* **929**, 148.

Schacter, B. A., Marver, H. S., and Meyer, U. A. (1972). Hemoprotein catabolism during stimulation of microsomal lipid peroxidation. *Biochim. Biophys. Acta*, **279**, 221.

Schwartz, R. S. *et al.* (1987). Protein 4·1 in sickle erythrocytes. Evidence for oxidative damage. *J. Biol. Chem.* **262**, 15666.

Scott, J. A. *et al.* (1987). Free radical-mediated depolarization in renal and cardiac cells. *Biochim. Biophys. Acta* **899**, 76.

Scott, G. (Ed.) (1985). *Developments in polymer stabilisation*, volume 7. Elsevier, London and New York.

Sevanian, A. and Hochstein, P. (1985). Mechanisms and consequences of lipid peroxidation in biological systems. *Ann. Rev. Nutr.* **5**, 365.

Shimizu, T., Kondo, K., and Hayaishi, O. (1981). Role of prostaglandin endoperoxides in the serum thiobarbituric acid reaction. *Arch. Biochem. Biophys.* **206**, 271.

Sies, H. (1985). *Oxidative stress*, Chapters 3, 10, 12, 13, 14. Academic Press, New York.

Siu, G. M. and Draper, H. H. (1982). Metabolism of malonaldehyde *in vivo* and *in vitro*. *Lipids* **17**, 349.

Smith, L. L. (1987). Cholesterol autoxidation 1981–1986. *Chem. Phys. Lipids* **44**, 87.

Stocker, R., Glazer, A. N., and Ames, B. (1987). Antioxidant activity of albumin-bound bilirubin. *Proc. Natl. Acad. Sci USA* **84**, 5918.

Stocks, J. *et al.* (1974). Assay using brain homogenate for measuring the antioxidant activity of biological fluids. *Clin. Sci.* **47**, 215.

Stocks, J. *et al.* (1974). The inhibition of lipid autoxidation by human serum and its relation to serum proteins and α-tocopherol. *Clin. Sci.* **47**, 223.

Subczynski, W. K. and Kusumi, A. (1985). Detection of oxygen consumption during very early stages of lipid peroxidation by ESR nitroxide spin probe method. *Biochim. Biophys. Acta* **821**, 259.

Suwa, K., Kimura, T., and Schaap, A. P. (1977). Reactivity of singlet molecular oxygen with cholesterol in a phospholipid membrane matrix. A model for oxidative damage of membranes. *Biochem. Biophys. Res. Commun.* **75**, 785.

Takahashi, K. and Cohen, H. J. (1986). Selenium-dependent glutathione peroxidase protein and activity: immunological investigations on cellular and plasma enzymes. *Blood* **68**, 640.

Takahashi, K., Newberger, P. E., and Cohen, H. J. (1986). Glutathione peroxidase protein. Absence in selenium deficiency states and correlation with enzymatic activity. *J. Clin. Invest.* **77**, 1402.

Takahashi, M. and Asada, K. (1983). Superoxide permeability of phospholipid membranes and chloroplast thylakoids. *Arch. Biochem. Biophys.* **226**, 558.

Tan, K. H. *et al.* (1984). Inhibition of microsomal lipid peroxidation by glutathione and glutathione transferases B and AA. *Biochem. J.* **220**, 243.

Tan, K. H. *et al.* (1988). Detoxication of DNA hydroperoxide by glutathione transferases and the purification and characterization of glutathione transferases of the rat liver nucleus. *Biochem. J.* **254**, 841.

Tappel, A. L. (1979). Measurement of and protection from *in vivo* lipid peroxidation. In *Biochemical and clinical aspects of oxygen* (W. S. Caughey, Ed.). Academic Press, New York.

Tauber, A. I., Fay, J. R., and Marletta, M. A. (1984). Flavonoid inhibition of the human neutrophil NADPH-oxidase. *Biochem. Pharmacol.* **33**, 1367.

Terelius, Y. and Ingelman-Sundberg, M. (1986). Metabolism of *n*-pentane by ethanol-inducible cytochrome P-450 in liver microsomes and reconstituted membranes. *Eur. J. Biochem.* **161**, 303.

Thompson, J. A. and Wand, M. D. (1985). Interaction of cytochrome P-450 with a hydroperoxide derived from butylated hydroxytoluene. *J. Biol. Chem.* **260**, 10637.

Thompson, S. and Smith, M. T. (1985). Measurement of the diene conjugated form of linoleic acid in plasma by high performance liquid chromatography: a questionable non-invasive assay of free radical activity? *Chem–Biol. Interact.* **55**, 357.

Thurnham, D. I. *et al.* (1986). The use of different lipids to express serum tocopherol: lipid ratios for the measurement of vitamin E status. *Ann. Clin. Biochem.* **23**, 514.

Tibell, L. *et al.* (1987). Expression of human extracellular superoxide dismutase in Chinese hamster ovary cells and characterization of the product. *Proc. Natl. Acad. Sci. USA* **84**, 6634.

Tipton, C. L. *et al.* (1987). Cholesterol hydroperoxides inhibit calmodulin and suppress atherogenesis in rabbits. *Biochem. Biophys. Res. Commun.* **146**, 1166.

Toth, K. M. *et al.* (1984). Intact human erythrocytes prevent hydrogen peroxide-mediated damage to isolated perfused rat lungs and cultured bovine pulmonary artery endothelial cells. *J. Clin. Invest.* **74**, 292.

Triggs, W. J. and Willmore, L. J. (1984). *In vivo* lipid peroxidation in rat brain following intracortical Fe^{2+} injection. *J. Neurochem.* **42**, 976.

Ursini, F., Maiorino, M., and Gregolin, C. (1985). The selenoenzyme phospholipid hydroperoxide glutathione peroxidase. *Biochim. Biophys. Acta* **839**, 62.

Valenzeno, D. P. (1987). Photomodification of biological membranes with emphasis on singlet oxygen mechanisms. *Photochem. Photobiol.* **46**, 147.

Van Kuijk, F. J. G. M. *et al.* (1986). Occurrence of 4-hydroxyalkenals in rat tissues determined as pentafluorobenzyl oxime derivatives by gas chromatography-mass spectrometry. *Biochem. Biophys. Res. Commun.* **139**, 144.

Videla, L. A. *et al.* (1984). Changes in oxygen consumption induced by *t*-butyl hydroperoxide in perfused rat liver. *Biochem. J.* **223**, 879.

Vladimirov, Y. A. (1980). Lipid peroxidation in mitochondrial membrane. *Adv. Lipid Res.* **17**, 173.

Wade, C. R. and Van Rij, A. M. (1985). *In vitro* lipid peroxidation in Man as measured by the respiratory excretion of ethane, pentane and other low-molecular-weight hydrocarbons. *Anal. Biochem.* **150**, 1.

Wahlgren, N. G. and Lindqvist, C. (1987). Haem derivatives in the cerebrospinal fluid after intracranial haemorrhage. *Eur. Neurol.* **26**, 216.

Wayner, D. D. M., Burton, G. W., and Ingold, K. U. (1986). The antioxidant efficiency of vitamin C is concentration-dependent. *Biochim. Biophys. Acta* **884**, 119.

Wayner, D. D. M. *et al.* (1985). Quantitative measurement of the total, peroxyl radical-trapping antioxidant capability of human blood plasma by controlled peroxidation. *FEBS Lett.* **187**, 33.

Wefers, H. (1987). Singlet oxygen in biological systems. *Bioelectrochem. Bioenerg.* **18**, 91.

Weiss, R. H. and Estabrook, R. W. (1986). The mechanism of cumene hydroperoxide-dependent lipid peroxidation. The function of cytochrome P-450. *Arch. Biochem. Biophys.* **251**, 348.

Weiss, R. H., Arnold, J. L., and Estabrook, R. W. (1987). Transformation of an arachidonic acid hydroperoxide into epoxyhydroxy and trihydroxy fatty acids by liver microsomal cytochrome P-450. *Arch. Biochem. Biophys.* **252**, 334.

Willmore, L. J., Triggs, W. J., and Gray, J. D. (1986). The role of iron-induced hippocampal peroxidation in acute epileptogenesis. *Brain Res.* **382**, 422.

Wills, E. D. (1964). The effect of inorganic iron on the thiobarbituric acid method for the determination of lipid peroxides. *Biochim. Biophys. Acta* **84**, 475.

Wills, E. D. (1966). Mechanisms of lipid peroxide formation in animal tissues. *Biochem. J.* **99**, 667.

Wills, E. D. (1969). Lipid peroxide formation in microsomes. The role of non-haem iron. *Biochem. J.* **113**, 325.

Wills, E. D. (1971). Effect of lipid peroxidation on membrane-bound enzymes of the endoplasmic reticulum. *Biochem. J.* **123**, 983.

Wills, E. D. (1980). Effect of antioxidants on lipid peroxide formation in irradiated synthetic diets. *Intl. J. Radiat. Biol.* **37**, 403.

Yagi, K. (Ed.) (1982). *Lipid peroxides in biology and medicine*. Academic Press.

Yamamoto, Y. *et al.* (1987). Detection and characterization of lipid hydroperoxides at picomole levels by high-performance liquid chromatography. *Anal. Biochem.* **160**, 7.

Yamoshoji, S. and Kajimoto, G. (1983). Antioxidant effect of caeruloplasmin on microsomal lipid peroxidation. *FEBS Lett.* **152**, 168.

Zaleska, M. M. and Floyd, R. A. (1985). Regional lipid peroxidation in rat brain in vitro: possible role of endogenous iron. *Neurochem. Res.* **10**, 397.

5

Protection against radical damage: systems with problems

In previous chapters we have described the problems faced by several different organs and tissues in coping with oxygen, and the ways in which they can deal with oxygen radicals and lipid peroxidation. For example, the lung is exposed to the highest concentration of oxygen of any body tissue and it uses SOD, catalase, and glutathione peroxidase for protection (Chapter 3). Human erythrocytes have to keep their membranes intact for 120 days in the face of a constant flux of O_2^- from haemoglobin, even though they cannot readily replace membrane lipids damaged by peroxidation. Hence they rely on α-tocopherol, SOD, catalase, and glutathione peroxidase (Chapters 3 and 4). The swim-bladder of at least one fish has a very high SOD activity to enable it to cope with high oxygen concentrations (Chapter 3). Even the earthworm *Eisenia foetida* increases its respiratory rate with temperature and also its SOD activity, presumably to cope with the increased O_2^- generation at higher temperatures.

The purpose of this chapter, however, is to focus on two systems that have exceptionally difficult problems—the chloroplasts of higher plants, and the mammalian eye. These very different systems have a lot in common.

5.1 The chloroplasts of higher plants

The chloroplasts present in the leaves of higher plants can be seen under the electron microscope to be bounded by an outer *envelope* consisting of two membranes separated by an electron-translucent space of about 10 nm (Fig. 5.1). The envelope encloses the *stroma* of the chloroplast, in which floats a complex internal membrane structure. The stroma is an aqueous solution containing various low-molecular-mass compounds plus a high concentration of proteins, most of which are the enzymes necessary to convert carbon dioxide into carbohydrate by a complex metabolic pathway known as the *Calvin cycle*. For its operation the Calvin cycle requires ATP and NADPH. The first enzyme in the Calvin cycle catalyses reaction of the 5-carbon sugar ribulose 1,5-bisphosphate with carbon dioxide to form two molecules of phosphoglyceric acid. It is called ribulose bisphosphate carboxylase.

The internal membrane structure of the chloroplast, which floats in the

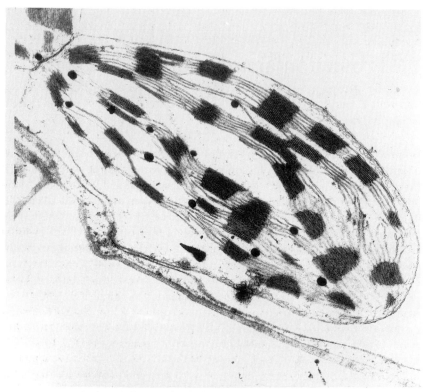

Fig. 5.1.   Electron micrograph of chloroplasts in the leaves of higher plants.

stroma, is extremely complex. In the electron micrograph shown in Fig. 5.1, two distinct features may be recognized, i.e. regions of closely stacked membranes known as *grana* that are interconnected by a three-dimensional network of membranes known as the *stroma thylakoids*. The membranes of this complicated internal structure contain the photosynthetic pigments (mainly chlorophylls a and b, green pigments that absorb light in the blue and red regions of the spectrum, together with *carotenoids* and *xanthophylls*, yellow pigments absorbing blue light). They function to produce the NADPH and, by the process of *photophosphorylation*, the ATP needed to drive CO_2 fixation by the Calvin cycle in the stroma. The structure of chlorophylls a and b is described in Chapter 2.

Absorption of light energy by chlorophyll or other pigment molecules causes electrons to move into higher energy states (*excited singlet states*). Absorption of blue light by chlorophyll results in formation of a more excited state (second excited state) than does absorption of red light, but the second excited state loses its excess energy as heat extremely quickly, so that absorption of either red or blue light effectively produces the same first excited state of the chlorophyll molecules. Energy so absorbed can be lost by re-emission of light to give fluorescence; it can be lost as heat or it can be transferred to another molecule, e.g. an adjacent chlorophyll, the first molecule returning to the ground state, while the second becomes excited.

Certain special molecules of chlorophyll a in the thylakoid membrane (*reaction centre chlorophylls*) can, when excited, lose an electron to a neighbouring electron acceptor (A), producing a (chlorophyll$^+$ A$^-$) pair. This initial charge separation is the basic reaction of photosynthesis. Light energy absorbed by other chlorophyll a or chlorophyll b molecules can be transferred quite efficiently in the thylakoid membrane to the reaction centre chlorophylls. Indeed, each reaction centre chlorophyll is associated with a *light-harvesting array* of other pigment molecules that can channel energy to it.

Two different types of reaction centre, known as P-700 and P-680, can be detected in the chloroplasts of higher plants. Each contains chlorophyll a and is served by its own light-harvesting pigment system; the complexes containing P-700 and its harvesting pigments are referred to as photosystems I and those containing P-680 as photosystems II. Stroma thylakoids are especially rich in photosystems I whereas the grana are enriched in photosystems II. Electrons ejected from the P-680 reaction centre of a photosystem II are accepted by a molecule of unknown structure, usually referred to as 'Q'. From Q electrons pass 'down' an electron transport chain that contains plastoquinone, plastocyanin, and cytochrome f, and they eventually enter photosystems I to replace the electrons lost on excitation of the P-700 reaction centres (Fig. 5.2). The molecule which accepts excited electrons from photosystems I has not been conclusively identified, but it contains iron (not in the form of a haem ring) and sulphur. From this primary acceptor, electrons pass to other iron–sulphur proteins bound to the membranes, and then onto a soluble (stromal) iron–sulphur protein, known as *ferredoxin*. Soluble ferredoxin from higher plants contains two iron ions and two atoms of sulphur per molecule, and it acts as a one-electron acceptor. Reduced ferredoxin can then, in the presence of a reductase enzyme, donate electrons to NADP$^+$ to give NADPH (Fig. 5.3).

The electrons ejected from the photosystem II reaction centres are replaced by the splitting of water molecules, accompanied by oxygen evolution. The detailed chemistry of the water-splitting reaction is not understood, but a 'charge-accumulating mechanism' is present in photosystem II. It stores up to four units of positive charge, corresponding to loss of four electrons from photosystem II. When fully 'charged', it can then take four electrons from two water molecules to generate a molecule of oxygen. This charge accumulator is usually designated as 'S' (like 'Q', we don't know what it is) so we can write:

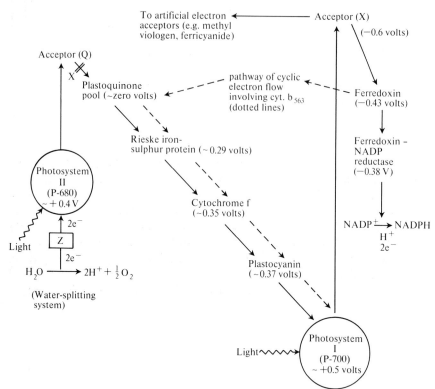

Fig. 5.2. The electron transport chain of the chloroplast. This current scheme of photosynthetic electron transport is often known as the Hill–Bendall scheme or the 'Z'-scheme. The sequence of carriers is presented as an energy diagram using redox potentials (shown in parentheses in volts). The lower the redox potential of a substance, the better is its electron-donating capacity. Hence a component with a negative redox potential is theoretically capable of donating electrons to another component with a less negative, zero, or positive redox potential. The redox potentials of the components are obtained by titrating them with reagents of known redox potential. X represents the site of action of the commonly used photosynthetic inhibitor DCMU. Q is a molecule of unknown structure, plastoquinone is a diphenol that resembles coenzyme Q (Chapter 3), cytochrome f is a typical cytochrome, and plastocyanin is a copper protein. Electrons ejected from photosystem II are replaced by the splitting of water. The acceptor Q passes its electrons into an electron transport chain, in which a plastoquinone pool is involved. Like coenzyme Q in mitochondria, oxidized plastoquinone (a quinone) can accept one electron to form a semiquinone or two to form a diphenol. This 'pool' of plastoquinone molecules may collect electrons from several different photosystems II and pass them onto an iron-sulphur protein (the Rieske iron-sulphur protein) and hence onto cytochrome f and plastocyanin. As electrons flow from Q to plastocyanin, sufficient energy is released to drive the synthesis of ATP from ADP and phosphate (*photophosphorylation*). Plastocyanin donates electrons to PSI to replace those ejected from the reaction centre.

$$S \xrightarrow{-1e^-} S^+ \xrightarrow{-1e^-} S^{2+} \xrightarrow{-1e^-} S^{3+} \xrightarrow{-1e^-} S^{4+}$$

$$S^{4+} + 2H_2O \longrightarrow 4H^+ + S + O_2$$

The transition metal manganese is intimately involved in the production of oxygen, but exactly what it does is not yet clear.

5.1.1 *The problems*

Chloroplasts are especially prone to oxygen-toxicity effects, firstly because their internal oxygen concentration in the light will always be greater than that in the surrounding atmosphere, due to oxygen production in photosystems II. Secondly, the lipids present in the chloroplast envelope, and thylakoids contain a high percentage of polyunsaturated fatty acids, and are thus very susceptible to peroxidation. Thirdly, we saw in Chapter 2 that illuminated chlorophyll can sensitize the formation of singlet oxygen, which is especially damaging to membrane lipids (Chapter 4). When chlorophyll absorbs light energy, it enters an excited singlet state. Most of this energy is

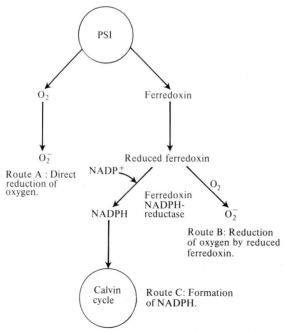

Fig. 5.3.   Routes of electron flow from reduced photosystem I. Reduced ferredoxin reduces oxygen to O_2^- as well as passing electrons onto $NADP^+$. The electron acceptor of PSI itself slowly reduces oxygen to O_2^-.

transferred to the reaction centre chlorophylls, and some can be lost as fluorescence, as explained previously. However, if energy is not dissipated by these mechanisms, *intersystem crossing* may occur to generate the triplet state of chlorophyll, which is the form that can transfer energy to O_2 and generate singlet oxygen. Prolonged illumination of chloroplast thylakoids *in vitro* causes marked lipid peroxidation. Indeed, when polyunsaturated fatty acids are simply mixed with chlorophyll and the mixture illuminated, they are rapidly peroxidized. The deterioration of isolated, illuminated chloroplast thylakoids is an extremely complicated process. In addition to lipid peroxidation and the formation of reactive aldehydes by peroxide decomposition, there is actual hydrolysis of lipids to release fatty acids. Both esterified and released fatty acids undergo peroxidation, and the released fatty acids themselves cause membrane damage and inhibit photosynthesis. Lipid hydrolysis is due to the action of lipase enzymes which normally show little activity in chloroplasts, but seem to be 'unmasked' during membrane deterioration.

Since the trapping of light by chlorophyll causes formation of excited states, then production of singlet oxygen would be expected, since there is plenty of oxygen around. If operation of the electron transport chain were prevented, the reaction-centre chlorophylls could not dispose of their excitation energy, nor could they accept energy from their harvesting pigments and so singlet O_2 formation should be greatly increased. Indeed, about 50 per cent of all known herbicides act by inhibiting the electron transport chain. They include monuron (CMU), ioxynil, and atrazine. Their inhibition of electron transport stops light-dependent generation of ATP and NADPH and so prevents the plant from fixing CO_2. In green plant tissues exposed to light, this inhibition is followed by carotenoid destruction, chlorophyll bleaching, and membrane deterioration accompanied by increased lipid peroxidation.

The urea and triazine classes of herbicide inhibitors of electron transport bind to a thylakoid protein of relative molecular mass 32 000. This protein is involved in the transfer of electrons between the primary electron acceptor of photosystem II ('Q') and a secondary acceptor ('B'), from which the electrons enter the 'pool' of plastoquinone molecules (Fig. 5.2). It is hence called the Q_B protein. When chloroplasts are illuminated, the Q_B protein turns over extremely rapidly, many times faster than other thylakoid proteins. Hence a major part of the chloroplast machinery for protein synthesis is used to make new Q_B protein. Both degradation and synthesis of Q_B stop in the dark. It has been *proposed* that the Q_B protein is rapidly inactivated in the light because reduced quinones formed at its active site can react with O_2 to form radicals that attack essential amino-acid residues, but no clear evidence supporting this proposal has yet been obtained.

Fourthly, the electron-transport chain of chloroplasts, like that of mitochondria and the endoplasmic reticulum (Chapter 3), can 'leak' electrons onto oxygen. Isolated illuminated chloroplast thylakoids slowly take up oxygen in the absence of added electron acceptors. This was first observed by A. H. Mehler in the USA, and is hence often referred to as the *Mehler reaction*. It appears to result from the reduction of oxygen to O_2^- by the electron acceptors of photosystems I. Addition of the stromal protein ferredoxin increases the amount of oxygen uptake, since it is reduced by photosystems I (PSI) much more quickly than is oxygen and the reduced ferredoxin can then itself reduce oxygen:

$$Fd_{red} + O_2 \rightarrow O_2^- + Fd_{ox}$$

In vivo, however, reduced ferredoxin also passes electrons onto $NADP^+$ via ferredoxin $- NADP^+$ reductase (Fig. 5.2). Thus, electrons from photosystems I can pass through at least three routes, as shown in Fig. 5.3, of which route C is preferred. If the supply of $NADP^+$ were limited, the rate of electron flow along pathway C would be expected to be decreased, and more O_2^- should be made by route B and, to a lesser extent, by route A (Fig. 5.3). In some experiments, however, isolated intact chloroplasts have been shown to reduce some oxygen even at fairly low light intensities.

Lastly, oxygen has an effect on the reaction catalysed by ribulose biphosphate carboxylase. One of the chemical intermediates formed during the action of the enzyme on ribulose bisphosphate is capable of reacting with molecular oxygen as shown in Fig. 5.4. As a result of this *oxygenase* activity of the enzyme, ribulose bisphosphate becomes converted into phosphoglyceric acid and phosphoglycollic acid. The phosphoglycollate so produced is further metabolized by a series of reactions known as the *photorespiratory pathway*, a metabolic pathway that eventually converts two molecules of phosphoglycollate into one molecule each of carbon dioxide and phosphoglycerate, which can re-enter the Calvin cycle. Hence, carbon dioxide previously fixed into the Calvin cycle intermediates is lost and has to be refixed, with expenditure of more energy. Both oxygen and carbon dioxide compete for the enzyme intermediate (Fig. 5.4). The affinity of the carboxylase for carbon dioxide is much greater than that for oxygen, but at elevated oxygen concentrations, CO_2 fixation into the Calvin cycle is decreased and CO_2 loss by photorespiration is increased. At some O_2/CO_2 ratios, a point will be reached at which the leaf is no longer achieving any net CO_2 fixation, and therefore plant growth is completely inhibited. At high O_2/CO_2 ratios there may be actual loss of carbon from the plant as photorespiration exceeds CO_2 fixation into the Calvin cycle: a situation which will result in death of the plant if continued for a long period.

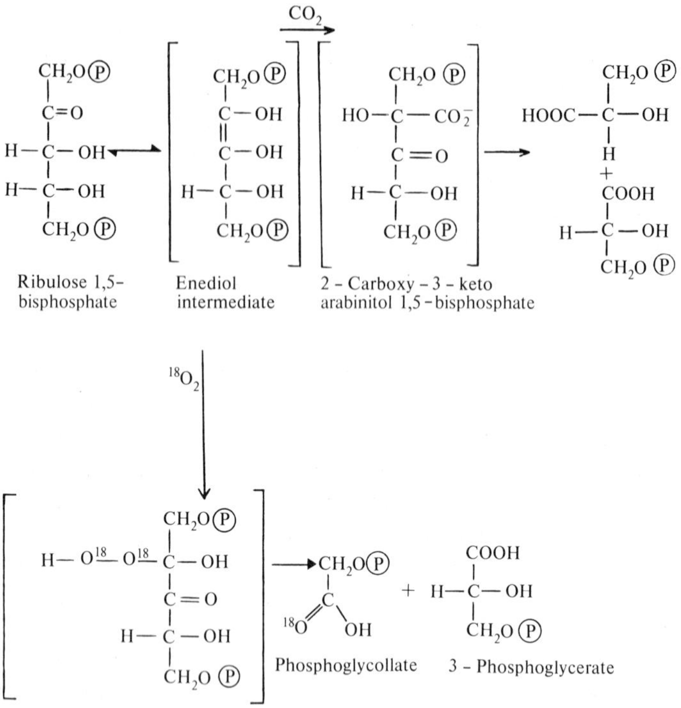

Fig. 5.4. Chemistry of ribulose bisphosphate carboxylase.

5.1.2   The solutions

Oxidant scavengers

The thylakoid membranes are especially rich in α-tocopherol, which interferes with the chain reaction of lipid peroxidation (Chapter 4) and directly scavenges singlet oxygen. The products of reaction of tocopherol with singlet oxygen include tocopherylquinone, which is found in chloroplasts and, as might be expected, increases in amount when they are illuminated. As with animal tissues, however, it must not be assumed that α-tocopherol in plants serves *only* as a protection against membrane damage. Other roles are perhaps suggested by the fact that a tocopherol oxidase enzyme, whose activity is under hormonal control, has been detected in many plant tissues, although this enzyme has so far resisted purification, and its subcellular location is unknown.

In animal tissues, α-tocopherol has been suggested to be the major, if not the only, lipid-soluble, chain-breaking antioxidant present (Chapter 4). Whether this is also true of the thylakoid membrane, and other plant membranes, remains to be established. For example, many plants accumulate 'secondary products' such as flavonoids and flavonols. A number of these products exert powerful antioxidant actions *in vitro*, such as scavenging of O_2^- and inhibition of lipid peroxidation. Examples include silymarin, (a 3-oxyflavone isolated from the thistle *Silybum marianum*), quercetin, rutin, and kaempferol. However, whether they act as antioxidants *in vivo* is unknown. For example, many plant secondary products are located in the central vacuole of cells and not in the organelles, such as chloroplasts, that are subject to oxidative stress.

Carotenoids, which are important constituents of chloroplast membranes, quench singlet oxygen extremely rapidly and can therefore help to protect chlorophyll and membranes against damage. Carotenoids are quickly destroyed on illumination of thylakoids *in vitro*, presumably as they absorb any singlet oxygen formed. Their protective function *in vivo* is well illustrated by mutant strains of maize which do not synthesize them: illumination of such plants under aerobic conditions causes rapid bleaching of chlorophylls and destruction of chloroplast membranes. Illumination under anaerobic conditions causes much less damage, since singlet oxygen cannot then be generated. Similar destructive effects of illumination in the presence of oxygen are seen in normal plants in which carotenoid biosynthesis has been inhibited by certain herbicides, such as aminotriazole, pyriclor, and norflurazon. The carotenoids present in chloroplasts are two main types: the carotenes (e.g. β-carotene, whose structure is shown in Fig. 5.5) and xanthophylls, which are oxygen-containing derivatives of carotenes. Carotenoids are also able to absorb energy from, and so diminish the concentration of, those excited states of chlorophyll that lead to singlet oxygen formation. Hence they have a dual role: decreasing the formation of singlet oxygen *in vivo*, and helping to remove any that does happen to be formed. They may also react directly with peroxy and alkoxy radicals, and so interfere with the chain reaction of lipid peroxidation.

Carotenoids are absorbed through the human gut, and might also act as antioxidants in humans. For example, large doses of β-carotene diminish the photosensitivity of some patients with abnormal porphyrin accumulation

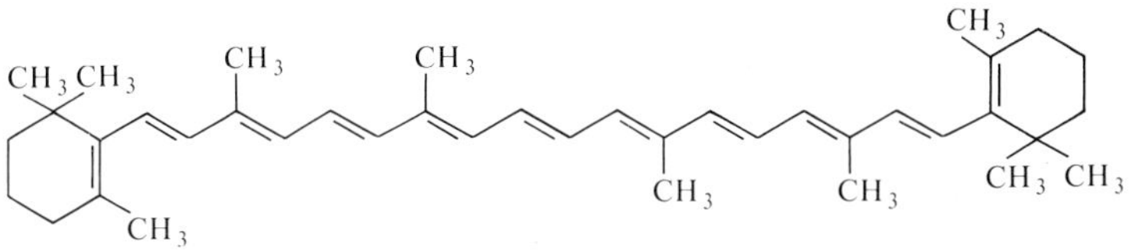

Fig. 5.5. The structure of β-carotene. Carotenes are orange-yellow pigments. The excited states of chlorophyll can transfer energy onto carotene, which dissipates it harmlessly and thus reduces chlorophyll-sensitized formation of singlet oxygen. Carotenes are also good quenchers of any singlet oxygen formed.

(Chapter 2), and it has been proposed that a high intake of β-carotene in Man may offer some protection against the development of cancer (Chapter 8).

Enzymes

Superoxide produced in the chloroplast is dealt with by a superoxide dismutase enzyme. Israeli scientists have shown that the SOD activity of developing tomatoes correlates with the resistance of the fruits to the 'sunscald' damage induced by exposure to heat and high light intensities (a nuisance for growers in hot countries who want to sell their tomatoes). This observation indicates that SOD plays an important protective role in illuminated pigment systems. In the blue-green alga *Anabaena cylindrica*, nitrogen fixation is carried out in specialized cells called heterocysts which lack the oxygen-evolving photosystems II. Heterocyst SOD activity is much lower than that of the photosynthetic cells. In spinach chloroplasts, a CuZnSOD is present, some of which is bound to the thylakoids, and the rest free in the stroma. No MnSOD or FeSOD has been detected in spinach chloroplasts, although there have been reports of a MnSOD in the chloroplasts of some other plants (possibly due to contamination, of the chloroplast fraction assayed, by mitochondria), and the FeSOD in *Brassica campestris* leaves (Chapter 3) is found in the chloroplast.

Careful work by Professor Asada's group in Japan has shown that the hydrogen peroxide produced by illuminated chloroplasts is derived from O_2^- arising from photosystem I and ferredoxin. Like all CuZnSODs, the chloroplast enzyme is inactivated on prolonged exposure to hydrogen peroxide. If this were permitted, then O_2^- could react with H_2O_2 to give OH˙ in a catalysed Haber–Weiss reaction (Chapter 3) if suitable transition metal ions are present. Very little is known about the availability of metal ions 'catalytic' for OH˙ formation and lipid peroxidation in chloroplasts. Chloroplasts often contain ferritin, from which O_2^- could conceivably mobilize iron ions (Chapters 2 and 3). Several iron proteins are damaged by a molar excess of H_2O_2, with release of 'catalytic' iron ions. Examples include mammalian haemoglobin (Chapter 3) and the leghaemoglobin of soybean root nodules. However, Puppo and the authors find that plant ferredoxins are not degraded by H_2O_2. Instead, they react with H_2O_2 to form a 'reactive species'. The chemical nature of this is presently unknown, except that it is not OH˙.

Hydrogen peroxide can also inhibit the Calvin cycle by inactivating the enzymes fructose bisphosphatase and sedoheptulose bisphosphatase. Thus it is imperative that the illuminated chloroplast disposes of hydrogen peroxide, yet these organelles contain no catalase or glutathione peroxidase. Both ascorbate and glutathione are, however, involved in removing hydrogen peroxide, since chloroplasts contain an ascorbate peroxidase activity, and GSH can reduce dehydroascorbate back to ascorbate. Figure 5.6 summarizes this 'ascorbate–glutathione cycle', a scheme first proposed in the authors'

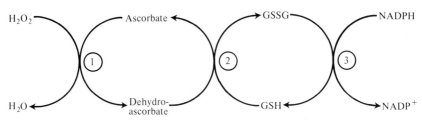

Fig. 5.6. An 'ascorbate–glutathione cycle' in chloroplasts. Enzymes involved: 1, Ascorbate peroxidase; 2, dehydroascorbate reductase (Chapter 2); 3, glutathione reductase. The first product of oxidation of ascorbate by ascorbate peroxidase is probably the semidehydroascorbate radical. Two such radicals then react in a disproportionation reaction to form ascorbate and dehydroascorbate, (i.e. 2 semi-dehydroascorbate → ascorbate + dehydroascorbate). NADH-dependent mechanisms for reducing semidehydroascorbate have also been described in chloroplasts.

laboratory and greatly strengthened by the discovery of ascorbate peroxidase in chloroplasts by D. Groden and E. Beck in Germany. Purified spinach ascorbate peroxidase is a monomer of relative molecular mass about 30 000, and it appears to contain haem. Ascorbate and GSH also scavenge OH˙ and singlet oxygen, and ascorbate can react with O_2^- (Chapter 2), so they may help to protect against these species as well. The chloroplast stroma contains GSH at concentrations in the range 1–4 mM and ascorbic acid at concentrations of 10–20 mM and in both cases the ratios of reduced to oxidized form (GSH/GSSG and ascorbate/dehydroascorbate) are kept high under both light- and dark-conditions. Thus direct scavenging of radicals by ascorbate is quite likely in view of the high concentrations present. Addition of hydrogen peroxide to chloroplasts in the light causes transient oxidation of both GSH and of ascorbate, as would be expected (Fig. 5.6). Ascorbate might also reduce tocopheryl radicals at the surface of the thylakoid membrane, so regenerating a-tocopherol for further use in protection against lipid peroxidation (Chapter 4). Ascorbate peroxidase purified from the alga *Euglena* can reduce not only hydrogen peroxide but also some artificial organic hydroperoxides. These enzymes might therefore be capable of acting on lipid peroxides *in vivo*, and further experiments are required on this point.

H_2O_2-dependent selenium-containing glutathione peroxidase activity is not generally found in plant tissues, but some GSH transferases are present and might be able to act on hydroperoxides. Japanese scientists have identified an enzyme system in the microsomal fraction of pea seeds that catalyses hydroxylation of some organic compounds (e.g. indole or phenol) in the presence of hydroperoxides such as linoleic acid hydroperoxide or even hydrogen peroxide. The overall reaction may be written as follows, where SH is the substrate and ROOH the peroxide:

$$SH + ROOH \rightarrow SOH + ROH.$$

Cytochrome P-450 might be involved in this reaction.

Chloroplasts also possess a number of systems that can 'repair' oxidative damage. These include methionine sulphoxide reductase (Chapter 3, Section 3.4.1) and the thioredoxin system. The protein thioredoxin has an essential disulphide bridge that can be reduced by an NADPH-dependent mechanism in illuminated chloroplasts to give a dithiol protein, thioredoxin—$(SH)_2$. Reduced thioredoxin can re-activate some enzymes inactivated by oxidants, such as the oxidized form of the Calvin cycle enzyme fructose bisphosphatase. Not until all these protective and repair systems are overloaded should toxic effects appear.

In summary, superoxide, hydrogen peroxide, singlet oxygen, the hydroxyl radical (OH˙), and lipid peroxides are species whose formation is damaging to plant tissues and must therefore be carefully controlled. Chloroplasts are especially affected by them because of a high internal oxygen concentration, the presence of molecules which can reduce oxygen to O_2^- (e.g. reduced ferredoxin), and the presence of pigments which can sensitize the formation of singlet oxygen. To allow their continued functioning, chloroplasts have multiple protective mechanisms. The main ones are SOD (which removes O_2^- and hence prevents formation of OH˙ from it), ascorbic acid (which reacts rapidly with O_2^-, OH˙, and singlet oxygen and helps remove hydrogen peroxide by the ascorbate peroxidase reaction), reduced glutathione (which reacts rapidly with OH˙ and singlet oxygen, protects enzyme —SH groups, and helps to regenerate ascorbate from dehydroascorbate), α-tocopherol (which inhibits the chain-reaction of lipid peroxidation and rapidly scavenges singlet oxygen), and carotenoids (which decrease formation of singlet oxygen by absorbing excess excitation energy from chlorophyll by direct transfer, and which also quench singlet oxygen).

Antioxidants and pollutants
The antioxidant mechanisms play a key part in the ability of plants to resist certain environmental pollutants. Thus fumigation of pea plants with mixtures of sulphur dioxide, nitrogen dioxide, and ozone at low concentrations caused rises in ascorbate peroxidase and glutathione reductase activities. Since the activity of these enzymes and the amounts of their substrates also appear to vary seasonally in plants, pollution resistance may well depend upon time of year, among other factors.

Photoinhibition and chilling injury
Exposure of leaves, or isolated chloroplasts, to light intensities greatly in excess of those required to saturate photosynthesis, produces marked reductions in photosynthetic capacity. This observation is known as *photo-*

inhibition. In both leaves and in isolated chloroplasts, photoinhibition is especially severe if illumination is performed in the absence of CO_2. Early studies identified photosystem II as the major site of damage in photo-inhibited chloroplasts, and showed that the process required O_2. Small protective effects of SOD and singlet O_2 scavengers have been reported. Excessive destruction of the Q_B protein, due to damage by oxygen-derived species (Section 5.1.1), might contribute to the phenomenon of photoinhibition. Catalase in leaf peroxisomes may also be inactivated at high light intensities. It has been suggested that the function of the metabolic pathway of photorespiration, which at first sight seems pointless in that it releases CO_2 from the plant which then has to be re-fixed into the Calvin cycle with expenditure of more energy, is actually to ensure that some CO_2 is always present within the leaf to minimize damage by excessive formation of O_2^- and singlet O_2.

Both high and low temperatures cause more-severe damage to higher plants in the light than they do in the dark, e.g. 'sunscald' damage to tomatoes has already been considered. At the other extreme, many temperate plant species show a depressed photosynthesis (*chilling injury*) after exposure to low temperature for one or two days, especially if they are illuminated. The involvement of photo-oxidant reactions has been suggested and there may be decreases in chloroplast SOD activity during the chilling process. A strain of the green alga *Chlorella ellipsoidea* resistant to chilling injury showed greater SOD activity than did the chilling-sensitive strain. Increasing the SOD activity of this *Chlorella* strain by treating it with paraquat (Chapter 6) also conferred upon it increased resistance to chilling injury. Chilling injury seems to particularly affect photosystem II activity and excessive damage to the Q_B protein may again be involved.

5.2   The eye

Retrolental fibroplasia (more usually now called 'retinopathy of prematurity') is a condition that can lead to blindness; it is a complication of the use of elevated oxygen concentrations in incubators for premature babies (Chapter 1). Although its incidence can be controlled by restricting the oxygen concentrations used, there is evidence that the weakest babies need a high oxygen concentration if they are to survive at all. Fortunately, there is now some evidence that administration of large doses of a-tocopherol (Chapter 4) to such babies lessens the risk of their developing severe retrolental fibroplasia and (possibly) brain haemorrhage. The latter is a common finding in premature babies who die within the first week of life and can lead to handicap in surviving babies.

Cataract, defined as a clinically-significant opacity of the lens, is one of the leading causes of blindness in the world. The lens is surrounded by a capsule, and held in place behind the iris by ligaments (Fig. 5.7). The anterior of it is covered by a single layer of epithelial cells which are metabolically highly active and undergo cell division, elongation, and development to form the *lens fibres*. Newly-formed fibres continuously push the older ones towards the centre of the lens to form the so-called *lens nucleus*. No blood vessels enter the lens, and it derives food and oxygen by diffusion through the aqueous and vitreous humours (Fig. 5.7). The epithelium actively transports sodium ions out of the lens, and potassium ions into it, using an ion-transporting ATPase enzyme, which is especially sensitive to damage by O_2^--generating systems.

We saw in Chapter 2 that the lens contains sensitizers of singlet oxygen formation, and singlet oxygen can damage and cross-link lens proteins.

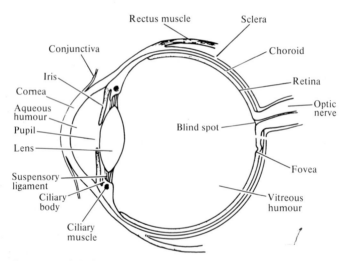

Fig. 5.7.   Structure of the human eye. The sclera is a tough fibrous coat, inside which is the choroid layer containing a network of blood vessels that supply food and oxygen to the eye. The choroid is highly pigmented. The aqueous humour is a watery fluid whereas the vitreous humour is more jelly-like. Pressure of these fluids outward on the sclera maintains the shape of the eye. The transparent cornea, lens, and conjunctiva obtain food and oxygen by diffusion from the aqueous humour—they contain no blood vessels. The lens is suspended in position by ligaments. The retina contains two types of light-sensitive cells: rods and cones. Only the cones are sensitive to coloured light, but the rods are more responsive to dim light (hence the perception of colour decreases as light intensity decreases). Any marked variation in the pressure of the humours can alter the blood flow in the eye. For example, *ocular hypertension* (too much pressure) can restrict blood flow and cause ischaemic damage; *glaucoma* is the name of a group of diseases in which this occurs.

Ionizing radiation often induces cataract, perhaps by increased formation of hydroxyl and superoxide radicals. Indeed, a considerable amount of evidence has accumulated implicating prolonged exposure to ultraviolet radiation as a factor in lens ageing and cataract formation. Proteins isolated from human cataractous lenses contain considerable quantities of methionine sulphoxide residues (Chapter 3). UV-induced degradation of tryptophan, a reaction that produces O_2^- (Chapter 3), has been reported in lens proteins.

The vitreous humour (Fig. 5.7) contains hyaluronic acid, which is attacked and depolymerized on exposure to O_2^--generating systems (Chapter 3) causing loss of viscosity. This reaction is due to O_2^--dependent formation of hydroxyl radicals (Chapter 3) in the presence of traces of iron salts. Indeed, introduction of iron salts into the eye causes severe problems; this can happen as a result of penetration by iron objects or by intravitreal haemorrhage. Both can cause severe visual impairment, sometimes leading to blindness. The lipids present in the membranes of retinal rod cells (Fig. 5.7) contain a high percentage of polyunsaturated fatty-acid side-chains, and are thus susceptible to lipid peroxidation. The pigment they contain, rhodopsin, can sensitize the formation of singlet oxygen (Chapter 2); and exposure of frog retina to light of the wavelengths absorbed by rhodopsin *in vitro* has been shown to induce lipid peroxidation. Lipid peroxides are increased in the rabbit retina after exposure to elevated oxygen concentrations or to ionizing radiation; and injection of preformed lipid peroxides into the eyes of rabbits causes severe retinal damage. Dietary deficiencies of selenium and vitamin E in rats cause marked loss of polyunsaturated fatty acids and accumulation of peroxidation products in the retinal pigment epithelium.

Therefore the eye has a lot of potential problems, and one would expect a corresponding degree of protection. Indeed, the concentration of GSH in the lens of several species is as high as that in the liver, being especially concentrated in the epithelium. GSH may serve to protect the thiol-groups of the lens proteins known as *crystallins*, preventing them from aggregating together to form opaque clusters. The GSH/GSSG ratio is normally kept high by the activity of a lens glutathione reductase, which obtains NADPH by operation of the pentose phosphate pathway in the lens (Chapter 3). Glaucoma (defined in the caption for Fig. 5.7), by decreasing the supply of glucose to the lens, interferes with this pathway and so restricts NADPH supply. Both glutathione peroxidase and catalase activities are present in the lens, as is methionine sulphoxide reductase (Chapter 3). Glutathione-S-transferase enzymes have also been detected in bovine lens but they do not use lipid peroxides as substrates, in contrast to the liver enzymes (Chapter 4). Bovine retina, on the other hand, does contain a glutathione-S-transferase that can act on lipid peroxides.

Feeding aminotriazole to rabbits to inhibit catalase raises the concentration of hydrogen peroxide in the lens and induces cataracts; so, at least in this animal, the glutathione peroxidase alone cannot cope with the normal rates of H_2O_2-generation. By contrast, cataract does not develop in mice with an inborn deficiency in catalase activity. The concentration of hydrogen peroxide in the aqueous humour of human patients with cataract is significantly higher than normal, and a common feature of nearly all types of cataract is that lens GSH concentrations fall.

It has been suggested that the ability to synthesize GSH decreases with age in the human lens, which would predispose to the development of cataract if the lens was subjected to oxidative stress. Calvin *et al* in the USA reported that very young (9–12 day old) mice injected with buthionine sulphoximine, an inhibitor of GSH synthesis (Chapter 3), rapidly developed dense cataracts.

Ascorbic acid is present at high concentrations in the lens, cornea, and aqueous humour of man, monkey, and many other animals (e.g. 1–2 millimoles per litre in human aqueous humour). Its ability to scavenge O_2^-, OH˙, and singlet oxygen may be of importance although, on the other hand, it has been suggested that light-induced degradation of ascorbate in the aqueous humour may be a source of hydrogen peroxide that has to be removed by glutathione peroxidase and catalase. Ascorbate has been observed to protect the lens ion-transporting ATPase enzyme against damage by a O_2^--generating system *in vitro*. Consistent with a protective role for ascorbate *in vivo* is the observation that nocturnal animals (e.g. cats) have lower ascorbate concentrations in the eye. Rat lens has virtually no ascorbate.

Russell *et al.* in the USA showed that incubation of human lens crystallins with ascorbate in the presence of iron ions produced oxidative changes to the protein resembling those seen with aging in the human lens. Generation of reactive species such as OH˙ by iron/ascorbate mixtures (Chapter 3) may explain the damage produced, and why introduction of iron ions into the eye is so damaging. Under normal conditions, such reactions may be minimized by the presence in aqueous and vitreous humors of an iron-binding protein similar to transferrin; transferrin-bound iron is much less effective in stimulating radical reactions than are low-molecular-mass iron complexes (Chapter 3). Lactoferrin is also present in tear fluid.

Superoxide dismutase is present in all parts of the eye, but problems in the assay of eye tissue have made comparison of the SOD activities in the different parts rather difficult. Rat lens has a higher SOD activity than human lens, and so ascorbate may play a greater role in scavenging O_2^- in the latter case. The SOD activities of eye tissues are inhibited by cyanide, and thus can be attributed to the copper–zinc enzymes (Chapter 3). CuZnSODs are susceptible to inactivation by hydrogen peroxide. Indeed, feeding aminotriazole to rabbits (see above) causes a decrease in lens CuZnSOD activity, probably as a result of the accumulated hydrogen peroxide. Finally, rod outer segments and retinal pigment epithelium are especially rich in vitamin E, and the retina is usually damaged in vitamin-E-deficient animals. Feeding extra vitamin E decreases the severity of the cataracts seen after aminotri-azole is administered to rabbits.

Thus the eye, like the chloroplast, has to put up with the problems caused by light, and relies heavily on ascorbate, GSH, and tocopherol as protective mechanisms, whilst methionine sulphoxide reductase may help regenerate methionine in proteins after oxidation by singlet oxygen (Chapter 3). On the other hand, the increased content of methionine sulphoxide residues known to be present in proteins from cataractous lenses is not due to any defect in this enzyme, since a comparison of normal and cataractous human lens showed similar activities.

Carotenoids may be important in protection of light receptors against singlet oxygen in some living organisms. In the compound eye of the housefly a carotenoid is present in large amounts (4–10 molecules per rhodopsin molecule). Carotenoid has also been reported in the lateral eye of the crab *Limulus*. The corneas of puffer fishes are clear in the dark but become yellow in the light, owing to the migration of carotenoid pigment in chromatophore cells.

5.2.1 *Diabetic cataract*

Diabetes mellitus, a disease marked by elevated blood glucose (hypergly-caemia) and urinary glucose excretion, is caused by faulty production of, or tissue response to, insulin. In the most common form of diabetes (type II, non-insulin-dependent) blood insulin is often elevated, yet tissue response to it is subnormal; there is 'peripheral resistance'. In the rarer type I (juvenile onset diabetes), insulin secretion is absent or impaired. Both types of diabetic are prone to develop complications that include atherosclerosis, retinal damage, cataract, and neuropathy. These complications appear to relate to the elevated blood glucose concentrations and it has thus been proposed that glucose, in excessive amounts, is toxic. Indeed, several routes by which excessive glucose and other sugars may be toxic have been proposed. All of these may involve oxidative stress.

Aldose reductase
An enzyme, aldose reductase, which converts glucose to the poly-alcohol sorbitol:

$$H^+ + \underset{\text{aldehyde}}{RCHO} + NADPH \rightarrow \underset{\text{alcohol}}{RCH_2OH} + NADP^+$$

appears to be implicated in the rapidly-developing cataract seen in diabetic rodents. Inhibitors of this enzyme prevent sorbitol accumulation and cataract development. Why excessive activity of aldose reductase should be toxic is unknown but it may be caused by an excessive osmotic stress placed on the lens, since sorbitol accumulates to high concentrations. Alternatively, high aldose reductase activity may place a burden on cellular NADPH (a co-substrate for glutathione reductase) and make the lens vulnerable to oxidative damage. Cataracts from human diabetics do not appear to contain osmotically-active concentrations of sorbitol, but there is some evidence to suggest that individuals who take dietary vitamin E suffer less often from cataracts. Levels of GSH are decreased in cataractous lenses, suggesting that they have been exposed to an oxidative insult of some sort (see above).

Non-enzymatic glycation
Glucose is also considered to be toxic by virtue of its ability to behave chemically as an aldehyde, although most glucose in solution exists as non-aldehydic ring structures. Aldehydes, such as those produced by lipid peroxidation (Chapter 4), are reactive substances and can bind to protein and DNA. Glucose in its straight-chain form is an aldehyde and very slowly reacts with proteins and DNA, modifying them in a process referred to as *non-enzymic glycation* (Fig. 5.8). Protein which has been glycated *in vitro* is conformationally-altered and acquires a brown colour. In fact these reactions are referred to as *Maillard browning*. The reactions with glucose are very slow, however, and are only considered relevant in the case of proteins with a very slow turnover, such as those which are found in the lens and may have been present for many decades. Glycation is detectable in many proteins from human diabetics and glycated haemoglobin is used as an index of blood glucose control over the previous month or so. Glucose also modifies CuZnSOD in the erythrocyte, decreasing its activity, and this may account for the lower SOD activity in the blood of some diabetics.

Monosaccharide oxidation
An alternative view is that sugars such as glucose produce free radicals and other oxidizing substances directly. Monosaccharides can oxidize, catalysed by trace amounts of iron and copper ions, and so produce hydroxyl radicals, superoxide, hydrogen peroxide, and toxic carbonyls (Fig. 5.9). This process of monosaccharide oxidation can lead to protein damage by free radicals and by the covalent binding of the carbon-products of the process (Fig. 5.9). Both these reactions and those shown in Fig. 5.8 may participate in the browning reactions. Furthermore, free radicals fragment and conformationally-alter protein in a way reminiscent of the changes seen in protein when it is exposed to high concentrations of glucose. Oxidizing monosaccharides can also stimulate peroxidation of lipids *in vitro*, and it has been claimed that elevated concentrations of lipid peroxidation products are found in the plasma of human diabetics. Some doubt has been cast on the validity of the methods used to measure the peroxidation, however.

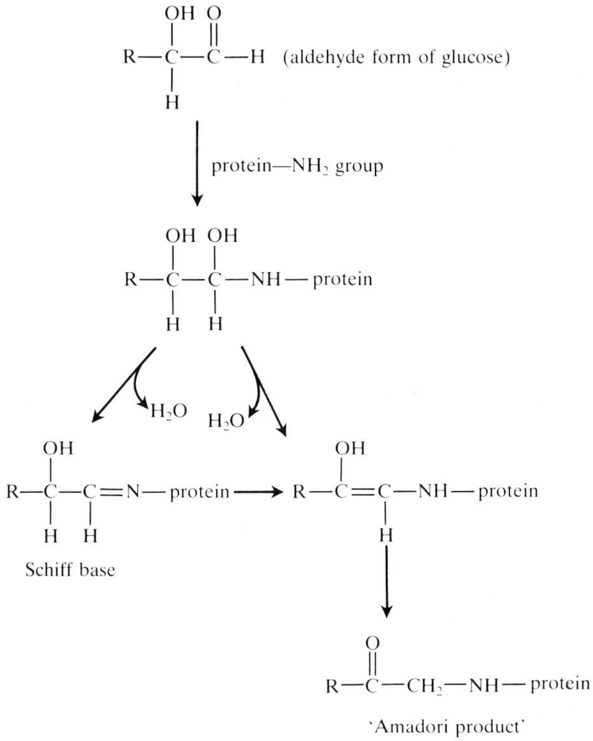

Fig. 5.8. 'Browning reactions' of glucose (and other monosaccharides) with proteins. These are slow chemical reactions, producing coloured products.

Oxidative stress in diabetes

Oxidative stress certainly appears to be an important feature of the diabetic complications. Apart from the claimed increase in lipid peroxide, diabetics have lower concentrations of erythrocyte GSH, consume ascorbate faster, have higher concentrations of dehydroascorbate in their plasma and lower levels of vitamin E in their platelets. It has, in fact, been suggested that diabetes and its complications may be related to abnormal levels of tissue oxidative stress. Many drugs which induce diabetes, such as alloxan and streptozotocin, act by producing concentrations of peroxides greater than can be tolerated by the islets of Langherhans, since these are relatively-poor in glutathione peroxidase (see Chapter 6).

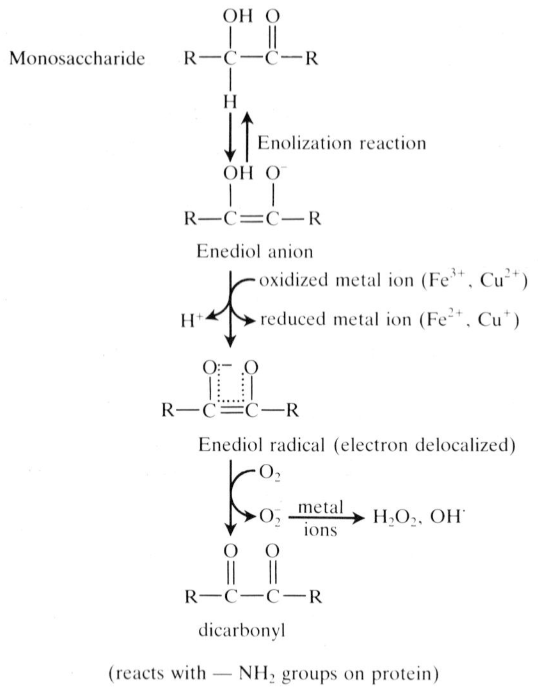

$$
\begin{array}{c}
\qquad\qquad\text{OH}\;\;\text{O} \\
\qquad\qquad\;\;|\quad\;\;\| \\
\text{Monosaccharide}\qquad \text{R}-\text{C}-\text{C}-\text{R} \\
\qquad\qquad\;\;| \\
\qquad\qquad\;\;\text{H}
\end{array}
$$

↓↑ Enolization reaction

$$
\begin{array}{c}
\text{OH}\;\;\text{O}^- \\
\;\;|\quad\;\;| \\
\text{R}-\text{C}=\text{C}-\text{R}
\end{array}
$$

Enediol anion

H⁺ ⤢ oxidized metal ion (Fe^{3+}, Cu^{2+})
reduced metal ion (Fe^{2+}, Cu$^+$)

$$
\begin{array}{c}
\text{O}^{.-}\;\;\text{O} \\
|\quad\quad| \\
\text{R}-\text{C}=\text{C}-\text{R}
\end{array}
$$

Enediol radical (electron delocalized)

O$_2$
O$_2^-$ $\xrightarrow[\text{ions}]{\text{metal}}$ H$_2$O$_2$, OH$^-$

$$
\begin{array}{c}
\text{O}\quad\;\;\text{O} \\
\|\quad\;\;\| \\
\text{R}-\text{C}-\text{C}-\text{R}
\end{array}
$$

dicarbonyl

(reacts with — NH$_2$ groups on protein)

Fig. 5.9. Involvement of radicals and metal ions in the oxidation of monosaccharides such as glucose. Dicarbonyls can react with amino groups of proteins to form coloured products. Attack of OH˙ produced in the above reactions upon further monosaccharide molecules can yield hydroxyalkyl radicals that decompose to produce more dicarbonyls.

5.3 Further reading

The effect of deficiency of vitamins E and A on the retina (1980). *Nutr. Rev.* **38**, 386.

Ahmad, M. U. *et al.* (1988) Oxidative degradation of glucose adducts to protein *J. Biol. Chem.* **263**, 8816.

Andley, U. P. (1988). Photodamage to the eye. *Photochem. Photobiol.* **46**, 1057.

Arai, K. *et al.* (1987). Increase in the glucosylated form of erythrocyte Cu–Zn-superoxide dismutase in diabetes and close association of the nonenzymatic glucosylation with the enzyme activity. *Biochim. Biophys. Acta* **924**, 292.

Asada, K. (1987). Production and scavenging of active oxygen in photosynthesis. In *Photoinhibition* (D. J. Kyle, C. B. Osmond, and C. J. Arntzen Eds) p. 227. Elsevier, Amsterdam.

Bhuyan, K. C. and Bhuyan, D. K. (1979). Mechanisms of cataractogenesis induced by 3-amino 1,2,4 triazole. In *Biochemical and clinical aspects of oxygen* (W. S. Caughey, Ed.), p. 785. Academic Press, New York.

Brownlee, M., Cerami, A., and Vlassara, H. (1988). Advanced glycosylation end products in tissue and the biochemical basis of diabetic complications. *New Engl. J. Med.* **318,** 1315.

Calvin, H. I., Medvedovsky, C., and Worgul, B. V. (1986). Near-total glutathione depletion and age-specific cataracts induced by buthionine sulfoximine in mice. *Science* **233,** 533.

Critchley, C. (1988). The molecular mechanism of photoinhibition—fact and fiction. *Aust. J. Pl. Physiol.* **15,** 27.

Demmig, B. *et al.* (1987). Photoinhibition and zeaxanthin formation in intact leaves. A possible role of the xanthophyll cycle in the dissipation of excess light energy. *Plant Physiol.* **84,** 218.

Elstner, E. F. (1982). Oxygen activation and oxygen toxicity. *Ann. Rev. Plant Physiol.* **33,** 73.

Finer, N. N. *et al.* (1982). Effect of intramuscular vitamin E on frequency and severity of retrolental fibroplasia. A controlled trial. *Lancet* **i,** 1087.

Gillham, D. J. and Dodge, A. D. (1986). Hydrogen-peroxide-scavenging systems within pea chloroplasts. A quantitative study. *Planta* **167,** 246.

Gillham, D. J. and Dodge, A. D. (1987). Chloroplast superoxide and hydrogen peroxide scavenging systems from pea leaves: seasonal variations. *Plant Sci.* **50,** 105.

Halliwell, B. (1984). *Chloroplast metabolism: the structure and function of chloroplasts in green leaf cells* (Revised edition). Clarendon Press, Oxford.

Halliwell, B. (1987). Oxidative damage, lipid peroxidation and antioxidant protection in chloroplasts. *Chem. Phys. Lipids,* **44,** 327.

Katz, M. L. *et al.* (1982). Effects of antioxidant nutrient deficiency on the retina and retinal pigment epithelium of albino rats: a light and electron microscopic study. *Exp. Eye Res.* **34,** 339.

Kirschfeld, K. (1982). Carotenoid pigments: their possible role in protecting against photo-oxidation in eyes and photoreceptor cells. *Proc. Roy. Soc. Lond.* **B216,** 71.

Krinsky, N. I. and Deneke, S. M. (1982). Interaction of oxygen and oxyradicals with carotenoids. *J. Natl. Cancer Inst.* **69,** 205.

Kuizenga, A., van Haeringen, N. J., and Kijlstra, A. (1987). Inhibition of hydroxyl radical formation by human tears. *Invest. Ophthalmol. Vis. Sci.* **28,** 305.

Law, M. Y., Charles, S. A., and Halliwell, B. (1983). Glutathione and ascorbic acid in spinach chloroplasts. The effect of hydrogen peroxide and of paraquat. *Biochem. J.* **210,** 899.

Mathews-Roth, M. M. (1987). Photoprotection by carotenoids. *Fed. Proc.* **46,** 1890.

McGahan, M. C. and Fleisher, L. N. (1986). A micromethod for the determination of iron and total iron-binding capacity in intraocular fluids and plasma using electrothermal atomic absorption spectroscopy. *Anal. Biochem.* **156,** 397.

Mehlhorn, H. *et al.* (1987). Induction of ascorbate peroxidase and glutathione reductase activities by interactions of mixtures of air pollutants. *Free Radical Res. Commun.* **3,** 193.

Nakano, Y. and Asada, K. (1987). Purification of ascorbate peroxidase in spinach chloroplasts: its inactivation in ascorbate-depleted medium and reactivation by monodehydroascorbate radical. *Plant Cell Physiol.* **28,** 131.

Osswald, W. F. and Elstner, E. F. (1987). Investigations on spruce decline in the Bavarian forest. *Free Radical Res. Commun.* **3**, 185.

Overbaugh, J. M. and Fall, R. (1985). Characterization of a selenium-independent glutathione peroxidase from *Euglena gracilis. Plant Physiol.* **77**, 437.

Palma, J. M., Sandalio, L. M., and del Rio, L. A. (1986). Manganese superoxide dismutase and higher plant chloroplasts: a reappraisal of a controverted cellular localization. *J. Plant Physiol.* **125**, 427.

Rathbun, W. B. (1986). Activity of glutathione synthesis enzymes in the rhesus monkey lens related to age: a model for the human lens. *Curr. Eye Res.* **5**, 161.

Rathbun, W. B., Kuck, J. F. R., and Kuck, K. D. (1986). Glutathione metabolism in lenses of Emory and cataract-resistant mice: activity of five enzymes. *Curr. Eye Res.* **5**, 189.

Russell, P. *et al.* (1987). Aging effects of vitamin C on a human lens protein produced *in vitro. FASEB J.* **1**, 32.

Schmidt, A. and Kunert, K. J. (1986). Lipid peroxidation in higher plants. The role of glutathione reductase. *Plant. Physiol.* **82**, 700.

Sinha, S. *et al.* (1987). Vitamin E supplementation reduces frequency of periventricular haemorrhage in very preterm babies. *Lancet* **i**, 466.

Spector, A. *et al.* (1982). Lens methionine sulphoxide reductase *Biochem. Biophys. Res. Commun.* **108**, 429.

Srivastava, S. K. (Ed.) (1982). *Red blood cell and lens metabolism.* Elsevier/North-Holland Biomedical Press, Amsterdam.

Steffens, G. J. *et al.* (1986). Primary structure of Cu–Zn superoxide dismutase of *Brassica oleracea* proves homology with corresponding enzymes of animals, fungi and prokaryotes. *Biol. Chem. Hoppe–Seyler* **367**, 1007.

Stephens, R. J. *et al.* (1988). Vitamin E distribution in ocular tissues following long-term dietary depletion and supplementation as determined by microdissection and gas chromatography-mass spectrometry. *Exp. Eye Res.* **47**, 237.

Takahama, U. (1983). Redox reactions between kaempferol and illuminated chloroplasts. *Plant Physiol.* **71**, 598.

Tanaka, K. *et al.* (1985). O_3 tolerance and the ascorbate-dependent H_2O_2 decomposing system in chloroplasts. *Plant Cell Physiol.* **26**, 1425.

Thornalley, P. *et al.* (1984). The autoxidation of glyceraldehyde and other simple monosaccharides under physiological conditions catalysed by buffer ions. *Biochim. Biophys. Acta* **797**, 276.

Torel, J., Cillard, J., and Cillard, P. (1986). Antioxidant activity of flavonoids and reactivity with peroxy radical. *Phytochemistry* **25**, 383.

Varma, S. D., Kumar, S., and Richards, R. D. (1979). Light-induced damage to ocular lens cation pump. Prevention by vitamin C. *Proc. Natl. Acad. Sci. USA* **76**, 3504.

Wolf, G. (1982). Is dietary β-carotene an anti-cancer agent? *Nutr. Rev.* **40**, 257.

Wolff, S. P. and Dean, R. T. (1987). Glucose autoxidation and protein modification. The potential role of 'autoxidative glycosylation' in diabetes. *Biochem. J.* **245**, 243.

6

Free radicals and toxicology

As a result of the 'Spanish cooking-oil crisis', which is discussed later in this chapter, radicals such as superoxide, and the enzyme superoxide dismutase, were mentioned on BBC television and in up-market British newspapers. As usual, the Americans were well ahead of us in that for some time it has been possible to buy superoxide dismutase tablets (sometimes with added catalase) in health food stores to protect oneself against ageing (with total disregard for the action of digestive enzymes on dietary proteins).

To return to the serious subject matter of this chapter, we have seen that free radicals can often damage cells. Any agent that could abstract a hydrogen atom from a membrane lipid and initiate peroxidation would obviously be unpleasant, and some toxic compounds that can do this are considered later in this chapter. The increase in the activities of superoxide dismutase and H_2O_2-removing enzymes that occurs when many organisms are exposed to elevated oxygen concentrations (Chapter 3) implies that the amounts of these enzymes present normally in some tissues are sufficient only to cope with the usual rates of O_2^-- and H_2O_2-generation. Hence any compound that increased O_2^-- and H_2O_2-production in such tissues at normal oxygen concentrations would be expected to be toxic.

Depending on the cell system under study, the primary target of such increased oxidant stress can vary. Thus DNA is an important early target of damage when H_2O_2 is added to mammalian cells (Chapter 2, Section 2.5.3). Excessive DNA damage is associated with the depletion of ATP and NAD^+, falls in GSH/GSSG ratios and increases in 'free' intracellular Ca^{2+} ions, which are normally maintained at sub-micromolar concentrations in most cells. Oxidant stress can damage the enzymes that normally maintain low intracellular Ca^{2+}, e.g. plasma membrane Ca^{2+}-extrusion systems and the ATP-dependent Ca^{2+}-pump of the endoplasmic reticulum. A number of oxidants cause Ca^{2+} release from mitochondria. Increases in intracellular 'free' Ca^{2+} might lead to increased availability of transition metal ions (by the mechanism suggested in Fig. 6.1) and hence to increased rates of lipid peroxidation. Indeed, we have already seen (Chapter 4) that lipid peroxidation often proceeds more rapidly in injured tissues, even if the damage was not initially by a free-radical mechanism. GSH, as a substrate of selenium-containing glutathione peroxidase, plays an important role in scavenging H_2O_2 (Chapter 3); GSH is also involved in protecting against lipid peroxida-

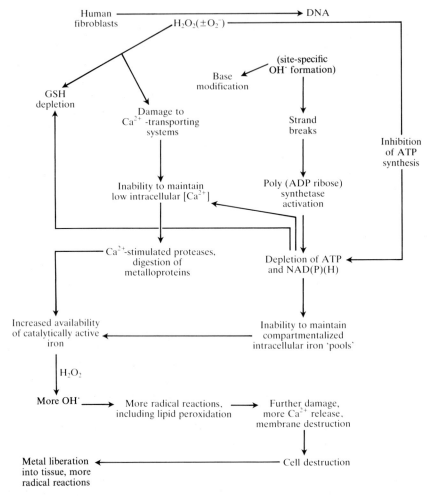

Fig. 6.1. Interacting mechanisms of oxidant damage: the action of H_2O_2 on human fibroblasts. H_2O_2 can produce DNA damage (Chapter 2, Section 2.10; Chapter 3, Section 3.5.4), apparently by site-specific generation of hydroxyl radicals and/or by activation of Ca^{2+}-dependent nuclease enzymes. H_2O_2 can also directly inactivate some enzymes involved in ATP synthesis, such as the glycolytic enzyme glyceraldehyde 3-phosphate dehydrogenase (Chapter 2, Section 2.10).

tion (Chapter 4). Some toxins are metabolized by GSH-dependent reactions (e.g. the opioid morphine) and excessive quantities of these toxins can so deplete cellular GSH concentrations that free-radical reactions, such as lipid peroxidation, accelerate as a consequence of the GSH depletion.

Thus 'free-radical reactions' have been implicated in the mechanism of action of many xenobiotics, but published reports often do not take sufficient care to distinguish between compounds for which free-radical formation is directly responsible for the toxicity, and other compounds in which the free-radical formation is a late stage in the process of cell injury. When increased oxidant formation is a primary toxic mechanism, it is also important to ask what is the first target of oxidant damage—is it DNA, proteins, or membrane lipids? Answering these questions is difficult because mechanisms of cell damage are inter-related (Fig. 6.1), and it is often hard to disentangle what is going on. Recent studies suggest that low levels of oxidant stress also inhibit communication between adjacent cells.

However, let us examine what is known about the mechanism of action of several xenobiotics.

6.1 Bipyridyl herbicides

We saw in Chapter 5 that a number of herbicides act by inhibiting electron transport in chloroplasts or by interfering with carotenoid synthesis.

The bipyridyl herbicides *paraquat* and *diquat* act in a rather different way. *Bipyridyl* means that the structure contains two pyridine rings, aromatic rings in which one carbon atom is replaced by a nitrogen atom. The rings may be joined either by their number-2 carbon atoms, or by their number-4 carbon atoms (the N counting as atom number-one), to give 2,2′-bipyridyl or 4,4′-bipyridyl respectively. The dash is used to indicate an atom in the second ring. In paraquat, a methyl group is attached to each nitrogen, giving a full chemical name as 1,1′-*dimethyl*-4,4′-*dipyridylium ion* (Fig. 6.2). Each N gains a positive charge because it has four bonds and is thus utilizing its lone pair of electrons (see Appendix I). In diquat, the two nitrogen atoms are joined by an ethylene group, to give 1,1′-*ethylene*-2,2′-*dipyridylium ion* (Fig. 6.2). Paraquat is usually manufactured as a salt with chloride (Cl^-) ion, and diquat with bromide (Br^-) ion. Upon chemical reduction, 4,4′-bipyridyl compounds form coloured solutions; and a number of such compounds, under the general name 'viologens', have been used for this reason to study the chemistry of reduction and oxidation reactions. Hence paraquat is sometimes called *methyl viologen*. The colour is produced by addition of one electron to the compounds to form a stable (in the absence of oxygen) radical that has a characteristic ESR spectrum, and that absorbs visible light (λ_{max} in the visible range is at 603 nm for paraquat). The extra electron is delocalized over both ring structures with partial neutralization of the positive charge on each nitrogen-atom.

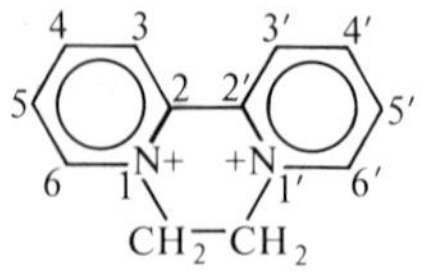

Diquat dibromide

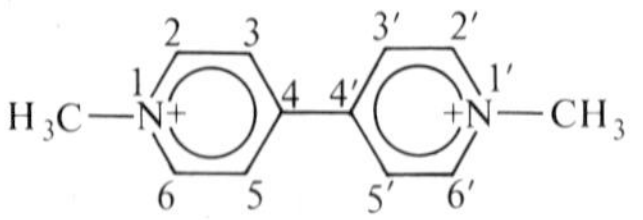

Paraquat dichloride

Fig. 6.2.   Structures of the herbicides diquat and paraquat.

6.1.1   *Paraquat toxicity to plants*

The bipyridyl herbicides, discovered by Imperial Chemical Industries in England, are extremely valuable because of their broad toxicity to a wide range of plants, and because they are completely inactivated on contact with soil, being tightly bound to clay minerals and destroyed by various micro-organisms. Hence a field of weeds can be sprayed with these herbicides and, after the dead plants are cleared, replanting can be done at once. It was soon observed that the killing of green plants by either paraquat or diquat is greatly accelerated by light and is slowed if oxygen is removed from the environment of the plant by flushing with nitrogen gas. Later work showed that both paraquat and diquat cross the chloroplast envelope easily, and can accept electrons from the non-haem-iron proteins associated with the photosystem I and also from the flavin at the active site of ferredoxin–NADP reductase (Chapter 5), in both cases becoming reduced to their radical forms. If these experiments were carried out in the absence of oxygen, the bipyridyl radicals could be identified in herbicide-treated chloroplasts by means of their absorption and ESR spectra. On admission of oxygen (and, of course, chloroplasts normally have a high internal oxygen concentration because of its production during photosynthesis) the radicals disappear because they react with oxygen extremely rapidly. If BP^{2+} is used to represent the herbicides, the reactions may be written as:

$$BP^{2+} \xrightarrow[\text{chain (1 electron)}]{\text{electron-transport}} BP^{\cdot +}$$

$$BP^{\cdot +} + O_2 \rightarrow BP^{2+} + O_2^- \quad k_2 = 7.7 \times 10^8 \, \text{M}^{-1}\text{s}^{-1} \quad \text{(for paraquat)}$$

Hence the superoxide radical, O_2^-, is formed. Thus treatment of illuminated chloroplasts *in vitro* with paraquat or diquat leads to a rapid uptake of oxygen as the herbicides are continuously reduced and re-oxidized. The O_2^- is presumably converted into hydrogen peroxide by chloroplast superoxide dismutase (Chapter 5). Since chloroplasts contain no catalase, hydrogen peroxide is dealt with by the ascorbate–glutathione cycle (Chapter 5), but as Table 6.1 shows, GSH and ascorbate are quickly oxidized and inactivation of Calvin cycle enzymes such as fructose bisphosphatase occurs, so that CO_2-fixation stops. An additional reason for inhibition of CO_2-fixation is that diversion of electrons from photosystem I onto the herbicides will decrease the supply of NADPH both for the Calvin cycle, and for glutathione reductase activity. In studies upon whole flax leaves, A. D. Dodge in England observed that inhibition of CO_2-fixation is followed by leakage of ions from the leaves, and accumulation within them of thiobarbituric acid-reactive material, indicative of lipid peroxidation. Analysis of total lipids extracted from the leaves showed destruction of many fatty-acid side-chains. Electron microscopy revealed deterioration and breakdown of a number of cellular membranes, including the chloroplast thylakoids and the tonoplast, the membrane which surrounds the central vacuole of the plant cell and is often very close to the envelopes of chloroplasts. The central vacuole contains a number of hydrolytic enzymes, and often accumulates organic acids to a high concentration; so release of its contents into the rest of the cell will potentiate the damage. Similar studies upon bean-leaves showed that paraquat caused decreased membrane fluidity and increased lipid destruction, accompanied by accumulation of TBA-reactive material. Comparable results have been obtained with several green algae. It does not, of course, follow that lipid peroxidation is the primary cause of paraquat-induced leaf death, but it certainly does not help. Inhibition of CO_2-fixation is a much earlier phenomenon than is detectable lipid peroxidation.

That these damaging effects are related to an increased production of O_2^- and H_2O_2 in the presence of paraquat or diquat is shown by a number of experiments. Paraquat induces synthesis of MnSOD in the green alga *Chlorella*. Some strains of ryegrass are relatively resistant to paraquat even though it is still taken up at a normal rate by the plants, and the leaves of the resistant plants have been shown to contain significantly more superoxide dismutase (SOD) and catalase activities than those from sensitive plants. Although leaf catalase is located in peroxisomes rather than in chloroplasts (Chapter 5), H_2O_2 is a fairly non-polar molecule and some of it can presumably diffuse across the chloroplast envelope to be dealt with in peroxisomes, which are often seen to be closely associated with chloroplasts in electron micrographs of leaf sections. Paraquat-resistant strains of *Conyza bonariensis* have been reported to have increased SOD, ascorbate peroxidase, and glutathione reductase activities within their chloroplasts, although they may take up less paraquat into the chloroplasts than the wild-type strain. Addition of paraquat to old cultures of the blue-green alga *Gloeocapsa*

Table 6.1. Effect of paraquat on illuminated spinach chloroplasts

Time after paraquat addition (minutes)	GSH (mmol l^{-1})	GSSG (mmol l^{-1})	GSH/GSSG ratio	[Ascorbate] in reduced form (mmol l^{-1})	Fructose bisphosphatase activity (units mg^{-1} chlorophyll)
0	6.5	0.25	26	13	62
2	3	3.2	0.9	8	20
5	3.1	1.5	2.1	6	8
10	3.0	0.8	3.8	4	0

Intact spinach-leaf chloroplasts were isolated and treated with paraquat in the light. The stromal contents of GSH, GSSG, and ascorbate were measured, as was the activity of fructose bisphosphatase, an enzyme essential to the Calvin cycle. Such rapid changes are not seen if paraquat is added to darkened chloroplasts, since the electron-transport chain is not then active to reduce paraquat. Data from *Biochem. J.* **210**, 899–903.

inhibited its ability to fix nitrogen, an effect that could be mimicked by addition of H_2O_2. Paraquat and diquat are also toxic to several non-green plant tissues. For example, they cause increased membrane permeability and accumulation of TBA-reactive material in some fungi such as *Aspergillus niger* and *Mucor hiemalis*. If this is due to the increased production of O_2^- and/or of H_2O_2, then reduction of the herbicides must be achieved *in vivo*. A number of flavoprotein enzymes, including glutathione reductase, have been shown to be capable of reducing paraquat and/or diquat; if they can penetrate into the active sites of the enzymes, these herbicides seem able to take electrons from the flavin ring and then, in the presence of oxygen, to generate O_2^-.

6.1.2 *Toxicity to bacteria*

Perhaps the most detailed studies of the role of oxygen radicals in paraquat toxicity have been performed with the bacterium *Escherichia coli*, largely by Hassan and Fridovich in the USA. Low concentrations (0.1–1.0 micromolar) of paraquat halt the growth of *E. coli* (a bacteriostatic effect), whereas much higher (> 100 micromolar) concentrations are required to kill the cells.

Paraquat added to a culture of *E. coli* is taken up by the cells and rapidly reduced. Under anaerobic conditions, the bipyridyl radical can be detected by observing its absorption spectrum. Extracts of *E. coli* cells have been observed to reduce paraquat if NADPH is added, a reaction that is catalysed by a soluble flavoprotein 'diaphorase' enzyme. In the presence of air, as would be expected, the radical disappears, and O_2^- is generated. Addition of paraquat to aerobically-grown *E. coli* induces a rapid synthesis of MnSOD activity, the same enzyme induced upon exposure of the bacteria to elevated oxygen concentrations (Chapter 3). Catalase activity increases as well. *E. coli* cells whose SOD activity had been increased by paraquat treatment were more resistant to elevated O_2 than normal cells and, vice versa, cells with raised SOD due to previous exposure to increased oxygen concentrations were more resistant to the toxic effects of paraquat. If induction of MnSOD by the bacteria was prevented, either by adding puromycin (an inhibitor of protein synthesis) or by a poor growth medium, then the toxicity of paraquat to the cells was greatly increased. Hassan and Fridovich further observed that addition of SOD to the growth medium offered some protection to *E. coli* against damage by paraquat. This at first sight is surprising, since O_2^- cannot cross the cell membrane and the SOD protein cannot enter the bacterial cells. The scientists were able to show, however, that just as paraquat can easily enter the cells, some of the paraquat radical can leak out of them and react with oxygen in the surrounding medium to give O_2^-, causing extracellular damage. The amount of leakage is inversely proportional to the intracellular oxygen concentration. Paraquat, like oxygen, has been shown to cause mutations in those *Salmonella typhimurium* strains that are used to test for mutagenic ability in the 'Ames test'. *S. typhimurium* cells containing high SOD activities were more resistant to the toxic and mutagenic effects of paraquat than were cells with normal SOD activity.

O. R. Brown's laboratory in the USA has shown that addition of paraquat to *E. coli* under aerobic conditions causes inhibition of the dihydroxyacid dehydratase enzyme involved in the biosynthesis of branched-chain amino acids. Providing amino acids to the growth medium partially relieves the inhibitory effect of paraquat on bacterial growth, i.e. its bacteriostatic effect. Addition of nicotinamide, a precursor of NAD^+, gives further relief. These results are very similar to the effects of high pressure oxygen on *E. coli* (Chapter 1) and they suggest that the actions of hyperbaric O_2, and the bacteriostatic action of paraquat, are mediated by a common mechanism. This is presumably an increased generation of O_2^- and H_2O_2 *in vivo*, especially as dihydroxyacid dehydratase is known to be inactivated by O_2^- (Chapter 3).

However, the lethal effects of high paraquat concentrations upon *E. coli* are not prevented by amino-acid supplementation, which means that other targets within the cell can also be attacked by O_2^- and/or H_2O_2. The mutagenicity of paraquat in the Ames test suggests that one of these targets is DNA, especially as some *E. coli* strains deficient in DNA repair mechanisms show an enhanced sensitivity to paraquat. It may be that the DNA damage is mediated by OH˙ formation at or close to the bacterial DNA, since iron or copper ions have been reported to aggravate the toxicity of paraquat to *E. coli*.

As mentioned in Chapter 3, several other compounds cause intracellular O_2^- generation and an increase in the synthesis of MnSOD, catalase, and/or peroxidase activities when they are added to *E. coli* under aerobic conditions. These include the antibiotic streptonigrin, juglone, menadione, pyocyanine (a pigment produced by *Pseudomonas aeruginosa*), and methylene blue. Their toxicity also appears to be mediated by increased generation of O_2^- since, for example, strains of *E. coli* with elevated SOD activity are resistant to the effect of streptonigrin. Some of these compounds are discussed further in Section 6.6.

6.1.3   *Toxicity to animals*

The main problem in the agricultural use of paraquat and diquat is that they are poisonous to several animal species, including fish, rat, mouse, cat, dog, sheep, cow, and man. Many cases of children drinking herbicides, carelessly put in lemonade or other soft-drink bottles, have been reported, and paraquat is sometimes used in suicide attempts. Oral intake of paraquat first results in local effects—irritation of the mouth, throat, and oesophagus, and sometimes vomiting and diarrhoea. Fortunately, gut absorption of bipyridyl herbicides is fairly slow, and life may often be saved by washing out the stomach and intestines repeatedly with saline solutions. Administration of suspensions of clays that absorb the herbicides (e.g. bentonite or Fullers earth), and dialysis of blood are often carried out as well.

The major lethal effect of paraquat in animals and humans is damage to the lungs. The membranous pneumocyte cells that line the alveoli (sometimes called type I alveolar cells) begin to swell and are eventually destroyed, an effect accompanied by oedema, capillary congestion, and a mild inflammation. Type II cells, or granular pneumocytes, are also damaged. The synthesis of surfactant, which lowers the surface tension of the lung linings and allows expansion, is decreased. As a result of both this and the oedema, gas exchange is hindered. In animals which survive, the damaged lung tissue is replaced by inelastic fibres that cause permanent interference with lung expansion. This major damaging effect on the lung is due to the fact that the granular pneumocytes, and probably several other lung cell types, actively accumulate paraquat, i.e. it is taken up even against a concentration gradient. The diamino compound putrescine ($H_3\overset{+}{N}(CH_2)_4\overset{+}{N}H_3$) has been shown to block this uptake in isolated rat lung slices. Other tissues, including liver and kidney, are damaged more slowly by paraquat. Large doses of diquat also affect the lung but it is not the major target tissue. In most diquat-exposed animals the intestines are especially attacked, becoming distended, and ceasing their normal peristaltic movements.

The special action of paraquat on the lungs means that poisoning can occur not only by oral intake but also by inhaling paraquat droplets from crop spraying. Paraquat has allegedly caused lung damage in the USA to 'pot'-smokers who obtained their marijuana from paraquat-treated Mexican plants. Bipyridyl herbicides can be slowly absorbed from the skin and they produce local irritation upon contact, and interfere with fingernail growth if present at high concentrations. Inhalation of the solid powders can cause nose-bleeds.

Microsomal fractions from several animal tissues, including lung, have been observed to reduce paraquat in the presence of NADPH and the paraquat radical can then react with oxygen to form $O_2^{\cdot-}$. It seems likely that paraquat can take electrons from the flavoprotein NADPH cytochrome P-450 reductase, since antibody against this enzyme inhibits paraquat reduction by microsomes *in vitro*. Intravenous injection of paraquat into rats causes a rapid activation of the pentose phosphate pathway in lung. NADPH is required for the biosynthesis of fatty acids, (needed to replace damaged membrane lipids), for surfactant biosynthesis, and for the glutathione reductase reactions. Fatty acid synthesis is decreased in the paraquat-treated lung, and there is an accumulation of 'mixed disulphides' formed between protein —SH groups and oxidized glutathione (see Chapter 3). Presumably hydrogen peroxide formed from O_2^- in the superoxide dismutase reaction is acted upon by glutathione peroxidase in the cytosol to form GSSG, which cannot be reduced quickly enough by glutathione reductase, and so combines with proteins. Feeding of nicotinamide has been reported to decrease the toxic effects of ingested paraquat in rats, an effect reminiscent of that in *E. coli* (see above).

The damaging effects of paraquat on lung tissue *in vivo*, and on isolated lung cells, are greatly potentiated at high oxygen concentrations, illustrating the importance of the reaction of the paraquat radical with O_2 *in vivo*. Selenium-deficient rats are more sensitive to paraquat poisoning than are animals fed on a normal diet, which suggests that the selenoprotein glutathione peroxidase has some protective role. There have been a few reports that injection of superoxide dismutase into animals ameliorates the symptoms of paraquat poisoning. Other scientists have not found this effect, however, and even when seen, it is fairly small. Injected superoxide dismutase cannot enter lung cells, and it is rapidly cleared from the body by the kidneys; so a large effect would not be expected. It may be that in lung, as in bacteria (see above), some of the paraquat radical can diffuse out of the cells and generate O_2^- in the surrounding tissue fluid. SOD would, of course, offer protection against this. *Intracellular* SOD may be an important protective mechanism, however, since the toxicity of paraquat to rat kidney cells in culture could be inversely correlated to their content of SOD after paraquat treatment.

Introduction of extra SOD into Chinese hamster ovary (CHO) cells by a 'scrape-loading' technique protected them against the loss of colony-forming ability caused by treatment with paraquat, and paraquat-resistant human cancer cells (HeLa cells) were found to have an increased activity of CuZn- and Mn-SODs. Further, injection of mice with diethyldithiocarbamate to inhibit CuZn-SOD (Chapter 2) greatly enhances paraquat toxicity. This inhibitor decreases lung SOD activity rapidly, being followed by a slower drop in glutathione peroxidase activity. Treatment of adult rats with bacterial endotoxin increases catalase, superoxide dismutase, glucose 6-phosphate dehydrogenase, and glutathione peroxidase activities in the lung; and these rats are not only more resistant to elevated oxygen concentrations but also to paraquat.

6.1.4   *Paraquat, lipid peroxidation, and hydroxyl radical formation*

Addition of paraquat to a mouse lung microsomal fraction in the presence of NADPH stimulated the peroxidation of membrane lipids, as observed by Bus *et al.* in the USA. However, attempts by other workers to extend these observations have led to a mixture of results. For example, pretreatment of mice with the antioxidant *N,N'*-diphenyl-*p*-phenylenediamine (Chapter 4) prevented this stimulatory effect being seen in microsome fractions subsequently isolated from the animals, yet it did not decrease the toxic effects of paraquat to the animals. Other workers have not observed any stimulation of lipid peroxidation by paraquat in microsomal fractions. On the other hand, light emission induced by infusion of *tert*-butylhydroperoxide into the perfused rat lung, and the lung content of TBA-reactive material, are increased by addition of paraquat to the perfusing medium, as is the production of both ethane and GSSG by the perfused rat liver. However, the ethane content of expired air from whole rats is only slightly increased by paraquat administration, and the basal light emission from perfused lung (i.e. in the absence of infused hydroperoxide) is not increased by paraquat infusion. It seems therefore that whereas paraquat can increase the rate of lipid peroxidation *in vivo* to some extent under certain circumstances, as would be expected if production of O_2^- and hydrogen peroxide is increased (Chapter 4), an increased rate of lipid peroxidation is *not* the primary damaging effect of paraquat. This is not a surprising conclusion; lipid peroxidation is not the only mechanism of cell injury by oxidant stress (Chapter 4, Section 4.9). It seems more likely that tissue damage by paraquat occurs first, mediated by increased formation of O_2^- and H_2O_2, and then damaged cells may sometimes peroxidize more rapidly.

The ability of paraquat to cause DNA damage in *E. coli*, and the potentiation of its toxicity by iron salts (Section 6.1.2), suggests that $OH^•$ radical may be involved. Treatment of Chinese hamster fibroblasts with paraquat has also been reported to produce chromosome damage. The paraquat radical ($PQ^{•+}$) is a very powerful reducing agent, much more so than O_2^-. Hence it can reduce Fe^{3+} complexes directly (A), or it can react with O_2 to form O_2^- (B):

A $Fe^{3+}\text{-complex} + PQ^{•+} \rightarrow PQ^{2+} + Fe^{2+}\text{-complex}$

B $PQ^{•+} + O_2 \rightarrow O_2^- + PQ^{2+}$

      $Fe^{3+}\text{-complex} + O_2^- \rightarrow Fe^{2+}\text{-complex} + O_2$

The Fe^{2+} complex can then react with H_2O_2 to form $OH^•$

      $Fe^{2+}\text{-complex} + H_2O_2 \rightarrow OH^• + OH^- + Fe^{3+}\text{-complex}$

The balance between route A and route B for generation of reduced iron complexes will depend upon the O_2 concentration in the system (the higher the O_2 concentration, the more favoured will be route B over route A). The presence of paraquat in a cell may also contribute to *providing* the metal ion promoters needed for $OH^•$ formation, since not only O_2^- (Chapter 3) but also $PQ^{•+}$ can reductively mobilize iron ions from the storage protein ferritin.

The hydroxyl radical may not be the only highly-reactive species generated in systems containing paraquat, a reducing system, oxygen, and iron ions. Elstner and Youngman in Germany suggested that a powerfully oxidizing species not identical to OH˙ could be formed; they called it 'crypto-OH''. Sutton and Winterbourn in New Zealand have more recently made similar suggestions that an iron(IV) species (ferryl) may be formed. As has already been discussed in Chapters 3 and 4, the existence of such 'alternative species' in biological systems seems very likely, but it is difficult to prove that they are responsible for producing the biological damage that is usually attributed to OH˙. The failure of certain OH˙ scavengers to protect against damage does not necessarily mean that the damage is not mediated by OH˙, since 'site-specific' OH˙ generation is very difficult to protect against by scavengers (see Chapter 3 for a detailed discussion on this point). Despite earlier claims, it now seems very unlikely that direct reaction of $PQ^{˙+}$ and H_2O_2 (i.e. in the absence of metal ions) produces a powerful oxidizing species of any kind.

The iron-chelating agent desferrioxamine (Chapter 3) has been used to 'probe' for a role of iron ions in mediating paraquat toxicity to animals *in vivo*. Some authors have reported protective effects, but others have claimed an exacerbation of paraquat toxicity. Desferrioxamine prevents O_2^--dependent OH˙ formation because it binds Fe^{3+} very tightly. and the Fe^{3+} complex cannot be reduced by O_2^-. However, it is possible that the highly-reducing $PQ^{˙+}$ radical *can* slowly reduce the Fe^{3+}–desferrioxamine complex, leading to Fe^{2+} formation and OH˙ generation. In any case, the toxic actions of desferrioxamine and its relatively-slow entry into cells (Chapters 3 and 8) suggest that it would not be a suitable therapy for use in the treatment of paraquat poisoning in humans.

Presumably the toxicity of bipyridyl herbicides to tissues other than the lung is mediated by their reduction and subsequent oxidation. As shown in Table 6.2, NADPH-cytochrome P-450 reductase is widely distributed in animal tissues. Even in tissues without this enzyme, reduction may be achieved by other flavoproteins. Feeding diquat to rats causes cataract formation, possibly due to its reduction by lens glutathione reductase. The haemolysis sometimes seen in paraquat-poisoned humans might be due to a reduction of this compound by erythrocyte glutathione reductase.

6.2   Alloxan and streptozotocin

Injection of alloxan into animals causes degeneration of the β-cells in the islets of Langerhans of the pancreas. Since these cells synthesize the hormone insulin, alloxan is often used to induce diabetes in experimental animals. Its structure is indicated below. Two-electron reduction of it gives *dialuric acid*:

Table 6.2. Distribution of cytochrome P-450 and its reductase in microsomal fractions obtained by centrifugation of different rat tissues

Organ examined	NADPH–cytochrome P-450 reductase	Cytochrome P-450
Liver (adult)	46.1 ± 5.0	497 ± 40
Liver (foetal)	3.7 ± 0.4	Not detected
Liver (new born)	18.9 ± 1.4	210
Kidney	21.2 ± 2.5	76 ± 11
Lung†	18.1 ± 2.4	41 ± 14
Brain	8.6 ± 2.0	30 ± 14
Spleen	5.5 ± 0.8	6
Testis	7.9 ± 0.8	46 ± 7
Adrenal	41.5 ± 6.6	604 ± 7
Stomach	11.1 ± 0.9	—
Small intestine	46.9 ± 5.8	Not detected
Mammary gland (late pregnant)	13.0 ± 0.9	—

†Tissues such as the lung are histologically very complex and the cytochrome P-450 and reductase may be concentrated in only one or two cell types.

Data were abstracted from Benedetto *et al. Biochim. Biophys. Acta* **677**, 363–72 (1981). Cytochrome P-450 reductase units are given in nmol min^{-1} mg^{-1} protein and P-450 is given in picomoles mg^{-1} protein. The absolute values should not be taken too seriously since microsomal fractions are very heterogeneous.

alloxan   dialuric acid

An intermediate radical, formed by one-electron reduction of alloxan, also exists. Dialuric acid is unstable in aqueous solution and undergoes oxidation, eventually to alloxan, that is accompanied by reduction of oxygen to O_2^-. Dialuric acid oxidation depends upon the presence of traces of transition metal ions, and it leads to generation of not only O_2^- but also H_2O_2, and OH· by Fenton-type reactions. Indeed, solutions of dialuric acid have been observed to stimulate lipid peroxidation *in vitro*, and to inhibit the growth of several strains of bacteria.

Hence, any body tissue that can take up and reduce alloxan will be at risk from the resulting oxidative stress. When alloxan is injected into rats, it accumulates in the islets of Langerhans and in the liver, but not in other body tissues. Whereas the liver contains high activities of superoxide dismutase, catalase, and glutathione peroxidase, the activities of these enzymes in the islet cells are moderate by comparison. Further, isolated islet cells are capable of reducing alloxan to dialuric acid at a high rate. Reduction of alloxan in the β-cell may involve a thioredoxin protein, normally involved in insulin synthesis, which is converted into its reduced (dithiol) form by NADH or NADPH. Reduced thioredoxin then converts alloxan to dialuric acid. It is probably this combination of fast radical generation and inadequate antioxidant defence that makes β-cells so sensitive to alloxan.

Addition of alloxan to isolated islet cells causes membrane damage and cell death, effects which can be decreased by addition of superoxide dismutase, catalase, compounds that can react with the hydroxyl radical, including mannitol and dimethylsulphoxide (Chapter 2), or the iron chelator DETA-PAC, which inhibits a key step in the iron-catalysed Haber–Weiss reaction (Section 3.5.4). Extensive DNA strand breaks have been observed in the islet cells of alloxan-treated rats. This DNA strand breakage, which is accompanied by an increased activity of poly(ADP-ribose) synthetase, may result from generation of OH˙ by transition metal complexes bound upon, or close to, the cell's DNA.

That these observations upon isolated islet cells are relevant to the action of alloxan *in vivo* is shown by a number of experiments. Grankvist *et al.* in Sweden found that injection of CuZnSOD, attached to a high-molecular-mass polymer to reduce its clearance by the kidneys, into mice protects them against the diabetes normally produced by a subsequent injection of alloxan. Heikkila and Cabbat in the USA observed a similar protective effect of injected DETAPAC whereas EDTA, which does not prevent the iron-catalysed Haber–Weiss reaction, did not protect. Dimethylurea, a powerful scavenger of OH˙, can also protect mice *in vivo*, as can dimethylsulphoxide, thiourea, or the alcohols methanol, ethanol, propanol, or butanol. It must be noted, however, that the effect of butanol is ambiguous since it produces an elevation of blood glucose concentration that can itself offer protection against alloxan. Injection of alloxan into rats deficient in vitamin E increased their exhalation of pentane and ethane, an effect that was decreased by prior injection of the OH˙ scavenger, mannitol. These effects were not seen in normally fed rats, however. The anti-cancer drug ICRF-159, a derivative of EDTA, has also been observed to protect against alloxan, possibly by chelating iron salts and preventing OH˙-production.

As is the case with paraquat (Section 6.1), there are conflicting reports on the ability of desferrioxamine to protect animals against the diabetogenic action of alloxan. Some scientists have reported protective effects in short-term experiments, but others find that desferrioxamine exacerbates alloxan-induced damage to β-cells *in vivo*. The chelator 1,10-phenanthroline, which decreases DNA damage in mammalian cells treated with H_2O_2 (Chapter 2), has been reported to decrease the severity of alloxan-induced diabetes in rats.

Another drug that has been used to induce diabetes in experimental animals is streptozotocin, a nitrosourea compound produced by *Streptomyces achromogenes* (Fig. 6.3). Streptozotocin induces DNA strand breakage in islet cells. Injection of it into rats has been observed to decrease the CuZnSOD activity of retina, erythrocytes, and islet cells, but not that in other tissues. Vitamin E has been reported to decrease the diabetogenicity of streptozotocin in rats. Injected SOD has been claimed by some scientists to decrease the diabetogenic effects of streptozotocin, but others have not confirmed this, and much more work is required on its mechanisms of action. The effect may depend very much on the exact timing of the SOD injection. Streptozotocin may well damage DNA directly by a chemical reaction of its nitroso group with the DNA, rather than by increasing oxygen radical formation. In any case, excessive DNA damage can lead to poly(ADP-ribose) synthetase activation and depletion of cellular NAD^+ to an extent that is lethal to the cell (Chapter 2). Okamoto and co-workers in Japan have proposed that this NAD^+ depletion is the mechanism of streptozotocin toxicity to the β-cells. (It might also contribute to alloxan toxicity.) They observed that β-cell tumours (insulinomas) could result from the combined administration of streptozotocin and inhibitors of poly(ADP-ribose) synthetase to rats. Thus, although inhibition of the enzyme prevented NAD^+ depletion and β-cell death, some of the DNA damage was not correctly repaired and led to a cancerous state (Fig. 6.4). These observations are fully consistent with the view (Chapter 2) that poly(ADP-ribose) synthetase

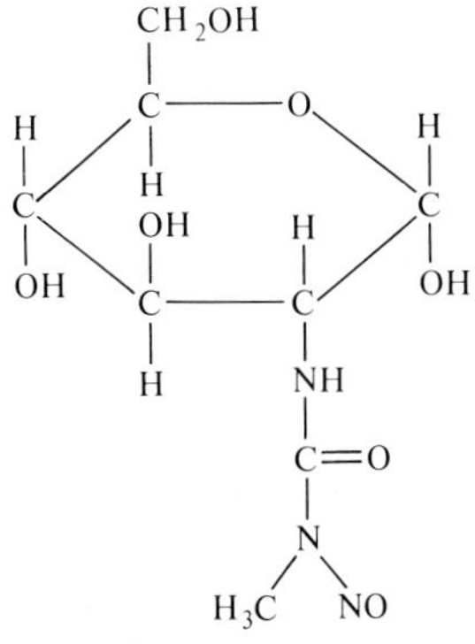

Fig. 6.3. Structures of streptozotocin. The full chemical name of the drug is *N*-methylnitrosocarbamylglucosamine. The α-form of the drug is shown; in the β-form the arrangement of the $-H$ and $-OH$ groups on carbon 1 (next to the oxygen in the ring) is reversed. The drug is normally a mixture of α- and β-forms.

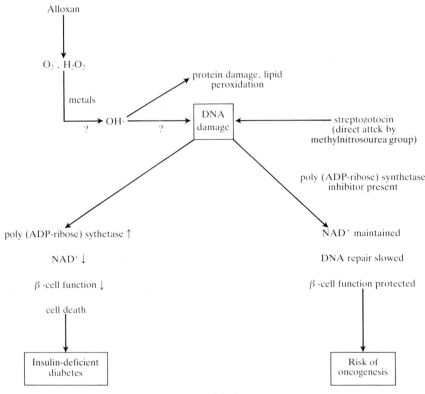

Fig. 6.4. Proposed mechanism of the diabetogenic action of streptozotocin (and possibly alloxan). Adapted from Okamoto, *Bioessays.* **2**, 19 (1985), with the kind permission of Professor H. Okamoto. Inhibitors of poly(ADP-ribose) synthetase used in such experiments include benzamide, 3-aminobenzamide, and theophylline.

activation and NAD^+ depletion may function to kill cells with excessive DNA damage, so restricting the occurrence of harmful mutations.

6.3 Substituted dihydroxyphenylalanines and other phenolic compounds

As Fig. 6.5 shows, DOPA, DOPAMINE, adrenalin, and noradrenalin are derivatives of the aromatic amino acid phenylalanine and they are often referred to collectively as *catecholamines*. As described in Chapter 3, the oxidations of adrenalin and noradrenalin can produce O_2^- and H_2O_2 in a very complicated series of reactions; DOPA and DOPAMINE do the same. In all cases the rate of oxidation is greatly accelerated by the presence of transition-metal ions, and it produces not only oxygen-derived species, but

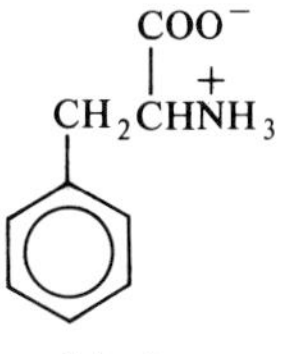

Phenylalanine

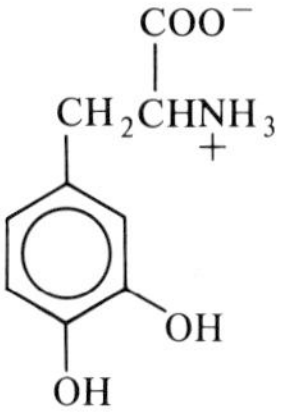

3,4-Dihydroxyphenylalanine
(DOPA)

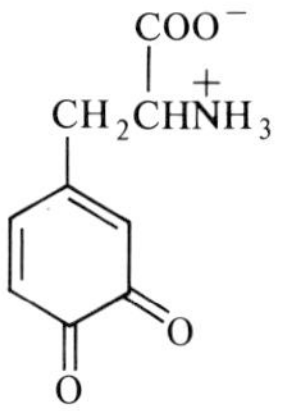

DOPA quinone

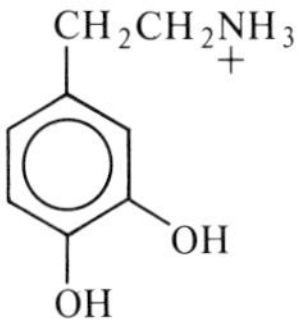

DOPAMINE

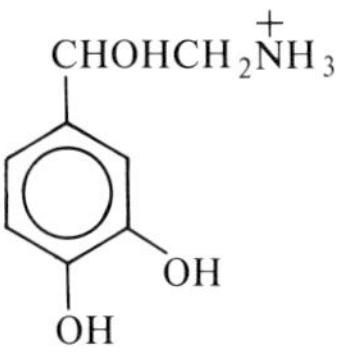

Noradrenalin

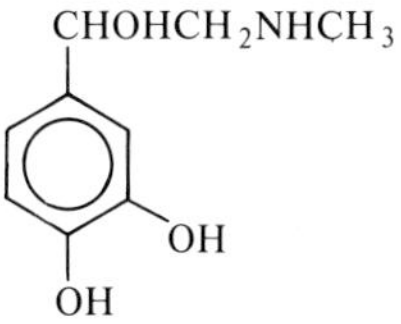

Adrenalin

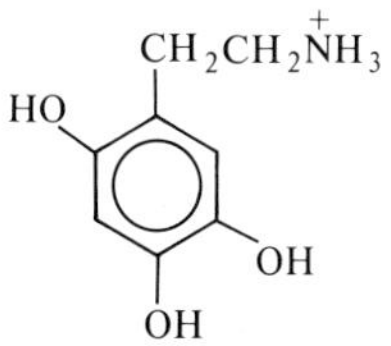

6-Hydroxydopamine

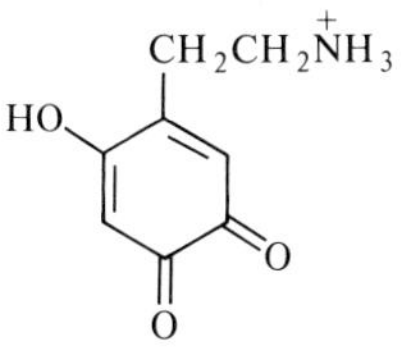

...*ortho*-quinone

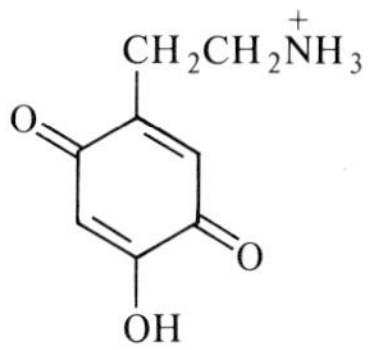

...*para*-quinone

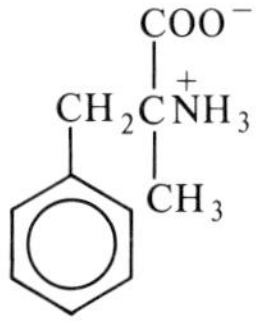

Methyldopa

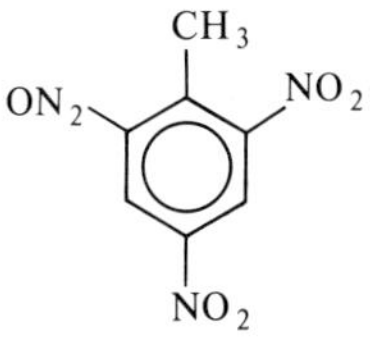

Trinitrotoluene (TNT)

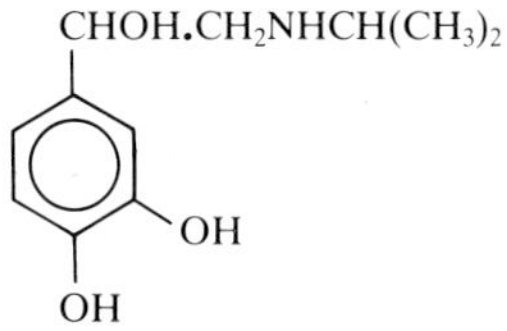

Isoproterenol

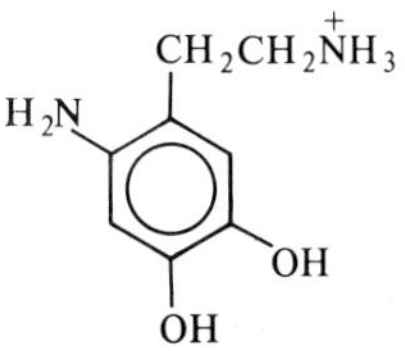

6-Aminodopamine

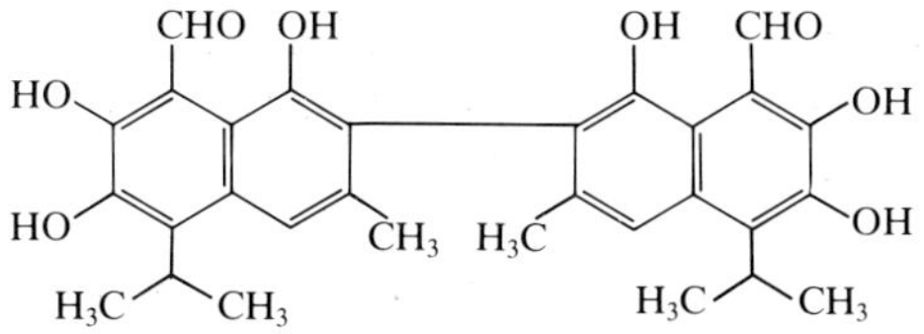

Gossypol

Fig. 6.5. Structures of dopa derivatives and some other aromatic compounds.

also quinones (e.g. DOPA quinone, Fig. 6.5) and semiquinones. These can combine with various cellular constituents to form covalent bonds, usually with thiol groups. Hence they can deplete cellular GSH concentrations. GSH plays a key role in drug metabolism (Chapter 3) and antioxidant defence systems, so its depletion can render the cell more sensitive to oxidant stress.

Treatment of animals or humans with large amounts of manganese compounds decreases the amount of dopamine in some parts of the brain, leading to symptoms that include muscular tremors. Mn(III) complexes readily oxidize catecholamines *in vitro*. The disorder 'manganese madness' or *locura manganica* has been observed in the miners of manganese ores in parts of northern Chile. *Locura manganica* in its later stages has a superficial clinical resemblance to Parkinson's disease, although the most extensive damage done to brain tissues is to the striatum and pallidum, not to the substantia nigra, which is the main site of cell death in Parkinson's disease (Section 6.12).

Catecholamines are important physiological regulators of cardiac contractility and metabolism, yet abnormally large doses of adrenalin, noradrenalin, or the synthetic catecholamine isoproterenol (Fig. 6.5) have been shown to produce an 'infarct-like' necrosis of heart muscle in experimental animals. Dhalla *et al.* in Canada have proposed that catecholamine oxidation products could contribute to this myocardial damage, especially as the heart seems sensitive to oxidant stress (Chapter 2, Section 2.4.3). The polyphenolic compound *gossypol* (Fig. 6.5), which has been clinically tested as an antifertility agent in men, exerts its spermicidal action by several mechanisms, one of which might be a slow oxidation reaction to give radical species and H_2O_2. Spermatozoa are very sensitive to oxidants and are easily inactivated by lipid peroxides (Chapter 4, Section 4.2.3).

Whereas the rate of oxidation of adrenalin and other phenols at physiological pH would be expected to be slow at the concentrations of transition metals likely to be present *in vivo* (Chapter 2), the related compounds 6-hydroxydopamine and 6-aminodopamine (Fig. 6.5) oxidize much more rapidly. During the oxidation, semiquinones and quinones are formed (Fig. 6.5 shows the two quinones that are produced by 6-hydroxydopamine) and spin-trapping experiments have shown the production of both superoxide and hydroxyl radicals. Formation of OH˙ during 6-hydroxydopamine oxidation *in vitro* is decreased by desferrioxamine, suggesting that it is formed by iron-catalysed reactions. The superoxide radical participates in the oxidation of further molecules, and hence addition of superoxide dismutase decreases the observed rate of oxidation of 6-hydroxydopamine *in vitro*. Indeed, this has been made the basis of an assay for SOD activity (Chapter 2).

When injected into the brains of experimental animals, both 6-hydroxydopamine and 6-aminodopamine cause a rapid and specific damage to catecholamine nerve terminals. Because of this, 6-hydroxydopamine is widely used as a research tool in experiments investigating the physiological roles of such nerve terminals. The selective action seems to be due to the specific uptake of these compounds by catecholamine neurones, and it is tempting to attribute the toxicity to increased formation of O_2^-, hydrogen peroxide, and OH˙ *in vivo*. The reaction of semiquinones and quinones with —SH groups on proteins might also produce damaging effects. Indeed, Graham *et al.* in the USA showed that the toxicity of various phenolic compounds to neuroblastoma cells in culture could be correlated with their rate of oxidation.

Figure 6.5 additionally shows the structure of methyldopa, a drug used to lower elevated blood pressure in humans. One of its side-effects is an impairment of liver metabolism. Incubation of liver microsomes with methyldopa in the presence of NADPH causes a covalent binding of the drug to the microsomal membranes that can be inhibited either by GSH or by superoxide dismutase. It seems that O_2^- produced by microsomes when NADPH is present (Chapter 3) can accelerate the oxidation of methyldopa, and the quinones and semiquinones formed can attack —SH groups in the microsomal proteins (hence the protection by GSH).

It has recently been suggested that the toxicity of the hydrocarbon benzene to animals and humans is due to its conversion *in vivo* into various phenols by the cytochrome P-450 system. Some diphenols and triphenols so produced can undergo oxidation to give quinones, semiquinones, and oxygen radicals. Benzene particularly damages the bone marrow. The explosive trinitro-toluene (TNT) has been reported to lead to O_2^- production when incubated with rat liver microsomes plus NADPH. Its structure is shown in Fig. 6.5.

6.4 Cigarette smoke

Soot from urban air, smoke from burning material, car exhaust fumes, and cigarette smoke all contain considerable numbers of free radicals (not counting the biradical oxygen molecule itself) that can be detected using ESR methods. Both cellulose and glass-fibre filters have been used by Pryor's group in the USA, and others, to trap radicals in the tar present in cigarette smoke. The tar in each puff of smoke contains about 10^{14} radicals. Most of them are highly stable and persist on the filters for hours. Figure 6.6 shows their ESR spectra. Gas-phase cigarette smoke radicals have been studied using the spin-trapping technique (Chapter 2) and Fig. 6.6(c) shows a typical spectrum obtained with phenyl-*tert*-butylnitrone (PBN). The gas phase contains about 10^{15} radicals per puff, both peroxy radicals (ROO˙) and, to a lesser extent, carbon-centred radicals being present. The smoke can be drawn as much as 180 cm down a glass tube without a significant decrease in radical concentration, which has interesting implications for what happens in the lungs. Fresh cigarette smoke contains high concentrations of the gaseous

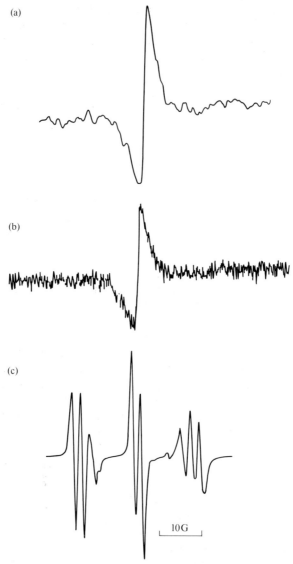

(a)

(b)

(c)

10G

Fig. 6.6. Detection of radicals in cigarette smoke by electron spin resonance. (a) ESR signal from glass-wool filter after smoke from four cigarettes has been drawn through it. The g-value is 2.002. (b) ESR signal from a glass-fibre filter after smoke from four cigarettes has been drawn through it. The g-value of the centre of the spectrum is 2.002. (c) ESR spectrum of the spin adduct formed by the trapping of radicals in cigarette smoke with PBN in benzene. By courtesy of Prof. W. A. Pryor.

compound nitric oxide (NO), which can react with oxygen to form nitrogen dioxide (NO_2). NO_2 is itself a radical and is capable of attacking compounds with $\diagup C{=}C\diagdown$ bonds. Hydrocarbons containing one or two double bonds are present in cigarette smoke and may react with nitrogen dioxide in a number of ways, e.g.

$$NO \xrightarrow{\ O_2\ } NO_2 \xrightarrow[\text{hydrocarbons}]{\text{unsaturated}} R^{\bullet} \xrightarrow{\ O_2\ } ROO^{\bullet} \xrightarrow{\ NO\ } RO^{\bullet} + NO_2$$

$\qquad$ slow reaction $\qquad\qquad\qquad\qquad\qquad$ peroxy radicals

A second reaction is of the type:

$$NO_2 + {-}HC{=}CH{-}CH_2{-} \rightarrow HNO_2 + {-}\overset{\cdots\cdots\overset{\bullet}{\cdots}\cdots}{CH{-}CH{-}CH}{-}$$

(a radical $\qquad\qquad\qquad\qquad\qquad$ nitrous acid $\quad$ (unpaired electron delocalized
itself) $\qquad\qquad\qquad\qquad\qquad\qquad\qquad\qquad\qquad$ between 3 carbon atoms)

The various radicals formed can then combine with each other to give non-radical products, so that a steady-state concentration of radicals is established. Nitrogen dioxide can abstract a hydrogen atom from linoleic and linolenic acids, and has been shown to stimulate the peroxidation of membrane lipids both *in vitro* and *in vivo*. For example, exposure of rats to air polluted with nitrogen dioxide increases the amounts of conjugated dienes subsequently detected in lung extracts, and also increases the rats' exhalation of ethane gas (Chapter 4). Clearly, the presence of nitrogen dioxide in cigarette smoke and in the gases produced during combustion of many nitrogen-containing compounds gives considerable potential for damage. Further, both nitric oxide and nitrogen dioxide *may* react with hydrogen peroxide to produce hydroxyl radicals (this has been reported in the gas phase but not in aqueous solution):

$$NO + H_2O_2 \rightarrow HNO_2 + OH^{\bullet}$$
$$NO_2 + H_2O_2 \rightarrow HNO_3 + OH^{\bullet}$$

Such reactions are feasible *in vivo*; cigarette smoke may contain H_2O_2, it increases the number of neutrophils in the lung and may activate them to produce O_2^{-} and H_2O_2, and it also increases the number of pulmonary macrophages as well as stimulating them to produce H_2O_2 (further discussion of oxidant production by neutrophils and macrophages can be found in Chapter 7, Section 7.3). The nicotine component of cigarette smoke has been claimed to enhance oxidant production by phagocytes, by making them more responsive to stimuli that activate the respiratory burst (Section 7.3). Cigarette smoke contains traces of metal ions (e.g. cadmium, iron, copper);

the last two metal ions can react with H_2O_2 to form $OH^{\cdot}$. It is known that alveolar macrophages from smokers contain more iron than normal.

Among the many damaging effects of $OH^{\cdot}$ generated in the lung is its ability to attack and inactivate a_1-*antiprotease*, sometimes called a_1-*antitrypsin*. This is an acute phase glycoprotein produced in the liver (Chapter 4) and released into extracellular fluids, and it serves to inhibit several proteolytic enzymes. Among them is *elastase*, an enzyme released by activated neutrophils. If elastase action is not inhibited, then it will hydrolyse the protein elastin within the lung. Elastin is the major component of elastic fibres, which can stretch to several times their length and then rapidly return to their starting length when tension is released. It is very important in allowing the correct expansion and contraction of the lung. An inborn deficiency of a_1-antiprotease in humans often causes a disease called *emphysema*, in which lung capacity is lost. Statistical evidence indicates that, even in humans with normal a_1-antiprotease concentrations, smoking predisposes to the development of emphysema. The most-popular explanation of this is that smoking produces a localized inactivation of a_1-antiprotease within the lung, allowing elastase released from neutrophils to remain active (smoke enhances neutrophil entry into lung). Lung lavage fluids from chronic smokers sometimes (but by no means always) show sub-normal antiprotease activity. Macrophages from some animals also contain elastase activity (much less than neutrophils). However, macrophage elastase is not inhibited by a_1-antiprotease. Hence the increased number of macrophages in the lungs of smokers, together with their activation by smoke to release enzymes, might represent an additional mechanism of elastin hydrolysis, if macrophage elastase is released in humans.

Various oxidants generated in cigarette smoke can inactivate a_1-antiprotease. These may include peroxynitrates, formed by reaction of NO_2 with peroxy radicals:

$$ROO^{\cdot} + NO_2 \rightarrow ROONO_2$$

as well as $OH^{\cdot}$. Oxidants damage a_1-antiprotease by oxidizing a methionine residue, essential for its function, into methionine sulphoxide. Modern genetic engineering techniques have been used to synthesize forms of a_1-antiprotease that lack the oxidant-sensitive methionine residue yet are still active in inhibiting elastase, and their use in the prevention of emphysema development in a_1-antiprotease-deficient humans has been proposed. Another approach has been to develop elastase inhibitors for therapeutic use; one of the most promising elastase inhibitors is *Eglin c*, a small peptide (relative molecular mass 8100) extracted from the bloodsucking leech *Hirudo medicinalis*.

An aqueous solution of cigarette smoke added to human lung cells has been reported to produce chromosome damage; damage was inhibited considerably by catalase and slightly by SOD. It may be that peroxides in the smoke cross the plasma membrane and damage DNA by site-specific OH$^{\cdot}$ generation (Chapters 2 and 3). Aqueous extracts of cigarette smoke also stimulate peroxidation of low-density lipoproteins, which might be relevant to the link between smoking and atherosclerosis (Chapter 8, Section 8.2).

Cigarette *tar* contains many complex aromatic compounds (over 3000 have been identified) and the stable free radicals present are mostly semiquinones held in a polymeric matrix. Aqueous extracts of cigarette tar generate O_2^{-} and peroxides and have been shown to damage isolated DNA. Inhibition of the DNA damage by the chelating agent DETAPAC suggests that OH$^{\cdot}$ generation is involved, presumably promoted by the metal ions present in cigarette smoke (see above). Some of the aromatic hydrocarbons present in cigarette tar are carcinogenic, and others promote the action of carcinogens (carcinogens and promoters are further considered in Chapter 8, Section 8.8.3).

6.4.1 *Lung defences against cigarette smoke*

Cigarette smoking predisposes to emphysema, lung cancer, heart disease, and atherosclerosis and the best defence is not to do it. We also advise you to keep well away from the 'sidestream' smoke produced by other smokers; this also contains carcinogens and damaging radicals.

The fluid which normally lines the alveoli of the lung seems to have little of such antioxidant proteins as SOD, selenium–glutathione peroxidase, catalase, caeruloplasmin, or transferrin. Alveolar surfactant is easily damaged by oxidants, and its synthesis is inhibited, so some protection must be present. Alveolar lining fluid, like lung in general, is very rich in ascorbic acid, which may directly scavenge several oxidants (Chapter 3, Section 3.2.1). Indeed, smoking may deplete both vitamin C and vitamin E in body fluids, perhaps predisposing to atherosclerosis (Chapter 8).

The authors have proposed that tracheobronchial mucus, which reacts with H_2O_2, OH$^{\cdot}$, oxides of nitrogen, and ozone, may help to protect the lung against cigarette smoke by scavenging many of the oxidants present before the smoke enters the alveoli. As scavenging occurs, the mucus is degraded, but it is continuously replenished by the mucus-secreting cells of the trachea. Irritation of these cells by any mechanism causes them to increase mucus production; mild oxidant stress can act as an irritant in this way.

6.5 Other air pollutants

Oxides of nitrogen are produced not only in cigarette smoke but also in motor vehicle exhausts and in the combustion of most biological materials.

322 *Free radicals and toxicology*

As well as promoting oxidative reactions (Section 6.4) NO_2 dissolves in water to give a highly acidic solution:

$$2NO_2 + H_2O \rightarrow HNO_3 + HNO_2$$
nitric acid nitrous acid.

Ozone (O_3) is also a common air pollutant; it performs an essential function in the higher levels of the atmosphere (Chapter 1), but is a nuisance if formed at lower levels (it can arise by photochemical reactions between nitrogen oxides and traces of hydrocarbons in the air: motor vehicle exhausts are a good source of the necessary gas mixture). Ozone is a powerful oxidizing agent, can lead to stimulation of lipid peroxidation (Chapter 4) and may cause OH˙ production in aqueous solution (Chapter 2). Ozone depletes ascorbic acid, vitamin E, and GSH in the lung, and can inactivate a_1-antiprotease by oxidizing the essential methionine residue of this protein (Section 6.4). Depletion of ascorbic acid and GSH may be due to direct chemical reaction with O_3, whereas depletion of vitamin E may be due to ozone-accelerated lipid peroxidation. Pre-exposure of adult rats to traces of O_3 (0.8 ppm for 7 days) caused increases in lung SOD, catalase, glutathione peroxidase, and glucose 6-phosphate dehydrogenase activities; these animals were then more resistant to the toxic effects of hyperoxia (Chapter 3). O_3 also increases synthesis of SOD and catalase by *E. coli*.

Another air pollutant found in urban areas is sulphur dioxide (SO_2), a colourless choking gas formed by the combustion of fuels containing sulphur (e.g. low-grade coals). It contains no unpaired electrons, and thus is not a free radical. Sulphur dioxide dissolves in water to reversibly form the sulphite and bisulphite ions, also producing an acidic solution:

$$SO_2 + H_2O \rightleftarrows H_2SO_3 \rightleftarrows H^+ + HSO_3^- \rightleftarrows H^+ + SO_3^{2-}$$
sulphurous acid   bisulphite ion   sulphite ion.

In addition to inhalation of SO_2, sulphite ion can enter the human body directly because it is used as a preservative in wine and in several foods. Much of it can be oxidized to sulphate (SO_4^{2-}) by the enzyme sulphite oxidase. However, some SO_3^{2-} may be available to stimulate free-radical reactions.

In one of the earliest uses of SOD to probe a radical reaction, McCord and Fridovich in the USA studied the ability of sulphite ions in solution to undergo oxidation. Sulphite ions are unstable and react with oxygen to eventually form sulphate ions, SO_4^{2-}. They observed that the oxidation of SO_3^{2-} in aqueous solution at pH 7.0 was not slowed by addition of SOD. However, the presence of EDTA decreased the rate threefold and changed the mechanism of the oxidation, in that SOD was then a powerful inhibitor of it. It appears that under certain circumstances, O_2^- can participate in sulphite oxidation. Asada in Japan showed that illuminated chloroplasts can accelerate the oxidation of SO_3^{2-}, an effect inhibited by SOD and presumably due to the production of O_2^- by electron-acceptors of photosystem I (Chapter 5). Young poplar leaves have more SOD activity than old poplar leaves, and are more resistant to the toxic effects of sulphur dioxide, whereas spraying spinach leaves with diethyldithiocarbamate to inhibit CuZnSOD increases their sensitivity to sulphur dioxide. These observations suggest that an interaction of SO_3^{2-} and O_2^- is relevant *in vivo*. Indeed, exposure of poplar leaves to concentrations of sulphur dioxide too low to cause visible injury induces a striking increase in leaf SOD activity. Growth of the green alga *Chlorella sorokiniana* in the presence of sulphite caused an increase in SOD activity within the cells and increased their resistance to paraquat (Section 6.1). Also, a paraquat-resistant strain of *Conyza boriensis* (with elevated SOD) is more tolerant to SO_2 than is the wild-type.

Ascorbic acid and GSH also play an important role in protecting plants against the toxicity produced by exposure to O_3, to oxides of nitrogen, or to SO_2. Both SO_2 and NO_2 contribute to the low pH of the 'acid rain' that is devastating trees in some European countries. Acid rain causes damage to plants not only by its low pH and the presence of oxidants, but also by converting insoluble complexes of toxic metals in the soil (e.g. aluminium) into soluble metal ions.

Horseradish peroxidase, among its many other reactions, will catalyse the one-electron oxidation of sulphite ions (Chapter 2), to form sulphite radical $SO_3^{\cdot-}$. Exposure of plants to SO_2-polluted air causes accumulation of hydrogen peroxide within chloroplasts, destruction of chlorophyll and carotenoids, and membrane damage, accompanied by the accumulation of TBA-reactive material and increased production of ethane gas. Indeed, SO_3^{2-} has been shown to stimulate the peroxidation of linoleic and linolenic acid emulsions, microsomes, and lipid extracts from rat lungs, probably by the interaction of a sulphite radical with preformed lipid hydroperoxides and/or by abstraction of hydrogen atoms (Chapter 4). Radicals such as $SO_3^{\cdot-}$, $SO_4^{\cdot-}$ (sulphate radical) and $SO_5^{\cdot-}$ (peroxysulphate radical) are known to be formed during sulphite oxidation: $SO_4^{\cdot-}$ and $SO_5^{\cdot-}$ are highly reactive. Inhalation of excess SO_2 might lead to lung damage by radical reactions as well as by low pH.

6.6 Haemolytic and anti-malarial drugs

We saw in Chapter 3 that the haemoglobin molecule acts as a source of O_2^- *in vivo*, producing the Fe(III)-form methaemoglobin. Methaemoglobin has been known for some time to catalyse certain peroxidase-like reactions *in vitro*, such as the oxidation of benzidine by hydrogen peroxide. Reactions of this type have often been used to detect blood in faeces, and during forensic examinations. Oxyhaemoglobin can even catalyse a hydroxylation of the aromatic compound aniline ($C_6H_5NH_2$) in the presence of NADPH.

6.6.1 *Hydrazines*

A number of drugs are known that can penetrate to the O_2-binding site of the haemoglobin molecule, and react with it. One of the most studied is phenylhydrazine and its derivative acetylphenylhydrazine (Fig. 6.7). Injection of these compounds into animals causes haemolysis and the bone marrow responds by putting immature erythrocytes into the circulation. Indeed reticulocytes, the precursors of erythrocytes, are often obtained for studies of cellular maturation and differentiation by injecting animals with phenylhydrazine and removing reticulocyte-rich blood several days later.

Phenylhydrazine and its derivatives slowly oxidize in aqueous solution to form O_2^- and hydrogen peroxide, a reaction catalysed by traces of transition-metal ions. The first stages in the oxidation can probably be represented by the equations below, in which M^{n+} represents the metal ions and Ph symbolizes the benzene ring to which the rest of the molecule is attached:

$$Ph\!-\!NH\!-\!NH_2 + M^{n+} \rightarrow H^+ + Ph\!-\!NH\!-\!NH^{\boldsymbol{\cdot}} + M^{(n-1)+}$$

$$Ph\!-\!NH\!-\!NH^{\boldsymbol{\cdot}} + O_2 \rightarrow H^+ + O_2^- + PhN\!=\!NH$$

Addition of SOD slows down the oxidation process.

However, Goldberg and Stern in the USA showed that the damage done by phenylhydrazine to erythrocytes is not prevented by SOD. They discovered that methaemoglobin, acting as a peroxidase, oxidizes phenylhydrazine in the presence of hydrogen peroxide. Oxyhaemoglobin also oxidizes phenylhydrazine, but hydrogen peroxide is not required in this reaction. These oxidase and peroxidase reactions of haemoglobin form a phenylhydrazine product that can react with oxygen to give O_2^-, and can also lead to formation of the phenyl radical. The reactions may be written, in a simplified form, as:

$$Ph\!-\!NH\!-\!NH_2 \xrightarrow{\text{haemoglobin}} PhN\!=\!NH$$
$$\text{phenyldiazine}$$

$$Ph\!-\!N\!=\!NH + O_2 \longrightarrow O_2^- + PhN\!=\!N^{\boldsymbol{\cdot}} + H^+$$
$$\text{phenyldiazine radical}$$

$$Ph\!-\!N\!=\!N^{\boldsymbol{\cdot}} \longrightarrow Ph^{\boldsymbol{\cdot}} + N_2$$
$$\text{phenyl radical}$$

$$Ph^{\boldsymbol{\cdot}} \xrightarrow[\text{abstraction}]{H^{\boldsymbol{\cdot}}} Ph\!-\!H$$
$$\text{benzene}$$

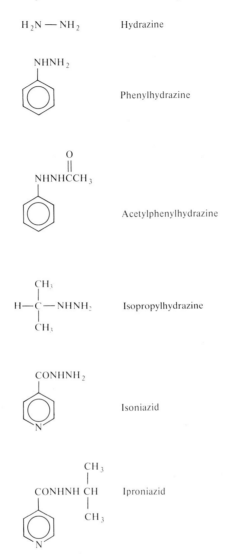

$H_2N \longrightarrow NH_2$ Hydrazine

Phenylhydrazine

Acetylphenylhydrazine

Isopropylhydrazine

Isoniazid

Iproniazid

Fig. 6.7. Structures of some haemolytic drugs.

Hence the end-products of the oxidation are benzene and nitrogen gas. Of these various species, the most damaging seems to be the phenyldiazine radical, which can denature the haemoglobin molecule, and stimulate peroxidation of membrane lipids, causing eventual haemolysis. The haem group is converted into a green product, both cleavage of the ring, and addition of phenyl groups to it, taking place. Oxidative denaturation of haemoglobin forms intracellular precipitates called *Heinz bodies*. There is evidence that damaged haemoglobin can be acted upon and broken down by the protease system in erythrocytes that recognizes abnormal proteins (Chapter 3, Section 3.4.3). It has been pointed out that reticulocytes obtained from phenylhydrazine-treated animals may themselves have suffered some membrane damage.

Although O_2^- and hydrogen peroxide are not required in the initial reaction of oxyhaemoglobin and phenylhydrazine, they are involved in the subsequent decomposition and precipitation reactions. Winterbourn and Carrell in New Zealand have conducted detailed studies on the interaction of acetylphenylhydrazine with oxyhaemoglobin *in vitro*. Addition of catalase to remove hydrogen peroxide decreased the rate of disappearance of oxyhaemoglobin, but addition of SOD actually increased it. Both ascorbate and GSH decreased the reaction, possibly by directly scavenging intermediate radicals (Chapter 3) such as $Ph-N{=}N^\cdot$, $Ph^\cdot$, and $Ph-NH-NH^\cdot$. Hence, these compounds, as well as the H_2O_2-removing enzymes catalase and glutathione peroxidase, would offer some protection to the erythrocyte *in vivo*. Consistent with this, the toxic effects of phenylhydrazine to mice are increased if they have been maintained on a selenium-deficient diet to lower glutathione peroxidase activity. Erythrocytes from patients with Fanconi's anaemia, in which SOD activities are decreased but glutathione peroxidase and catalase activities are normal (Chapter 8), do not show an increased susceptibility to haemolysis by acetylphenylhydrazine, consistent with the lack of protection by SOD in *in vitro* experiments.

Hydrazine derivatives are widely used in industry, as rocket fuels, and in medicine. For example, hydralazine is used to treat high blood pressure, but it has several side-effects, and prolonged use of it can produce symptoms of rheumatoid arthritis and of the autoimmune disease lupus erythematosus (Chapter 8). Like other hydrazines, this compound oxidizes *in vitro* in the presence of transition-metal ions to form O_2^-, hydrogen peroxide, and nitrogen-centred radicals. These may well participate in the side-effects of the drug. The metabolism of the anti-tubercular drug *isoniazid* and of the antidepressant *iproniazid* (Fig. 6.7) has been reported to produce hydrazine derivatives and free radicals, which again might contribute to their side-effects. For example, iproniazid can give rise to the hepatotoxic compound isopropylhydrazine (Fig. 6.7). An antioxidant often used in the rubber industry, *N*-isopropyl-*N'*-phenyl-*p*-phenylenediamine, can cause rapid oxidation and denaturation of both pure haemoglobin and also of the protein in erythrocytes. The reaction is about 40 times faster than that induced by phenylhydrazine. Thus antioxidants designed by chemists are not always antioxidants in biological systems.

6.6.2 *Sulphur-containing haemolytic drugs*

Diphenyl disulphide administered orally to rats causes erythrocyte destruction. It may be reduced by GSH to form thiophenol:

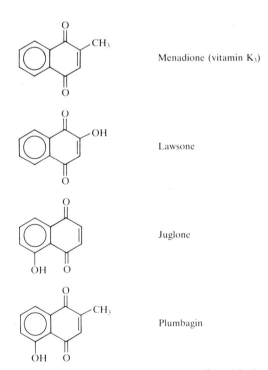

Thiophenol is oxidized by oxyhaemoglobin with formation of methaemo-globin, thiyl (RS˙) radicals, O_2^-, and H_2O_2; these radicals may account for its haemolytic action. Thiyl radicals combine with oxygen to yield species such as $RSO_2˙$ and $RSO˙$, that might do biological damage.

6.6.3 *Quinones*

Quinones (Fig. 6.8) are widely distributed in nature, being involved in many biochemical reactions (e.g. ubiquinone and plastoquinone, Chapters 3 and 5), used as dyes (e.g. lawsone, found in henna), and sometimes as drugs (e.g. the 'anti-cancer' quinones; Chapter 8). They are also present in cigarette smoke tar (Section 6.4) and their widespread industrial use guarantees our exposure to a wide range of quinones as environmental pollutants. We have

Fig. 6.8. Some simple quinones. Lawsone is a coloured pigment found in henna. Juglone is exuded by the roots of walnut trees and has been suggested to prevent germination of other plant seeds in the vicinity of the tree. Plumbagin and juglone have been used to increase SOD activity in *E. coli* (section 6.1).

already seen that quinone metabolites may mediate the toxicity of some compounds, e.g. 6-hydroxydopamine (Section 6.3).

Quinones can be toxic by at least two mechanisms. First, they or their semiquinones can often react with the —SH groups on essential molecules such as proteins and GSH (ubiquinone and plastoquinone cannot do this). Second, like paraquat, they may create 'oxidative stress' by redox cycling, i.e.

$$\text{Quinone} \xrightarrow[\text{1e}^-\ \text{reduction}]{\text{cellular reducing system}} \text{semiquinone}$$

For example, *plumbagin* and *juglone*, redox cycling quinones, are powerful inducers of MnSOD activity in *E. coli* (Chapter 3). In animal cells, both mitochondria and endoplasmic reticulum can catalyse one-electron reduction of quinones.

Reaction of the synthetic quinone menadione (Fig. 6.8) with oxyhaemoglobin causes oxidation and precipitation of the protein. It has been proposed that menadione first reacts with oxyhaemoglobin to give a semiquinone (SQ$^{\cdot-}$):

$$\text{Hb(Fe}^{2+}\text{)O}_2 + \text{Q} \rightleftarrows \underset{\text{methaemoglobin}}{\text{Hb(Fe}^{3+}\text{)}} + \text{O}_2 + \text{SQ}^{\cdot-}$$

The semiquinone can both re-reduce methaemoglobin and can also convert oxygen to the superoxide radical. In addition, menadione can react directly with —SH groups on the haemoglobin molecule itself. If these are written as protein—SH, then the reaction product is:

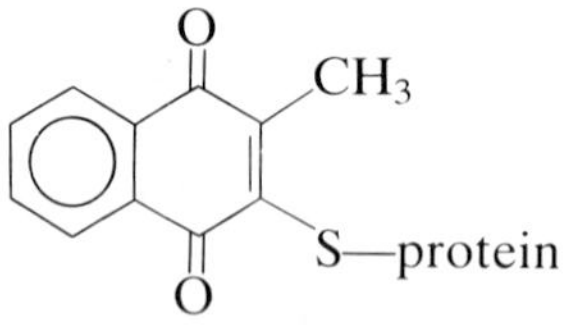

Treatment of isolated erythrocytes with diethyldithiocarbamate to inhibit CuZnSOD has been shown to accelerate the rate of haemolysis induced by *naphthoquinone-2-sulphonate*, a derivative of menadione in which the —CH$_3$ group is replaced by a sulphonate group to make the compound more soluble in water and aid experimentation. Diethyldithiocarbamate also potentiates the toxicity of menadione to liver cells. This is because a number of enzyme systems in liver, including NADPH-cytochrome P-450 reductase, catalyse the one-electron reduction of quinones into semiquinones that then react with oxygen to give O$_2^-$. Liver also contains high activities of the enzyme *diaphorase*, which, by contrast, catalyses a two-electron reduction of quinones into stable hydroquinones at the expense of NADH or NADPH. Hochstein and Ernster have suggested that the physiological function of diaphorase is as a 'quinone reductase', decreasing formation of O_2^- *in vivo* by removing quinones and thus preventing their reduction to semiquinones by the other enzyme systems. Consistent with this hypothesis, inhibition of diaphorase in isolated hepatocytes by the anti-coagulant drug *dicoumarol* increases the toxicity of menadione to these cells.

In hepatocytes, high concentrations of menadione rapidly deplete GSH, increase formation of O_2^- and H_2O_2 by redox cycling and cause the formation of numerous small 'blebs' (similar to those shown in Fig. 4.9) on the membrane surface. Bleb formation in menadione-treated hepatocytes seems to be associated with increases in intracellular 'free' Ca^{2+} concentrations. Intracellular free Ca^{2+} is normally very low (< 0.1 µmoles per dm^3) but menadione can increase it by inactivating the Ca^{2+}-sequestering mechanism (Ca^{2+}-ATPase) of the endoplasmic reticulum (probably by oxidizing essential —SH groups on the protein), by causing mitochondria to release calcium or by interfering with plasma membrane systems that pump Ca^{2+} out of cells. A rise in intracellular free Ca^{2+} activates Ca^{2+}-dependent proteases, which may play some role in bleb formation. Menadione, plumbagin, and several other quinones are mutagenic to bacteria, suggesting that DNA damage can also occur (Fig. 6.1). The ability of menadione to induce DNA strand breaks in rat hepatocytes has been described.

6.6.4 *Oxidant stress, favism, and malaria*

The haemolytic effects of menadione *in vivo* are usually minor unless the patient has an inborn defect in one of the erythrocyte defence mechanisms, the commonest being a deficiency in glucose 6-phosphate dehydrogenase activity (Chapter 3, Section 3.1.1). Most patients with a partial deficiency of this enzyme show little haemolysis normally, but it can be induced, often severely, by a wide range of drugs and even by ingestion of certain foods such as the broad bean *Vicia faba*. This latter condition, known as favism, is common in certain Mediterranean countries such as Sardinia and Greece, in the Middle East, and parts of south-east Asia. Its distribution therefore follows that of glucose 6-phosphate dehydrogenase deficiency, but not all patients deficient in this enzyme are sensitive to the bean, for an unknown reason. The erythrocytes of patients with favism show a rapid fall in $NADPH/NADP^+$ ratios and GSH concentrations very soon after ingesting the bean, resulting in impaired functioning of glutathione peroxidase and glutathione reductase. The chemicals that cause this effect are the pyrimidine derivatives, vicine and convicine, which are present at about 0.5 per cent of the total weight of the bean. These can be hydrolysed by β-glucosidase enzymes to give 'aglycone' products (Fig. 6.9) that react rapidly with oxygen

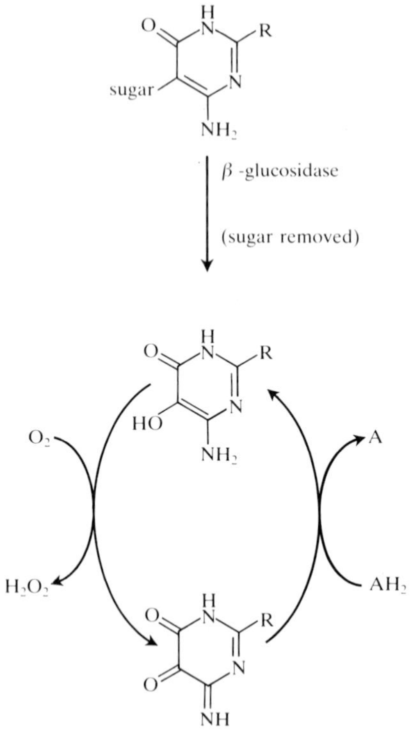

Free radicals and toxicology

Fig. 6.9. Redox cycling of the aglycones derived by hydrolysis of convicine (R = OH) and vicine (R = NH$_2$) from *Vicia faba* seeds. The aglycone of convicine is called *isouramil*, that of vicine is called *divicine*. Diagram by courtesy of Prof. G. Rotilio. AH$_2$ signifies a reducing system (such as GSH).

to form hydrogen peroxide and an oxidized form that can be re-reduced by GSH, thus leading to more H$_2$O$_2$-generation.

Drugs that trigger haemolysis in glucose 6-phosphate dehydrogenase deficient patients include the antimalarials primaquine and pamaquine, as summarized in Table 6.3. Mixtures of primaquine with NADPH and with oxyhaemoglobin have been shown to produce hydrogen peroxide *in vitro*, an observation perhaps relevant to its haemolytic effects. Inborn defects in other erythrocyte enzymes can also cause problems, although these defects occur much more rarely. The oxidation and precipitation of several abnormal haemoglobins, which produces O$_2^-$, is accelerated by certain drugs (Table 6.3). Infections frequently initiate haemolytic crises in carriers of unstable haemoglobins, which may sometimes be due to their faster denaturation during the elevated body temperatures that occur in fever. Some of the denaturation products of haemoglobin might be able to catalyse an interac-

Table 6.3. Inborn deficiencies of erythrocyte enzymes in relation to drug-induced haemolysis

Abnormality	Prevalence	Usual clinical feature	Drugs inducing haemolysis	Normal clinical use of drug
Glucose-6-phosphate dehydrogenase deficiency (X-linked, recessive)	Very common in patients in tropical or Mediterranean areas or their descendants	Some RBC damage often detected in laboratory tests but severe haemolysis *in vivo* very rare	Fava beans Furazolidone Nitrofurantoin Nitrofurazone Pamaquine Primaquine Sulphonamides	— (See Section 6.10) Antimalarials Antibacterial
Glutathione peroxidase deficiency (recessive)	Very rare	Often none, sometimes severe haemolysis, infertility	Sulphonamides Nitrofurantoin	Antibacterial (Section 6.10)
Glutathione reductase deficiency (recessive)	Very rare, but lack of riboflavin in the diet can also decrease the activity (enzyme has FAD at the active site). The drug BCNU, used in cancer chemotherapy, is a powerful inhibitor of glutathione reductase and can cause erythrocyte damage	Often none, sometimes severe haemolysis	Sulphonamides	Antibacterial
Abnormal haemoglobins (e.g. Hb Torino, Hb Shepherd's Bush, Hb Peterborough, Hb Zurich)	Rare	Often some haemolysis seen related to protein instability. Drug administration can cause a severe haemolytic crisis	Sulphonamides	

Data are largely taken from the article by Gaetani, G. F. and Luzzatto, L. (1980). In *Pseudo-allergic reactions. Involvement of drugs and chemicals*, Volume 2 (eds P. Dukor *et al.*), S. Karger, Basle, Switzerland.

tion of O_2^- and H_2O_2 to form hydroxyl radicals, as can any iron ions released from the denatured proteins.

The prevalence of glucose 6-phosphate dehydrogenase deficiency and sickle-cell anaemia has been related to their ability to confer some protection against the malarial parasite (Section 3.1.1), since malaria-infected erythrocytes already appear to be under an oxidant stress, perhaps as a consequence of parasite metabolism. Malarial parasites are damaged by systems generating O_2^- and H_2O_2, and these oxidants are used by macrophages (Chapter 7, Section 7.3) to attack such parasites. Parasites are also damaged by lipid peroxides and cytotoxic aldehydes (Chapter 4, Section 4.2.3). Clark in Australia has shown that injection of alloxan or *tert*-butylhydroperoxide (Chapter 4) into malaria-infected mice kills a large number of the parasites. The effect of alloxan can be overcome by pre-treatment of the mice with desferrioxamine, an inhibitor of iron-dependent formation of OH˙ and iron-dependent decomposition of peroxides.

However, it should be noted that, under slightly different experimental conditions, desferrioxamine and other chelators can themselves suppress the growth of malarial parasites, probably by depriving them of the iron that they need. When grown in isolated human erythrocytes, the malarial parasite *Plasmodium falciparum* suffers complete growth inhibition at only $30\,\mu M$ concentrations of desferrioxamine. Phenylhydrazine and divicine also decrease parasitaemia in malaria-infected mice, and the anti-parasite action of divicine can be prevented by desferrioxamine.

Thus any defect in protection of erythrocytes against oxidative damage, or an increased rate of free-radical production *in vivo*, should favour the eradication of malarial parasites. Clark *et al.* in Australia have made the interesting suggestion that much of the pathology associated with malarial infection may be host-dependent, as the host over-enthusiastically tries to kill the malarial parasites without focusing clearly enough on the target.

Several other protozoan parasites are sensitive to oxidant attack, which raises prospects for the design of new drugs. *Trypanosoma brucei*, for example, is unable to synthesize haem and lacks catalase activity. It can be killed by increasing its rate of intracellular oxidant generation, e.g. by adding menadione, and killing is accelerated by the addition of free haem. Several nitro-compounds have anti-parasite action (Section 6.10). Iron-chelators could conceivably be used as anti-parasite agents.

6.7 Ethanol

Most people drink solutions of ethanol (ethyl alcohol) as alcoholic drinks, and even in teetotallers it is formed in small amounts by gut bacteria. Ethanol is very soluble both in water and in organic solvents, and can cross cell membranes readily. It penetrates the blood–brain barrier and affects the central nervous system. One beneficial effect that it has, apart from the social ones, is to partially protect experimental animals against alloxan-induced diabetes (Section 6.2). It may do this by scavenging the highly reactive hydroxyl radical to form a much less reactive hydroxyethyl radical:

$$CH_3CH_2OH + OH^{\cdot} \rightarrow CH_3{}^{\cdot}CHOH + H_2O$$

Indeed, this reaction is made use of in spin-trapping experiments with DMPO (Chapter 2).

Ethanol absorbed into the body is mainly metabolized in the liver by an alcohol dehydrogenase enzyme to form the aldehyde ethanal (acetaldehyde):

$$CH_3CH_2OH + NAD^+ \rightarrow NADH + H^+ + CH_3CHO$$

Smaller quantities are oxidized by the peroxidatic action of catalase in peroxisomes (Chapter 3) and the MEOS system, which involves cytochrome P-450 and, possibly, hydroxyl radicals. Prolonged intake of ethanol increases the activity of the MEOS system, largely by stimulating synthesis of the 'ethanol-inducible' cytochrome P-450 (Chapter 2). Prolonged intake of excessive amounts of ethanol by humans causes severe damage to many tissues, especially the liver, which may become cirrhotic. Hepatic iron overload is common in patients with alcoholic liver diseases.

In England, it is a criminal offence to drive a motor vehicle with a blood ethanol concentration greater than 80 mg per 100 ml, corresponding to a concentration of 17.4 millimoles per litre. Some 'heavy' drinkers get themselves up to 300 mg per 100 ml of blood, or 65.2 mmol l^{-1}. Such high concentrations have been shown in experimental animals to affect antioxidant protective systems. Injection of large doses of ethanol into rats has been reported to decrease the SOD activity measurable in homogenates of their brains by about 25 per cent, and an even greater decrease is produced by repeated injection. Similarly, exposure of cultures of neuronal or glial cells from rats, mice, hamsters, and chicks to 100 mM ethanol *in vitro* decreased their SOD activity. Whether these small changes are sufficient to cause biological damage has yet to be determined. By contrast, other scientists have reported increases in SOD activity in animals treated with ethanol for prolonged periods. Both large doses of ethanol, and smaller doses given repeatedly, have been shown to increase lipid peroxidation in the livers of rats and baboons, as followed by accumulation of conjugated dienes or ethane production. Large doses of ethanol have also been observed to cause a significant decrease in the GSH concentrations of liver and kidney cells of rats, but not in other tissues. Comparable falls in GSH are seen in mice and baboons. Such a drop in GSH might either cause more lipid peroxidation or merely be a consequence of it in view of the reactions catalysed by glutathione peroxidase (Chapter 4). Neither does it follow that an induction of lipid peroxidation is the mechanism by which ethanol damages the liver, since increased peroxidation often accompanies damage caused by other means (Chapter 4). There is some evidence that the damage is done by ethanal rather than ethanol itself. Ethanal is metabolized mainly by an aldehyde dehydrogenase enzyme that converts it into ethanoic acid (acetic acid, CH_3COOH), but is possible that some of it could be acted upon by the enzyme aldehyde oxidase in the liver, which is known to produce O_2^-. In the liver of alcoholics, mitochondria, which contain the alcohol dehydrogenase enzyme, are particularly damaged. Ethanal may also react chemically with GSH and so decrease its concentration.

The potential relevance of free radicals to chronic ethanol effects in man is shown by the observation of American scientists that red blood cells from alcoholics have increased SOD content. Curiously, this effect is only seen with black alcoholics and not white ones, for an unknown reason. There is suggestive, but as yet not convincing evidence, that increased oxidant damage causes the pathology related to excess ethanol consumption. Analysis of the mechanism of tissue damage by ethanol in chronic alcoholics is made more difficult by the fact that their diet is often inadequate and may lack enough vitamin E, selenium, or polyunsaturated fatty acids.

6.8 Paracetamol and phenacetin

Paracetamol (sometimes called acetaminophen) is a mild pain-killer that has found increasing use in recent years as a substitute for aspirin. Unlike aspirin, it does not irritate the stomach lining and it appears fairly safe in the recommended dosage. At high doses, however, it is acutely toxic to both the liver and kidneys, and poisoning by paracetamol overdosage in suicide attempts is becoming increasingly common. The paracetamol derivative, phenacetin, is also a mild pain-killer, but it is no longer widely used because of the risk of kidney damage.

Paracetamol is a substrate for the cytochrome P-450 system, and phenacetin is converted into paracetamol *in vivo* by removal of the ethyl group. The action of the P-450 system on paracetamol produces a highly-reactive quinoneimine (Fig. 6.10), which can attack cell membrane proteins and other proteins by combining with —SH groups. It also causes rapid depletion of GSH. Indeed, protection against the hepatotoxicity of paracetamol is afforded by giving sulphur-containing compounds such as *N*-acetylcysteine or methionine. These may act by maintaining intracellular GSH concentrations.

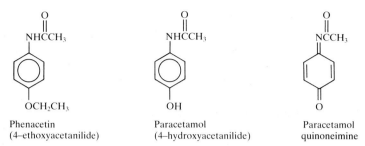

Phenacetin
(4–ethoxyacetanilide)

Paracetamol
(4–hydroxyacetanilide)

Paracetamol
quinoneimine

Fig. 6.10. Structure of paracetamol, phenacetin, and paracetamol quinoneimine.

Some evidence for paracetamol-induced lipid peroxidation in the livers of animals has been reported, but there is no good evidence that lipid peroxidation causes paracetamol hepatotoxicity. Instead, the GSH depletion and tissue injury caused by paracetamol may lead to increased rates of lipid peroxidation.

6.9 Halogenated hydrocarbons

6.9.1 *Carbon tetrachloride and chloroform*

One of the first reactions encountered by any student of organic chemistry is that of the hydrocarbon gas methane (CH_4) with chlorine or bromine vapour. The reaction only proceeds in the presence of ultraviolet light, which provides sufficient energy to cause homolytic fission of the covalent bond in the halogen molecule. The reaction then proceeds as a typical radical chain reaction, i.e.

Initiation: $\qquad Cl_2 \xrightarrow{\text{uv}} Cl^{\cdot} + Cl^{\cdot}$

Propagation reactions: $\qquad Cl^{\cdot} + CH_4 \longrightarrow CH_3^{\cdot} + HCl$

$\qquad CH_3^{\cdot} + Cl_2 \longrightarrow CH_3Cl + Cl^{\cdot}$

$\qquad Cl^{\cdot} + CH_3Cl \longrightarrow HCl + {}^{\cdot}CH_2Cl$

$\qquad {}^{\cdot}CH_2Cl + Cl_2 \longrightarrow {}^{\cdot}CH_2Cl_2 + Cl^{\cdot}, \text{ etc. ....}$

Termination reactions: $\qquad CH_3^{\cdot} + Cl^{\cdot} \longrightarrow CH_3Cl$

$\qquad CH_3^{\cdot} + CH_3^{\cdot} \longrightarrow C_2H_6$

Similar reactions occur with bromine, but much more slowly. Hence the methane is successively converted into chloromethane (CH_3Cl), dichloromethane (CH_2Cl_2), trichloromethane ($CHCl_3$, often called 'chloroform'), and tetrachloromethane (CCl_4, often called 'carbon tetrachloride'). Both $CHCl_3$ and CCl_4 are liquids at room temperature. Chloroform was the first anaesthetic to be employed in surgery, by the Scottish doctor Sir James Young Simpson in 1847, but its liver-damaging properties were quickly discovered. Carbon tetrachloride is a constituent of many dry-cleaning fluids because of its ability to easily dissolve greasy stains. Its lipid solubility allows it to cross cell membranes rapidly, and any tetrachloromethane taken into the body is quickly distributed to all organs. However, its main toxic effects are shown on the liver (and it is much more toxic than chloroform) although there is some injury to other tissues. Even a single small dose of tetrachloromethane to rats produces fat accumulation in the liver due to a blockage in synthesis of the lipoproteins that carry triglyceride away from this organ. The normal structure of the liver cell endoplasmic reticulum as seen under the electron microscope is distorted, hepatic protein synthesis slows down, and the activity of enzymes located in the endoplasmic reticulum, such as glucose 6-phosphatase and the P-450 system, rapidly declines, as does the ability of the reticulum to bind Ca^{2+} ions. Hence rises in intracellular Ca^{2+} concentrations may occur (Fig. 6.11). The nuclear membrane is attacked more slowly. Eventually there is necrosis of liver cells in the central areas of the organ. Yet incubation of microsomal fractions with tetrachloromethane *in vitro* does not directly inactivate these enzymes. As a result of detailed work by the groups of Recknagel in the USA, Slater in England, and Comporti in Italy, it has been concluded that the effects produced by tetrachloromethane are due to the fact that it is a substrate for the cytochrome P-450 system which is, of course, especially concentrated in the liver (Table 6.2). If microsomes from rat liver are exposed to tetrachloromethane *in vitro*, nothing happens until NADPH is provided to allow NADPH-cytochrome P-450 reductase to operate. There is then a rapid peroxidation of microsomal lipids accompanied by the inactivation of the above enzymes, and destruction of cytochrome P-450 itself. Microsomal fractions isolated from CCl_4-treated rats also show increased peroxidation. Fatty-acid side-chains attached to phosphatidylserine (Chapter 4) are especially attacked, perhaps because this lipid is adjacent to P-450 *in vivo*. Phenobarbital increases the activity of the P-450 system in liver (Chapter 3). Phenobarbital-treated rats or sheep are more susceptible to tetrachloromethane and show more peroxidation in liver microsomal fractions isolated from them. Administration of antioxidants such as vitamin E, promethazine, propyl gallate, and GSH, or of the compound SKF-525A, which inhibits microsomal drug metabolism, decreases CCl_4-toxicity in animals.

    As a result of these observations, it was suggested that tetrachloromethane is metabolized by the P-450 system to give the trichloromethyl radical (CCl_3·), a carbon-centred radical:

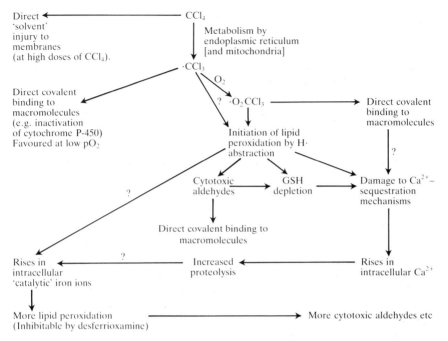

Fig. 6.11. Mechanisms of hepatotoxicity of CCl_4. Several possible links are not shown, e.g. O_2CCl_3 could react with GSH and contribute to GSH depletion. In organic solvents, O_2^- can attack CCl_4 to produce CCl_3O_2 and other products (Chapter 2, section 2.9): whether this contributes to the hepatotoxicity of CCl_4 is uncertain (see *J. Biol. Chem.* **263**, 12224 [1988]).

$$CCl_4 \xrightarrow[\text{P-450 system}]{\text{1 electron}} {}^{\cdot}CCl_3 + Cl^-$$

Mitochondria may also be able to form $^{\cdot}CCl_3$ from CCl_4. Spin-trapping experiments upon isolated rat liver microsomes in the presence of **NADPH** and **tetrachloromethane** have been carried out by several groups. Addition of the spin-trap, phenyl-*tert*-butylnitrone (PBN, see Chapter 2) to the system gives the ESR signal expected from reaction of the trap with $CCl_3^{\cdot}$

Stable nitroxyl radical

Conversion of tetrachloromethane into $CCl_3\dot{}$ appears to be brought about by the cytochrome P-450 molecule itself (after reduction by the NADPH-cytochrome P-450 reductase), although it is possible that the reductase might also interact with tetrachloromethane. Studies with PBN in perfused rat liver exposed to CCl_4 have also trapped $CCl_3\dot{}$, as well as the carbon dioxide anion radical, $\dot{}CO_2{}^-$, which appears to arise from $CCl_3\dot{}$. The $CCl_3\dot{}$ radical has also been trapped in the livers of mice given both CCl_4 and PBN.

The trichloromethyl radical might combine directly with biological molecules, causing covalent modification (Fig. 6.11), as well as abstracting hydrogen from membrane lipids, setting off the chain reaction of lipid peroxidation (Chapter 4). Products of peroxidation are known to inhibit protein synthesis and the activity of certain enzymes (Chapter 4). Indeed, liver microsomal fractions from CCl_4-treated rats show an increased amount of protein-bound carbonyl compounds and CCl_4-treated rats exhale more pentane gas than normal, indicative of increased lipid peroxidation *in vivo*. The rats also exhale chloroform vapour; trichloromethane would be expected to be produced by the combination of a trichloromethyl radical and a hydrogen atom abstracted from a membrane lipid. These effects are more marked if the rats have been fed on diets deficient in vitamin E. Increased ethane production has also been observed in isolated liver cells treated with tetrachloromethane, together with increased light emission, and accumulation of TBA-reactive material. Peroxidation occurred rapidly and was seen before the onset of loss of cell viability. The antioxidant promethazine decreased peroxidation in the liver cells and prevented loss of glucose 6-phosphatase activity, but not the loss of P-450 itself. Probably cytochrome P-450 can be directly attacked by $CCl_3\dot{}$ or other radical species formed, whereas the inactivation of glucose 6-phosphatase is brought about by products of lipid peroxidation. Supporting this conclusion is the observation that hepatocytes isolated from rats pre-treated with vitamin E were less sensitive to killing by CCl_4 than hepatocytes from control rats. Inactivation of glucose 6-phosphatase after treatment of tocopherol-enriched hepatocytes with CCl_4 was slowed, but destruction of cytochrome P-450 was not. This P-450 destruction means that CCl_4 metabolism is a self-limiting event.

Despite the mass of evidence reviewed above, some questions have been raised about peroxidation of membrane lipids induced by $CCl_3\dot{}$ as an explanation of CCl_4-toxicity. Pulse radiolysis studies of the reactivity of the trichloromethyl radical have been carried out by Willson's group in England, whereupon it was found that its most rapid reaction is with molecular oxygen to form the *trichloromethylperoxy radical*:

$$\dot{}CCl_3 + O_2 \rightarrow CCl_3O_2\dot{} \quad (k_2\ 3.3 \times 10^9\ \mathrm{M^{-1}s^{-1}})$$

$CCl_3O_2\dot{}$ reacts much more rapidly with arachidonic acid ($k = 6 \times 10^6$ at pH 7), promethazine, ascorbate, thiol compounds, and the tyrosine and tryptophan residues of proteins than does the trichloromethyl radical, and so $CCl_3O_2^{\cdot}$ would seem a more likely candidate for a damaging species. Spin-trapping experiments have so far failed to reveal its presence in CCl_4-treated microsomes, however, although its high reactivity will make it correspondingly more difficult to detect. Formation of the trichloromethylperoxy radical could explain why small amounts of phosgene gas ($COCl_2$) are produced by CCl_4-treated microsomes. It could simply arise by the reactions below:

$$CCl_3O_2^{\cdot} + lipid-H \rightarrow CCl_3O_2H + lipid^{\cdot}$$
$$CCl_3O_2H \rightarrow COCl_2 + HOCl$$

It is possible that the local oxygen concentration in different parts of the liver influences whether $CCl_3^{\cdot}$ itself covalently binds to biological molecules or combines with O_2 to form $CCl_3O_2^{\cdot}$ which should be a much better initiator of lipid peroxidation (Fig. 6.11).

The 'radical stress' imposed on the liver by CCl_4 may lead to rises in intracellular Ca^{2+} and some depletion of GSH. Younes and Siegers in West Germany reported that pre-treatment of mice with desferrioxamine before CCl_4 decreased the hepatotoxicity and depressed the CCl_4-induced exhalation of ethane, taken as an index of lipid peroxidation *in vivo* (Chapter 4). This suggests that the lipid peroxidation initiated by free-radical metabolites of CCl_4 is made worse by release of iron ions, a possible mechanism being illustrated in Fig. 6.11.

Chloroform
Chloroform ($CHCl_3$—trichloromethane) is no longer used as an anaesthetic, but it is still widely employed in industry as a solvent, and small amounts are sometimes added to cough mixtures and mouth-washes, although this is no longer encouraged. Trichloromethane is much less damaging to the liver than is tetrachloromethane, and induces lipid peroxidation in isolated liver microsomes at a much lower rate. This may be because the energy required to cause homolytic fission of trichloromethane to produce the trichloromethyl radical is greater than with tetrachloromethane. Consistent with this argument, compounds in which homolytic fission is easier such as bromo-trichloromethane ($BrCCl_3$) induce peroxidation even more rapidly than does CCl_4 (Table 6.4). None the less, liver damage is still a significant problem in humans exposed to chloroform vapour. The toxicity appears to be due to metabolism by the P-450 system, in that inducers of this system potentiate $CHCl_3$-toxicity in animals, and there is covalent binding of the molecule to cell membranes. Phosgene is produced during $CHCl_3$-metabolism.

Table 6.4.   Lipid peroxidation induced by halogenated hydrocarbons

Reaction	Energy needed for reaction (kcal mole^{-1})	Relative rate of lipid peroxidation
$CCl_4 \rightarrow Cl^{\cdot} + {}^{\cdot}CCl_3$	68	100
$CHCl_3 \rightarrow H^{\cdot} + {}^{\cdot}CCl_3$	90	7
$BrCCl_3 \rightarrow Br^{\cdot} + {}^{\cdot}CCl_3$	49	3650

The rate of peroxidation induced in rat-liver microsomes in the presence of NADPH was measured by the thiobarbituric acid method. Results are expressed relative to the stimulatory effect of CCl_4. Data were abstracted from Slater, T. F. and Sawyer, B. C., *Biochem. J.* **123**, 805–14 (1971).

6.9.2   Other halogenated hydrocarbons

Bromoethane and bromobenzene

1,2-Dibromoethane (ethylene dibromide, $CH_2Br.CH_2Br$) is widely used as a 'lead scavenger' in petrol, as an industrial solvent, and in agriculture. Unfortunately it is mutagenic and carcinogenic, causing especial damage to the liver and kidneys. Free-radical reactions may contribute to this damage: bromoethane can be metabolized by conjugation with GSH to form a mercapturic acid (Chapter 3) or by cytochrome P-450-dependent oxidation to give bromoacetaldehyde via intermediate free radicals. Toxic amounts of dibromoethane cause rapid GSH depletion which is *followed* by lipid peroxidation. Although, unlike the case of CCl_4, lipid peroxidation is not a primary mechanism of tissue injury by bromoethane, it does aggravate the damage to some extent. The aromatic hydrocarbon bromobenzene (C_6H_5Br) also produces rapid GSH depletion in the liver, again leading to lipid peroxidation. In experiments with mice, Trolox C (Chapter 4) offered some protection against liver necrosis induced by toxic doses of bromobenzene, and desferrioxamine was also protective. Rises in intracellular 'free' Ca^{2+} would also be expected in bromobenzene-treated hepatocytes (e.g. Fig. 6.11).

Halothane

Halothane is a frequently used inhalation anaesthetic. It has the structure:

$$F_3C-CHBrCl$$

It normally has no significant side-effects but it has been observed to produce liver damage in a few patients, for unknown reasons. Incubation of halothane with rat liver microsomes in the presence of NADPH and the spin-trap PBN (see above) resulted in the formation of an ESR signal. The same signal was observed if the spin trap was fed in an oil emulsion to rats which were allowed to inhale halothane, killed, and the liver lipids extracted and placed in the ESR spectrometer.

Incubation of halothane with liver microsomes has been observed to produce reactive metabolites that can bind chemically to the membranes and might be able to stimulate lipid peroxidation under aerobic conditions. One action of cytochrome P-450 on halothane is to cause formation of the radical $F_3C\dot{C}HCl$; this is favoured at low O_2 concentrations. Formation of this radical could account for the observation that the expired air of rabbits or humans exposed to halothane has been shown to contain chlorotrifluoroethane and chlorodifluoroethene. They could be formed by the following reactions; the C—Br bond is the weakest carbon–halogen bond in halothane.

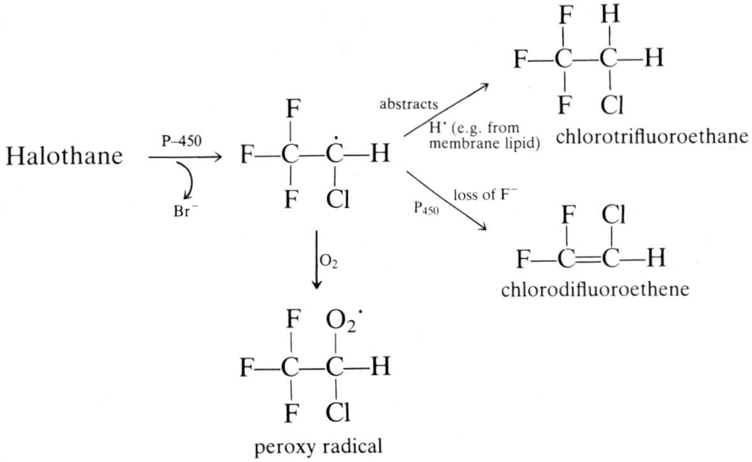

Formation of chlorodifluoroethene from the radical may be catalysed by cytochrome P-450 donating an electron to remove fluorine as F^-. The $F_3C\dot{C}HCl$ radical might also combine with oxygen to form a reactive peroxy radical. It therefore seems that radical reactions are important in halothane toxicity, but the reasons for the variability in observed effects in human patients have yet to be established.

6.10 Hypoxic cell sensitizers and related compounds

We saw in Chapter 2 that absence of oxygen decreases the sensitivity of cells to ionizing radiation. Such an observation is particularly relevant in the treatment of large cancerous tumours by radiotherapy. Often, as such tumours grow, areas within them no longer receive an adequate blood supply and thus becomes short of oxygen (hypoxic). Whilst radiation treatment may destroy most of the tumour, the hypoxic cells are more resistant, and can serve as 'nuclei' for subsequent regrowth. There has therefore been some interest in various therapies combining increased oxygen exposure with radiation (Chapter 1), and a greater interest in various drugs that, like oxygen, make hypoxic cells more sensitive to ionizing radiation. Such drugs are collectively known as *hypoxic cell sensitizers*.

In 1973, it was proposed that metronidazole (Fig. 6.12) might be useful as a hypoxic cell sensitizer. Metronidazole, under the trade name *Flagyl®*, was originally introduced as an effective treatment for diseases caused by

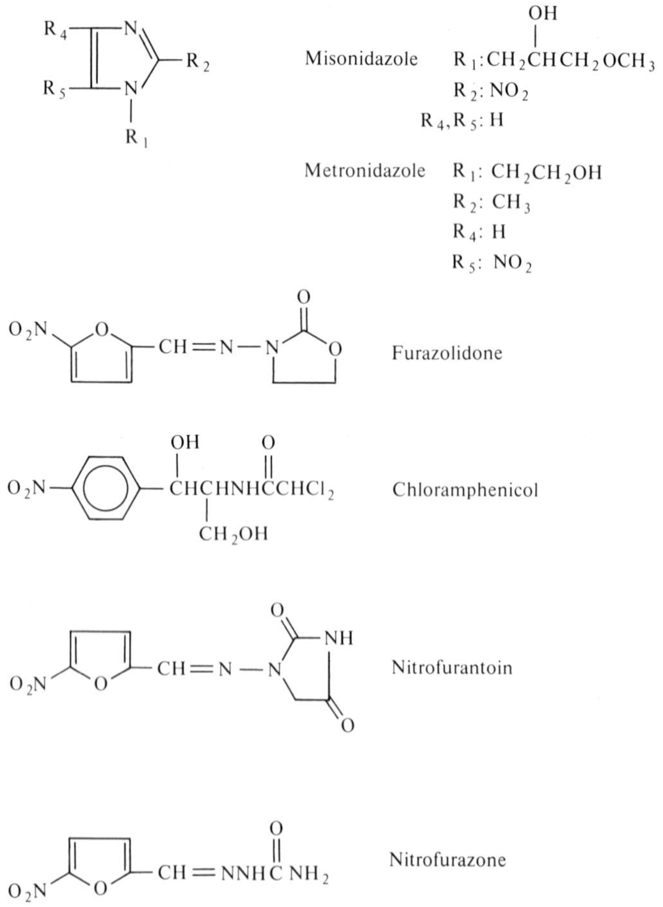

Fig. 6.12. Structure of hypoxic cell-sensitizers and some other nitro-compounds.

anaerobic protozoa such as *Trichomonas vaginalis* and *Giardia lamblia*. Subsequent work showed that it is also effective against *Trypanosoma cruzi*, which causes *Chagas' disease*, an infection particularly common in South America that produces recurrent fever, with damage to the heart and gastro-intestinal system. Metronidazole was later found to be effective against a wide range of anaerobic bacteria. Exposure of *E. coli*, plant tissues, and animal cells in culture to metronidazole was observed to increase their susceptibility to ionizing radiation under anaerobic conditions, and experiments on tumour-bearing animals confirmed this effect *in vivo*. A number of other imidazole compounds have been introduced for chemotherapy of bacterial and protozoal infections, although Flagyl® is still very popular. By contrast, although metronidazole has been used experimentally to supplement radiation treatment in some human cancer patients, several other sensitizers are more promising, especially *misonidazole* (Fig. 6.12). Unfortunately, a side-effect of this drug—damage to peripheral nerves—limits the amounts that can safely be given to patients. There is therefore considerable research to find less toxic sensitizers and/or drugs which minimize the neurotoxicity.

As discussed in Chapter 3, one reason for the effect of oxygen in potentiating radiation damage seems to be prevention of the repair of organic radicals i.e.

$$RH \xrightarrow{\text{radiation}} R^{\cdot} \text{ (radical production from cell components)}$$

$$R^{\cdot} \xrightarrow[\text{ascorbate}]{\text{GSH or}} RH \text{ (repair)}$$

$$R^{\cdot} + O_2 \longrightarrow RO_2^{\cdot} \text{ (damage fixation).}$$

Metronidazole and other hypoxic cell sensitizers might act in a similar way to oxygen, forming either a radical adduct (A) or a radical anion (B) that cannot easily be repaired:

In addition, it has been suggested that cell sensitizers might interfere with normal recombination reactions between radiation-produced radicals.

Metronidazole, misonidazole, and related compounds are reduced *in vivo* to form radical species, and this seems to be the basis of their toxicity to anaerobic bacteria. It may also explain their toxicity to anaerobic animal cells, an effect seen in addition to radiation sensitization. Reduction of these drugs has been demonstrated *in vitro* using a number of enzyme systems such as xanthine oxidase, the hydrogenase complex of *T. vaginalis* (in which reduced ferredoxin is probably the electron-donating component), and with microsomal fractions from several animal tissues, apparently by the action of NADPH-cytochrome P-450 reductase. Exposure of DNA to reduced metronidazole causes strand-breakage in the molecule, especially at thymidine residues. The reduction products also bind to microsomal membranes under anaerobic conditions and react with GSH.

Probably the first stage in reduction is formation of a radical anion $RNO_2{}^{\cdot-}$ (compound B above). The ability of this to react with GSH could potentiate radiation damage by interfering with an important protective mechanism:

$$RNO_2{}^{\cdot-} + GSH \rightarrow RNO + GS^{\cdot} + OH^-$$

Further reduction of $RNO_2{}^{\cdot-}$ to nitroso compounds (RNO) and to hydroxylamines (RNHOH) has been observed *in vivo*, although which of these species actually causes the DNA damage has not been established. The nitro-drug furazolidone (Fig. 6.12), used in veterinary medicine, is also reduced *in vivo*.

The covalent binding of metronidazole-derived products to microsomes in the presence of NADPH is not seen if oxygen is present, since $RNO_2{}^{\cdot-}$ reacts quickly to form the superoxide radical:

$$RNO_2{}^{\cdot-} + O_2 \rightarrow RNO_2 + O_2{}^-$$

It is possible that increased $O_2{}^-$-formation by this reaction might account for some of the observed side-effects of hypoxic cell sensitizers on normal aerobic tissues, although there is no direct evidence for this.

Figure 6.12 shows the structure of nitrofurantoin, an antibacterial drug that produces lung damage as a side-effect. It has been suggested that this may be due to reduction of the drug by lung endoplasmic reticulum:

$$X-NO_2 \xrightarrow[\text{electron}]{\text{one}} X-NO_2{}^{\cdot-}$$

followed by re-oxidation of the nitro-radical to produce $O_2{}^-$ and hence hydrogen peroxide. Consistent with this, feeding chicks on selenium-deficient diets to decrease glutathione peroxidase activity greatly potentiates the toxicity of nitrofurantoin. Similar reactions have been suggested to account for the toxicity of nitrofurazone (Fig. 6.12) and its derivatives to *Trypanosoma cruzi*. In the absence of oxygen the nitro-radicals seem to undergo disproportionation to form highly-reactive nitroso compounds:

$$2XNO_2^{\cdot -} + 2H^+ \rightarrow X-N{=}O + XNO_2 + H_2O$$

XNO may then be further reduced to a hydroxylamine, XNHOH.

Trypanosoma cruzi may be particularly sensitive to oxidants because it lacks catalase and glutathione peroxidase, although it may operate an 'ascorbate-glutathione cycle', of the type found in plants (Chapter 5), for the removal of H_2O_2.

6.11   Antibiotics

Antibiotics are substances produced by living organisms which, when present at low concentrations, are antagonistic to the growth and life of other micro-organisms. Antibiotics have a very wide range of chemical structures and recent work has shown that many can participate in the formation of oxidants. Thus, the tetracyclines are sensitizers of singlet oxygen formation (Chapter 2) and complexes formed by binding iron or copper ions to tetracyclines (which have established metal-chelating ability) can generate $OH^{\cdot}$ from H_2O_2. On the other hand, tetracyclines are powerful *scavengers* of hypochlorous acid, an oxidant produced by neutrophil myeloperoxidase (Chapter 8). Hence their effects at sites of inflammation will be complex, and it must never be assumed that an action of an antibiotic *in vivo* is necessarily mediated by its anti-bacterial activity. Indeed, many antibiotics are now known to bring about morphological and ultrastructural changes in bacteria at concentrations well below their established minimum inhibitory concentrations (MIC).

Most anticancer antibiotics undergo redox cycling to produce $O_2^{\cdot -}$ and H_2O_2 *in vivo* (Chapter 8). Rifamycin SV, an antibiotic often used in the treatment of tuberculosis, can oxidize in the presence of transition metal ions to give a quinone (called *rifamycin S*), with intermediate formation of a semiquinone, and $O_2^{\cdot -}$

$$QH_2 + O_2 \underset{\text{metal ions}}{\overset{}{\rightleftharpoons}} QH^{\cdot} + O_2^{\cdot -} + H^+$$

rifamycin SV   semiquinone

$$QH^{\cdot} + O_2 \rightleftharpoons Q + O_2^{-} + H^{+}$$

$$QH^{\cdot} + \text{metal ion} \longrightarrow Q + H^{+} + \text{reduced metal ion}$$

Kono in Japan has suggested that the bactericidal action of rifamycin SV may involve increased oxidant generation within the bacteria, as well as prevention of bacterial RNA synthesis. Rifamycin shows many biological and chemical properties similar to those of the copper chelator 1,10-phenanthroline. Both have been shown to mediate $OH^{\cdot}$-dependent damage to DNA, lipids, or carbohydrates in the presence of transition metal ions (e.g. Chapter 2, Section 2.4.3).

Free-radical reactions have been suggested to be involved in the toxic side-effects produced by several antibiotics. Thus aminoglycosides such as gentamicin damage the kidneys and increase lipid peroxidation in this organ. The cephalosporin antibiotic cephaloridine has a similar effect. Whether the lipid peroxidation is an early, causative, stage in the kidney injury or a consequence of tissue injury (Fig. 6.1) is as yet uncertain. For example, vitamin E administration to rats depressed gentamicin-induced lipid peroxidation in the kidney, but did not prevent the kidney damage. However, desferrioxamine has been reported to diminish gentamicin-induced kidney damage in rats.

The antibiotic chloramphenicol, sometimes used to treat humans, is a nitro-compound (Fig. 6.13). Chloramphenicol is active against a wide range of bacteria, but its use is severely restricted since it depresses the action of the bone marrow and sometimes produces severe and irreversible anaemia. Its action has often been attributed to its ability to inhibit protein synthesis, but this may not be the whole story since it also induces DNA damage *in vivo*. Chloramphenicol in which the nitro group has been reduced *in vitro* will damage isolated DNA. Further, microsomal fractions from rats pretreated with phenobarbital convert chloramphenicol into a number of products, at least one of which can combine covalently with microsomal protein. However, the microsome-catalysed activation involves the terminal $-CHCl_2$ group rather than the nitro-group (Fig. 6.13). The exact relation of these effects to the observed bone-marrow damage has yet to be determined. The β-lactam antibiotics (penicillins and cephalosporins) have been shown in the authors' laboratories to oxidatively damage DNA and deoxyribose in the presence of iron and copper salts; the damaging species has properties consistent with its being $OH^{\cdot}$. The lipophilic macrocyclic polypeptide polymyxin antibiotics, and bacitracin, are able to stimulate lipid peroxidation when present in membrane lipids. The group of antifungal antibiotics known as polyenes (containing numerous double bonds) are active because they bind to sterols in the membranes of fungal and mammalian cells. This group

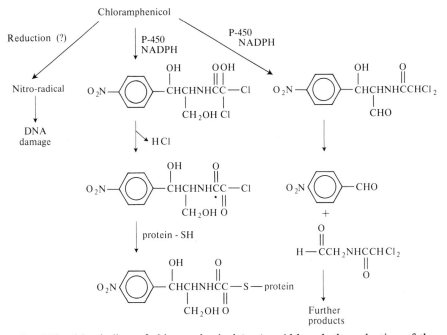

Fig. 6.13. Metabolism of chloramphenicol *in vivo*. Although the reduction of the nitro-group of chloramphenicol can be achieved *in vitro* and the product so obtained damages DNA, it has not been established how it could be reduced *in vivo*.

includes antibiotics such as amphotericin, candidicin, natamycin, and nystatin. Their polyunsaturated structures gives them a propensity to oxidize with the formation of peroxy radicals and numerous aldehyde fragments, some of which are TBA-reactive.

As the above shows, it must be emphasized that many antibiotics do not act solely as pharmacological agents blocking key enzymes and metabolic functions of micro-organisms, but that some also possess the ability to redox cycle and produce reactive intermediates capable of damaging the molecular structure of cells. This action of antibiotics can be called a 'phago-mimetic' effect and may prove to be an important component of their biological activities.

6.12   MPTP and Parkinson's disease

We saw in Chapter 4 (Section 4.9) that the brain is prone to undergo oxidant damage because of its relatively-low content of antioxidant enzymes and its high content of iron, that becomes easily released when cells are injured and cannot be safely bound because cerebrospinal fluid has no significant iron-binding capacity. Thus oxidation of 6-hydroxydopamine produces significant damage to catecholamine nerve terminals (Section 6.3), severe vitamin E deficiency in humans produces neurological symptoms (Chapter 4) and iron-stimulated lipid peroxidation may mediate some of the brain and spinal cord degeneration that occurs after trauma (Chapter 4). There is a high concentration of ascorbic acid in the grey and white matter of the brain, and ascorbate concentrations in the cerebrospinal fluid (CSF) are about ten times those in the plasma. Hence release of metal ions into CFS creates potentially-damaging ascorbate/metal ion mixtures (Chapter 3).

Because of the sensitivity of the nervous system to oxidative damage, there have been many proposals that oxidants are involved in the pathology of neurodegenerative diseases. One such disease is *Parkinson's disease*, first described by James Parkinson in 1817. It usually appears in middle to old age (rarely before 50) and the person begins to have increasing problems in controlling movement. Head and limbs shake; eating, speaking, and dressing become more and more difficult.

Parkinson's disease attacks a group of cells (the *substantia nigra*) in the upper part of the brainstem; the word 'nigra' derives from the fact that the cells are rich in a black pigment, *neuromelanin*. The substantia nigra sends nerve fibres to a structure called the *striatum* at the base of the brain; the terminals of these fibres secrete the neurotransmitter dopamine (Fig. 6.5), which helps the striatum to control movement. The progressive death of nigral cells means that less and less dopamine is available, which explains why Parkinsonian patients benefit from therapy with L-DOPA, a precursor of dopamine (Fig. 6.5). (Cells also die in another part of the brain, the *locus coeruleus*, leading to lowered noradrenalin concentrations.) The possibility of grafting foetal dopamine-producing cells into the brain is also being tested. However, L-DOPA treatment produces many distressing side-effects, and it is possible that metabolism of excess dopamine by the *monoamine oxidase* enzymes in the brain produces too much H_2O_2:

$$RCH_2NH_2 + O_2 + H_2O \xrightarrow{MAO} RCHO + NH_3 + H_2O_2$$

An illustration of this mechanism of neurotoxicity may be provided by *5,7-dihydroxytryptamine*. This was developed as an experimental agent to destroy serotonin-containing neurones, but it also attacks catecholamine neurones. The action of 5,7-dihydroxytryptamine is blocked by inhibitors of MAO, and evidence exists to suggest that it may involve generation of excess H_2O_2 with subsequent formation of hydroxyl radicals.

Two major clues have emerged as to the cause of Parkinson's disease. Until recently, the native Chamorro Indians of Guam and Rota, islands in the Western Pacific, were more than 50 times more likely than usual to develop Parkinson's symptoms and they also developed them at an earlier age than usual. This high incidence of Parkinsonism has been declining since the 1950s, as the Chamorros have adopted an increasingly 'US-style' life. It has been proposed that a neurotoxin present in extracts from nuts of the 'false sago palm' (*Cycas circinalis*), formerly a common part of the Chamorro diet but now consumed less and less often, could cause the disease. The neurotoxin appears to be an amino acid, β-N-methylamino-L-alanine (a-amino-β-methyl aminopropionic acid). It has been suggested that it might kill cells by producing excessive and sustained firing, leading to enhanced entry of Ca^{2+} and consequent cell damage. Another proposal has been that the high levels of aluminium in the soil of Guam could contribute to neurodegeneration (Section 6.13).

A second clue about the origin of Parkinson's disease came from the USA. In 1982, clinicians at hospitals in California were surprised by a sudden influx of young patients with Parkinsonian symptoms. These patients were found to have used 'heroin' from a drug pusher in northern California, but this heroin substitute was contaminated by *1-methyl-4-phenyl-1,2,3,6-tetra-hydropyridine* (MPTP). The structure of this compound is shown in Fig. 6.14, and it bears some resemblance to paraquat (Fig. 6.2). MPTP crosses the blood–brain barrier and is oxidized by an isoenzyme of monoamine oxidase (*monoamine oxidase B*), which is largely located in the mitochondria of glial cells (the non-neuronal cells of the brain that apparently help nerve cells to function). Oxidation produces MPDP$^+$, which then forms MPP$^+$ by a mechanism not yet understood (Fig. 6.14). The substantia nigra actively accumulates MPP$^+$, so it is possible that MPP$^+$ formed in glial cells and released by them is taken up by nigral cells, a process that may be accelerated by the binding of MPP$^+$ to neuromelanin. The MPP$^+$ kills nigral cells,

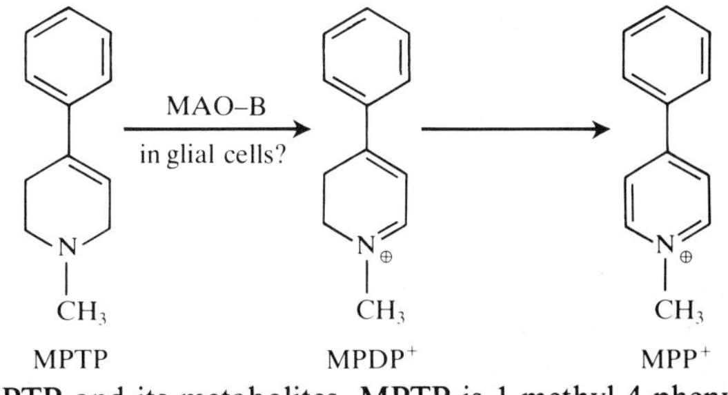

Fig. 6.14. MPTP and its metabolites. MPTP is 1-methyl-4-phenyl-1,2,3,6-tetrahy-dropyridine. It is oxidized to MPDP$^+$ (1-methyl-4-phenyl-2,3-dihydropyridine ion) which forms MPP$^+$, the 1-methyl-4-phenylpyridinium ion. Inhibitors of monamine oxidase, such as pargyline or deprenyl (specific for MAO-B) diminish damage by MPTP. How MPDP$^+$ is converted into MPP$^+$ is not completely clear.

producing Parkinsonian symptoms. How precisely the MPP^+ produces nigral cell death is undecided; suggestions that it does so by a free-radical mechanism and/or by GSH depletion do not seem supported by experimental evidence. Neither MPTP nor MPP^+ appears to undergo rapid 'redox cycling' in the way demonstrated for paraquat and quinones. A more likely explanation is that MPP^+ accumulates in mitochondria and blocks the electron transport chain, so starving the cells of ATP.

MPTP contains a pyridine ring, as do many industrial chemicals, such as herbicides (Fig. 6.2). This naturally raises the possibility that Parkinson's disease is caused by an environmental toxin, as is further suggested by studies of the Chamorros on Guam and Rota. Humans are born with a limited number of nigrostriatal cells and this number declines with age (8–10 per cent per decade). Depletions of 70–75 per cent seem to be the minimum for Parkinsonian symptoms to appear. Hence an environmental insult to the nigra in early life, combined with the normal age-related death of nigral cells, might be sufficient to produce Parkinson's disease in late life. It must also be pointed out that, even if free-radical reactions are not involved in the origin of Parkinson's disease, they could occur as a result of the tissue damage and make it worse (Chapter 4). Pall and his colleagues in England have made the interesting observation that CSF from Parkinsonian patients often contains elevated concentrations of copper, but not of manganese or iron. The substantia nigra of Parkinsonian patients has also been reported to be enriched in iron, suggesting that iron-dependent radical reactions might contribute to disease progression and/or to the side-effects of therapy with L-DOPA.

6.13 Aluminium and Alzheimer's disease

Aluminium is the most abundant metal in the Earth's crust, and we are constantly exposed to it. It is slowly leached from aluminium cookware and cans, ingested from beverages such as tea and many processed foods, sprayed on skin as a constituent of deodorants, injected in some vaccines where aluminium salts are used in the vaccine preparation, and consumed in antacids containing the hydroxide $Al(OH)_3$.

Aluminium has a fixed valency of 3, so that aluminium compounds release the Al^{3+} ion in aqueous solution. Animal studies reveal that the gastrointestinal tract presents a formidable barrier to the entry of aluminium but since high doses of $Al(OH)_3$ administered orally to rats or to humans (as antacids) have succeeded in raising tissue and urine aluminium levels, the barrier is clearly not impervious. The *possibility* that inhaled aluminium compounds (e.g. in deodorants) might enter sensory neurones of the olfactory epithelium and spread into the brain has also been raised. Some dietary organic acids, such as citrate, have been shown to accelerate uptake of aluminium from the gastrointestinal tract, and absorbed Al^{3+} can bind to the iron transport protein transferrin. Aluminium ions can also be deposited within the iron-storage protein ferritin.

Until the early 1970s, the possible toxicity of Al was not considered. However, with the development of long-term haemodialysis therapy for patients with kidney malfunction, it was discovered that many patients maintained by dialysis develop a serious neurological syndrome, called *dialysis encephalopathy*. The finding of high concentrations of aluminium in grey matter from the brains of patients dying from this disease led to the suggestion that dialysis encephalopathy is caused by aluminium intoxication. Dialysis encephalopathy is also associated with demineralization of the bones, leading to increased risk of fracture. Subsequent work has amply confirmed the role of aluminium in causing these conditions and aluminium still represents a significant health hazard to dialysis patients in certain parts of the world.

More recent work has implicated aluminium in the pathogenesis of senile dementia. The dementing condition known as *Alzheimer's disease* affects 10–15 per cent of individuals aged over 65, and perhaps 20 per cent over the age of 80. Early manifestations are a gradually-developing loss of memory, combined with growing confusion and disorientation, leading eventually to severe mental disability. Two of the pathological features of Alzheimer-type dementia, first observed by Alois Alzheimer in 1906, are the presence of *neurofibrillary tangles* and *senile plaques* in the brain (Fig. 6.15). The tangles are fibrous masses within affected neurones in several brain regions and they largely consist of pairs of filaments, each about 10 nanometres in diameter, helically twisted around each other with a cross-over roughly every 80 nanometres (*paired helical filaments*). Senile plaques are localized areas of degenerating and frequently-swollen axons/synaptic terminals (neurites) and glial cells often surrounding a core of amyloid protein (defined in Chapter 8, Section 8.7.7); many of these neurites contain the paired helical filaments described above. Plaques are most common in the regions of the brain known as the amygdala, the hippocampus, and the neocortex, but they can also occur elsewhere. Amyloid protein is also deposited in blood vessel walls in the 'Alzheimer' brain; this protein seems identical with the amyloid of plaque cores. Amyloid protein deposits contain a truncated product of a much larger protein, possibly a cell-surface receptor. Neurofibrillary tangles of the Alzheimer type are also found in adults with Down's syndrome (Chapter 8) and in cases of Parkinsonian dementia on Guam (Section 6.12).

In Alzheimer's disease, correlations exist between the extent of dementia and the number of plaques present, and also with decreased synthesis of the neurotransmitter acetylcholine in the brain. It must be noted that neuro-fibrillary tangles and plaques also occur, to a much more limited degree, in non-demented old people.

Fig. 6.15. Section of brain from a patient with Alzheimer's disease showing senile plaques (arrows) and neurofibrillary tangles (arrowheads). Holmes staining (silver impregnation) × 250. Photography by courtesy of Dr B. H. Anderton and Dr Jean-Pierre Brion. Amyloid protein deposits contain a truncated cell-surface receptor and a protease inhibitor, α_1-antichymotrypsin.

A significant percentage of Alzheimer patients (25–40 per cent of cases) seem to have *some* genetic predisposition to the disease. A rare dominantly-inherited form of Alzheimer's disease also exists; genetic studies on this rare familial form show that the gene defect is on chromosome 21. The gene coding for the amyloid protein in plaques and vessels also maps to chromosome 21, but is not the same gene. In Down's syndrome (Chapter 8) an extra copy of chromosome 21 is present, which might explain why these patients accumulate brain plaques and tangles at an accelerated rate.

Interest in the role of aluminium in Alzheimer's disease arose when it was reported that the cores of senile plaques are enriched in aluminium and silicon. Guam Indians (Section 6.14) with Alzheimer's disease also showed increased intracellular deposition of calcium, silicon, and aluminium in neurofibrillary tangle-bearing neurones. Is excessive aluminium accumulation over a lifetime the *cause* of Alzheimer's disease? The authors do not think so. Although aluminium can induce the formation of intra-neuronal tangles when injected into cats, ferrets, and rabbits, the fibrils produced are not identical to those of Alzheimer's disease. No such tangles can be formed in primates, and no unusual incidence of Alzheimer's disease has been reported amongst workers in aluminium mines or smelting plants. Dialysis encephalopathy patients do not show neurofibrillary tangles, but only have the type of intra-neuronal tangles found in experimental animals after injection of aluminium around the brain. It may be that Alzheimer's disease leads to an impaired blood–brain barrier that allows increased amounts of aluminium to reach the central nervous system. Since the neurotoxicity of aluminium is well-established from the studies upon dialysis patients, it follows that aluminium accumulation, even if not causative for Alzheimer's disease, could worsen the neurological damage being done. Hence part of the symptoms suffered by Alzheimer's patients may be related to aluminium accumulation, and therapy aiming to remove excess aluminium from such patients (as well as measures to prevent further accumulation) might prove useful. The iron-binding agent desferrioxamine chelates Al^{3+}, and it has been used in the treatment of dialysis encephalopathy. However, its stability constant for Al^{3+} (about 10^{25}) is several orders of magnitude smaller than that for Fe^{3+} (about 10^{31}), and so prolonged treatment with desferrioxamine may lead to anaemia and other side-effects (Chapter 8). Fortunately, chelating agents more specific for Al^{3+} are being developed.

6.13.1 *Aluminium and lipid peroxidation*

What is the mechanism of the neurotoxicity of Al^{3+}? Several suggestions have been made, including interference with Ca^{2+} uptake mechanisms and inhibition of the enzyme dihydropteridine reductase, which catalyses the NADPH-dependent reduction of dihydrobiopterin to tetrahydrobiopterin, a cofactor required for biosynthesis of tyrosine (and hence of dopamine, adrenalin, and noradrenalin) from phenylalanine. The authors found that Al^{3+} ions cannot themselves stimulate lipid peroxidation, which is not surprising because of their fixed valency. However, if peroxidation in liposomes, erythrocytes, or microsomes is stimulated by adding Fe^{2+} ions (Chapter 4), the simultaneous addition of Al^{3+} greatly accelerates the peroxidation rate. It may be that Al^{3+} ions bind to membranes and cause a subtle rearrangement of membrane lipids that aids the propagation of lipid peroxidation. This action of Al^{3+} might contribute to its neurotoxic properties, since the brain is sensitive to free-radical reactions.

Friedreich's ataxia
Friedreich's ataxia, an autosomal recessive disease, is the most common form of hereditary ataxia (loss of control over voluntary movements), affecting some 1–2 per 100 000 of the population in Europe. Onset usually occurs between the ages of 8 to 16 but always before the age of 25 years. Pathological changes involve neural and heart muscle degeneration, although the underlying cause of the disease remains unknown. The neuro-

logical features of vitamin E deficiency are similar to those seen in Friedreich's ataxia, and skin fibroblasts from patients with Friedreich's ataxia show increased sensitivity to ionizing radiation. However, increased free-radical damage due to decreased tissue levels of vitamin E is unlikely to account for the increased sensitivity to radiation, since Muller and his colleagues in London have shown that serum vitamin E and vitamin E: cholesterol ratios in Friedreich's ataxia are similar to those of control subjects. The molecular defect in this disease appears to lie on a gene located on chromosome 9.

6.14   Toxicity of other metals

Increased rates of free-radical reactions have often been suggested to contribute to the toxicity of high doses of several other metal ions, including lead (Pb^{2+}), cobalt (Co^{2+}), mercury (Hg^{2+}), nickel (Ni^{2+}), molybdenum (Mo^{3+}), and cadmium (Cd^{2+}). While increased lipid peroxidation has often been demonstrated in isolated cells exposed to these metals, or in tissues from animals poisoned by metals, this peroxidation may be a consequence of tissue injury and GSH depletion caused by the metals rather than an early contributor to the metal toxicity. However, the occurrence of peroxidation will not improve the health of the tissue. Several metal ions, including Ni^{2+}, are carcinogenic in high doses.

Many metal ions have been shown to accelerate free-radical reactions *in vitro*. Thus Hg^{2+}, Pb^{2+} and, under certain circumstances, Cd^{2+} and Ca^{2+} ions, have been found to accelerate iron-stimulated lipid peroxidation *in vitro*. Low concentrations of Cd^{2+} ($\sim 10^{-5}$ M) and Hg^{2+} ($\sim 10^{-9}$ M) have been reported to stimulate O_2^- production by activated phagocytes, although higher concentrations diminish it by injuring the cells. Similar effects have been claimed for Al^{3+}. Several transition metal ions can react with H_2O_2 to form hydroxyl radical or a species that resembles it; these include Co^{2+}, Cr^{2+}, and titanium (Ti^{3+}) ions. Indeed, a mixture of a Ti^{3+} salt and H_2O_2 was introduced by Dixon and Norman in 1962 in England as a convenient laboratory source of OH˙.

$$Ti^{3+} + H_2O_2 \rightarrow Ti(IV) + OH\cdot + OH^-$$

Both vanadate (Chapter 3) and molybdate accelerate oxidation of NADH by O_2^-.

6.15 Sporidesmin

Sporidesmin (Fig. 6.16) is a product of the fungus *Pithomyces chartarum*, which often occurs in New Zealand pastures. Ingestion of fungal material by sheep or other ruminants leads to the photosensitivity disease 'facial eczema'. The sporidesmin attacks the biliary epithelium, producing inflammation and necrosis that eventually lead to cessation of bile flow, deposition of phylloerythrin in the animal's skin, and photodamage due to singlet oxygen formation (Chapter 2). The mechanism of sporidesmin toxicity is unknown; Munday (1987) in New Zealand has suggested that generation of O_2^-, H_2O_2, OH^-, and sulphur-centred radicals by oxidation of reduced sporidesmin, in the presence of traces of transition metal ions, may be involved.

6.16 The 'Spanish cooking-oil' syndrome

In September 1981, Dr J. M. Tabuenca reported on a new syndrome that has caused at least 350 deaths and affected over 20 000 other people in Spain up to the end of 1982. The symptoms of the disease in its acute stage are respiratory distress, fever, headache, itching, nausea, and sometimes muscular pains and neurological disorders. Lung damage has been responsible for most of the early deaths. When followed up 5 months later, about 10 per cent of patients had developed muscular wasting and weakness, and another 49 per cent had suffered some muscular impairment. All sufferers from the disease were found to have consumed 'olive oil' sold by door-to-door salesmen. Analysis showed that the oil in question had never seen an olive

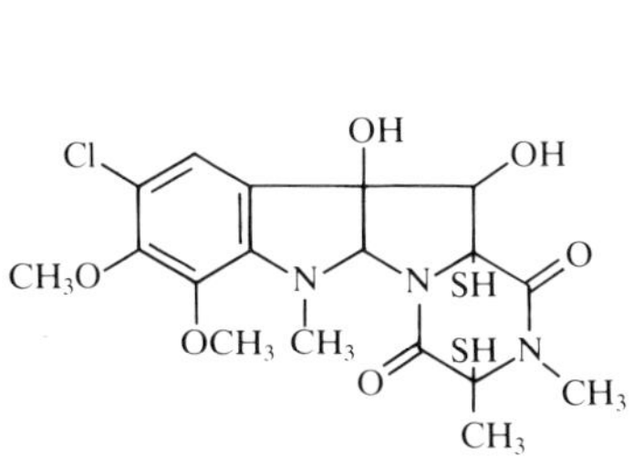

Fig. 6.16. Structure of sporidesmin.

tree, and was basically oil obtained from seeds of the rape plant *Brassica napus*. Rapeseed oil is normally used as an industrial lubricant. Such oil imported into Spain is treated with aniline (phenylamine, $C_6H_5NH_2$) to deter its consumption; and it appears that attempts had been made to refine the aniline out of the rapeseed oil. During this attempted purification, a number of chemical changes took place, resulting in the formation of various toxins. The disease differs from that of simple poisoning by aniline itself.

It has been suggested that part of the pathology of the disease may be related to peroxidation of membrane lipids induced by one or more of the toxins. This does not, of course, mean that peroxidation causes the damage, since increased lipid peroxidation often accompanies cell damage caused in other ways (Chapter 4). The single good thing to come out of this affair is the setting-up of an international society for free-radical research to combine the expertise of chemists, physicists, biochemists, biologists, and clinicians in studying free-radical reactions. (Further information about this international society may be obtained from the Secretary, SFRR Europe, Research Division, North East Wales Institute, Deeside, Clwyd, CH5 4BR, Wales.)

6.17 Ultraviolet light

In contrast to X-rays and γ-rays, UV radiation does not deposit sufficient energy in water molecules to ionize them. However, UV light can cause the homolytic fission of H_2O_2 and so generate hydroxyl radicals (Chapter 2, Section 2.10), and of chlorine molecules to generate chlorine radicals (Section 6.9.1). Some other molecules can absorb UV-light and enter a short-

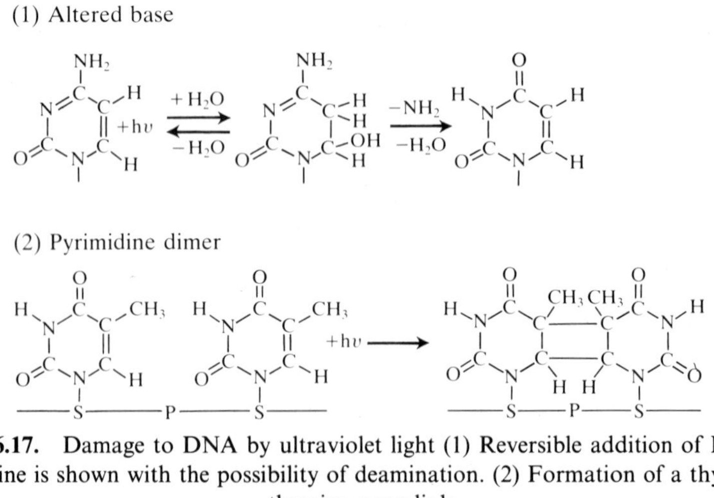

(1) Altered base

(2) Pyrimidine dimer

Fig. 6.17. Damage to DNA by ultraviolet light (1) Reversible addition of H_2O to cytosine is shown with the possibility of deamination. (2) Formation of a thymine–thymine cross-link.

lived excited state, in which they are more reactive chemically. Thus in DNA, absorption of UV light in the wavelength range 200–300 nm by thymine or cytosine creates excited states that can react with water to form pyrimidine hydrates, or with an adjacent pyrimidine to produce cross-links (thymine–thymine, thymine–cytosine, or cytosine–cytosine). The dimers are chemically stable, but the hydrates are not and usually lose water again to re-form the original pyrimidine. Alternatively, the cytosine hydrate can lose its amino group in a deamination reaction (Fig. 6.17). Defects in the ability to repair UV-induced lesions in DNA are seen in some human diseases (Chapter 8).

6.18 Further reading

Abraham, C.R. and Potter, H. (1989). Alzheimer's disease. *Biotechnology* **7**, 147.

Ahr, H. J. *et al.* (1982). The mechanism of reductive dehalogenation of halothane by liver cytochrome P_{450}. *Biochem. Pharmacol.* **31**, 383.

Albano, E. and Tomasi, A. (1987). Spin trapping of free radical intermediates produced during the metabolism of isoniazid and iproniazid in isolated hepatocytes. *Biochem. Pharmacol.* **36**, 2913.

Altmann, P. *et al.* (1987). Serum aluminium levels and erythrocyte dihydropteridine reductase activity in patients on hemodialysis. *New Engl. J. Med.* **317**, 80.

Anderton, B., Brion, J. P., and Power, D. (1988). The protein constituents of paired helical filaments and senile plaques in Alzheimer's disease. *ISI Atlas of Science, Biochemistry* **1**, 81.

Archibald, F. S. and Tyree, C. (1987). Manganese poisoning and the attack of trivalent manganese upon catecholamines. *Arch. Biochem. Biophys.* **256**, 638.

Augusto, O. *et al.* (1986). Hydroxyl radical formation as a result of the interaction between primaquine and reduced pyridine nucleotides. Catalysis by hemoglobin and microsomes. *Arch. Biochem. Biophys.* **244**, 147.

Bagley, A. C., Krall, J., and Lynch, R. E. (1986). Superoxide mediates the toxicity of paraquat for Chinese hamster ovary cells. *Proc. Natl. Acad. Sci. USA* **83**, 3189.

Banda, M. J. *et al.* (1988). a_1-Proteinase inhibitor is a neutrophil chemoattractant after proteolytic inactivation by macrophage elastase. *J. Biol. Chem.* **263**, 4481.

Beloqui, O. I. and Cederbaum, A. I. (1985). Microsomal interactions between iron, paraquat and menadione: effect on hydroxyl radical production and alcohol oxidation. *Arch. Biochem. Biophys.* **242**, 187.

Borg, D. C. and Schaich, K. M. (1986). Prooxidant action of desferrioxamine: Fenton-like production of hydroxyl radicals by reduced ferrioxamine. *J. Free Radical Biol. Med.* **2**, 237.

Calderbank, A. (1968). The bipyridylium herbicides. *Adv. Pest. Control. Res.* **8**, 127.

Candy, J. M. *et al.* (1986). Aluminosilicates and senile plaque formation in Alzheimer's disease. *Lancet* **i**, 354.

Carrell, R. W. (1986). a_1-Antitrypsin: molecular pathology, leukocytes and tissue damage. *J. Clin. Invest.* **78**, 1427.

Casini, A. F. *et al.* (1986). Lipid peroxidation and cellular damage in extrahepatic tissues of bromobenzene-intoxicated mice. *Am. J. Pathol.* **123**, 520.

Casini, A. F. *et al.* (1987). Lipid peroxidation, protein thiols and calcium homeostasis in bromobenzene-induced liver damage. *Biochem. Pharmacol.* **36**, 3689.

Cheeseman, K. H. *et al.* (1985). Biochemical studies on the metabolic activation of halogenated alkanes. *Environ. Health Perspect.* **64**, 85.

Chevion, M. *et al.* (1982). The chemistry of favism-inducing compounds. The properties of isouramil and divicine and their reaction with glutathione. *Eur. J. Biochem.* **129**, 405.

Clark, I. A., Hunt, N. H., and Cowden, W. B. (1986). Oxygen-derived free radicals in the pathogenesis of parasitic disease. *Adv. Parasitol.* **25**, 1.

Clark, I. A. *et al.* (1987). Toxicity of certain products of lipid peroxidation to the human malaria parasite *Plasmodium falciparum*. *Biochem. Pharmacol.* **36**, 543.

Cohen, G. (1978). The generation of hydroxyl radicals in biological systems. Toxicological aspects. *Photochem. Photobiol.* **28**, 669.

Cohen, G. (1983). The pathobiology of Parkinson's disease: biochemical aspects of dopamine neuron senescence. *J. Neurol. Transmission* suppl. **19**, 89.

Comporti, M. (1985). Lipid peroxidation and cellular damage in toxic liver injury. *Lab. Invest.* **53**, 599.

Cox, C. D. (1986). Role of pyocyanin in the acquisition of iron from transferrin. *Infect. Immunol.* **52**, 263.

Cox, F. E. G. (1983). Oxidant killing of the malaria parasites. *Nature* **302**, 19.

Cross, C. E., Halliwell, B., and Allen, A. (1984). Antioxidant protection: a function of tracheobronchial and gastrointestinal mucus. *Lancet*, **i**, 1328.

Dahlin, D. C. *et al.* (1984). *N*-acetyl-*p*-benzoquinone imine: a cytochrome P_{450}-mediated oxidation product of acetaminophen. *Proc. Natl. Acad. Sci. USA* **81**, 1327.

D'Amato, R. J. *et al.* (1987). Evidence for neuromelanin involvement in MPTP-induced neurotoxicity. *Nature* **327**, 324.

De Flora, S. *et al.* (1988). Influence of DT diaphorase on the mutagenicity of organic and inorganic compounds. *Carcinogenesis* **9**, 611.

Del Villano *et al.* (1980). Elevated superoxide dismutase in black alcoholics. *Science*, **207**, 911.

Dianzani, M. U. (1985). Lipid peroxidation in ethanol poisoning: a critical reconsideration. *Alcohol and Alcoholism* **20**, 161.

Di Monte, D. *et al.* (1986). Comparative studies on the mechanisms of paraquat and 1-methyl-4-phenylpyridine (MPP^+) toxicity. *Biochem. Biophys. Res. Commun.* **137**, 303.

Do Campo, R. and Moreno, S. N. J. (1986). Free radical metabolism of anti-parasitic agents. *Fed. Proc.* **45**, 2471.

Ehhalt, D. H. (1987). Free radicals in the atmosphere. *Free Radical Res. Commun.* **3**, 153.

Eizirik, D. L. *et al.* (1986). 1,10 phenanthroline, a metal chelator, protects against alloxan- but not streptozotocin-induced diabetes *J. Free Radical Biol. Med.* **2**, 189.

Eklow-Lastbom, L. *et al.* (1986). Effects of oxidative stress caused by hyperoxia and diquat. A study in isolated hepatocytes. *Free Radical Res. Commun.* **2**, 57.

Eklow-Lastbom, L., Moldeus, P. and Orrenius, S. (1986). On the mechanisms of glutathione depletion in hepatocytes exposed to morphine and ethylmorphine. *Toxicology* **42**, 13.

Fagan, J. M., Waxman, L., and Goldberg, A. L. (1986). Red blood cells contain a pathway for the degradation of oxidant-damaged hemoglobin that does not require ATP or ubiquitin. *J. Biol. Chem.* **261**, 5705.

Farr, S. B., Natvig, D. O., and Kogoma, T. (1985). Toxicity and mutagenicity of plumbagin and the induction of a possible new DNA repair pathway in *Escherichia coli*. *J. Bacteriol.* **164**, 1309.

Farrington, J. A. *et al.* (1973). Bipyridylium quaternary salts and related compounds. Pulse radiolysis studies of the reaction of paraquat radical with oxygen. Implications for the mode of action of bipyridyl herbicides. *Biochim. Biophys. Acta* **314**, 372.

Fernandez-Pol, J. A., Hamilton, P. D., and Klos, D. J. (1982). Correlation between the loss of the transformed phenotype and an increase in SOD activity in a revertant subclone of sarcoma virus-infected mammalian cells. *Cancer Res.* **42**, 609.

Fischer, V. *et al.* (1985). Free radical metabolites of acetaminophen and a dimethylated derivative. *Environ. Health Perspect.* **64**, 127.

Forni, L. G. *et al.* (1983). Reactions of the trichloromethyl and halothane-derived peroxy radicals with unsaturated fatty acids: a pulse radiolysis study. *Chem.–Biol. Interact.* **45**, 171.

Frank, L. (1981). Prolonged survival after paraquat. Role of the lung antioxidant enzyme systems. *Biochem. Pharmacol.* **30**, 2319.

Fuerst, E. P. *et al.* (1985). Paraquat resistance in *Conyza*. *Plant Physiol.* **77**, 984.

Ganrot, P. O. (1986). Metabolism and possible health effects of aluminium. *Environ. Health Perspect.* **65**, 363.

Garruto, R. M. and Yase, Y. (1986). Neurodegenerative disorders of the Western Pacific: the search for mechanisms of pathogenesis. *Trends Neurosci.* **12**, 368.

Gee, P. and Davison, A. J. (1985). Effect of scavengers of oxygen free radicals on the anaerobic oxidation of 6-hydroxydopamine by H_2O_2. *Biochim. Biophys. Acta* **838**, 183.

Giamalva, D., Church, D. F., and Pryor, W. A. (1985). A comparison of the rates of ozonation of biological antioxidants and oleate and linoleate esters. *Biochem. Biophys. Res. Commun.* **133**, 773.

Gilsanz, V. (1982). Late features of toxic syndrome due to denatured rapeseed oil. *Lancet* **i**, 335.

Goldberg, B., Stern, A., and Peisach, J. (1976). The mechanisms of superoxide anion generation by the interaction of phenylhydrazine with hemoglobin. *J. Biol. Chem.* **251**, 3045.

Goldstein, R. S. *et al.* (1986). Biochemical mechanisms of cephaloridine nephrotoxicity: time and concentration dependence of peroxidative injury. *Toxicol. Appl. Pharmacol.* **83**, 261.

Graham, D. G. *et al.* (1978). Autoxidation versus covalent binding of quinones as the mechanism of toxicity of dopamine, 6-hydroxydopamine, and related compounds towards C1300 neuroblastoma cells *in vitro*. *Mol. Pharmacol.* **14**, 644.

Grankvist, K. *et al.* (1982). Thioredoxin and thioredoxin reductase in pancreatic islets may participate in diabetogenic free radical reactions. *Biochem. Biophys. Res. Commun.* **107**, 1412.

Grankvist, K. and Marklund, S. L. (1983). Opposite effects of two metal-chelators on alloxan-induced diabetes in mice. *Life Sci.* **33**, 2535.

Grankvist, K., Marklund, S., and Taljedal, I. B. (1981). Superoxide dismutase is a prophylactic against alloxan diabetes. *Nature* **294**, 158.

Halliwell, B. (1987). Oxidants and human disease: some new concepts. *FASEB J.* **1**, 358.

Halliwell, B. and Gutteridge, J. M. C. (1985). Oxygen radicals and the nervous system. *Trends Neurosci.* **8**, 22.

Harper, D. B. and Harvey, B. M. R. (1978). Mechanisms of paraquat tolerance in perennial ryegrass. Role of superoxide dismutase, catalase and peroxidase. *Plant Cell Environ.* **1**, 211.

Hassan, H. M. and Fridovich, I. (1979). Paraquat and *E. coli*. Mechanism of production of extracellular superoxide radical. *J. Biol. Chem.* **254**, 10846.

Hebbel, R. P. (1985). Auto-oxidation and a membrane-associated "Fenton reagent": a possible explanation for development of membrane lesions in sickle erythrocytes. *Clin. Haematol.* **14**, 129.

Heikkila, R. E. and Cabbat, F. S. (1982). The prevention of alloxan-induced diabetes in mice by the chelator detapac. Suggestion of a role for iron in the cytotoxic process. *Experientia* **38**, 378.

Heppner, G. D. *et al.* (1988). Antimalarial properties of orally active iron chelators. *Blood*, **72**, 358.

Hill, H. A. O. and Thornalley, P. J. (1983). The effect of spin-traps on phenylhydrazine-induced haemolysis. *Biochim. Biophys. Acta*, **716**, 249.

Horton, J. K. *et al.* (1986). Paraquat uptake into freshly isolated rabbit lung epithelial cells and its reduction to the paraquat radical under anaerobic conditions. *Mol. Pharmacol.* **29**, 484.

Jackson, R. M. and Frank, L. (1984). Ozone-induced tolerance to hyperoxia in rats. *Am. Rev. Resp. Dis.* **129**, 425.

Janoff, A., Pryor, W. A., and Bengali, Z. H. (1987). Effect of tobacco smoke components on cellular and biochemical processes in the lung. *Am. Rev. Resp. Dis.* **136**, 1058.

Jay, M., Kojima, S. and Gillespie, M. N. (1986). Nicotine potentiates superoxide anion generation by human neutrophils. *Toxicol. Appl. Pharmacol.* **86**, 484.

Jefcoate, C. R. E., and Norman, R. O. C. (1968). Electron spin resonance studies. Part XIV. Hydroxylation. Part III. Reactions of anisole, acetanilide, fluorobenzene and some phenols with the titanium(III)-hydrogen peroxide system. *J. Chem. Soc.* **(B)**, 48.

Kappus, H. (1987). Oxidative stress in chemical toxicity. *Arch. Toxicol.* **60**, 144.

Keeling, P. L., Smith, L. L., and Aldridge, W. N. (1982). The formation of mixed disulphides in rat lung following paraquat administration. Correlation with changes in intermediary metabolism. *Biochim. Biophys. Acta* **716**, 249.

Kera, Y., Sippel, H. W., Penttila, K. E. and Lindros, K. O. (1987). Acinar distribution of glutathione-dependent detoxifying enzymes. Low glutathione peroxidase activity in perivenous hepatocytes. *Biochem. Pharmacol.* **36**, 2003.

Kitzler, J. and Fridovich, I. (1986). The effects of paraquat on *Escherichia coli*: distinction between bacteriostasis and lethality. *J. Free Radical Biol. Med.* **2**, 245.

Komulainen, H. and Bondy, S. C. (1988). Increased free intracellular Ca^{2+} by toxic agents: an index of potential neurotoxicity? *Trends Pharm. Sci.* **9**, 154.

Kono, Y. (1982). Oxygen enhancement of bactericidal activity of rifamycin SV on *Escherichia coli* and aerobic oxidation of rifamycin SV to rifamycin S catalysed by manganous ions: the role of superoxide. *J. Biochem.* **91**, 381.

Korbashi, P. *et al.* (1986). Iron mediates paraquat toxicity in *Escherichia coli*. *J. Biol. Chem.* **261**, 12472.

Krall, J. *et al.* (1988). Superoxide mediates the toxicity of paraquat for cultured mammalian cells. *J. Biol. Chem.* **263**, 1910.

La Cagnin, L. B. (1988). The carbon dioxide anion radical adduct in the perfused rat liver: relationship to halocarbon-induced toxicity. *Mol. Pharmacol.* **33**, 351.

Lachocki, T. M. *et al.* (1988). Persistent free radicals in the smoke of common household materials: biological and clinical implications. *Environ. Res.* **45**, 127.

Langston, J. W., Irwin, I., and Ricaurte, G. A. (1987). Neurotoxins, Parkinsonism and Parkinson's disease. *Pharmac. Ther.* **32**, 19.

Lau, S. S. and Monks, T. J. (1988). The contribution of bromobenzene to our current understanding of chemically-induced toxicities. *Life Sci.* **42**, 1259.

Lind, C., Hochstein, P., and Ernster, L. (1982). DT-diaphorase as a quinone reductase: a cellular control device against semiquinone and superoxide radical formation. *Arch. Biochem. Biophys.* **216**, 178.

Long, R. M. and Moore, L. (1987). Cytosolic calcium after carbon tetrachloride, 1,1-dichloroethylene, and phenylephrine exposure. Studies in rat hepatocytes with phosphorylase a and quin 2. *Biochem. Pharmacol.* **36**, 1215.

Macdonald, T. L. and Martin, R. B. (1988). Aluminium ion in biological systems. *Trends Biochem. Sci.* **13**, 15.

Marklund, S. L. *et al.* (1985). Superoxide dismutase isoenzymes in normal brains and in brains from patients with dementia of Alzheimer type. *J. Neurol. Sci.* **67**, 319.

Mason, R. P. and Chignell, C. F. (1982). Free radicals in pharmacology and toxicology—selected topics. *Pharmacol. Rev.* **33**, 189.

McLachlan, D. R. C. (1986). Aluminium and Alzheimer's disease. *Neurobiology of Aging* **7**, 525.

McMurray, C. H. and Rice, D. A. (1981). Toxic edible oils. *Nature* **293**, 332.

Miners, J. O., Drew, R., and Birkett, D. J. (1984). Mechanisms of action of paracetamol protective agents in mice *in vivo*. *Biochem. Pharmacol.* **33**, 2995.

Misra, H. P. and Fridovich, I. (1976). The oxidation of phenylhydrazine. Superoxide and mechanism. *Biochemistry*, **15**, 681.

Moller, L. (1986). Consequences of cadmium toxicity in rat hepatocytes: mitochondrial dysfunction and lipid peroxidation. *Toxicology* **40**, 285.

Moorhouse, C. P. *et al.* (1985). Cobalt(II) ion as a catalyst of hydroxyl radical and possible 'crypto-hydroxyl' radical formation under physiological conditions. Differential effects of hydroxyl radical scavengers. *Biochim. Biophys. Acta* **843**, 261.

Morris, P. L. *et al.* (1982). A new pathway for the oxidative metabolism of chloramphenicol by rat liver microsomes. *Drug Metab. Disposition* **10**, 439.

Morrison, H. M. (1987). The proteinase-antiproteinase theory of emphysema: time for a reappraisal? *Clin. Sci.* **72**, 151.

Mottley, C., Trice, T. B., and Mason, R. P. (1982). Direct detection of the sulfur trioxide radical anion during the horseradish peroxidase–hydrogen peroxide oxidation of sulfite (aqueous sulfur dioxide). *Mol. Pharmacol.* **22**, 732.

Munday, R. (1987). Studies on the mechanism of toxicity of the mycotoxin Sporidesmin. V. Generation of hydroxyl radical by sporidesmin. *J. Appl. Toxicol.* **7**, 17.

Munday, R. and Manns, E. (1985). Toxicity of aromatic disulphides. III. *In vivo* haemolytic activity of aromatic disulphides. *J. Appl. Toxicol.* **5**, 414.

Nakayama, T. *et al.* (1985). Cigarette smoke induces DNA single-strand breaks in human cells. *Nature* **314**, 462.

Neta, P. and Huie, R. E. (1985). Free-radical chemistry of sulfite. *Environ. Health Perspect.* **64**, 209.

Neuberger, J. (1988). Halothane hepatitis. *ISI Atlas Science, Pharmacology* **1**, 309.

Nicotera, P. *et al.* (1986). The formation of plasma membrane blebs in hepatocytes exposed to agents that increase cytosolic Ca^{2+} is mediated by the activation of a non-lysosomal proteolytic system. *FEBS Lett.* **209**, 139.

Nicotera, T. M. *et al.* (1985). Induction of superoxide dismutase, chromosomal aberrations and sister-chromatid exchanges by paraquat in Chinese hamster fibroblasts. *Mutat. Res.* **151**, 263.

Nohl, H. (1986). Quinones in biology: functions in electron transfer and oxygen activation. *Adv. Free Radical Biol. Med.* **2**, 211.

Pacht, E. R. *et al.* (1986). Deficiency of vitamin E in the alveolar fluid of cigarette smokers. Influence on alveolar macrophage cytotoxicity. *J. Clin. Invest.* **77**, 789.

Pall, H. S. *et al.* (1987). Raised cerebrospinal-fluid copper concentration in Parkinson's disease. *Lancet* **ii**, 238.

Pellack-Walker, P. and Blumer, J. L. (1986). DNA damage in L5178YS cells following exposure to benzene metabolites. *Mol. Pharmacol.* **30**, 42.

Peiser, G. D., Lizada, M. C. C., and Yang, S. F. (1982). Sulfite-induced lipid peroxidation in chloroplasts as determined by ethane production. *Plant Physiol.* **70**, 994.

Perl, D. P. and Good, P. F. (1987). Uptake of aluminium into central nervous system along nasal-olfactory pathways. *Lancet* **i**, 1028.

Pestana, A. and Munoz, E. (1982). Anilides and the Spanish toxic oil syndrome. *Nature* **298**, 608.

de Peyster, A. *et al.* (1984). Oxygen radical formation induced by gossypol in rat liver microsomes and human sperm. *Biochem. Biophys. Res. Commun.* **118**, 573.

Picardo, M. *et al.* (1987). Mechanism of antitumoral activity of catechols in culture. *Biochem. Pharmacol.* **36**, 417.

Pollack, S. *et al.* (1987). Desferrioxamine suppresses *Plasmodium falciparum* in aotus monkeys. *Proc. Soc. Exp. Biol. Med.* **184**, 162.

Preston, A. M. (1985). Modification of streptozotocin-induced diabetes by protective agents. *Nutr. Res.* **5**, 435.

Pryor, W. A., Dooley, M. M., and Church, D. F. (1986). The mechanism of the inactivation of human a-1-proteinase inhibitor by gas-phase cigarette smoke. *Adv. Free Radical Biol. Med.* **2**, 161.

Quinlan, G. J. and Gutteridge, J. M. C. (1987). Oxygen radical damage to DNA by rifamycin SV and copper ions. *Biochem. Pharmacol.* **36**, 3629.

Quinlan, G. J. and Gutteridge, J. M. C. (1988). Oxidative damage to DNA and deoxyribose by β-lactam antibiotics in the presence of iron and copper salts. *Free Radical Res. Commun.* **5**, 149.

Quinlan, G. J. *et al.* (1988). Action of lead(II) and aluminium(III) ions on iron-stimulated lipid peroxidation in liposomes, erythrocytes and rat liver microsomal fractions. *Biochim. Biophys. Acta* **962**, 196.

Rabinowitch, H. D. and Fridovich, I. (1985). Growth of *Chlorella sorokiniana* in the presence of sulfite elevates cell content of superoxide dismutase and imparts resistance towards paraquat. *Planta* **164**, 524.

Ramsammy, L. S. *et al.* (1987). Failure of inhibition of lipid peroxidation by vitamin E to protect against gentamicin nephrotoxicity in the rat. *Biochem. Pharmacol.* **36**, 2125.

Reif, D. W. *et al.* (1988). Effect of diquat on the distribution of iron in rat liver. *Toxicol. Appl. Pharmacol.* **93**, 506.

Ribarov, S. R., Benov, L. C., and Benchev, I. C. (1984). $HgCl_2$ increases the methaemoglobin prooxidant activity. Possible mechanism of Hg^{2+}-induced lipid peroxidation in erythrocytes. *Chem–Biol. Interact.* **50**, 111.

Rich, E. A. (1988). Iron metabolism by smokers' alveolar macrophages: protective or problematic in the lung? *J. Lab. Clin. Med.* **111**, 598.

Richmond, R. and Halliwell, B. (1982). The formation of hydroxyl radicals from the paraquat radical cation, demonstrated by a highly-specific gas chromatographic technique. The role of superoxide radical anion, hydrogen peroxide and glutathione reductase. *J. Inorg. Biochem.* **17**, 95.

Richter, C. and Frei, B. (1988). Ca^{2+} release from mitochondria induced by prooxidants. *Free Radical Biol. Med.* **4**, 365.

Rona, G. (1985). Catecholamine cardiotoxicity. *J. Mol. Cell. Cardiol.* **17**, 291.

Roth, E. F. *et al.* (1986). Pathways for the reduction of oxidized glutathione in the *Plasmodium falciparum*-infected erythrocyte: can parasite enzymes replace host red cell glucose-6-phosphate dehydrogenase? *Blood* **67**, 827.

Ruch, R. J. and Klaunig, J. E. (1988). Inhibition of mouse hepatocyte intercellular communication by paraquat-generated oxygen free radicals. *Toxicol. Appl. Pharmacol.* **94**, 427.

Sagai, M. and Ichinose, T. (1987). Lipid peroxidation and antioxidative protection mechanism in rat lungs upon acute and chronic exposure to nitrogen dioxide. *Environ. Health Perspect.* **73**, 179.

Seeger, W. *et al.* (1985). Alteration of alveolar surfactant function after exposure to oxidative stress and to oxygenated and native arachidonic acid *in vitro*. *Biochim. Biophys. Acta* **835**, 58.

Shaaltiel, Y. *et al.* (1988). Cross tolerance to herbicidal and environmental oxidants of plant biotypes resistant to paraquat, sulfur dioxide, and ozone. *Pestic. Biochem. Physiol.* **31**, 13.

Shapiro, A, *et al.* (1982). *In vivo* and *in vitro* activity by diverse chelators against *Trypanosoma brucei*. *J. Protozool.* **29**, 85.

Sherlock, S. (1986). The spectrum of hepatotoxicity due to drugs. *Lancet* **ii**, 440.

Shimazaki, K. *et al.* (1980). Active oxygen participation in chlorophyll destruction and lipid peroxidation in SO_2-fumigated leaves of spinach. *Plant Cell Physiol.* **21**, 1193.

Singal, P. K. *et al.* (1982). Role of free radicals in catecholamine-induced cardiomyopathy. *Can. J. Physiol. Pharmacol.* **60**, 1390.

Singer, T. P., Trevor, A. J., and Castagnoli, N. (1987). Biochemistry of the neurotoxic action of MPTP. *Trends Biochem. Sci.* **12**, 266.

Sinha, B. K. (1987). Activation of hydrazine derivatives to free radicals in the perfused rat liver: a spin-trapping study. *Biochim. Biophys. Acta* **924**, 261.

Smith, C. E., Stack, M. S., and Johnson, D. A. (1987). Ozone effects on human neutrophil proteinases. *Arch. Biochem. Biophys.* **253**, 146.

Smith, M. T. *et al.* (1985). Quinone-induced oxidative injury to cells and tissues. In *Oxidative stress* (H. Sies, Ed.). Academic Press, London.

Sokol-Anderson, M. *et al.* (1988). Role of cell defense against oxidative damage in the resistance of *Candida albicans* to the killing effect of amphotericin B. *Antimicrob. Agents Chemother.* **32**, 702.

Spencer, P. S. *et al.* (1987). Guam amyotrophic lateral sclerosis—Parkinsonism-dementia linked to a plant excitant neurotoxin. *Science* **237**, 465.

Srivastava, L. M., Bora, P. S., and Bhatt, S. D. (1982). Diabetogenic action of streptozotocin. *Trends Pharm. Sci.* September issue, 376.

Stacey, N. H. and Kappus, H. (1982). Cellular toxicity and lipid peroxidation in response to mercury. *Toxicol. Appl. Pharmacol.* **63**, 29.

Stacey, N. H. and Klaassen, C. D. (1981). Inhibition of lipid peroxidation without prevention of cellular injury in isolated rat hepatocytes. *Toxicol. Appl. Pharmacol.* **58**, 8.

Starkey, B. J. (1987). Aluminium in renal disease: current knowledge and future developments. *Ann. Clin. Biochem.* **24**, 337.

Stocker, R. *et al.* (1985). Oxidative stress and protective mechanisms in erythrocytes in relation to *Plasmodium vinckei* load. *Proc. Natl. Acad. Sci. USA* **82**, 548.

Stockley, R. A. (1983). Proteolytic enzymes, their inhibitors and lung diseases. *Clin. Sci.* **64**, 119.

Sunderman, F. W. (1986). Metals and lipid peroxidation. *Acta Pharmacol. Toxicol.* suppl. **59**, 248.

Sutton, H. C., Vile, G. F., and Winterbourn, C. C. (1987). Radical driven Fenton reactions—evidence from paraquat radical studies for production of tetravalent iron in the presence and absence of ethylenediaminetetraacetic acid. *Arch. Biochem. Biophys.* **256**, 462.

Tabuenca, J. M. (1981). Toxic–allergic syndrome caused by injection of rapeseed oil denatured with aniline. *Lancet* **ii**, 567.

Takasawa, S. *et al.* (1986). Novel gene activated in rat insulinomas *Diabetes*, **35**, 1178.

Thomas, C. E. and Aust, S. D. (1986). Reductive release of iron from ferritin by cation free radicals of paraquat and other bipyridyls. *J. Biol. Chem.* **261**, 13064.

Thor, H. *et al.* (1982). The metabolism of menadione (2-methyl-1,4-naphthoquinone) by isolated hepatocytes. A study of the implications of oxidant stress in intact cells. *J. Biol. Chem.* **257**, 12419.

Tomasi, A. *et al.* (1987). Free-radical metabolism of carbon tetrachloride in rat liver mitochondria. A study of the mechanism of activation. *Biochem. J.* **246**, 313.

Vennerstrom, J. L. and Eaton, J. W. (1988). Oxidants, oxidant drugs and malaria. *J. Med. Chem.* **31**, 1269.

Walker, P. D. and Shah, S. V. (1988). Evidence suggesting a role for hydroxyl radical in gentamicin-induced acute renal failure in rats. *J. Clin. Invest.* **81**, 334.

Wardman, P. (1984). Radiation chemistry in the clinic: hypoxic cell radiosensitizers for radiotherapy. *Radiat. Phys. Chem.* **24**, 293.

Wasil, M., Halliwell, B., and Moorhouse, C. P. (1988). Scavenging of hypochlorous acid by tetracycline, rifampicin and some other antibiotics. A possible antioxidant action of rifampicin and tetracycline? *Biochem. Pharmacol.* **37**, 775.

Whiteside, C. and Hassan, H. M. (1987). Induction and inactivation of catalase and superoxide dismutase of *Escherichia coli* by ozone. *Arch. Biochem. Biophys.* **257**, 464.

Willey, J. C. *et al.* (1987). Biochemical and morphological effects of cigarette smoke condensate and its fractions on normal human bronchial epithelial cells *in vitro*. *Cancer Res.* **47**, 2045.

Willson, R. L. (1977). Metronidazole (Flagyl®) in cancer radiotherapy: a historical introduction. In *Metronidazole: Proceedings*. Montreal May 26–28th 1976, Excerpta Medica.

Winterbourn, C. C. (1985). Free-radical production and oxidative reactions of haemoglobin. *Environ. Health Perspect.* **64**, 321.

Winterbourn, C. C., French, J. K., and Claridge, R. F. C. (1979). The reaction of menadione with haemoglobin. Mechanism and effect of superoxide dismutase. *Biochem. J.* **179**, 665.

Yonei, S. *et al.* (1986). Mutagenic and cytotoxic effects of oxygen free radicals generated by methylviologen (paraquat) on *Escherichia coli* with different DNA-repair capacities. *Mutat. Res.* **163**, 15.

Younes, M. and Siegers, C. P. (1985). The role of iron in the paracetamol- and CCl_4-induced lipid peroxidation and hepatotoxicity. *Chem–Biol. Interact.* **55**, 327.

7

Free radicals as useful species

We have already seen a number of cases in which free-radical reactions are employed in living systems for useful purposes, such as the role of highly reactive ferryl species at the active sites of cytochrome P-450 and peroxidase enzymes, and the possible involvement of hydroxyl radicals in the small amount of ethanol oxidation *in vivo* that is achieved by the microsomal ethanol-oxidizing system (Chapter 2). In the present chapter we will examine some other examples in detail.

7.1 Reduction of ribonucleosides

Deoxyribonucleotides, the precursors of DNA, are formed *in vivo* by the reduction of ribonucleoside diphosphates in the presence of an enzyme called *ribonucleoside diphosphate reductase*. This enzyme catalyses replacement of the $2'$ —OH group on the ribose sugar ring by a hydrogen atom (Fig. 7.1) and the overall reaction may be written:

$$\text{Ribonucleoside diphosphate} + R\!\!\begin{array}{c} \diagup \text{SH} \\ \diagdown \text{SH} \end{array} \longrightarrow$$

$$\text{H}_2\text{O} + \text{deoxyribonucleoside diphosphate} + R\!\!\begin{array}{c} \diagup \text{S} \\ \diagdown | \\ \text{S} \end{array}$$

where $R(SH)_2$ is a dithiol compound. In most cases $R(SH)_2$ is *thioredoxin*, a small protein with two cysteine —SH groups in close proximity. The oxidized thioredoxin (RS_2) is re-reduced by a reductase enzyme at the expense of NADPH. However, in *Escherichia coli* and in calf thymus tissue, not only thioredoxin but also a related protein called *glutaredoxin* can donate electrons to ribonucleoside diphosphate reductase. Oxidized glutaredoxin is reduced at the expense of GSH, the GSSG so produced being itself reduced by glutathione reductase (Fig. 7.1).

In *E. coli* and in higher cells the ribonucleoside diphosphate reductase enzyme contains two different subunits, B1 and B2, in a 1:1 stoichiometry.

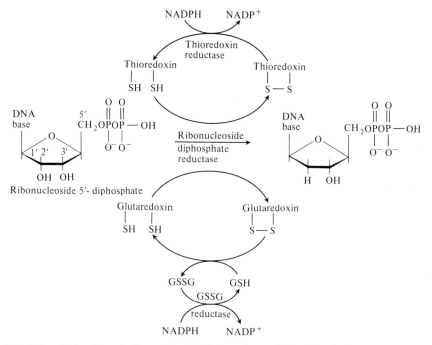

Fig. 7.1. Reduction of ribonucleoside diphosphates. The thioredoxin system operates in all cells containing the ribonucleoside diphosphate reductase activity. In addition, glutaredoxin is present in *E. coli* and calf thymus, and perhaps in other tissues. The relative importance of these two reducing systems *in vivo* is not clear.

The two catalytic sites of the enzyme are made up from parts of each subunit. B1 provides substrate binding sites and also contains —SH groups which react with the substrate during catalysis, donating hydrogen. The oxidized groups are then re-reduced at the expense of thioredoxin or glutaredoxin (Fig. 7.1). Subunit B2 contains 2 moles of Fe(III) ion, and gives an ESR signal that has been identified as arising from a tyrosine radical produced by loss of one electron from a tyrosine residue in the amino-acid chain. The presence of the tyrosine radical is closely-linked to the presence of iron; the radical is lost on removal of iron and re-formed upon incubation of the enzyme with a Fe(II) salt and a thiol compound in the presence of oxygen.

The function of the iron, as it binds to the protein, is both to generate the tyrosine radical (during oxidation from Fe^{2+} to Fe^{3+}) and then to stabilize the radical by some continued interaction with it. The positive charge is delocalized over the whole of the benzene ring structure of the tyrosine. This tyrosine radical is known to participate in the catalytic mechanism of the enzyme, but details have yet to be established. *Hydroxyurea* ($H_2NCON-HOH$), a widely used inhibitor of ribonucleoside diphosphate reductase (and hence of DNA synthesis in cells) appears to act by reacting with the tyrosine radical.

If ribonucleoside diphosphate reductase that has been inactivated by treatment with hydroxyurea is incubated with an extract of *E. coli* in the presence of NAD(P)H, a flavin (riboflavin, FMN, or FAD), O_2, and a thiol compound, the tyrosine radical re-appears and enzyme activity is restored. Fractionation of the *E. coli* extract shows that three proteins are essential for restoration of activity. One is an NAD(P)H-dependent flavin reductase, which catalyses the reaction:

$$NAD(P)H + H^+ + flavin_{ox} \rightarrow NAD(P)^+ + flavin_{red}$$

The reduced flavin may lead to reduction of Fe(III) on subunit B2, as a first step in a series of reactions leading to one-electron oxidation of the tyrosine to generate the tyrosine radical.

Reduced flavins, however, can donate an electron to O_2, to form O_2^- (Chapter 3). Superoxide can damage B2, probably by producing more-reactive species such as $OH^{\cdot}$ (Chapter 3). Thus the second protein required to restore B2 activity is superoxide dismutase, which removes O_2^- and so protects B2. The function of the third protein is unknown at present.

A number of bacteria and algae contain a different ribonucleoside diphosphate reductase in which the reaction involves a deoxyadenosyl cobalamin radical instead of a tyrosine radical. *Cobalamin*, otherwise known as vitamin B_{12}, consists of a *corrin* ring structure with a central cobalt ion. Like the haem ring, the corrin ring is based on four pyrrole units (Fig. 7.2). The radical involved in the reductase reaction seems to be formed by a homolytic cleavage of the bond between the cobalt and the $-CH_2$ group in deoxyadenosylcobalamin (Fig. 7.2).

7.2   Oxidation, carboxylation, and hydroxylation reactions

7.2.1   *Dioxygenases*

As we saw in Chapter 3, the enzyme indoleamine dioxygenase, which catalyses cleavage of the indole rings of tryptophan, tryptamine, and serotonin, has been conclusively shown to involve O_2^- in its reaction mechanism; and it is inhibited by superoxide dismutase. Indeed, decreasing the SOD activity of rabbit intestinal cells by treatment with diethyldithiocarbamate increases the rate of action of the dioxygenase *in vivo*. Similarly, the fungal enzymes, nitropropane dioxygenase and galactose oxidase, are prevented from functioning by high concentrations of SOD. If these enzymes indeed use O_2^- *in vivo*, then any SOD present in the cells containing them cannot be completely effective in removing O_2^- or else the enzymes would

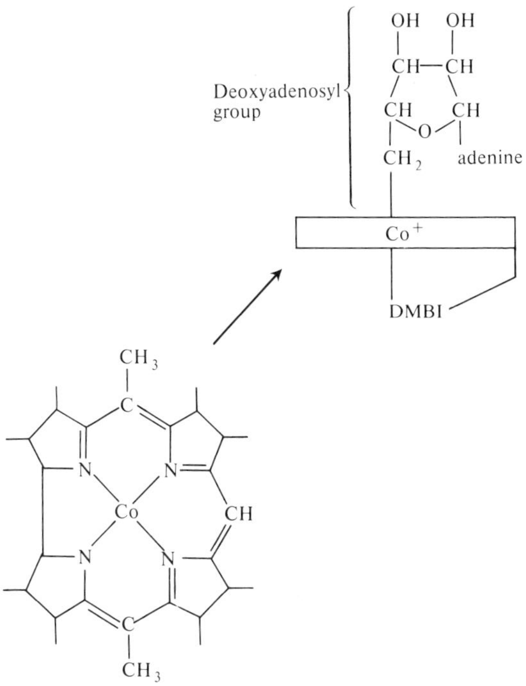

Fig. 7.2. Vitamin B_{12} and its derivatives. Co is cobalt, a transition metal (see Appendix). The corrin ring is represented as a flat plate, four N-atoms being co-ordinated to the cobalt. The fifth co-ordination position is used to attach a derivative of dimethylbenzimidazole (DMBI, structure not shown) which is also attached to a side chain of the corrin ring. The sixth co-ordination position of the cobalt can be occupied by a number of groups, such as cyanide (CN^-) ion to give *cyanocobalamin*. The cyanide is introduced during the isolation procedure and is not present *in vivo*. The cobalt atom in cobalamin can have a $+1$, $+2$, or $+3$ oxidation state. In hydroxycobalamin, OH^- occupies the sixth co-ordination position and the cobalt is in the Co^{3+} state. This form, called B_{12a}, is reduced by a flavoprotein enzyme to $B_{12r}(Co^{2+})$ which is reduced by a second flavoprotein to give $B_{12s}(Co^+)$. B_{12s} is the substrate for a reaction with ATP that yields 5'-deoxyadenosyl cobalamin, whose structure is shown. Impaired absorption of cobalamin from the human diet results in pernicious anaemia.

not be able to function. Perhaps they are located in parts of the cell that have low SOD activities.

7.2.2 *Pyruvate-metabolizing enzymes*

In animal cells, conversion of pyruvate produced by glycolysis into acetyl-coenzyme A (acetylSCoA) is catalysed by a multienzyme complex, known as pyruvate dehydrogenase. However, in many bacteria and in some parasitic protozoa of the trichomonad group (e.g. *Trichomonas vaginalis*) the reaction is brought about by other enzymes, such as pyruvate–ferredoxin oxidoreductase:

$$HSCoA + pyruvate + 2\ ferredoxin_{ox} \rightarrow acetylSCoA + CO_2 + 2\ ferredoxin_{red}.$$

In trichomonas parasites, this enzyme is located in organelles known as *hydrogenosomes*. The reduced ferredoxin can then be used to reduce H^+ ions to hydrogen gas, a reaction accompanied by ATP synthesis.

Long-lived free radicals have been detected by ESR during the operation of bacterial and trichomonad enzymes. It appears that, after decarboxylation of pyruvate, one electron is transferred from it onto an iron–sulphur cluster on the enzyme, from which the electron eventually moves to ferredoxin. A *possible* mechanism is shown in Fig. 7.3 from which it may be seen that both carbon-centred (from pyruvate) and sulphur-centred (from coenzyme A) radicals could be involved. In the presence of O_2, the reduced iron–sulphur cluster might react to form O_2^-; reduced ferredoxin also reduces O_2 to O_2^- (Chapter 5). O_2^- could participate in the inactivation of the enzyme that is observed under aerobic conditions.

Another pyruvate-metabolizing enzyme found in many bacteria is pyruvate–formate lyase, which catalyses a non-oxidative conversion of pyruvate to acetyl-coenzyme A:

$$pyruvate + HSCoA \rightarrow acetylSCoA + formate.$$

It has been suggested that the activity of this enzyme may involve a free radical derived from an amino-acid residue on the enzyme.

7.2.3 Hydroxylases

There is no clear evidence for the participation of *free* superoxide and hydroxyl radicals in the mechanism of action of any hydroxylase enzyme, although iron–oxygen radical complexes such as ferryl species are important at the active sites of cytochrome P-450 and peroxidase. Similarly, the lysyl and prolyl hydroxylases involved in the synthesis of collagen (Chapter 2) require three things for their action: an Fe(II) ion at the active site, a reducing agent (ascorbic acid), and the compound 2-oxoglutarate. The first stage of the hydroxylation catalysed by these enzymes may be the attack of an Fe^{2+} —O_2 complex at the active site upon the carbonyl group of 2-oxoglutarate to give an intermediate which then hydroxylates the substrate. A complex such as Fe^{2+}—O_2 would have considerable resonance contribution from Fe^{3+}—O_2^-, and so the initial step might well be regarded as nucleophilic attack by a

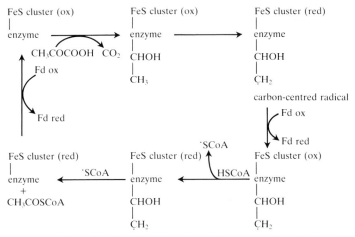

Fig. 7.3. A *possible* mechanism for the action of pyruvate–ferredoxin oxidoreductase, showing the involvement of carbon-centred and sulphur-centred radicals (many details remain to be established). HSCoA-coenzyme A.

form of 'bound-superoxide'. Hydroxylation is not inhibited by SOD because *free* O_2^- is not involved in the catalytic mechanism, but it can be inhibited by low-molecular-weight copper complexes with SOD activity, which can presumably penetrate to the active site and react with 'bound-superoxide'. Great caution should be exercised in the use of such complexes however, since they often act as unspecific enzyme inhibitors.

7.2.4 Carboxylation of glutamic acid

The synthesis of prothrombin and factors VII and IX, essential constituents of the blood coagulation system, requires the attachment of carboxyl (—COOH) groups to a number of glutamic acid residues to give *γ-carboxylglutamic acid* residues in the polypeptide chains.

polypeptide   polypeptide

Carboxylation is achieved by an enzyme located in the endoplasmic reticulum of the liver cells, and it requires the fat-soluble vitamin K_1 for its action. As Fig. 7.4 shows, this vitamin is a quinone and it can be enzymically reduced by liver microsomal fractions to a semiquinone form at the expense of NADPH. This semiquinone, or possibly the fully-reduced hydroquinone form of vitamin K_1, is required for the carboxylation reaction. The semiquinone can react reversibly with O_2 to form the quinone and O_2^-.

It has been observed that carboxylation of glutamate by isolated liver microsomes is inhibited by SOD and by copper chelates that can scavenge O_2^-. This may implicate a 'superoxide-like' species in the catalytic mechanism of the carboxylase. Another possibility is that SOD, by removing O_2^-, 'drains away' the required semiquinone by causing the equilibrium:

$$\text{semiquinone} + O_2 \rightleftharpoons \text{quinone} + O_2^-$$

to move towards the right (Chapter 3).

7.3 Phagocytosis

In the late 1800s, the Russian scientist Metchnikoff observed the engulfment of bacteria by cells from the bloodstream of animals. This process is called *phagocytosis*—the cell 'flows around' the foreign particle and encloses it in a plasma membrane vesicle that becomes internalized into the cytoplasm of the phagocytic cell. Most of the phagocytic cells in the human bloodstream are *neutrophils* (Table 7.1), which have a multilobed nucleus (hence they are called *polymorphonuclear* cells) and a large number of cytoplasmic granules, which are of several different types (Fig. 7.5). The *primary* or *azurophil* granules contain the enzymes myeloperoxidase and lysozyme, several proteases (e.g. elastase), and a number of 'granular cationic proteins'. The *specific* or *secondary granules* contain a protein that binds vitamin B_{12} (*cobalophilin*), lysozyme, and may also contain lactoferrin, an iron-binding protein similar to transferrin. Like transferrin itself, lactoferrin can bind two moles of Fe^{3+} ion per mole of protein (Chapter 2). A whole series of lysosomal enzymes is probably housed within the so-called *tertiary granules*.

When human tissue is injured, an *acute inflammatory response* develops, characterized by swelling, warmth, pain, reddening, and partial immobilization. The arterioles in and around the injured area relax, so that the capillary

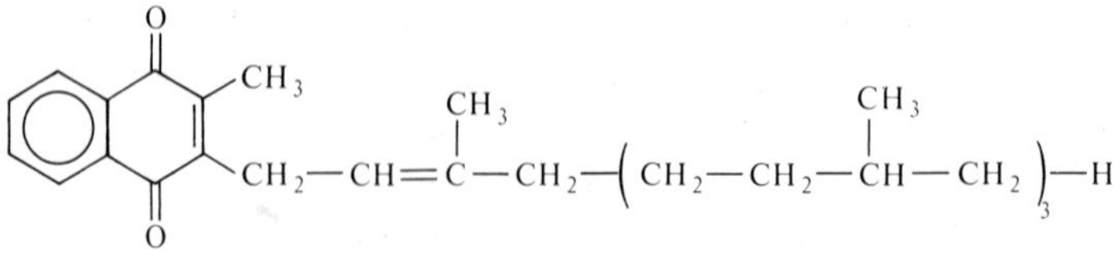

Fig. 7.4.   Vitamin K_1 quinone.

Table 7.1. Distribution of leucocytes ('white blood cells') in human blood

Type of cell	Normal number per μl (10^{-6} l) of blood	Function
Polymorphonuclear leucocytes ('polymorphs')		
neutrophils	2500–7500	Phagocytosis (see text).
eosinophils	40–400	Allergic and hypersensitivity responses (see text).
basophils	0–100	Hypersensitivity reactions (see text).
Monocytes	200–800	Precursors of macrophages.
Lymphocytes	1500–3500	Immune response (see Appendix II).

Polymorphs, monocytes, and lymphocytes differ in their precursor cells, morphology, and function. An abnormally high content of white cells in the blood (above 11 000 per μl) is called *leucocytosis*, whilst a decrease below 4000 per μl is termed *leucopenia* or, since neutrophils constitute such a high proportion of white cells, *neutropenia*. Eosinophils differ from neutrophils (Fig. 7.4) in having larger cytoplasmic granules and often a nucleus with only two lobes. Monocytes are the precursors of macrophages in sites of inflammation. Most eosinophils reside in the tissues rather than circulating in the blood.

network becomes engorged with blood (hence the heat and redness). The permeability of the blood vessel walls increases so that more fluid leaks out, causing oedema. This fluid is rich in protein. As they enter the inflamed area, neutrophils often stop on the endothelial cells lining the blood vessels, a phenomenon known as *pavementing* or *margination*. Neutrophils then push out cytoplasmic pseudopodia and squeeze through the gaps between endothelial cells, crossing the vessel wall and entering the inflamed tissue. The migration appears to be produced by a number of compounds which are formed in the inflamed area and attract the neutrophils (*chemotactic* factors). Such factors are numerous and include products of complement activation.

At a later stage of inflammation, monocytes (Table 7.1) leave the circulation and enter the inflamed area. These cells are less actively mobile and phagocytic than neutrophils, but once in the inflamed area they undergo differentiation and change into *macrophages* (Figs. 7.6 and 7.7), which involves increase in their content of lysosomal enzymes, metabolic activity, motility, and phagocytic and microbicidal capacity. Macrophages are larger than neutrophils and do not usually have a lobed nucleus. Apart from their formation during inflammation, they can be found in the lymphatic system,

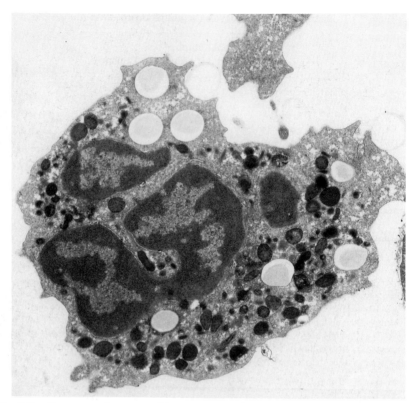

Fig. 7.5. Structure of a neutrophil as seen under the electron microscope. The cell is phagocytosing opsonized latex beads (the white circles). Some are being taken up by the cell flowing around them and others are already present within the cell in vacuoles (× 18,000). Photograph by courtesy of Prof. A. W. Segal.

Fig. 7.6. Human pulmonary macrophages. The cells were obtained from the lungs of healthy volunteers. A, C, and E are from non-smokers; B, D, and F from smokers. A and B are light-micrographs. The cells vary from 15–50 μm in diameter. C and D are electron micrographs (× 6175); and E and F are scanning electron micrographs which show the surface of the cells. Large numbers of vesicles are present. Cigarette smoke causes marked changes in the cells, which contain characteristic 'smokers inclusions', these include 'needle-like' or 'fibre-like' structures. These needle-like structures consist of kaolinite, an aluminium silicate present in cigarette smoke that the macrophages can engulf. Cigarette smoking increases the number of alveolar macrophages (Chapter 6, Section 6.4). The cell surface membrane of macrophages is highly ruffled. Photographs by courtesy of Prof. W. G. Hocking.

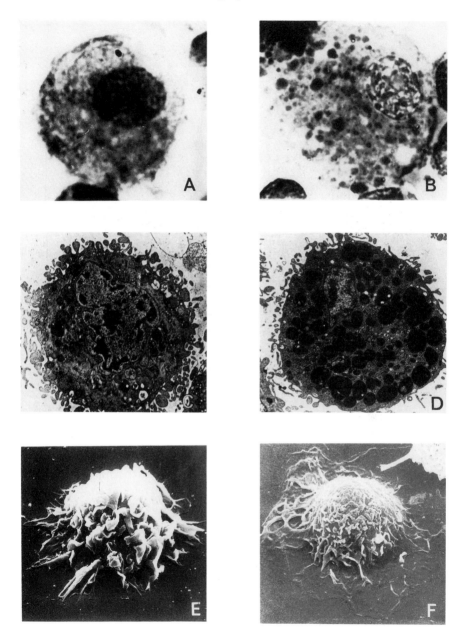

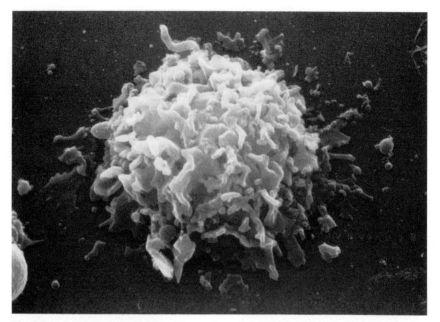

Fig. 7.7. A spleen macrophage. The surface of the cell is shown by scanning electron microscopy (× 15 000). Note the intensely ruffled surface, characteristic of an activated macrophage. Photograph by courtesy of Dr W. Dawson.

spleen (Fig. 7.7), and as scattered cells in connective tissue and in the lungs. Indeed, the *alveolar* macrophages lie on the alveolar walls, and are a major defence of the lung against inhaled bacteria and other particles (Fig. 7.6). Pulmonary macrophages can be found in other parts of the lung as well. The Kupffer cells, which form part of the lining of the liver sinusoids, are also macrophages. Macrophages can phagocytose bacteria and dead cells, e.g. degenerated erythrocytes. They can also ingest large amounts of insoluble material and retain it for months or even years (Fig. 7.6).

The fluid leaking into the inflamed area contains various antibodies which can bind to bacteria. Both neutrophils and macrophages have surface receptors that can recognize immunoglobulin G antibodies and the C3b component of reacted complement. Coating of bacteria with such host-derived proteins, known collectively as *opsonins*, enables the neutrophils and macrophages to recognize them, although a few bacterial strains have coats that can be recognized directly. Once the bacteria are engulfed, various cytoplasmic granules fuse with, and hence mix their contents with, the vacuole containing the engulfed particle (*phagocytic vacuole*). The engulfed particles are then killed and, if possible, digested within the phagocytic vacuole, the digestion products eventually being expelled from the cell.

After the bacteria causing the lesion have been destroyed (or, if injury was caused by another mechanism such as heat or chemicals, after the insult ceases) there is usually reversal of the inflammatory changes. The vessel walls regain their normal permeability. Most of the emigrated neutrophils probably die, and the fragments are phagocytosed by macrophages. During phagocytic activity, neutrophils and macrophages release lysosomal enzymes into the surrounding fluid, where they contribute to the digestion of inflammatory debris.

If the bacteria are not completely eliminated, however, or the tissue injury continues, chronic inflammation can result. There is formation of vascular granulation tissue, which matures into fibrous tissue. Bacteria likely to cause chronic inflammation include those responsible for syphilis and tuberculosis; and it is often induced by silica dust (quartz) or silicate fibres (e.g. asbestos) inhaled into the lungs. The fibrous tissue produced during chronic inflammation causes loss of function. Chronic inflammatory lesions are especially rich in macrophages, several of which may fuse together to give *giant cells* with multiple nuclei. The various hydrolytic enzymes secreted by macrophages play an important role in tissue damage during chronic inflammation. The function of giant cells is not clear, but they retain some phagocytic ability.

7.3.1 *The bacterial killing mechanism*

Superoxide, hydrogen peroxide, and hydroxyl radical
Most studies on the biochemistry of phagocytosis have been carried out upon neutrophils and macrophages, especially alveolar macrophages since they are the only macrophages that can readily be obtained from humans, by the technique of bronchoalveolar lavage. Resting neutrophils consume little oxygen since they rely mainly on glycolysis for ATP production and are rich in stored glycogen. By contrast, macrophages possess mitochondria and so make use of oxidative phosphorylation and show more oxygen consumption. At the onset of phagocytosis, however, both cell types show a marked increase in oxygen uptake that is not prevented by cyanide, and so is unrelated to mitochondrial electron transport. The accelerated O_2 uptake is often called the *respiratory burst* of the cells, although that is rather a bad name since it is unrelated to mitochondrial respiration. The increase in O_2 uptake during the respiratory burst can be to ten or twenty times the 'resting' O_2 consumption of neutrophils, but is less marked in macrophages. At the same time, there is an increased consumption of glucose by the cells, which is fed into the pentose phosphate pathway as shown by the increased release of $^{14}CO_2$ from [1-^{14}C] glucose. Triggering of the respiratory burst depends on some kind of membrane perturbation, possibly a decrease in membrane potential, since not only opsonized bacteria but also small latex beads (Fig. 7.5), opsonized zymosan (a preparation of yeast cell walls), and certain

chemicals can induce it. These chemicals include *phorbol myristate acetate*, a co-carcinogen (see Chapter 8), high concentrations of fluoride ion, certain small peptides such as *N*-formylmethionylleucylphenylalanine (fmet-leu-phe), and concanavalin A, a lectin (lectins are plant proteins that bind with high affinity to carbohydrate-containing substances, such as the glycoproteins found in cell surface membranes). Hence the act of phagocytosis itself is not required for the respiratory burst to occur. There is considerable person-to-person variability in the extent and time-course of O_2 uptake by neutrophils isolated from human blood and activated *in vitro*. Neutrophil responsiveness also varies when these cells are obtained from the same person at different times.

The exact size of the respiratory burst shown by macrophages depends on how, and from which tissue site, they were obtained. *Resident macrophages*, tissue macrophages that have not yet met foreign materials, are poorly active. More active macrophages can be obtained by infecting animals with micro-organisms such as bacillus Calmette–Guerin (BCG) or by injecting irritating materials such as thioglycollate. The antibacterial behaviour of the macrophages obtained by these different procedures is very variable and papers concerning, for example, the relative importance of oxygen radicals in killing by macrophages should be read with this in mind. Macrophage activation is partly achieved during infection by the release of macrophage-activating factors (*lymphokines*) from T lymphocytes sensitized to antigens from the infecting organism (see Appendix II for further explanation). Interferon γ appears to act as an important lymphokine.

Macrophages can be 'primed' to release more O_2^- upon activation: priming agents include interferon γ, bacterial endotoxin (lipopolysaccharide), and the bacterial cell wall component muramyl dipeptide. The priming agents themselves induce little or no O_2^- release, but they potentiate the response of the macrophage to stimuli that do activate the respiratory burst. Macrophages are complex multifunctional cells that are capable, under the right conditions, of secreting a wide range of substances into their environment—not only oxidants and proteases (e.g. collagenase and macrophage elastase, an enzyme that has properties very different from neutrophil elastase and is not inhibited by α_1-antiproteinase; Chapter 6, Section 6.4), but also enzyme inhibitors (e.g. the protease inhibitor α-macroglobulin), some components of complement, angiogenesis factor (a compound that stimulates the migration and proliferation of fibroblasts and endothelial cells), and interleukin I, a modulator of immune function (Chapter 4, Section 4.7.6). Macrophages also secrete *platelet-activating factor* (1-*O*-alkyl-2-acetyl-*sn* glyceryl-3-phosphorylcholine), a glyceryl ether lipid that stimulates platelets (small cells circulating in the blood in vast numbers) to aggregate and release various products, such as vasoactive amines. Platelets play an important part in the blood coagulation system (Section 7.6). Macrophages exposed to endotoxin secrete tumour-necrosis factor, TNF (Chapter 8, Section 8).

If the respiratory burst in neutrophils is prevented by placing them under an atmosphere of nitrogen gas, the killing of some bacterial strains, such as *Bacillus fragilis* and *Clostridium perfringens*, is not impaired. Many of these organisms are anaerobes, and the neutrophil would not be expected to rely upon an oxygen-dependent killing mechanism for species that it would be likely to encounter under conditions of low oxygen concentration. The killing is presumably achieved by the contents of the various vacuoles as they empty into the phagocytic vacuole. The enzyme lysozyme can attack and digest the cell walls of various bacteria, especially Gram-positive strains. Lysosomal enzymes might be able to attack some strains and a number of the 'granular cationic proteins' have bactericidal activity. For example, a protein of this type from rabbit and human neutrophils causes breakdown of the cell membranes of *E. coli* and *Salmonella typhimurium*. Lactoferrin can bind iron essential for the growth of certain bacteria, and it has been reported to kill a few bacterial strains directly.

However, the killing of many other bacterial strains by neutrophils is greatly decreased under anaerobic conditions. Indeed, patients suffering from *chronic granulomatous disease* (CGD), an inherited condition in which phagocytosis is normal but the respiratory burst is absent, show persistent and multiple infections, especially in skin, lungs, liver, and bones, by those bacterial strains whose killing by neutrophils requires oxygen (Table 7.2).

The oxygen uptake in both neutrophils and macrophages is due to the activation of an enzyme complex associated with the plasma membrane. The system oxidizes NADPH provided by the pentose phosphate pathway in the cytosol into $NADP^+$, the electrons being used to reduce oxygen to the superoxide radical, O_2^-, as was first shown by B. M. Babior's group in the USA. The overall equation can be written:

$$NADPH + 2O_2 \rightarrow NADP^+ + H^+ + 2O_2^-$$

Indeed, very severe inborn deficiencies of glucose 6-phosphate dehydrogenase, the first enzyme in the pentose phosphate pathway, also cause increased susceptibility to infections. The superoxide radical can be detected outside stimulated neutrophils and macrophages by its ability to reduce cytochrome c or nitro-blue tetrazolium, both reactions being inhibited by added SOD. Alternatively, O_2^- can be identified by spin-trapping experiments (Chapter 2). The NADPH oxidase complex is a flavoprotein, containing FAD. It is very likely that a b-type cytochrome is also involved in O_2^- generation. An electron transport chain may operate, as indicated below:

Table 7.2. Synopsis of chronic granulomatous disease

Incidence	1 in 1,000,000
Male:female ratio	6:1
Inheritance	80% X-linked
	rest mostly autosomal recessive
Carrier state	Seen in X-linked CGD in mothers and sisters, usually asymptomatic
Age of onset	Congenital, first symptoms < 1 year of age (80%), rarely adult onset
Clinical manifestations	1. Recurrent bacterial and fungal infections
	2. Sequelae of chronic inflammation (failure to thrive, anaemia, enlarged spleen etc.).
	3. Obstruction due to granuloma formation.
Diagnosis	Absence of O_2^- and H_2O_2 production by activated phagocytes
Organ involvement	1. Lung (80%)—pneumonia (*Staphylococcus aureus*, *Aspergillus*) abscess, chronic changes
	2. Lymph nodes (75%)—suppurative lymphadenitis (*S. aureus*), adenopathy
	3. Skin (65%)—infectious dermatitis, abscesses
	4. Liver (40%)—abscess (*Staphylococcus aureus*)
	5. Bone (30%)—osteomyelitis (*Serratia*)
	6. Gastrointestinal tract (30%)
	7. Sepsis and meningitis (20%)
	8. Genito-urinary (10%)
Prognosis	Variable—death in infancy to middle age
Treatment	Prophylactic antibiotics

Adapted from J. T. Curnutte and B. M. Babior, Clinically-significant phagocytic cell defects. In *Current clinical topics in infectious disease* **6**, 138 (1985).

The involvement of a quinone molecule in O_2^- generation has also been suggested, but the evidence is not compelling. The NADPH oxidase complex is arranged vectorially in the membrane so that electrons pass across it from the NADPH-oxidizing site on the inside to the external O_2-reducing site. Movement of electrons across the plasma membrane seems to be accompanied by outward movement of H^+ through proton-conducting channels in the membrane. The K_m for O_2 of the activated oxidase complex in rat neutrophils is within the range of the normal oxygen concentration in body fluids, and so it is possible that the amount of O_2^- produced *in vivo* will depend on O_2 supply.

The oxidase complex is defective in chronic granulomatous disease. Different forms of this condition exist, with different defects in the oxidase system. Patients with X-linked disease (Table 7.2) usually lack the b-type cytochrome, whereas most patients with autosomal recessive disease have a defect elsewhere in the oxidase complex.

Since the bacteria or other engulfed particles are 'wrapped up' in a plasma membrane vesicle in the phagocyte cytoplasm, they are exposed to a high flux of O_2^- (Fig. 7.8). In view of its low reactivity in aqueous solution (Chapter 3), it is unlikely that O_2^- itself is responsible for bacterial killing. It is often suggested that, at the acidic pH of the phagocytic vacuole, much O_2^- will exist as the more reactive $HO_2^{\cdot}$ (Chapter 2). However, Segal and his colleagues in England used a pH indicator dye (fluorescein) conjugated to bacteria to measure intra-vacuolar pH. Despite several problems with these studies (e.g. oxidation of dye by the myeloperoxidase-derived oxidant hypochlorous acid; discussed below), the results obtained suggested that a few minutes after phagocytosis of bacteria the intra-vacuolar pH of human neutrophils *rose* from 7.4 to about 7.75, and then fell slowly, reaching 6.0–6.5 after 2 hours. The rise in pH might be accounted for by consumption of protons as O_2^- dismutates to H_2O_2 in the phagocytic vacuole,

$$2O_2^- + 2H^+ \rightarrow H_2O_2 + O_2$$

If vacuoles are neutral to slightly alkaline initially, this would favour the operation of the NADPH oxidase, some neutral proteases, and some of the 'cationic antibacterial proteins'. An acidic pH value at later times would favour the operation of myeloperoxidase and hydrolytic enzymes with acidic pH optima. More of any O_2^- that was still being generated at that time would also exist as $HO_2^{\cdot}$ (Chapter 2).

Some strains of bacteria are quickly killed by hydrogen peroxide, which will be formed from O_2^- by the dismutation reaction (Chapter 3). The hydroxyl radical, $OH^{\cdot}$, formed from O_2^- and hydrogen peroxide by an iron-catalysed Haber–Weiss reaction may sometimes participate in killing, since dimethylsulphoxide, a scavenger of $OH^{\cdot}$ that can penetrate into cells (Chapter 3), can partially protect *Staphylococcus aureus* against killing by human neutrophils. During these experiments, the formation of methane, a product of the reaction between dimethylsulphoxide and $OH^{\cdot}$ (Chapter 2), was observed. This observation could mean that H_2O_2 is converted into $OH^{\cdot}$ in the phagocytic vacuole, and the $OH^{\cdot}$ then attacks the bacteria. Alternatively, it is possible that H_2O_2 penetrates into the bacteria and causes intracellular $OH^{\cdot}$ formation using iron ions inside the bacterium. The latter possibility seems more likely to the authors, since $OH^{\cdot}$ formed outside bacteria would presumably react with the cell wall, and so very large fluxes of $OH^{\cdot}$ would be required to reach and damage the plasma membrane. We have already commented (Chapter 3) on the variability of bacterial sensitivity to H_2O_2 as due both to the different efficiencies of H_2O_2-metabolizing systems within bacteria and the different intracellular availabilities of transition metal ions that can convert H_2O_2 into highly-reactive OH˙.

Can neutrophils themselves make the OH˙ radical from the O_2^- and H_2O_2 that they generate? Several scientists have detected OH˙ radical in suspensions of activated neutrophils by benzoate decarboxylation, oxidation of methional to ethene gas, or spin-trapping with DMPO (for an explanation of these techniques see Chapter 3). Other experiments using DMPO or aromatic hydroxylation (Chapter 3) as assays for OH˙ gave negative results unless an iron complex (such as Fe^{3+}–EDTA) was added to the reaction mixture. When neutrophils are prepared for studies of oxygen-radical formation, they are separated from their normal plasma environment and suspended in laboratory buffers. Many such buffers are contaminated with iron ions. Thus neutrophil-derived O_2^- and H_2O_2 can react with these contaminating iron ions to form OH˙. The authors believe that neutrophils themselves do not secrete an iron 'catalyst' of ˙OH formation: the O_2^- and H_2O_2 that they produce on activation of the respiratory burst (Fig. 7.8) will only lead to OH˙ formation detectable outside the neutrophil if the environment of the neutrophils contains catalytic metal complexes. Such complexes are not present in normal human plasma, but can become available at sites of inflammation, e.g. in the inflamed rheumatoid joint (Chapter 8). Thus Ward *et al.* in the USA found that the killing of cultured bovine pulmonary artery endothelial cells by activated human neutrophils was apparently mediated by OH˙, but that iron came from the endothelial cells and not from the neutrophils.

The iron-binding protein lactoferrin is secreted into the phagocytic vacuole of activated neutrophils, and also released from the cells. Lactoferrin is similar to transferrin in that it binds two moles of Fe(III) per mole of protein, but its affinity for iron is higher and it needs a pH more acidic than is required by transferrin (pH 2–3 for lactoferrin, pH 5–6 for transferrin) in order for iron to be released. Iron bound to the specific binding sites of either lactoferrin or transferrin is unable to promote OH˙ formation, and, since neutrophil lactoferrin has only a low percentage iron saturation *in vivo*, the authors have suggested that its secretion by neutrophils may help to *minimize* iron-dependent radical reactions, such as OH˙ generation, by binding any available iron ions at sites of inflammation, e.g. in the inflamed rheumatoid joint (Chapter 8). Lactoferrin iron could only promote OH˙ formation if it were released from the protein by some mechanism (Chapter 3).

The killing of *E. coli*, *Staphylococcus aureus*, and *Streptococcus viridans* by human neutrophils was decreased by addition of SOD or of catalase, attached to latex particles to aid their uptake into the phagocytic vacuole. H_2O_2 can enter bacteria and may promote intra-bacterial OH˙ generation. It is also possible that O_2^- and H_2O_2 could interact in the phagocytic vacuole to form OH˙ if a suitable transition metal complex was present; this could be

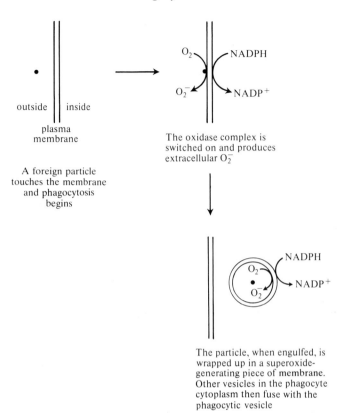

The oxidase complex is switched on and produces extracellular O_2^-

A foreign particle touches the membrane and phagocytosis begins

The particle, when engulfed, is wrapped up in a superoxide-generating piece of membrane. Other vesicles in the phagocyte cytoplasm then fuse with the phagocytic vesicle

Fig. 7.8. A schematic representation of the respiratory burst. The engulfed particles are exposed to a flux of O_2^- inside the phagocytic vacuole.

provided if bacteria damaged by other mechanisms (e.g. HOCl produced by myeloperoxidase) 'leaked' intracellular metal ions. However, unless a neutrophil-derived promoter of OH˙ generation can be identified, it seems unlikely that extra-bacterial OH˙ generation in the phagocytic vacuole is a primary mechanism of 'attack' upon phagocytosed bacteria by neutrophils.

Myeloperoxidase
Once the phagocytic vacuole is formed, fusion of it with other granules in the neutrophil cytoplasm empties, among other things, the enzyme myeloperoxidase over the engulfed particle. (Myeloperoxidase is not normally present in macrophages.) Myeloperoxidase is a green haem-containing enzyme that shows 'non-specific' peroxidase activity (Chapter 3) but it catalyses other reactions as well. In the presence of hydrogen peroxide, and chloride or iodide ions, myeloperoxidase can kill a number of bacteria and fungi *in vitro*.

Since much more Cl^- than I^- ion is present in the phagocyte cytoplasm, it has been suggested that myeloperoxidase can participate in the killing mechanisms by oxidizing Cl^- into hypochlorous acid, HOCl. Hypochlorous acid is highly reactive, being able to oxidize many biological molecules (especially —SH groups), and can itself kill a number of bacteria. It is known to react with hydrogen peroxide at alkaline pH values to form singlet oxygen $^1\Delta g$ (Chapter 2), but this reaction does not occur rapidly at pH 7.4.

$$HOCl + H_2O_2 \rightarrow O_2^* + Cl^- + H_2O + H^+.$$

Patients with an inborn deficiency of myeloperoxidase in their neutrophils show only minor decreases in resistance to infection, and so this enzyme must be of less importance in bacterial killing than is the respiratory burst. Thus Roos *et al.* in Holland studied the fate of *E. coli* cells engulfed by human neutrophils. The bacterial envelope was quickly perforated and bacterial enzymes were then inactivated. Perforation was much slower in neutrophils from patients with chronic granulomatous disease. However, neutrophils from myeloperoxidase-deficient patients showed a normal rate of perforation, but a slower rate of enzyme inactivation. Thus HOCl produced by myeloperoxidase might help to destroy bacteria after they have been irreversibly damaged, but does not, at least in this case, appear responsible for the initial damage to the bacteria.

Neutrophils from myeloperoxidase-deficient patients often kill bacteria more slowly than normal *in vitro*, although the difference seems physiologically insignificant. Myeloperoxidase may be of greater importance in protection against fungi such as *Candida albicans*. In neutrophils from patients with chronic granulomatous disease myeloperoxidase is present normally, but the absence of the respiratory burst means that it is not provided with the necessary hydrogen peroxide. Thus both antibacterial mechanisms fail to operate, accounting for the severe symptoms. Indeed, neutrophils from patients with CGD can destroy certain strains of bacteria that themselves release hydrogen peroxide, thus providing the myeloperoxidase with the means of operation.

Myeloperoxidase, like horseradish peroxidase (Chapter 3), reacts quickly with O_2^- to form an 'oxyferrous' enzyme, known as compound III:

$$enzyme - Fe(III) + O_2^- \rightarrow enzyme - Fe(II) - O_2.$$

Thus secretion of myeloperoxidase into the phagocytic vacuole might decrease the amount of O_2^- available as compound III is formed. *In vitro*, compound III reacts with H_2O_2 to form compound II, and O_2^- can regenerate myeloperoxidase in the Fe(III) form from compound II. The extent to which these various reactions occur in the phagocytic vacuole has yet to be established.

During phagocytosis, some myeloperoxidase is released from neutrophils and can react with H_2O_2 and Cl^- to form HOCl outside the cells. HOCl *in vitro* at low concentrations rapidly inactivates a_1-antiprotease, the main inhibitor of proteolytic enzymes, such as elastase, in extracellular fluids (Chapter 6, Section 4). HOCl inactivates by oxidizing a methionine residue essential for the action of a_1-antiprotease. The extent to which inactivation could happen *in vivo* depends on the environment of the neutrophil. Thus HOCl generated in plasma does not damage a_1-antiprotease, because the HOCl reacts preferentially with albumin, present at concentrations of 50–60 mg/ml (as compared to 1–3 mg/ml for a_1-antiprotease). In many other biological fluids (e.g. cerebrospinal fluid, synovial fluid, the fluid that lines the alveoli of the lung) albumin concentrations are much lower than in plasma, and any a_1-antiprotease present could be attacked by HOCl. Ascorbic acid is also oxidized by HOCl, and the authors have shown that the concentrations present in several extracellular fluids (50–150 μM in plasma and synovial fluid, and greater than this in alveolar lining fluid and cerebrospinal fluid) can protect a_1-antiprotease against inactivation by HOCl, until the concentration of ascorbate has been depleted. In the inflamed rheumatoid joint, ascorbate concentrations are sub-normal and inactivated a_1-antiprotease can be detected (Chapter 8). HOCl can also *activate* two enzymes present within neutrophils in a latent form—collagenase (which digests native collagen) and gelatinase (which attacks denatured collagen). HOCl can *inactivate* several other neutrophil enzymes, including lysozyme and the myeloperoxidase molecule itself. Thus the overall effect of myeloperoxidase secretion and HOCl generation at sites of inflammation depends on what else is present at the site in question, and is difficult to predict.

Human neutrophils, and several other human tissues (e.g. nerve tissue) are rich in the compound *taurine* (2-aminoethanesulphonic acid), which is an end-product of the metabolism of sulphur-containing amino acids. There have been several proposals that the biological role of taurine is to act as an antioxidant. However, the products of reaction of taurine with HOCl (taurine-*N*-chloramines), although less oxidizing than is HOCl itself, are still capable of inactivating a_1-antiprotease.

Light emission by phagocytes

Activated phagocytes produce a weak 'background' light emission; this can be greatly enhanced by addition of *luminol* or *lucigenin*. The background light emission may be due to production of O_2^- and H_2O_2 by the phagocytes, leading to the chemiluminescence that accompanies oxidant production

(Chapter 4, Section 4.6.6; Chapter 2, Section 2.8.3). By contrast, the enhanced light production in the presence of luminol appears largely to involve the myeloperoxidase–H_2O_2–Cl^- system. Under some experimental conditions (e.g. Fig. 7.9) this light production is biphasic: the first phase seems to depend on interaction of luminol with extracellular myeloperoxidase–H_2O_2–Cl^- and the second to be due to entry of luminol into the cells and its measurement of intracellular oxidant generation, presumably also involving myeloperoxidase. By contrast, the enhanced light emission produced by addition of lucigenin to activated neutrophils seems to involve O_2^- rather than the myeloperoxidase system.

Light emission by neutrophils in the presence of luminol or lucigenin has become widely used to measure the respiratory burst of isolated activated neutrophils, and has also been used to measure their activation in samples of body fluids (e.g. plasma) from patients with various diseases. However, it

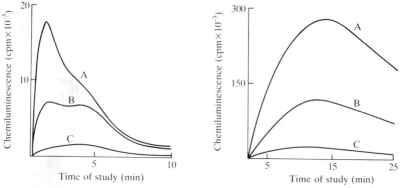

Fig. 7.9. Time course of luminol-dependent chemiluminescence from human neutrophils exposed to fmet-leu-phe (left-hand graph) or to phorbol myristate acetate (right-hand graph) at 37°C. A, 10^5 cells; B, 5×10^5 cells; C, 10^6 cells. Diagram adapted from *Agents and Actions*, **21**, 104 (1987) by courtesy of Dr C. Dahlgren. In the left-hand curve, the first peak is probably related to extracellular events (such as HOCl production by myeloperoxidase) and the second to intracellular events. Luminol has the structure   lucigenin the structure

should be noted that the time-course and extent of enhanced light emission by activated phagocytic cells depend on pH, temperature, the presence of compounds that might scavenge HOCl and thus depress the 'extracellular' part of luminol-dependent light emission (e.g. albumin in plasma), the stimulus used to activate the cells, and the cell density. Figure 7.9 shows how cell density and the nature of the activating stimulus (fmet-leu-phe or phorbol myristate acetate) can alter the kinetics and the amount of light produced by human neutrophils in the presence of luminol.

Some scientists have suggested that activated neutrophils generate singlet oxygen, but attempts to use various singlet oxygen scavengers (Chapter 2) in experiments have been frustrated by the fact that HOCl can oxidize a number of them directly, including DABCO, 2,5-diphenylfuran, histidine, and β-carotene. Formation of singlet oxygen by neutrophils *in vivo* cannot yet be regarded as rigorously proved.

Other phagocytes

Formation of O_2^- by a 'respiratory burst' has been observed not only in macrophages and neutrophils, but also in the microglial cells of the brain, the Kupffer cells of the liver, monocytes, basophils, mast cells, and eosinophils. Indeed, human monocytes show a more marked respiratory burst than do macrophages, although it is not as large as in neutrophils. Monocytes contain a myeloperoxidase-like enzyme which disappears as they differentiate into macrophages. Eosinophils and basophils are rare in the circulation (Table 7.1) but the former cells increase in number in allergic conditions, such as bronchial asthma and hay fever, and during infestations by some parasites, such as schistosomes. Eosinophils phagocytose bacteria very sluggishly, but they do seem to be important in defence against parasitic worms where, of course, the worm is much bigger than the eosinophil. Damage to the parasites can be achieved by oxygen radicals, by an eosinophil peroxidase activity, and by a 'highly basic protein' that attacks cell membranes. All of these are released when the eosinophils contact the parasite. Human eosinophils are especially active in O_2^- production.

Basophils and mast cells, which are present in connective tissue, are rich in histamine, and both may be activated by allergens which combine with immunoglobulin E antibody bound to their surface. Such activation not only produces O_2^- but also pharmacologically active products, including histamine, 5-hydroxytryptamine, and SRS (Section 7.6). These products are responsible for the symptoms produced when patients with allergies are exposed to the allergen that their bound IgE recognizes, such as grass pollen in the case of hay fever, or the house dust mite in many cases of bronchial asthma. Histamine has been reported to stimulate O_2^--production by human eosinophils.

7.3.2   *Significance of extracellular oxidant production by phagocytes*

The syndrome of chronic granulomatous disease shows that production of O_2^- and H_2O_2 in the phagocytic vacuole is essential for the killing of many (not all) strains of engulfed bacteria. Myeloperoxidase has a less important role.

However, activated neutrophils secrete myeloperoxidase into the surrounding medium, where it can lead to HOCl generation. Collagenase, gelatinase, elastase, lysozyme, several other hydrolytic enzymes, and lactoferrin are also released. Activated neutrophils, macrophages, monocytes, and eosinophils produce extracellular O_2^- and H_2O_2 upon activation, with the potential to generate OH˙ if the environment of the cells contains a suitable metal catalyst. Indeed, activated phagocytes are weakly mutagenic in the Ames test and H_2O_2 vapour, possibly arising from alveolar macrophages, has been detected in exhaled human breath.

Release of these various oxidants can help phagocytes to attack extracellular targets, such as opsonized cancer cells (Chapter 8) or parasitic worms, but they might also damage the surrounding environment, especially as extracellular fluids in the human body have only low activities of SOD or H_2O_2-removing enzymes (Chapter 4, Section 4.7.6). For example, by inactivating a_1-antiprotease, HOCl can allow elastase released from neutrophils to digest elastin. However, the interactions are very complex; we have seen that extracellular proteins such as albumin might scavenge HOCl. Both HOCl˙ and oxygen radicals can attack the neutrophils themselves, as well as enzymes released from them, e.g. HOCl inactivates lysozyme. Both oxidation of methionine residues in proteins to methionine sulphoxide, and DNA strand breakage, have been observed to occur within activated neutrophils. The DNA damage perhaps reflects diffusion of H_2O_2 into the nucleus and formation of OH˙ close to the DNA.

Neutrophils do have some defences against oxidants. They contain a CuZnSOD enzyme in the cytosol, and a MnSOD in the mitochondria, but these cannot offer protection against externally-generated O_2^-. Neutrophils also contain catalase and glutathione peroxidase activities, together with GSH at millimolar concentrations, which help to protect them against hydrogen peroxide. Methionine sulphoxide reductase and large amounts of ascorbic acid are also present. Indeed, neutrophils deficient in glutathione reductase activity are much more rapidly inactivated during phagocytosis than normal, presumably because hydrogen peroxide can easily diffuse out of the phagocytic vacuole into the cell cytoplasm, and GSH cannot be regenerated for glutathione peroxidase activity. The turnover of GSH in mouse peritoneal macrophages has been observed to increase during phagocytosis. Catalase activity seems to be less important in protection against hydrogen peroxide than the glutathione system, and indeed it varies considerably between species. For example, human or guinea-pig neutrophils contain much more catalase than cells from rats and mice.

Superoxide radical has been claimed to react with a component of normal human plasma to form a factor chemotactic for neutrophils, but a detailed investigation of the chemistry of this process has not been performed. By contrast, other leukotriene (Section 7.7) chemotactic factors produced by neutrophils can be *inactivated* by OH˙ or HOCl. In view of the complex interactions between all the oxidants, proteins and other factors present at sites of inflammation, it seems likely that tissue 'damage' will be limited during normal inflammation. The increased blood flow and vascular permeability that occurs at sites of inflammation will tend to replenish supplies of a_1-antiprotease and may permit more albumin than normal to enter the site of inflammation, leading to more HOCl scavenging. Concentrations of the acute-phase antioxidant protein caeruloplasmin are also increased during inflammation (Chapter 4), giving more antioxidant protection.

It is possible that oxygen radicals play some role in mediating the increased vascular permeability at sites of inflammation, although how important the effect of oxidants is remains to be determined. Thus Del Maestro in Sweden found that perfusion of a hamster cheek pouch with O_2^--generating systems caused increased leakage of material from the vascular network and also increased adherence of neutrophils to the vessel walls. Leakage was decreased by adding SOD or catalase to the perfusion medium.

However, if activation of phagocytic cells at a site of inflammation is excessively prolonged, or an abnormally large number of cells is activated at a particular site, then significant tissue damage might result. For example, a factor in the serum of patients with the skin disease psoriasis has been reported to increase oxygen-radical production by phagoctyes, which are known to accumulate in the epidermal lesions. Macrophages are present in the lesions of myelinated nerves seen in multiple sclerosis. Initial damage to the lungs by exposing animals to elevated oxygen concentrations has been reported to provoke an influx of neutrophils, which become activated and make the damage worse as O_2^- and hydrogen peroxide attack the alveolar cells. Pulmonary macrophages increase their SOD activity on exposure of animals to elevated O_2 (Chapter 3). Cigarette smoke has been reported to increase the production of hydrogen peroxide and O_2^- by pulmonary macrophages, an effect that may be involved in the development of smoker's emphysema (Chapter 6, Section 6.4). Silica particles and asbestos fibres can be phagocytosed by pulmonary macrophages. They then appear to cause rupture of the phagocytic vesicles, leading to release of proteolytic enzymes into the macrophage cytoplasm. If the macrophage is killed, the particles will be released to be taken up by other cells with the same effect. The proteases will thus be released into the surrounding lung tissue. Asbestos fibres have been shown to contain iron, some of which is in a form that can stimulate lipid peroxidation and OH˙ formation. This iron may be an integral part of the silicate crystal structure (e.g. in crocidolite asbestos, which has the overall formula $Na_2O \cdot Fe_2O_3$. $3FeO \cdot 8SiO_2$) or may be present as a contaminant on the surface of fibres, as with chrysotile asbestos ($3MgO \cdot 2SiO_2 \cdot 2H_2O$).

It is possible that penetration of asbestos fibres of a particular length and shape into cells leads to iron-dependent radical reactions that contribute to cancer development. Inhalation of asbestos fibres for prolonged periods causes not only lung fibrosis but also two types of cancer, mesothelioma and bronchogenic carcinoma. In addition, several mineral dust fibres, including asbestos, are capable of activating the respiratory burst of phagocytes. Gulamian and Van Wyk in South Africa have claimed that glass fibres also contain iron ions capable of promoting OH˙ generation from H_2O_2.

In 1968, it was discovered that within the first few minutes of blood dialysis, leucopenia (Table 7.1) occurs, which was traced to the fact that contact of plasma with cellophane in the dialyser causes activation of the complement system, leading to accumulation, aggregation, and activation of neutrophils and monocytes in the lung. This can cause damage by interfering with blood flow, and perhaps by producing oxygen radicals. Indeed, the adult respiratory distress syndrome (ARDS)—acute respiratory failure due to pulmonary oedema—is a frequent result of severe shock, tissue damage due to burns or accidents, or massive infections. The accumulation and activation of neutrophils in the lung plays an important part in the pathology of ARDS. Oxygen radicals are known to produce lung oedema *in vivo* (Chapter 3). Perhaps the most striking consequence of abnormal phagocyte action are seen in the autoimmune diseases. All the above diseases are considered in detail in Chapter 8.

7.4 Peroxidase and NADH oxidase enzymes

The action of peroxidases in the presence of hydrogen peroxide is important in several areas of metabolism in addition to the leucocyte killing mechanism. For example, formation of thyroid hormones involves iodination of the phenolic ring of tyrosine residues in the protein thyroglobulin, and coupling of the iodinated tyrosyls to form iodothyronines. Both these processes are catalysed by a *thyroid peroxidase* enzyme, which is apparently located in the endoplasmic reticulum of thyroid cells. The H_2O_2 required by thyroid peroxidase seems to be generated by a membrane-bound enzyme system that uses NADPH and requires Ca^{2+} ions. Salivary peroxidase, and lactoperoxidase, may have some anti-bacterial action (Chapter 3). The ability of peroxidase to oxidize phenols into quinones in the presence of hydrogen peroxide is made use of by bombardier beetles, which attack their enemies by spraying them with a hot quinone-containing fluid. A sac within the insect contains a 25 per cent aqueous solution of hydrogen peroxide plus 10 per cent hydroquinone, and the spray is generated when the contents of the sac are pushed into a reaction chamber containing catalase and peroxidase. The hydroquinones are explosively oxidized to quinones, causing the temperature to rise up to 100 °C. Figure 7.10 shows such a beetle in action.

The polymer lignin, a major constituent of wood, is derived by the oxidation and polymerization of phenols synthesized from the amino acids phenylalanine and tyrosine (Fig. 7.11). Oxidation of phenols to the phenoxy radicals which polymerize seems to be achieved by one or more peroxidase enzymes bound to the plant cell wall. The source of the hydrogen peroxide required for peroxidase action is not entirely clear, but a likely possibility is that the cell wall peroxidases simultaneously oxidize NADH. This reaction does not require the addition of hydrogen peroxide (Chapter 3), and the NADH oxidation actually generates both O_2^- and hydrogen peroxide, the latter being used by the peroxidase to oxidize phenols. Work by Elstner and others in Germany, and in the authors' laboratory, has shown the reaction mechanism to be complicated.

Breakdown of lignin can also involve peroxidases. Some wood-destroying fungi release both H_2O_2 and peroxidase enzymes extracellularly to aid lignin degradation. Thus the white-rot fungus *Phanerochaete chrysosporium* can oxidize 60–70 per cent of lignin to CO_2 and water, leaving the rest as small molecular fragments. This fungus produces at least two extracellular peroxidases. One of these is a haem enzyme that resembles horseradish peroxidase

Fig. 7.10. A bombardier beetle in action. Photograph by courtesy of Thomas Eisner and Daniel Aneshansley.

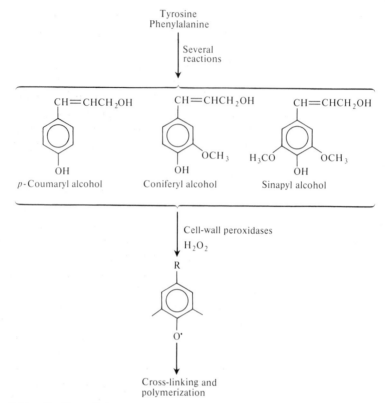

Fig. 7.11. Outline of the structure of lignin found in plant cell walls. Lignin is derived by the oxidation of various phenols into phenoxy radicals which then polymerize. The phenols are derived by adding —OH and —OCH₃ groups onto the benzene ring of the amino acids tyrosine or phenylalanine. Oxidation of phenols to phenoxy radicals is probably achieved by one or more peroxidase enzymes bound to the plant cell walls.

in its mechanism of action; it reacts with H_2O_2 to form a compound I that then performs one-electron oxidations of the substrate (Chapter 3). The substrate radicals undergo a wide variety of very complex non-enzymic cleavage reactions, some of which involve uptake of O_2 and formation of peroxy radicals.

The second peroxidase produced by this fungus is also a haem enzyme. It appears to use H_2O_2 to oxidize Mn^{2+} into Mn(III), which becomes chelated by an organic acid, such as lactate. This Mn(III) chelate is then responsible for the oxidative cleavage of lignin (the chemical reactivity of manganese ion complexes is discussed in Chapter 3, Section 3.5.5). The H_2O_2 required by these two peroxidases may arise, in part, from glucose oxidase and from the action of a cellobiose oxidase enzyme also secreted by the fungus. Cellobiose oxidase is a flavoprotein that attacks cellulose and generates superoxide (O_2^-), which can dismute to give H_2O_2.

Changing to a completely different system, the fertilization of sea-urchin eggs induces a rapid uptake of oxygen and a fertilization membrane containing cross-linked tyrosine residues is produced to stop further spermatozoa from entering. Cross-linking is catalysed by a peroxidase (*ovoperoxidase*) present in granules in the egg, and most of the extra oxygen uptake can be accounted for by synthesis of the hydrogen peroxide that it requires (a 'respiratory burst' with a novel purpose). Administration of oestrogen to female rats induces the synthesis of a uterine peroxidase activity that has been suggested to play a role in destruction of this hormone, and might additionally have an antibacterial effect in the uterine fluid. Whether this peroxidase is synthesized by the uterus itself, or is released by uterine eosinophils (it strongly resembles eosinophil peroxidase), is not entirely clear.

A peroxidase enzyme plays a role in the light emission from the boring mollusc *Pholas dactylus*. In most bioluminescent systems, light is produced when an enzyme (called a *luciferase*) acts on a low-molecular-mass substrate (*a luciferin*) to generate an excited state which then emits light as it decays. Detailed work by A. M. Michelson's group in France has shown, however, that *Pholas dactylus* luciferin is itself a protein, and can be induced to emit light when exposed to several systems generating oxygen radicals, such as Fe^{2+} in the presence of O_2, or a mixture of xanthine oxidase and hypoxanthine. The *Pholas* luciferase is a glycoprotein with peroxidase activity, but it contains copper ions and not a haem ring. Light emission from *Pholas* luciferin (often called *pholasin*) has been used as a highly-sensitive method for detecting oxidant production by activated phagocytic cells.

Luciferase from the earthworm *Diplocardia longa* (a luminescent earthworm several inches long found in Georgia, USA) has also been discovered to be a copper-containing protein with peroxidase activity. All the above systems must be provided with hydrogen peroxide *in vivo* and so the enzymes that normally dispose of hydrogen peroxide cannot be completely effective in their vicinity.

The plasma membranes of some animal cells other than phagocytes contain enzymes that can oxidize NADH and NADPH, including adipose tissue cells, erythrocytes, and renal brush border membranes. The electrons from NAD(P)H oxidation can be accepted *in vitro* by a variety of reagents such as ferricyanide or cytochrome c, but the electron acceptors *in vivo* have not been identified, and there is no clear evidence that the action of NADH or NADPH oxidases in these other tissues can lead to O_2^- generation. However, Mukherjee *et al.* in the USA have suggested that the action of insulin, in increasing glucose transport into adipocytes and decreasing breakdown of stored triglycerides, may be mediated by a stimulation of membrane NAD(P)H oxidase by the hormone, resulting in increased intra-cellular generation of hydrogen peroxide. Several laboratories have obtained evidence consistent with this proposal, which means that hydrogen peroxide would be acting as an intracellular second messenger of insulin action. However, it is probably premature to accept the proposal fully at this stage. By contrast, the NADH oxidase in renal brush border membranes seems to be involved in transport processes, and a similar role has been suggested for an NADH oxidase system detected in the plasma membrane of corn root cells. Our knowledge of electron-transport systems in plasma membranes is meagre, largely because of the difficulty in obtaining membrane fractions uncontaminated with mitochondria or endoplasmic reticulum. However, any such systems present might well contribute to oxygen radical production by cells, if only by minor 'leakage' of electrons to oxygen to form O_2^- in the way already described for mitochondrial and microsomal electron-transport chains (Chapter 3).

7.5   Fruit ripening and the 'wound response' of plant tissues

The ripening and senescence of fruits seems to be a controlled oxidative process. Studies on pears have shown that, as ripening proceeds, the concentration of free —SH groups in the fruit decreases, and there is an accumulation of hydrogen peroxide and of lipid peroxides. Ripening of pears, or the senescence of rice plant leaves, can be speeded up by treatments that stimulate formation of hydrogen peroxide within the tissue. As pears and bananas ripen, fluorescent products, that may arise by lipid peroxida-tion, accumulate within them. Membrane lipid fluidity decreases in cell and organelle membranes from many senescing plant tissues; this might occur as a result of lipid peroxidation and of the action of lipases that become activated during senescence. A role for leaf peroxidase activity has been suggested in the breakdown of cell walls, and in the degradation of chlorophyll to yellow and brown pigments during leaf senescence.

Many non-green plant tissues, such as tubers, fruits, and seeds, contain the enzyme lipoxygenase, an iron-containing protein which catalyses a direct reaction of polyunsaturated fatty acids with oxygen to give 13- and 9-hydroperoxides (Fig. 7.12). Low activities of lipoxygenase are sometimes present in green leaves. The first such enzyme purified was from soyabeans, and it is now known as *soybean lipoxygenase 1*, since at least two other lipoxygenases have been discovered in this plant (lipoxygenases-2 and -3). Lipoxygenase 1 has a pH optimum of 9 and its preferred substrate is linoleic acid, upon which it acts to produce almost entirely the 13,L-hydroperoxide (Fig. 7.12). Other lipoxygenases have different pH optima, substrate specifi-city and produce different ratios of products. For example, soybean lipoxy-

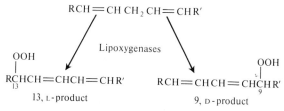

Fig. 7.12. Reactions catalysed by lipoxygenases. Purified lipoxygenases from plant tissue catalyse the peroxidation of fatty acids with a *cis,cis*-1,4-pentadiene structure. For example, linoleic acid may be converted into 13-L or 9-D hydroperoxy derivatives or both, depending on the enzyme. The reactions are highly stereospecific, L- and D- referring to the exact arrangements of groups in space around the carbon atom bearing the —OOH group.

genase-2 has a pH optimum of 6.8, prefers arachidonic acid as a substrate, and converts linoleic acid into approximately equal amounts of the 13- and 9-hydroperoxides.

Lipoxygenases contain one iron ion per molecule, and begin their action by abstracting a hydrogen atom from a fatty-acid substrate in a highly-stereospecific manner. Bond rearrangement and oxygen insertion follow. Often, the reactions that occur are more complicated and the hydroperoxide products interact further with the enzyme. The action of lipoxygenases on their substrates can produce a 'co-oxidation' of other added materials, such as thiols, carotenoids, and chlorophylls. Indeed, lipoxygenases are employed commercially to bleach wheat flour carotenoids during the bread-making process: as the carotenoids are oxidized they lose their yellow-orange colours. Chlorophyll is also bleached in the presence of lipoxygenase and a fatty-acid substrate. It seems that the action of lipoxygenase produces a peroxy radical that is normally reduced to a hydroperoxide but can also interact with an oxidizable co-substrate to cause damage. If L—H is the fatty-acid substrate and XH the co-oxidizable molecule the mechanism may be written:

$$L-H \xrightarrow{\text{enzyme}} L \xrightarrow{O_2} LOO \xrightarrow{\text{enzyme}} LOOH$$
$$\text{substrate} \qquad\qquad\qquad\qquad \text{product}$$

$$LOO + XH \longrightarrow LOOH + X$$

$$X + O_2 \longrightarrow \text{co-oxidation products.}$$

Oxidation of linoleic acid by soybean lipoxygenase isoenzymes 2 and 3 has been reported to be accompanied by singlet O_2 production, which might also contribute to co-oxidation reactions under certain circumstances. The singlet O_2 could arise by a Russell-type mechanism:

$$2 \,\text{\textbackslash CHOO}^{\cdot} \longrightarrow \,\text{\textbackslash CHOH} + \,\text{\textbackslash C}{=}\text{O} + {}^1\text{O}_2$$

Hence lipoxygenase action may cause damage to surrounding tissues. The subcellular location of lipoxygenase activity in plant tissues is variable, and it is usually found in several subcellular fractions. Indeed, the contamination of plant mitochondrial fractions by lipoxygenase activity has greatly confused studies of their oxygen uptake.

Many fruits, tubers, and seeds (and some leaves) respond to wounding or other tissue damage by initiating a series of complex biochemical reactions (Fig. 7.13). Firstly, a set of hydrolytic enzymes attacks membrane lipids and releases fatty acids, many of which are polyunsaturated. For example, linoleic and linolenic acids represent about 75 per cent of the total fatty-acid side-chains in potato tuber lipids. The released fatty acids are then acted upon by lipoxygenases, and the resulting hydroperoxides cleaved, both non-enzymically in the presence of metal complexes (Chapter 4) and by the action of 'cleavage enzymes'. This results in a wide variety of products, including volatile aldehydes and hydrocarbon gases, such as ethane and pentane. Some of the aldehydes so produced have characteristic smells which are responsible for the aroma of damaged plant tissues, such as sliced tomatoes and cucumbers (see legend to Fig. 7.13). The odour of crushed green leaves is sometimes caused by processes similar to these in Fig. 7.13. The reaction sequence shown in Fig. 7.13 can be extremely swift: over 30 per cent of the lipids in potato tuber slices are hydrolysed in less than 15 minutes even at 3 °C. Slicing a potato greatly increases its uptake of oxygen, both in the lipoxygenase reaction and in the subsequent metabolism of fatty acid products. Indeed, such enzyme oxidation of disrupted plant material can give rise to problems of rancidity and 'off flavour' during processing and storage,

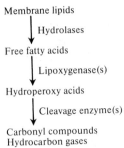

Membrane lipids

↓ Hydrolases

Free fatty acids

↓ Lipoxygenase(s)

Hydroperoxy acids

↓ Cleavage enzyme(s)

Carbonyl compounds
Hydrocarbon gases

Fig. 7.13. Pathway for the breakdown of membrane lipids induced on wounding plant tissues. The volatile products from the 9-hydroperoxides of 18:2 and 18:3 fatty acids are *cis*-3-nonenal and *cis*-3-, *cis*-6-nonadienal respectively. These are the two main components of the odour of sliced cucumbers. The 13-hydroperoxides give *hexanal* and *cis-3-hexenal*.

that have to be controlled by the use of antioxidants to scavenge lipid—OO˙ radicals (Chapter 4). Plant lipoxygenase activity can be inhibited by several antioxidants, such as propyl gallate and nordihydroguiaretic acid. These may act by scavenging peroxy radicals formed during the catalytic cycle (see above and Chapter 4).

Formation of lipid peroxides and aldehydes when plant tissues are damaged may play an important role in killing fungi and bacteria attempting to enter the wound, since several of these products have been shown to be toxic. Damage to certain plant tissues results in formation of *traumatic acid* or *wound hormone* (Fig. 7.14) a compound which induces proliferation of new cells. Traumatic acid is formed by oxidative degradation of polyunsaturated fatty acids by a process similar to that in Fig. 7.13. It seems that the 'wound response' of plant tissues is an example of controlled lipid peroxidation put to a useful purpose.

A lipoxygenase enzyme is found in reticulocytes and it has been suggested to be responsible for the degradation of mitochondria that occurs as these cells mature to form erythrocytes.

7.6 Eicosanoids: prostaglandins and leukotrienes

The prostaglandins and leukotrienes comprise a large and complex family of biologically-active lipids derived from polyunsaturated fatty acids by insertion of molecular oxygen (O_2) in enzyme-catalysed reactions. Oxygen insertion takes place by stereospecific free-radical mechanisms at the active sites of enzymes. Prostaglandins and leukotrienes have extremely potent and varied actions in the body; they are involved in regulation of numerous physiological processes and play a role in many diseases that involve tissue injury and inflammation. The prostaglandins, leukotrienes, and other related substances such as the thromboxanes and various families of hydroxy-fatty acids are often collectively referred to as *eicosanoids*, because they are synthesized from polyunsaturated fatty acids with twenty carbon atoms. The most important precursor of eicosanoids in humans is arachidonic acid (20:4;eicosatetraenoic acid).

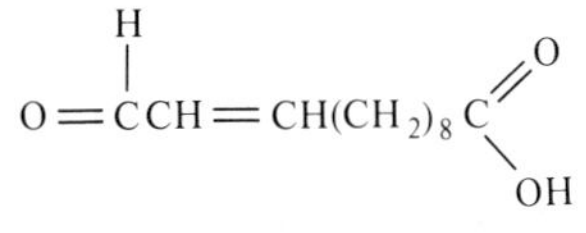

Fig. 7.14.   Structure of traumatic acid. The correct chemical name is 12-oxo-*trans*-10-dodecenoic acid. Both the aldehyde form (shown) and the dicarboxylic acid form produced by oxidation of the —CHO group are biologically active.

The eicosanoids act as local hormones. They are synthesized and released from cells in response to chemical or mechanical stimuli, interact with receptors on adjacent target cells, and are quickly inactivated, i.e. their actions are usually localized.

7.6.1 *Prostaglandins and thromboxanes*

Prostaglandins and thromboxanes (sometimes collectively called *prostanoids*) comprise those products of the metabolism of arachidonic acid and similar fatty acids (see below) in which insertion of oxygen leads to ring formation (Fig. 7.15). The term prostaglandin was coined in the 1930s to describe compounds present in semen that could affect blood pressure in animals and cause contraction of smooth muscle. In fact the term prostaglandin is a misnomer, because they do not arise from the prostate gland but from the seminal vesicle. This is the richest known source of prostaglandin-synthesizing enzymes, and seminal fluid is the only biological material in which prostaglandins accumulate in any significant concentration. However, almost every body tissue (with the possible exception of erythrocytes) is capable of making prostaglandins to some extent.

The first prostaglandins to be isolated (by Swedish scientists in the late 1960s) were the chemically-stable species PGE_2 and PGF_2 (Fig. 7.15), but better analytical techniques have since demonstrated a series of unstable prostanoids such as PGG_2, PGH_2, thromboxane A_2 (TXA_2), and PGI_2 (prostacyclin); all of these bicyclic compounds possess a chemically-unstable, oxygen-containing ring structure (Fig. 7.15).

Chemically, prostaglandins may be regarded as derivatives of *prostanoic acid*, a hypothetical C_{20} acid containing a saturated 5-membered ring structure (*cyclopentane* ring). There are several types of prostaglandins distinguished by the chemical nature and geometry of the groups attached to the ring (e.g. E, F, D) and by the number of double bonds in the side-chain (e.g. E_1, E_2, F_1, F_2 etc.); all these variations in chemical structure critically affect biological activity. Prostaglandins are synthesized *in vivo* from fatty acids that contain *cis* double bonds at positions 8, 11, and 14. The most important fatty acid of this type in humans is arachidonic acid ($C_{20:4}$, *cis* double bonds at 5, 8, 11, 14; see Chapter 4, Section 4.1), but others include

Fig. 7.15. Structure of prostanoic acid, and of thromboxanes and prostaglandins derived from arachidonic acid. Enzymes involved: 1, prostaglandin endoperoxide synthetase (cyclo-oxygenase); 2, glutathione-S-transferase; 3, prostaglandin endoperoxide E isomerase; 4, prostaglandin endoperoxide reductase; 5, prostacyclin synthetase; 6 and 7, prostaglandin endoperoxide: thromboxane A isomerase (thromboxane synthetase). Reactions 2 and 3 require glutathione. Prostaglandins E_3 and E_1 come from other fatty acids.

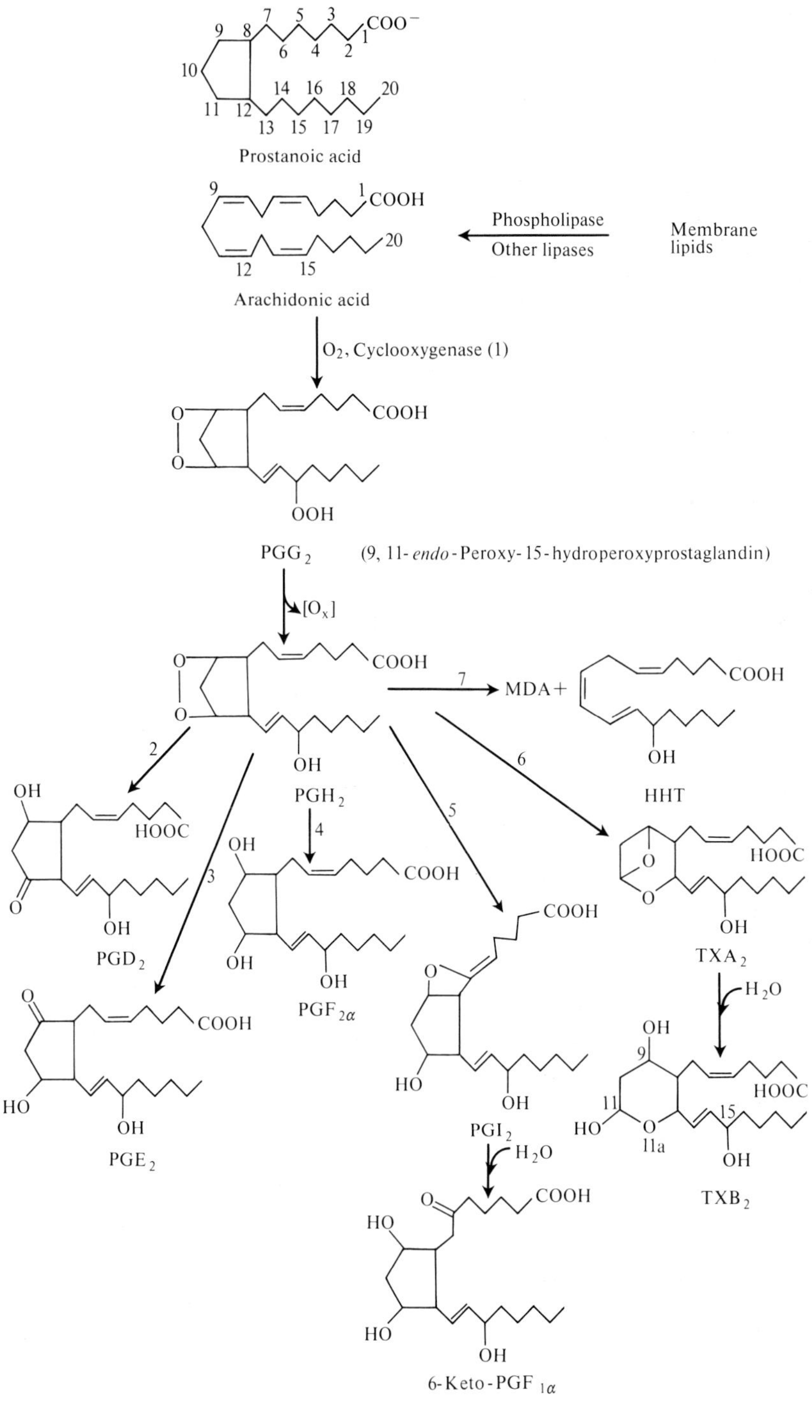

Prostanoic acid

Arachidonic acid

O_2, Cyclooxygenase (1)

PGG_2 (9, 11-*endo*-Peroxy-15-hydroperoxyprostaglandin)

$[O_x]$

PGH_2

MDA+

HHT

PGD₂

PGF₂α

TXA₂

H_2O

PGE₂

PGI₂

TXB₂

H_2O

6-Keto-PGF₁α

8,11,14-eicosatrienoic acid ($C_{20:3}$) and 5,8,11,14,17-eicosapentaenoic acid ($C_{20:5}$). This last fatty acid is unusually prevalent in the blood and membrane lipids of Eskimos, probably due to their fish-rich diet. It is thought that the relatively-low incidence of coronary heart disease shown by Eskimos might, in part, be due to the formation of anti-thrombotic prostanoids from this $C_{20:5}$ fatty acid. However, we will confine further discussion to products derived from arachidonic acid.

The first stage in the formation of prostaglandins (and other eicosanoids) is to provide a fatty-acid substrate. The main source of arachidonic acid is membrane lipids. Release of arachidonic acid from membrane phospholipids occurs by activation of phospholipase A_2 or by the concerted action of phospholipase C and diglyceride lipase, as part of the phenomenon of 'phosphatidylinositol turnover'. Free intracellular Ca^{2+} ions are important in activating these systems, especially phospholipase A_2, and thus agents increasing intracellular free Ca^{2+} concentrations are powerful triggers of eicosanoid production. Control over this vital early step can be exerted by a family of Ca^{2+}-binding anti-phospholipase proteins, including the *lipocortins*. Lipocortin was discovered in the early 1980s (and called *macrocortin*, or *lipomodulin*) and can be produced by cells under the action of glucocorticoid steroid hormones. This explains the powerful anti-inflammatory effect of pharmacological doses of these steroids in such conditions as rheumatoid arthritis (Chapter 8, Section 8.3.4); the increased synthesis of lipocortin stops many (but not all) cells from generating prostaglandins and other pro-inflammatory eicosanoids.

Once arachidonic acid is available, the enzyme *cyclo-oxygenase* (sometimes called *prostaglandin H synthetase*) acts upon it to form two endoperoxides, PGG_2 and PGH_2 (Fig. 7.14). Cyclo-oxygenase is present in almost all mammalian tissues, uses molecular oxygen and requires haem for its action. It can be inhibited by aspirin and other members of the group of drugs known as non-steroidal anti-inflammatory drugs (NSAIDs). This is further discussed in Chapter 8, Section 8.3.4.

If crude or purified preparations of cyclo-oxygenase are treated with GSH and glutathione peroxidase to remove traces of lipid peroxides, added arachidonic acid substrate is not immediately oxidized by the enzyme: there is a lag period before rates of oxidation reach maximum. This lag period may be shortened or abolished by adding PGG_2 or several other hydroperoxides to the system; effective ones include those formed by the action of lipoxygenase (Section 7.6.2) and during the non-enzymic peroxidation of arachidonic acid (Chapter 4). Indeed, commercially-available samples of arachidonic acid are peroxidized to variable extents and can therefore produce different initial rates of reaction when used in assays for cyclo-oxygenase activity. Cumene hydroperoxide, and even H_2O_2, can also activate cyclo-oxygenase, but higher concentrations are required. The spawning of abalones (herbivorous marine snails valued as a food source in certain parts of the world) can be induced by addition of 5 mM H_2O_2 to their sea-water, apparently by stimulation of prostaglandin synthesis.

Hence there is an intimate relationship between non-enzymic lipid peroxidation and prostaglandin metabolism, in that very efficient antioxidant protection will slow down prostaglandin synthesis, at least until sufficient PGG_2 is formed to activate cyclo-oxygenase maximally. It seems that a trace of hydroperoxide is required to react with iron(III) haem at the active site of the enzyme, forming a peroxy radical that can stereospecifically abstract a hydrogen atom from arachidonic acid to start off the process of PGG_2 formation. By contrast, an excess of lipid peroxides can inactivate cyclo-oxygenase. Lands *et al.* in the USA have suggested that the 'peroxide tone' of the cell (i.e. its content of lipid hydroperoxides, as determined by the balance between hydroperoxide generation by enzymic and non-enzymic mechanisms and the rate of hydroperoxide removal) can control the activity of cyclo-oxygenase and hence the rate of prostaglandin synthesis. The supply of arachidonic acid is also an important regulatory mechanism, as noted above.

PGG_2 is then converted into PGH_2 (Fig. 7.15) by a peroxidase that co-purifies with cyclo-oxygenase, since it is part of the same protein. This peroxidase is non-specific, in that it can reduce several different hydroperoxides whilst co-oxidizing a variety of other substances. It appears to operate by a cycle of reactions similar to that of horseradish peroxidase (Chapter 3, Section 3.1.1), i.e. reaction with hydroperoxide to give a compound I that can accept one electron to give a compound II. For example, during oxidation of PGG_2, this peroxidase activity can co-oxidize methional into ethene, oxidize adrenalin or diphenylisobenzofuran (yet another example of the lack of specificity of this compound as a 'singlet oxygen scavenger'; Chapter 2, Section 2.8.1), activate several aromatic amines to mutagenic products, convert the carcinogen benzpyrene into a quinone (Chapter 8, Section 8.8.3), cause the emission of light from luminol, and oxidize bisulphite ion (HSO_3^-) into sulphur trioxide anion (sulphite) radical (SO_3^{-}). The precise metabolic importance of most of these co-oxidations is unclear, however.

If the peroxidase activity of cyclo-oxygenase is not provided with an oxidizable substrate during conversion of PGG_2 to PGH_2, then the haem-associated ferryl species present in compound I, previously called '$[O_x]$', appears to break down and can cause inactivation of the cyclo-oxygenase itself. Thus appropriate concentrations of phenols, adrenalin, and other substrates can cause enhanced prostaglandin synthesis in some tissue systems, by preventing this self-inactivation. Some anti-inflammatory drugs, such as sulindac or 5-aminosalicylate, might exert some of their actions *in vivo* by acting as substrates for the peroxidase and removing $[O_x]$ (Chapter 8, Section 8.3.4).

Thus antioxidants of different types can have complex actions on prostaglandin synthesis, not only by regulating the 'peroxide tone' of the system but also by preventing self-inactivation of cyclo-oxygenase. High concentrations of antioxidants can also slow down cyclo-oxygenase by preventing formation of the traces of hydroperoxides that shorten the lag period. For example, variations in dietary vitamin E intake have been shown to affect eicosanoid production by platelets; increased dietary vitamin E intake in rats also leads to decreased activity of phospholipase A_2 in platelets.

Both PGG_2 and PGH_2 are unstable, having a half-life of only minutes under physiological conditions. They can, in theory, exert profound biological effects (including contraction of smooth muscle and stimulation of platelet aggregation) if they are not rapidly transformed into other products such as PGE_2, PGD_2, PGF_{2a}, TXA_2, and prostacyclin (Fig. 7.15). The precise fate of PGG_2 and PGH_2 depends on the enzymes present in each tissue. For example, platelets form predominantly TXA_2 and 12-hydroxy-5,8,10-heptadecatrienoic acid (HHT), whereas endothelial cells generate predominantly prostacyclin and mast cells make PGD_2. Whilst a detailed discussion of the tissue-specific complexities of prostaglandin biosynthesis is not appropriate here, it is worth saying a little about platelets and endothelial cells, since prostaglandins seem to play important roles in controlling blood vessel tone and the behaviour of platelets.

Platelets play a vital function in haemostasis to prevent blood loss from injured vessels; they rapidly adhere to collagen and basement membrane exposed when endothelial cells are damaged, and they can aggregate together. The processes of adhesion and aggregation are triggered by changes in the surface chemistry of platelets, and by the release from them of pro-aggregatory mediators. These include ADP (released from the 'dense granules' of the platelets) and TXA_2, which is formed upon platelet activation as a result of Ca^{2+}-dependent activation of phospholipase. Other mediators, such as platelet-activating factor, may also be important in certain species, as well as thrombin from the clotting blood. Platelets may also interact in a complex way with phagocytic cells during inflammation. For example, macrophages secrete platelet-activating factor (Section 7.3) and ADP released by platelets has been claimed to enhance O_2^- production by human neutrophils.

TXA_2 is unstable (half-life about 20 seconds under physiological conditions) and quickly forms the stable biologically-inactive thromboxane B_2 (TXB_2; Fig. 7.15). TXA_2 is a potent vasoconstrictor and platelet-aggregatory agent which aids the haemostatic process by assisting platelet clumping and decreasing blood flow. Its short half-life means that its action is localized. Low doses of aspirin can prolong bleeding time after vessel injury by decreasing TXA_2 production (due to cyclo-oxygenase inhibition) and it has been suggested that such low doses might be helpful in decreasing the incidence of thrombotic events in atherosclerosed arteries, thus reducing the risk of myocardial infarction (Chapter 8, Section 8.4.2). TXA_2 is also produced by activated neutrophils in some species.

The enzyme thromboxane synthetase (Fig. 7.15) cleaves PGH_2 to HHT plus a three-carbon compound, probably malondialdehyde (MDA). The function of HHT is not clear, but it has also been found in lung tissue and seminal vesicles. MDA reacts with thiobarbituric acid to give a pink chromogen. Prostaglandin endoperoxides decompose under the acid-heating conditions of the TBA test (Chapter 4, Section 4.6.8) to form MDA. Here application of the TBA test to tissues synthesizing prostaglandins might overestimate their MDA content. For example, the amount of 'TBA-reactive' material found in rabbit serum decreases if the animals are pretreated with aspirin to inhibit cyclo-oxygenase, suggesting that prostaglandins are a significant contributor to the observed TBA-reactivity. This should be borne in mind in attempts to use TBA-reactivity of body fluids as an index of lipid peroxidation *in vivo* in various disease states. Indeed, the generation of TBA-reactivity by lipid hydroperoxides may be due to the formation and decomposition of cyclic endoperoxides during the assay (Chapter 4, Section 4.6.8).

The endothelial cells which line blood vessels synthesize prostacyclin (PGI_2), which dilates blood vessels and is a powerful inhibitor of platelet aggregation. Its actions therefore oppose those of TXA_2, and it is likely that the ratio of thromboxane to prostacyclin is important in controlling platelet-vessel wall interactions and local blood flow. Prostacyclin is unstable (half-time of minutes), decomposing to give 6-keto-PGF_{1a} (Fig. 7.15). The capacity of vessel walls to synthesize prostacyclin *in vivo* seems less affected by aspirin than is the ability of platelets to synthesize TXA_2, so that controlled doses of aspirin may be useful in the prevention of unwanted thromboses, as mentioned above. Prostacyclin has other important effects, such as to potentiate pain and swelling at sites of inflammation, and to affect fluid transport in the kidney and gut. It is cytoprotective to the stomach mucosa, and decreases gastric-acid secretion. All these properties are shared by prostaglandins of the E-series, and probably depend on their binding to cell-membrane receptors, leading to activation of the enzyme adenylate cyclase and hence increased intracellular formation of cyclic AMP.

Formation of cyclic GMP from GTP by the enzyme guanylate cyclase appears to be stimulated by systems generating oxygen radicals, by prostaglandin endoperoxides and other fatty-acid hydroperoxides and by oxidation of thiols. However, the physiological significance of these effects is not yet clear. Several vasodilator drugs cause relaxation of blood vessels by acting on guanylate cyclase—examples are glyceryl trinitrate ('nitroglycerine') and nitroprusside. Endothelial-derived relaxing factor may also act in this way (Section 7.7).

7.6.2   *Leukotrienes and other lipoxygenase products*

In Section 7.5 we described the importance of lipoxygenase enzymes as initiators of 'controlled lipid peroxidation' in the wound response of plant tissues. In 1974, it was discovered that a lipoxygenase enzyme is active in platelets, and similar enzymes have since been studied in many mammalian tissues. Like plant lipoxygenases, the animal enzymes contain iron.

Platelet lipoxygenase acts upon arachidonic acid, its natural substrate, to form 12-hydroperoxy-5,8,11,14-eicosatetraenoic acid (12-HPETE). This is unstable and is quickly reduced to the 12-hydroxy derivative (12-HETE) *in vivo*. Reduction appears to be achieved by platelet glutathione peroxidase activity in the presence of GSH, since 12-HETE formation is decreased in the platelets of selenium-deficient rats.

Lipoxygenases present in other tissues introduce oxygen atoms at different places in the carbon chain (Fig. 7.16). Formation of 12-HETE has been observed in skin, 11-HETE in lung and mast cells, 5-HETE and 15-HETE in rabbit and human neutrophils, and various 5,12-DHETE S (specific stereoisomers of 5,12-dihydroxy-6,8,10,14-eicosatetraenoic acids) in phagocytes. Indeed, the lipoxygenase pathway for the metabolism of arachidonic acid in phagocytes appears to be more important biologically than is the cyclooxygenase route, and the resulting leukotriene products have important biological actions (see below). At concentrations effective against cyclo-

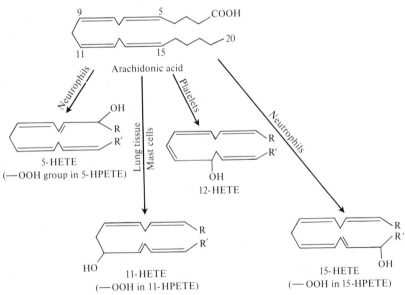

Fig. 7.16.   Conversion of arachidonic acid catalysed by mammalian lipoxygenases.

oxygenase, aspirin and many other NSAIDs do not inhibit lipoxygenase. It is therefore feasible for them to 'divert' the metabolism of arachidonic acid from cyclo-oxygenase to the lipoxygenase route. By contrast, anti-inflammatory steroids will inhibit both pathways by decreasing release of the fatty-acid substrate.

Animal lipoxygenases are inhibited by the antioxidant NDGA (Chapter 4, Section 4.7.4), which appears to act by reducing Fe(III) at the active site of the enzyme to Fe(II), so decreasing the turnover rate. Several phenolic antioxidants (e.g. some flavonoids; Chapter 4, Section 4) are also inhibitors of lipoxygenase, but the relationship between their ability to inhibit lipoxygenase and their antioxidant and iron-binding properties has not yet been fully worked out, although it seems very likely that some relationship exists. Lipoxygenase can also be inhibited by 5,8,11,14-eicosatetraynoic acid (ETYA), an analogue of arachidonic acid in which all the double bonds have been replaced by triple bonds, so preventing hydrogen atom abstraction at the enzyme active site. ETYA also inhibits cyclo-oxygenase.

The HPETE compounds are precursors of a range of chemicals with potent biological activity known as the *leukotrienes*. Leukotrienes differ structurally from prostanoids in that the leukotrienes have no cyclopentane ring. Leukotrienes have a conjugated triene structure (three double bonds separated from each other by single bonds; Fig. 7.17) and leukotrienes C_4, D_4, and E_4 also possess a peptide substituent containing one, two, or three amino-acid residues linked to the molecule by an atom of sulphur (Fig. 7.17).

Figure 7.17 shows the leukotrienes that arise from 5-HPETE, but others can be generated enzymically or non-enzymically from different HPETEs. LTA_4 is an unstable epoxide structure (Fig. 7.17) that is, in some tissues, enzymatically hydrolysed to LTB_4. This pathway is especially important in neutrophils, as LTB_4 is a powerful chemotactic agent for these cells and promotes aggregation, degranulation, and the 'respiratory burst'. Thus neutrophils themselves generate a mediator which attracts more such cells to a site of inflammation and promotes their activation. Human alveolar macrophages produce LTB_4 and other leukotrienes, as do lung mast cells. These mediators may contribute to the inflammation and bronchial constriction seen in allergic asthma (along with other mediators such as PGD_2, platelet-activating factor, and histamine). Activated phagocytes can also destroy leukotrienes; they can be oxidized and inactivated by HOCl produced by the myeloperoxidase system in neutrophils. If suitable metal ions are present, O_2^- and H_2O_2 produced by activated phagocytes can interact to form OH·, which can attack leukotrienes (and most other molecules).

Leukotrienes C_4 and D_4 are the principal components of 'slow-reacting substance A (SRS-A), a mixture of compounds produced when lung tissue taken from humans allergic to a given substance is perfused with that substance. Other tissues can also produce SRS-A, whose effects are to

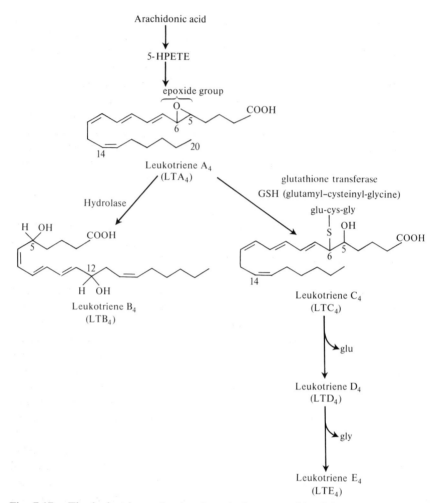

Fig. 7.17. The leukotrienes. Leukotriene A$_4$ is an epoxide (full name: 5,6-epoxy-7,9,11,14-eicosatetraenoic acid). It can be hydrolysed enzymatically into leukotriene B$_4$ (5,12-dihydroxy-6,*cis*-8,*trans*-10,*trans*-14,*cis*-eicosatetraenoic acid) or non-enzymically to other 5,12- and 5,6-dihydroxyacids. Only LTB$_4$ is shown for simplicity. LTA$_4$ is also converted by conjugation with GSH into LTC$_4$, which can be converted into LTD$_4$ and LTE$_4$ by successive removal of amino acids. LTC$_4$ and LTD$_4$ are responsible for the biological activity of most preparations of slow-reacting substance of anaphylaxis (SRS-A). Another group of leukotrienes can be formed by initial oxygenation of arachidonic acid at C$_{15}$. 'SRS-A' is a substance produced during allergic reactions that causes increased vascular permeability and, in lung, prolonged spasm of smooth muscle. Other hydroxylated compounds such as trihydroxy(20-OH-LTB$_4$) and dicarboxylate (20-COOH-LTB$_4$) derivatives have also been observed to be formed in neutrophils. The same enzyme (5-lipoxygenase) is thought both to generate 5-HPETE and to convert it into LTA$_4$.

increase vascular permeability and contract bronchial muscles in a slow and sustained way (unlike the rapid and faster-acting action of histamine). LTC_4 is formed by the conjugation of LTA_4 with GSH in the presence of a glutathione transferase enzyme. Removal of a glutamic acid residue by glutamyltransferase activity yields LTD_4, which can be further degraded to LTE_4. This latter compound appears less biologically active.

In the next few years much more information about the chemical nature and biological properties of leukotrienes will surely become available. For example, incubating neutrophils with 15-HPETE has been shown to generate two novel trihydroxytetraene metabolites called *lipoxins*. They result from the combined actions of 5- and 12-lipoxygenases. Lipoxins have been reported to inhibit the cytotoxic actions of natural killer cells (Chapter 8, Section 8.8.2) and to have properties that distinguish them from other leukotrienes or prostaglandins. It is interesting to note that *in vitro*, both 15-HPETE and 12-HPETE are substrates for the peroxidase activity of cyclo-oxygenase, indicating an interaction between the two pathways and non-enzymic lipid peroxidation (Section 7.6).

Before leaving this area, we wish to re-emphasize that polyunsaturated fatty acids such as arachidonate are easily peroxidized by non-enzymic methods (Chapter 4). Commercially-available samples may be extensively peroxidized, thus causing variable results in, for example, cyclo-oxygenase assays. If such fatty acids are included in a system generating oxygen radicals, then products can be formed not only by cyclo-oxygenase and lipoxygenase pathways but also by the attack of species such as hydroxyl radical. This has already caused confusion in studies of arachidonate metabolism by activated neutrophils, which can produce OH· if a suitable metal ion catalyst is available in the surrounding fluid. Fortunately, products formed by enzyme-catalysed reactions are highly stereospecific, whereas those from non-enzymic reactions usually are not, and determinations of stereochemical purity should be made before assuming that a novel oxidized product has been produced by the action of an enzyme. The rat liver microsomal cytochrome P-450 system has been claimed to hydroxylate arachidonic acid into 9-, 11-, 12-, and 15-HETEs, although it is unclear if this has any physiological importance.

7.7 Endothelial-derived relaxing factor (EDRF)

The endothelial cells that line blood vessels perform many important metabolic functions in addition to their role as a 'barrier'. These functions include synthesis of glycoproteins, interleukin-1, prostacyclin, 15-HETE, and platelet-activating factor. Control of the production of these substances depends on interactions between the endothelium and phagocytes, platelets, or components of plasma. Recently it has become clear that many neurohormones that cause blood vessels to relax do so by releasing an unstable non-eicosanoid mediator from endothelial cells, rather than by a direct action on the vascular smooth muscle. For example, both acetylcholine and bradykinin cause the release of the so-called 'endothelial-derived relaxing factor' (EDRF). EDRF is extraordinarily labile (half-life about 10 seconds), and also inhibits platelet aggregation. It is therefore another local hormone affecting the cardiovascular system.

In 1986, Gryglewski, Palmer, and Moncada, in England, observed that the stability of EDRF in isolated tissue preparations was increased by adding superoxide dismutase, implying that O_2^- is involved in its breakdown. Catalase had no effect, suggesting that O_2^- was not acting by leading to production of OH˙. This is a further example of the inter-relationship between oxygen radicals and other products formed at sites of injury. Evidence is now accumulating to suggest that EDRF is very similar, if not identical, to nitric oxide (NO), and that, like other nitro-vasodilators, it acts by stimulating guanylate cyclase (Section 7.6.1). How the NO is produced and its breakdown controlled remain to be established; the amino acid arginine may be a precursor. It is possible that the endothelium constantly produces O_2^- and NO *in vivo* in balanced quantities, so that changes in the production of either could modulate vessel tone.

The vascular endothelium seems to be very sensitive to oxidative damage. Thus low concentrations of H_2O_2 inhibit prostacyclin synthesis by cultured porcine endothelial cells, and the presence of xanthine oxidase activity within endothelium has been claimed to play an important role in reperfusion injury (Chapter 8, Section 8.4).

7.8 Further reading

Albrich, J. M. *et al.* (1986). Effect of the putative neutrophil-generated toxin hypochlorous acid on membrane permeability and transport systems of *Escherichia coli*. *J. Clin. Invest.* **78**, 177.

Aneshansley, D. J. *et al.* (1983). Thermal concomitants and biochemistry of the explosive discharge mechanism of some little known bombardier beetles. *Experientia* **39**, 366.

Babior, B. M. (1978). Oxygen-dependent microbial killing by phagocytes. *New Engl. J. Med.* **298**, 721 and 645.

Bass, D. A. and Szejda, P. (1979). Mechanisms of killing of newborn larvae of *Trichinella spiralis* by neutrophils and eosinophils. *J. Clin. Invest.* **64**, 1558.

Bhatnagar, R. *et al.* (1981). Superoxide release by zymosan-stimulated rat Kupffer cells *in vitro*. *Eur. J. Biochem.* **119**, 171.

Brennan, R. and Frenkel, C. (1977). Involvement of hydrogen peroxide in the regulation of senescence in pears. *Plant Physiol.* **59**, 411.

Briheim, G., Stendahl, O., and Dahlgren, C. (1984). Intra- and extracellular events in luminol-dependent chemiluminescence of polymorphonuclear leukocytes. *Infect. Immunol.* **45**, 1.

Britigan, B. E. and Cohen, M. S. (1986). Effects of human serum on bacterial competition with neutrophils for molecular oxygen. *Infect. Immunol.* **52**, 657.

Britigan, B. E., Cohen, M. S., and Rosen, G. M. (1987). Detection of the production of oxygen-centered free radicals by human neutrophils using spin trapping techniques: a critical perspective. *J. Leukocyte Biol.* **41**, 349.

Burke, T. M. and Wolin, M. S. (1987). Hydrogen peroxide elicits pulmonary arterial relaxation and guanylate cyclase activation. *Am. J. Physiol.* **252**, H721.

Cadenas, E., Daniele, R. P., and Chance, B. (1981). Low-level chemiluminescence of macrophages. *FEBS Lett.* **125**, 225.

Carpenter, M. P. (1981). Antioxidant effects on the prostaglandin endoperoxide synthetase product profile. *Fed. Proc.* **40**, 189.

Cech, P. and Lehrer, R. I. (1984). Phagolysosomal pH of human neutrophils. *Blood* **63**, 88.

Clark, R. A. and Borregaard, N. (1985). Neutrophils autoinactivate secretory products by myeloperoxidase-catalysed oxidation. *Blood* **65**, 375.

Clark, R. A. *et al.* (1981). Myeloperoxidase-catalysed inactivation of a_1-protease inhibitor by human neutrophils. *J. Biol. Chem.* **256**, 3348.

Colton, C. A. and Gilbert, D. L. (1987). Production of superoxide anions by a CNS macrophage, the microglia. *FEBS Lett.* **223**, 284.

Cuperus, R. A., Muijsers, A. O., and Wever, R. (1986). The superoxide dismutase activity of myeloperoxidase; formation of compound III. *Biochim. Biophys. Acta* **871**, 78.

Curnutte, J. T. and Babior, B. M. (1987). Chronic granulomatous disease. *Adv. Human Genetics* **16**, 229.

Dahlgren, C. (1988). Effects on extra- and intracellularly localized, chemoattractant-induced, oxygen radical production in neutrophils following modulation of conditions for ligand-receptor interaction. *Inflammation* **12**, 335.

Del Maestro, R. F., Planker, M., and Arfors, K. E. (1982). Evidence for the participation of superoxide anion radical in altering the adhesive interaction between granulocytes and endothelium *in vivo*. *Intl. J. Microcirc. Clin. Exp.* **1**, 105.

Deme, D. *et al.* (1985). NADPH-dependent generation of H_2O_2 in a thyroid particulate fraction requires Ca^{2+}. *FEBS Lett.* **186**, 107.

Dix, T. A., Kuhn, D. M., and Benkovic, S. J. (1987). Mechanism of oxygen activation by tyrosine hydroxylase. *Biochemistry* **26**, 3354.

Docampo, R., Moreno, S. N. J., and Mason, R. P. (1987). Free radical intermediates in the reaction of pyruvate-ferredoxin oxidoreductase in *Tritrichomonas foetus* hydrogenosomes. *J. Biol. Chem.* **262**, 12417.

Douglas, C. E., Chan, A. C., and Choy, P. C. (1986). Vitamin E inhibits platelet phospholipase A_2. *Biochim. Biophys. Acta* **876**, 639.

Edwards, J. E. Jr. *et al.* (1987). Neutrophil-mediated protection of cultured human vascular endothelial cells from damage by growing *Candida albicans* hyphae. *Blood* **69**, 1450.

Edwards, S. W., Hallett, M. B., and Campbell, A. K. (1984). Oxygen-radical formation during inflammation may be limited by oxygen concentration. *Biochem. J.* **217**, 851.

Ellis, J. A., Mayer, S. J., and Jones, O. T. G. (1988). The effect of the NADPH oxidase inhibitor diphenyleneiodonium on aerobic and anaerobic microbicidal activities of human neutrophils. *Biochem. J.* **251**, 887.

Filley, W. Y. *et al.* (1982). Identification by immunofluorescence of eosinophil granule major basic protein in lung tissues of patients with bronchial asthma. *Lancet* **ii**, 11.

Fletcher, M. P. and Seligmann, B. E. (1986). PMN heterogeneity: long-term stability of fluorescent membrane potential responses to the chemoattractant *N*-formyl-methionyl-leucyl-phenylalanine in healthy adults and correlation with respiratory burst activity. *Blood* **68**, 611.

Fliss, H., Weissbach, H., and Brot, N. (1983). Oxidation of methionine residues in proteins of activated human neutrophils. *Proc. Natl. Acad. Sci. USA* **80**, 7160.

Foerder, C. A., Klebanoff, S. J., and Shapiro, B. M. (1978). Hydrogen peroxide production, chemiluminescence and the respiratory burst of fertilisation. Interrelated events in early sea urchin development. *Proc. Natl. Acad. Sci. USA* **75**, 3183.

Fontecave, M., Graslund, A., and Reichard, P. (1987). The function of superoxide dismutase during the enzymatic formation of the free radical of ribonucleotide reductase. *J. Biol. Chem.* **262**, 12332.

Fontecave, M. *et al.* (1987). The stimulatory effect of asbestos on NADPH-dependent lipid peroxidation in rat liver microsomes. *Biochem. J.* **241**, 561.

Foote, C. S., Goyne, T. E., and Lehrer, R. I. (1983). Assessment of chlorination by human neutrophils. *Nature* **301**, 715.

Forman, H. J., Nelson, J., and Fisher, A. B. (1980). Rat alveolar macrophages require NADPH for superoxide production in the respiratory burst. *J. Biol. Chem.* **255**, 9879.

Galliard, T. (1978). Lipolytic and lipoxygenase enzymes in plants and their action in wounded tissue. In *Biochemistry of wounded plant tissues* (G. Kahl, Ed.), p. 155. Walter de Gruyter & Co, Berlin.

Gay, J. C. *et al.* (1986). Modulation of neutrophil oxidative responses to soluble stimuli by platelet-activating factor. *Blood* **67**, 931.

Glenn, J. K., Akileswaran, L., and Gold, M. H. (1986). Mn(II) oxidation is the principal function of the extracellular Mn-peroxidase from *Phanerochaete chrysosporium*. *Arch. Biochem. Biophys.* **251**, 688.

Goldenberg, H. (1982). Plasma membrane redox activities. *Biochim. Biophys. Acta* **694**, 203.

Gormley, I. P., Kowolik, M. J., and Cullen, R. T. (1985). The chemiluminescent response of human phagocytic cells to mineral dusts. *Br. J. Exp. Pathol.* **66**, 409.

Green, M. J., Hill, H. A. O., and Tew, D. G. (1987). The rate of oxygen consumption and superoxide anion formation by stimulated human neutrophils. *FEBS Lett.* **216**, 31.

Grisham, M. B. *et al.* (1984). Chlorination of endogenous amines by isolated neutrophils. *J. Biol. Chem.* **259**, 10404.

Gross, G. G., Janse, C., and Elstner, E. F. (1977). Involvement of malate, monophenols and the superoxide radical in hydrogen peroxide formation by isolated cell walls from horseradish. *Planta* **136**, 271.

Gryglewski, R. J., Palmer, R. M. J., and Moncada, S. (1986). Superoxide anion is involved in the breakdown of endothelium-derived vascular relaxing factor. *Nature* **320**, 454.

Gulumian, M. and Van Wyk, J. A. (1987). Hydroxyl radical production in the presence of fibres by a Fenton-type reaction. *Chem–Biol. Interact.* **62**, 89.

Gutteridge, J. M. C. and Kerry, P. J. (1982). Detection by fluorescence of peroxides and carbonyls in samples of arachidonic acid. *Br. J. Pharmacol.* **76**, 459.

Halliwell, B. (1977). Superoxide and hydroxylation reactions. In *Superoxide and superoxide dismutases* (A. M. Michelson, J. M. McCord, and I. Fridovich, Eds), p. 335. Academic Press, London.

Halliwell, B. (1978). Lignin synthesis: the generation of hydrogen peroxide and superoxide by horseradish peroxidase and its stimulation by manganese(II) and phenols. *Planta* **140**, 81.

Halliwell, B. (1988). Albumin—an important extracellular antioxidant? *Biochem. Pharmacol.* **37**, 569.

Hamers, M. N. *et al.* (1984). Kinetics and mechanism of the bactericidal action of human neutrophils against *Escherichia coli. Blood* **64**, 635.

Harrison, J. E., Watson, B. D., and Schultz, J. (1978). Myeloperoxidase and singlet oxygen: a reappraisal. *FEBS Lett.* **92**, 327.

Hatanaka, A., Kajiwara, T., and Sekiya, J. (1987). Biosynthetic pathway for C_6-aldehydes formation from linolenic acid in green leaves. *Chem. Phys. Lipids* **44**, 341.

Hayes, G. R. and Lockwood, D. H. (1987). Role of insulin receptor phosphorylation in the insulinomimetic effects of hydrogen peroxide. *Proc. Natl. Acad. Sci. USA* **84**, 8115.

Henderson, L. M., Chappell, J. B., and Jones, O. T. G. (1988). Internal pH changes associated with the activity of NADPH oxidase of human neutrophils. Further evidence for the presence of an H^+ conducting channel. *Biochem. J.* **251**, 563.

Henderson, W. R. and Klebanoff, S. J. (1983). Leukotriene production and inactivation by normal, chronic granulomatous disease and myeloperoxidase-deficient neutrophils. *J. Biol. Chem.* **258**, 13522.

Henry, J. P., Monny, C., and Michelson, A. M. (1975). Characterisation and properties of *Pholas* luciferase as a metalloglycoprotein. *Biochemistry* **14**, 3458.

Hocking, W. G. and Golde, D. W. (1979). The pulmonary-alveolar macrophage. *New Engl. J. Med.* **301**, 580 and 639.

Holme, E. *et al.* (1982). Does superoxide anion participate in 2-oxoglutarate-dependent hydroxylation? *Biochem. J.* **205**, 339.

Hume, D. A. *et al.* (1984). The mononuclear phagocyte system of the mouse defined by immunohistochemical localization of antigen F4/80: macrophages of endocrine organs. *Proc. Natl. Acad. Sci. USA* **81**, 4174.

Johnston, R. B. *et al.* (1975). The role of superoxide anion generation in phagocyte bactericidal activity. Studies with normal and chronic granulomatous disease leukocytes. *J. Clin. Invest.* **55**, 1357.

Johnston, R. B. Jr (1988). Monocytes and macrophages. *New Engl. J. Med.* **318**, 747.

412   *Free radicals as useful species*

Kalyanaraman, B. *et al.* (1982). The free radical formed during the hydroperoxide-mediated deactivation of ram seminal vesicles is hemoprotein-derived. *J. Biol. Chem.* **257**, 4764.

Kanabus-Kaminska, J. M. and Girardot, J. M. (1984). Inhibition of vitamin K-dependent carboxylase by metal ions and metal complexes: a reassessment. *Arch. Biochem. Biophys.* **228**, 646.

Kanofsky, J. R. and Axelrod, B. (1986). Singlet oxygen production by soybean lipoxygenase isozymes. *J. Biol. Chem.* **261**, 1099.

Kanofsky, J. R. *et al.* (1984). Biochemical requirements for singlet oxygen production by purified human myeloperoxidase. *J. Clin. Invest.* **74**, 1489.

Kanofsky, J. R. *et al.* (1988). Singlet oxygen production by human eosinophils. *J. Biol. Chem.* **263**, 9692.

Kemal, C. *et al.* (1987). Reductive inactivation of soybean lipoxygenase I by catechols: a possible mechanism for regulation of lipoxygenase activity. *Biochemistry* **26**, 7064.

Kerscher, L. and Oesterhelt, D. (1982). Pyruvate: ferredoxin oxidoreductase—new findings on an ancient enzyme. *Trends Biochem. Sci.* **7**, 371.

Kettle, A. J. and Winterbourn, C. C. (1988). Superoxide modulates the activity of myeloperoxidase and optimizes the production of hypochlorous acid. *Biochem. J.* **252**, 529.

Knappe, J. *et al.* (1984). Post-translational activation introduces a free radical into pyruvate formate-lyase. *Proc. Natl. Acad. Sci. USA* **81**, 1332.

Konze, J. R. and Elstner, E. F. (1978). Ethane and ethylene formation by mitochondria as indication of aerobic lipid degradation in response to wounding of plant tissues. *Biochim. Biophys. Acta* **528**, 312.

Lambert, A. M. *et al.* (1985). Spectral properties of the higher oxidation states of prostaglandin H synthase. *J. Biol. Chem.* **260**, 14894.

Locksley, R. M., Wilson, C. B., and Klebanoff, S. J. (1983). Increased respiratory burst in myeloperoxidase-deficient monocytes. *Blood*, **62**, 902.

Lutter, R. *et al.* (1984). Cytochrome b, flavins and ubiquinone-50 in enucleated human neutrophils (polymorphonuclear leukocyte cytoplasts). *J. Biol. Chem.* **259**, 9603.

Mack, A. J., Peterman, T. K., and Siedow, J. N. (1987). Lipoxygenase isozymes in higher plants: biochemical properties and physiological role. In *Isozymes: current topics in biological and medical research* **13**, 127.

Makins, R. *et al.* (1986). Stoichiometric conversion of oxygen to superoxide anion during the respiratory burst in neutrophils. *J. Biol. Chem.* **261**, 11444.

Marcus, A. J. *et al.* (1977). Superoxide production and reducing activity in human platelets. *J. Clin. Invest.* **59**, 149.

Matheson, N. R. and Travis, J. (1985). Differential effects of oxidizing agents on human plasma a_1-proteinase inhibitor and human neutrophil myeloperoxidase. *Biochemistry* **24**, 1941.

McKenna, S. M. and Davies, K. J. A. (1988). The inhibition of bacterial growth by hypochlorous acid. Possible role in the bactericidal activity of phagocytes. *Biochem. J.* **254**, 685.

Milks, L. C., Conyers, G. P., and Cramer, E. B. (1986). The effect of neutrophil migration on epithelial permeability. *J. Cell Biol.* **103**, 2729.

Monboisse, J. C. *et al.* (1987). Collagen activates superoxide production by human polymorphonuclear neutrophils. *Biochem. J.* **246**, 599.

Moore, P. K. (1985). *Prostanoids: pharmacological, physiological and clinical relevance.* Cambridge University Press, England.

Morita, Y. *et al.* (1986). Crystallization and properties of myeloperoxidase from normal human leukocytes. *J. Biochem. (Tokyo)* **99**, 761.

Mossman, B. T., Marsh, J. P., and Shatos, M. A. (1986). Alterations of superoxide dismutase activity in tracheal epithelial cells by asbestos and inhibition of cytotoxicity by antioxidants. *Lab. Invest.* **54**, 204.

Muchmore, D. B., Little, S. A., and de Haen, C. (1982). Counter-regulatory control of intracellular hydrogen peroxide production by insulin and lipolytic hormones in isolated rat epididymal fat cells: a role of free fatty acids. *Biochemistry* **21**, 3886.

Mukherjee, S. P. and Mukherjee, C. (1982). Similar activities of nerve growth factor and its homologue proinsulin in intracellular hydrogen peroxide production and metabolism in adipocytes. *Biochem. Pharmacol.* **31**, 3163.

Nakagawa, A., Nathan, C. F., and Cohn, Z. A. (1981). Hydrogen peroxide metabolism in human monocytes during differentiation *in vitro. J. Clin. Invest.* **68**, 1243.

Nakamura, M. *et al.* (1985). Thyroid peroxidase selects the mechanism of either 1- or 2-electron oxidation of phenols, depending on their substituents. *J. Biol. Chem.* **260**, 13546.

Nathan, C. F. *et al.* (1985). Administration of recombinant interferon γ to cancer patients enhances monocyte secretion of hydrogen peroxide. *Proc. Natl. Acad. Sci. USA* **82**, 8686.

Newburger, P. E. and Tauber, A. I. (1982). Heterogeneous pathways of oxygen radical production in human neutrophils and the HL-60 cell line. *Pediatr. Res.* **16**, 856.

Ohno, Y. and Gallin, J. I. (1985). Diffusion of extracellular hydrogen peroxide into intracellular compartments of human neutrophils *J. Biol. Chem.* **260**, 8438.

Palmer, R. M. J., Ashton, D. S., and Moncada, S. (1988). Vascular endothelial cells synthesize nitric oxide from L-arginine. *Nature* **333**, 664.

Petreccia, D. C., Nauseef, W. M., and Clark, R. A. (1987). Respiratory burst of normal human eosinophils. *J. Leukocyte Biol.* **41**, 283.

Rapoport, S. M. and Schewe, T. (1986). The maturational breakdown of mitochondria in reticulocytes. *Biochim. Biophys. Acta* **864**, 471.

Reichard, P. and Ehrenberg, A. (1983). Ribonucleotide reductase—a radical enzyme. *Science* **221**, 514.

Renganathan, V. and Gold, M. H. (1986). Spectral characterization of the oxidized states of lignin peroxidase, an extracellular heme enzyme from the white rot basidiomycete *Phanerochaete chrysosporium. Biochemistry*, **25**, 1626.

Repine, J. E., Fox, R. B., and Berger, E. M. (1981). Hydrogen peroxide kills *Staphylococcus aureus* by reacting with staphylococcal iron to form hydroxyl radical. *J. Biol. Chem* **256**, 7094.

Roberts, P. A., Knight, J., and Campbell, A. K. (1987). Pholasin—a bioluminescent indicator for detecting activation of single neutrophils. *Anal. Biochem.* **160**, 139.

Rosen, H. *et al.* (1987). Myeloperoxidase-mediated damage to the succinate oxidase system of *Escherichia coli. J. Biol. Chem.* **262**, 15004.

Rouzer, C. A. *et al.* (1982). Glutathione metabolism in resting and phagocytozing peritoneal macrophages. *J. Biol. Chem.* **257**, 2002.

Samuelsson, B. *et al.* (1987). Leukotrienes and lipoxins: structures, biosynthesis and biological effects. *Science* **237**, 1171.

Sasada, M., Pabst, M. J., and Johnston, R. B. (1983). Activation of mouse peritoneal macrophages by lipopolysaccharide alters the kinetic parameters of the superoxide-producing NADPH oxidase. *J. Biol. Chem.* **258**, 9631.

Segal, A. W. *et al.* (1981). The respiratory burst of phagocytic cells is associated with a rise in vacuolar pH. *Nature* **290**, 406.

Segal, A. W. *et al.* (1983). Absence of cytochrome b_{-245} in chronic granulomatous disease. *New Engl. J. Med.* **308**, 245.

Segal, A. W. *et al.* (1983). Iodination by stimulated human neutrophils. *Biochem. J.* **210**, 215.

Segal, A. W. *et al.* (1985). Stimulated neutrophils from patients with autosomal recessive chronic granulomatous disease fail to phosphorylate a M_r-44,000 protein. *Nature* **316**, 547.

Shimizu, T., Kondo, K., and Hayaishi, A. (1981). Role of prostaglandin endoperoxides in the serum thiobarbituric acid reaction. *Arch. Biochem. Biophys.* **206**, 271.

Stevens, P. and Hong, D. (1984). The role of myeloperoxidase and superoxide anion in the luminol- and lucigenin-dependent chemiluminescence of human neutrophils. *Anal. Biochem.* **30**, 135.

Sugahara, K. *et al.* (1986). Epithelial permeability produced by phagocytosing neutrophils *in vitro*. *Am. Rev. Resp. Dis.* **133**, 875.

Takemura, R. and Werb, Z. (1984). Secretory products of macrophages and their physiological functions. *Am. J. Physiol.* **246**, C1.

Thompson, J. E., Legge, R. L., and Barber, R. F. (1987). The role of free radicals in senescence and wounding. *New Phytol.* **105**, 317.

Till, G. O. *et al.* (1982). Intravascular activation of complement and acute lung injury. Dependence on neutrophils and toxic oxygen metabolites. *J. Clin. Invest.* **69**, 1126.

Vane, J. R., Gryglewski, R. J., and Botting, R. M. (1987). The endothelial cell as a metabolic and endocrine organ. *Trends Pharm. Sci.* **8**, 491.

Wagner, D. K., Collins-Lech, C., and Sohnle, P. G. (1986). Inhibition of neutrophil killing of *Candida albicans* pseudohyphae by substances which quench hypochlorous acid and chloramines. *Infect. Immunol.* **51**, 731.

Ward, P. A. *et al.* (1988). Platelet enhancement of O_2^- responses in stimulated human neutrophils: identification of platelet factor as adenine nucleotide. *Lab. Invest.* **58**, 37.

Washburn, R. G., Gallin, J. I., and Bennett, J. E. (1987). Oxidative killing of *Aspergillus fumigatus* proceeds by parallel myeloperoxidase-dependent and independent pathways. *Infect. Immunol.* **55**, 2088.

Wasil, M. *et al.* (1987). The antioxidant action of extracellular fluids. *Biochem. J.* **243**, 219.

Weiss, J. *et al.* (1985). Oxygen-independent intracellular and oxygen-dependent extracellular killing of *Escherichia coli S15* by human polymorphonuclear leukocytes. *J. Clin. Invest.* **76**, 206.

Weiss, S. J. and Peppin, G. (1986). Collagenolytic enzymes of the human neutrophil. *Biochem. Pharmacol.* **35**, 3189.

White, A. A., Karr, D. B., and Patt, C. S. (1982). Role of lipoxygenase in the O_2-dependent activation of soluble guanylate cyclase from rat lung. *Biochem. J.* **204**, 383.

Wharton, A. R., Montgomery, M. E., and Kent, R. S. (1985). Effect of hydrogen peroxide on prostaglandin production and cellular integrity in cultured porcine aortic endothelial cells. *J. Clin. Invest.* **76**, 295.

Yamazaki, I. (1987). Free radical mechanisms in enzyme reactions. *Free Radical Biol. Med.* **3**, 397.

Yazdanbakhsh, M. *et al.* (1986). Bactericidal action of eosinophils from normal human blood. *Infect. Immunol.* **53**, 192.

Zimmerman, D. C. and Coudron, C. A. (1979). Identification of traumatin, a wound hormone, as 12-oxo-*trans*-10-dodecenoic acid. *Plant Physiol.* **63**, 536.

8

Free radicals, ageing, and disease

8.1 General principles

We have seen that oxidants are of great importance in the mechanism of action of several toxins (Chapter 6). Their involvement in several disease states has already been considered, including porphyria (Chapter 2, Section 2.6), eye disorders such as cataract and retrolental fibroplasia (Chapter 5), iron and copper overload (Chapter 2, Section 2.4.3), Parkinson's disease (Chapter 6, Section 6.12), Alzheimer's disease (Chapter 6, Section 6.13), diabetes (Chapter 5 and Chapter 6, Section 6.2), emphysema (Chapter 6, Section 6.4), haemolytic diseases, and malaria (Chapter 6, Section 6.2). Table 8.1 lists some of the other conditions in which the involvement of oxidants has been suggested. Before going on to evaluate some of these proposals in detail, it is worth asking a few basic questions.

Does increased oxidant formation cause any human disease?
It is tempting to attribute the cardiomyopathy seen in Keshan disease (chronic selenium deficiency; Chapter 4, Section 4.7.5) to the lack of active glutathione peroxidase causing a failure to remove H_2O_2 and lipid peroxides at a sufficient rate *in vivo*. Although it is not strictly proven that this is the mechanism of tissue injury, patients suffering from Keshan disease do show very low selenium-dependent glutathione peroxidase activity in blood and organs. It is also possible that some cancers may originate as a result of DNA damage produced by free radicals (Section 8.8), and that the vessel weakness in retrolental fibroplasia is due to lipid peroxidation, since it can be ameliorated by vitamin E (Chapter 5). In most diseases, however, the increased oxidant formation is a consequence of the disease activity. For example, infiltration of a large number of neutrophils into a localized site, followed by activation of these cells to generate O_2^- and H_2O_2, can produce an intense localized oxidative stress. This appears to happen in rheumatoid arthritis (Section 8.2) and in some forms of the adult respiratory distress syndrome (Section 8.5). The question then arises as to what caused the phagocyte infiltration and activation in the first place.

In addition, many forms of tissue injury, including mechanical disruption, can cause increased free-radical reactions such as lipid peroxidation, by inactivating cellular antioxidants and by liberating reactive transition metal ions from their sites of sequestration within cells (Chapter 4, Section 4.9). Thus traumatic injury to the brain and spinal cord leads to increases in the availability of iron ions able to catalyse free-radical reactions, and often an associated increase in lipid peroxidation (Chapter 4, Section 4.9).

Does increased oxidant formation matter?
If increased oxidant formation is usually a consequence of disease, does it then make a significant contribution to the disease pathology or is it just an epiphenomenon? The answer probably differs in different diseases. For example, the increased lipid peroxidation seen after impact injury to the brain does seem to contribute to post-injury degeneration of the tissue (Chapter 4, Section 4.9). By contrast, the increased lipid peroxidation demonstrated in the wasted muscles of patients with muscular dystrophy may simply be a consequence of the tissue damage and make no contribution to it; there is no evidence that antioxidants are beneficial in this disease. To take a second example, human endothelial cells infected with *Rickettsia rickettsii* (the causative agent of Rocky Mountain spotted fever) show increased rates of lipid peroxidation, but inhibition of this peroxidation does not prevent the cellular injury caused by the organism.

Bearing these points in mind, let us now examine some diseases in which the evidence for an involvement of oxidants in the disease pathology is stronger than average. We will also attempt to evaluate critically the 'free-radical theory' of ageing.

8.2 Atherosclerosis

Cardiovascular disease is the chief cause of death in the USA and Europe. Many important forms of human cardiac disease, together with many cases of localized cerebral ischaemia ('stroke') are secondary to the condition of *atherosclerosis*, a disease of arteries that is characterized by a local thickening of the intima, or innermost part of the vessel. In general, three types of thickening are recognized. *Fatty streaks* are slightly-raised, yellow, narrow, longitudinally-lying areas. They are characterized by the presence of *foam cells*, lipid-laden distorted cells that can arise both from endogenous smooth muscle cells and from macrophages (formed from monocytes that entered the arterial wall from the blood). Fatty streaks probably serve as precursors of *fibrous plaques*. These are approximately-rounded raised lesions, usually off-white in colour and perhaps a centimetre in diameter. A typical fibrous plaque consists of a fibrous cap (composed mostly of smooth muscle cells and dense connective tissue containing collagen, elastin, proteoglycans, and basement membranes) covering an area rich in macrophages, smooth muscle cells, and T lymphocytes (Appendix II) and a deeper necrotic core which

Table 8.1. Clinical conditions in which the involvement of oxygen radicals has been suggested

Inflammatory-immune injury
Glomerulonephritis (idiopathic, membranous)
Vasculitis (hepatitis B virus, drugs)
Autoimmune diseases
Rheumatoid arthritis

Ischaemia—reflow states
Stroke/myocardial infarction/arrythmias
Organ transplantation
Inflamed rheumatoid joint
Frostbite
Dupuytren's contracture?

Drug and toxin-induced reactions
See Chapter 6

Iron overload
Idiopathic haemochromatosis
Dietary iron overload (Bantu)
Thalassaemia and other chronic anaemias
  treated with multiple blood transfusions
Nutritional deficiencies (kwashiorkor)

Red blood cells
Phenylhydrazine
Primaquine, related drugs
Lead poisoning
Protoporphyrin photoxidation
Malaria
Sickle cell anaemia
Favism
Fanconi's anaemia
Haemolytic anaemia of prematurity

Lung
Cigarette smoke effects
Emphysema
Hyperoxia
Bronchopulmonary dysplasia
Oxidant pollutants (O_3, NO_2)
ARDS (some forms)
Mineral dust pneumoconiosis
Asbestos carcinogenicity
Bleomycin toxicity
SO_2 toxicity
Paraquat toxicity

Gastrointestinal tract
Endotoxic liver injury
Halogenated hydrocarbon liver injury (e.g., bromobenzene, CCl_4, halothane)
Diabetogenic action of alloxan
Pancreatitis
NSAID-induced gastrointestinal tract lesions
Oral iron poisoning

Brain/nervous system/neuromuscular disorders
Hyperbaric oxygen
Vitamin E deficiency
Neurotoxins
Parkinson's disease
Hypertensive cerebrovascular injury
Neuronal ceroid lipofuscinoses
Allergic encephalomyelitis and other demyelinating diseases
Aluminium overload (Alzheimer's disease?)
Potentiation of traumatic injury
Muscular dystrophy
Multiple sclerosis

Alcoholism
including alcohol-induced iron overload and alcoholic myopathy

Radiation injury
Nuclear explosions
Accidental exposure
Radiotherapy
Hypoxic cell sensitizers

Ageing
Disorders of premature ageing

Heart and cardiovascular system
Alcohol cardiomyopathy
Keshan disease (selenium deficiency)
Atherosclerosis
Adriamycin cardiotoxicity

Kidney
Autoimmune nephrotic syndromes
Aminoglycoside nephrotoxicity
Heavy metal nephrotoxicity (Pb, Cd, Hg)

Eye
Cataractogenesis
Ocular haemorrhage
Degenerative retinal damage
Retinopathy of prematurity (retrolental fibroplasia)
Photic retinopathy

Skin
Solar radiation
Thermal injury
Porphyria
Hypericin, other photosensitizers
Contact dermatitis

ARDS—adult respiratory distress syndrome. NSAID—non-steroidal anti-inflammatory drug.

contains cellular debris, extracellular lipid deposits, and cholesterol crystals. In general, fibrous plaques begin to obstruct the arterial lumen. *Complicated plaques* are probably fibrous plaques that have been altered by necrosis, calcium deposition, bleeding, and thrombosis. Plaques cause disease by limiting blood flow to a region of an organ such as the heart or brain. A 'stroke' or 'heart attack' (myocardial infarction) occurs when the lumen of an essential artery becomes completely occluded, usually by a thrombus forming at the site of a plaque.

Fatty streaks are regularly present in the arteries of children on Western diets. In the disease *familial hypercholesterolaemia*, the accumulation of plaques is much increased and myocardial infarcts can be observed as early as 2 years of age. As explained in Chapter 4 (Section 4.5), very-low-density lipoproteins (VLDL) enter the circulation from the liver. Low-density lipoproteins (LDL) are largely produced from VLDL in the circulation by several processes, including removal of some triglyceride from VLDL as they circulate through the tissues. LDL are rich in cholesterol esters and supply cholesterol to body tissues; LDL bind to receptors on the surface of cholesterol-requiring cells, and are internalized, releasing cholesterol within the cells. In familial hypercholesterolaemia, the receptors are defective, so that blood cholesterol (and LDL) levels become very high. The increased accumulation of atherosclerotic plaques naturally attracted attention to the possible role of cholesterol in leading to atherosclerosis, especially as it had also been observed that healthy people with high blood cholesterol develop atherosclerosis at an accelerated rate.

The origin of atherosclerosis is uncertain, but a popular current theory is that it begins with damage, by some mechanism (probably haemodynamic), to the vascular endothelium. This could be followed by attachment of monocytes from the circulation, that develop into macrophages within the vessel wall. Activated monocytes and macrophages could injure neighbouring cells by secreting O_2^-, H_2O_2 and hydrolytic enzymes, and factors released by macrophages can stimulate the proliferation of smooth muscle cells. Macrophages also release platelet-stimulating factors (Chapter 7, Section 7.3) and adherence of platelets to injured endothelium could cause release of other agents that encourage proliferation of smooth muscle cells.

What roles could be played by oxidants in atherogenesis? First, as mentioned above, activation of macrophages or their monocyte precursors, e.g. in a fatty streak, could injure neighbouring cells and lead to more endothelial damage and damage to smooth muscle cells.

Second, normal macrophages possess some LDL receptors, but if LDL is peroxidized it is recognized by separate receptors known as the *acetyl-LDL receptors* or the *scavenger receptors*. LDL bound to these receptors is taken up with enhanced efficiency, so that cholesterol rapidly accumulates within the macrophage and may convert it into a foam cell. Arterial endothelial cells, smooth muscle cells, and macrophages are themselves known to be capable of oxidizing LDL so that macrophages will internalize faster. The modification process may involve the formation of derivatives of lysine residues of the protein moiety of LDL by lipid peroxidation products, such as cytotoxic aldehydes. LDL modification by these cells appears to require the presence of traces of iron or copper ions, whose origin has not yet been identified *in vivo*. Incubation of LDL with aqueous extracts of cigarette smoke has also been shown to cause oxidative modification, leading to increased LDL uptake by isolated macrophages.

Third, any lipid peroxides in LDL might contribute to the initial endothelial cell damage that is thought to start off the whole process, e.g. by worsening haemodynamic damage. For example, studies *in vitro* have shown that linoleic acid hydroperoxide increases the permeability of endothelial cell monolayers to macromolecules. Fourth, it has been suggested that peroxidized LDL might act as chemotactic factors for blood monocytes, encouraging their recruitment into an atherosclerotic lesion. Fifth, low concentrations of peroxides can accelerate cyclooxygenase- and lipoxygenase-catalysed reactions in endothelium and in any platelets present (Chapter 7, Section 7.6), leading to enhanced formation of eicosanoids. Oxidized LDL may also stimulate the production of eicosanoids by macrophages. There have been several suggestions that oxidation products of cholesterol might also be involved in atherogenesis; cholesterol is oxidized to a wide variety of products in peroxidizing lipid systems (Chapter 4) and oxidized cholesterol is known to be toxic to arterial smooth muscle cells.

It might be supposed, therefore, that elevated blood lipid concentrations could lead to elevated blood lipid peroxide concentrations, contributing to endothelial injury and accelerating the whole process of atherogenesis. Ca^{2+} accumulation in plaques may also be significant, since increases in intracellular Ca^{2+} are important damaging events during oxidant stress (Chapter 6, Section 6.1). Thus, *is* there more lipid peroxide in patients with atherosclerosis? The deposition of ceroid in atherosclerotic lesions is strongly suggestive that peroxidation is occurring within the plaque (Section 8), but studies upon plasma have not given clear-cut answers. One problem that has bedevilled studies on this point is that storage or mis-handling of plasma samples leads to peroxidation of lipoproteins because of release of low-molecular-mass copper complexes from caeruloplasmin (Chapter 2, Section 2). Several studies identifying 'cytotoxic factors' in the plasma of patients with various diseases have been lead astray by this artefact. It must be remembered that plasma contains powerful preventative (metal-ion-binding) and chain-breaking antioxidants (Chapter 4; Section 4) which limit lipid peroxidation; perhaps depletion or failure of these protective mechanisms is also involved in the pathogenesis of atherosclerosis.

If oxidants do indeed initiate atherosclerosis, or contribute to its pathology, then an increased intake of antioxidants (especially lipid-soluble chain-breaking antioxidants that accumulate in lipoproteins) might be expected to have a beneficial effect. Probucol, a drug used clinically to lower blood cholesterol levels, is a powerful antioxidant, as would be expected because its structure (Fig. 8.1) is similar to that of several phenolic antioxidants (Chapter 4, Table 4.4). American and Japanese scientists found that its anti-atherogenic effect in rabbits is far greater than expected from its cholesterol-lowering ability, suggesting that its antioxidant activity might also be biologically-relevant, by inhibiting the local oxidative modification of LDL in the arterial wall. LDL isolated from humans after probucol administration resist peroxidation *in vitro*, consistent with the above proposals.

8.3   Chronic inflammation and the autoimmune diseases

We saw in Chapter 7 that the acute inflammatory response is beneficial to the organism in that it deals with unwanted and potentially dangerous foreign particles such as bacteria. Inflammation is normally a self-limiting event. However, anything causing abnormal activation of phagocytes has the potential to provoke a devastating response. For example, the major biochemical feature of *gout* is an elevated concentration of uric acid in the blood. Inflammation of joints is triggered by the deposition within them of sodium urate crystals. These crystals can provoke inflammation by a variety of mechanisms, including the stimulation of a respiratory burst in neutrophils and production of leukotriene B_4, which will attract more neutrophils (Chapter 7). Gout can be treated with *allopurinol*, an inhibitor of the xanthine dehydrogenase activity that converts hypoxanthine and xanthine into urate *in vivo* (Chapter 3). Perhaps the most striking consequences of abnormal phagocyte actions, however, are seen in the *autoimmune diseases*.

The body has mechanisms to prevent formation of antibodies against its own components. Any failure of these mechanisms allows formation of *autoantibodies* that can bind to normal body-components, and provoke

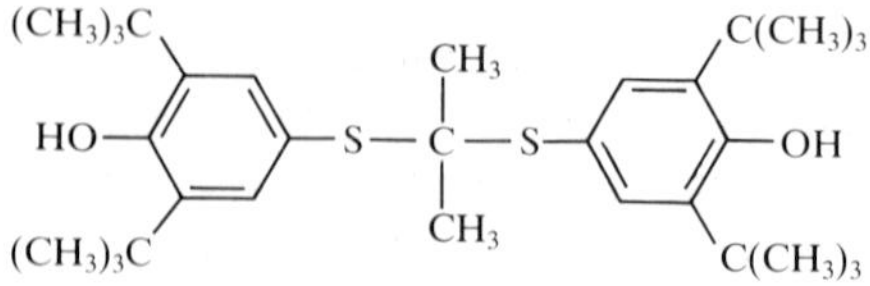

Fig. 8.1. Structure of probucol. The full chemical name is 4,4′-[(1-methylethylidene)*bis*(thio)]*bis*-[2,6-*bis*(1,1-dimethylethyl)phenol]. It has been used as a therapeutic agent to lower blood lipid concentrations. The phenolic—OH groups confer some antioxidant activity on the molecule.

attack by phagocytic cells. In some autoimmune diseases, only a single tissue is attacked. For example, in *Hashimoto's thyroiditis* infiltration of the thyroid gland by phagocytes is accompanied by tissue changes and fibrosis, and the presence of circulating antibodies against certain thyroid constituents, such as thyroglobulin. In *myasthenia gravis*, a neuromuscular disorder character- ized by weakness and fatigue of voluntary muscles, antibodies against the transmitter acetylcholine are present. *Chronic autoimmune gastritis*, in which antibodies to gastric parietal cells are present, is a third example.

More serious than these are the autoimmune diseases in which lesions are widespread and autoantibodies are present against many tissues. These diseases include *systemic lupus erythematosus, dermatomyositis,* and *autoim- mune vasculitis.* To take one example, systemic lupus affects mainly young women, and produces a wide variety of lesions involving the skin, kidneys, muscles, joints, heart, and blood vessels. A wide range of autoantibodies is present, including circulating antibodies to DNA and RNA, anti-erythrocyte antibodies, and antibodies to subcellular organelles and plasma proteins. The kidney lesions are probably due to deposition of immune complexes on the basement membranes of renal glomeruli, followed by complement activa- tion. Phagocytic cells have been identified in rat glomeruli, and respond to activation by mounting a respiratory burst. *Rheumatoid arthritis*, a disease characterized by chronic joint inflammation, especially in the hands and legs, has many features of an autoimmune disease, although the exact cause is unknown. The blood serum and joint fluid of rheumatoid patients often contains autoantibodies directed against immunoglobulin G, mostly against the F_C region of this protein. Rheumatoid factor autoantibodies can belong to any of the immunoglobulin classes. Clearly, something may cause IgG to become antigenic in rheumatoid arthritis; this may be because the carbohy- drate side-chains are different (with fewer galactose residues) and/or because IgG becomes modified by oxidants generated during the inflammatory process (see below).

In the normal synovial joint, articular cartilage covers the bone ends and both are enclosed by a synovial membrane. Articular cartilage is a dense tissue containing collagen fibres and cells (*chondrocytes*) embedded in a matrix of proteoglycan. The normal synovial lining of the joint exists as a thin fibrous tissue; it is made up of lining cells (*synoviocytes*) overlying fatty and fibrous material. Synoviocytes are of at least two types: type A cells have macrophage-like properties and type B may be secretory (it is uncertain how clear this distinction is). The synovial lining synthesizes the polymer *hyaluro- nic acid* (Fig. 8.2) and secretes it into the synovial fluid, where it is largely responsible for the viscosity of the fluid. The synovial lining also acts as a barrier to the free movement of proteins from plasma into the synovial fluid. It has been suggested that the macrophage-like cells may function to engulf debris produced as a result of 'wear and tear' in the joint.

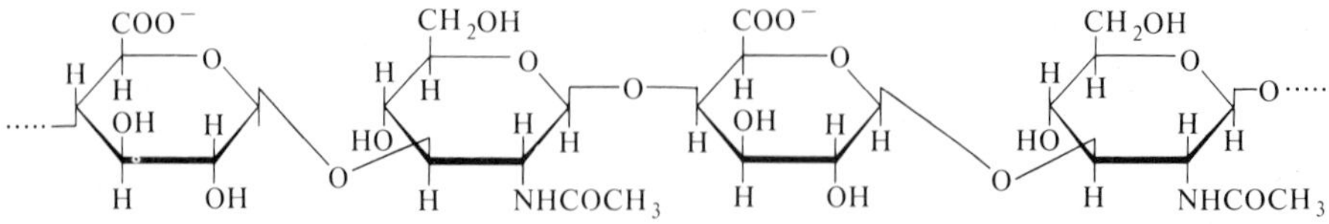

Fig. 8.2. Structure of hyaluronic acid. Hyaluronic acid is a long polymer formed by joining together alternately two different sugars: glucuronic acid (GA) and *N*-acetylglucosamine (NAG). The negative charge on the carboxyl groups of GA at physiological pH means that these subunits repel each other, so that the molecule extends out in solution. Hence solutions of hyaluronic acid are extremely viscous.

The onset of rheumatoid arthritis is usually slow. The synovial lining thickens and folds, and there is infiltration by blood vessels and chronic inflammatory cells. Its permeability rises, so that more proteins enter the joint from the plasma. Iron deposition is seen within the synovial cells; much of this iron occurs within ferritin and haemosiderin. In many patients with rheumatoid arthritis, the disease leads to destruction of the articular carti-lage, bone erosion, and impairment of joint function. Damage usually occurs from the periphery (where the synovial lining normally forms a junction with the articular cartilage) by the growth of the inflamed synovial lining over, and into, the cartilage. This 'overgrowing' tissue is called *pannus*; the word comes from the Latin for 'cloth', the pannus being said to resemble a reddish cloth spreading out over the cartilage structure. Pannus is often vascularized and contains iron deposits, and most cells in pannus are large and mono-nuclear, many having fibroblastic features.

The synovial fluid of the inflamed rheumatoid joint swarms with neutro-phils. Many of these are activated, since the fluid contains increased quantities of products released by activated phagocytic cells, including the enzyme lysozyme, the iron-binding protein lactoferrin, and stable prosta-glandins. Neutrophils are also present at the interface of cartilage and pannus. The viscosity of rheumatoid synovial fluid is much lower than normal; this is because the hyaluronic acid present (Fig. 8.2) has a much lower average molecular mass than in normal synovial fluid. The synovial membrane in rheumatoid patients may synthesize hyaluronic acid of a shorter chain length than usual, and hyaluronic acid can also be subject to radical-induced degradation in synovial fluid (Section 8.3.1). The cartilage wear particles, produced by increased friction in the joints, can activate neutrophils and make matters worse. Joint inflammation often accompanies other autoimmune diseases, such as systemic lupus.

Autoantibodies are present in the nonspecific inflammatory bowel diseases such as *Crohn's disease* and *ulcerative colitis*. The former is a recurrent inflammation and ulceration of the whole digestive tract, although it is often most severe in the lower part of the ileum, and in the colon and rectum, whereas in the latter disease the ulceration and inflammation affect the colon and rectum only. In both conditions autoantibodies to bowel components can be found in the blood serum. *Multiple sclerosis* has some features of an autoimmune disease.

Autoimmune diseases generally have active and quiescent phases, which makes the evaluation of medical treatment especially difficult. This should be borne in mind when assessing the effectiveness of any therapy, such as the use of superoxide dismutase injections, in other than a properly-controlled clinical trial over a long period of time. How autoimmune diseases arise is not known, although there is an inherited predisposition to them; and viral infections have often been suggested to be involved. Certain drugs can induce a condition resembling systemic lupus erythematosus: most significantly, hydralazine, isoniazid, chlorpromazine, and procainamide. A few other drugs have been shown to induce conditions resembling lupus, but much less frequently; these include penicillamine, a-methyldopa, and diphenylhydantoin. It has been suggested that reaction of drug-derived metabolites with normal tissues can produce products that behave as 'foreign antigens'. At least in the cases of hydralazine, a-methyldopa, penicillamine, and isoniazid (Chapter 6), this binding may involve radical reactions.

In 1982, Halliwell proposed that oxidants might attack normal biomolecules to create new antigens, and that this could be an origin of, or contribute to, autoimmunity. Clark *et al.* in Australia found that desferrioxamine, a chelating agent that inhibits iron-dependent radical reactions (Chapter 3), suppressed an experimental model of autoimmune disease (experimental allergic encephalomyelitis) in rats, although this does not necessarily mean that the desferrioxamine is acting by blocking radical reactions. For example, it might simply act by chelating iron necessary for the proliferation of pro-inflammatory cells. It has also been reported that exposure of DNA to a O_2^--generating system (hypoxanthine plus xanthine oxidase) causes the DNA to become antigenic when injected into animals, possibly because O_2^--dependent generation of DNA-damaging $OH^{\cdot}$ radicals leads to the formation of antigenic material. This observation may be of relevance to the presence of anti-DNA antibodies in systemic lupus erythematosus. Lunec *et al.* in England found that oxidants produced by activated phagocytic cells can alter the antigenic behaviour of immunoglobulin G, producing fluorescent protein aggregates that can themselves activate phagocytic cells and are pro-inflammatory in animals. Such radical-modified proteins could presumably provoke antibody (rheumatoid factor) formation. Oxidants produced by phagocytes that could modify IgG would include $OH^{\cdot}$ (provided that a source of iron was present to convert O_2^- and H_2O_2 produced by phagocytes into $OH^{\cdot}$) and hypochlorous acid, HOCl, from the myeloperoxidase system (Chapter 7, Section 7.3).

Autoantibodies are often demonstrable in small amounts in a few 'normal' members of the population, and their incidence increases with age. Postmortem examination of apparently-healthy people with such antibodies often shows minor lesions, too small to produce symptoms. Perhaps oxidative damage to tissues during the ageing process (Section 8.7) can create new antigens.

8.3.1 *Autoimmune diseases and phagocyte action*

Oxidants, prostanoids, leukotrienes, and hydrolytic enzymes produced by neutrophils, macrophages, and monocytes may all be important in mediating inflammation. Let us examine the evidence that oxidants play a biologically-significant role.

Superoxide dismutase, and other scavengers of oxygen radicals, have been observed to suppress inflammation in some animal model systems, such as the reversed passive Arthus reaction in skin. (The 'Arthus reaction' is the name given to a local inflammation that results when an antigen is injected into the skin of an animal that has a high level of circulating antibody against that antigen. It is largely mediated by neutrophils. It can also be observed if the antibody is injected into the animal's bloodstream rather than being formed by the animal itself; this is the passive Arthus reaction. In the reversed passive reaction, the antigen is injected intravenously, and the antibody locally.)

For example, injection of human serum into the bloodstream of rats, followed by injection of an antibody against it into the skin, causes swelling and heat, which is largely mediated by neutrophils. McCord's group in the USA showed that intravenous injection of SOD into the animals had little anti-inflammatory effect in this system, because SOD is cleared from the circulation within minutes by the kidneys. However, if clearance was prevented by binding the SOD to a high-molecular-mass polymer (e.g. Ficoll), there was a marked anti-inflammatory effect. Ficoll-bound SOD also had an inhibitory effect against inflammation induced by injecting carrageenan (an irritating substance derived from seaweed) into the feet of rats. In neither case did Ficoll-bound catalase have an anti-inflammatory effect. Native SOD was observed to protect against glomerulonephritis induced by the intravenous injection of preformed antigen–antibody complexes into mice. These become deposited in the kidneys and provoke inflammation. The effectivenes of native SOD in this system may be due to its rapid accumulation in the kidneys prior to removal from the body. McCord attributed the above results to the action of SOD in removing O_2^-, so preventing O_2^--dependent formation of a factor chemotactic for neutrophils. This explanation has been challenged, however. The lack of inhibition by Ficoll-bound catalase suggested that hydroxyl radical formation was not important.

Several other workers have confirmed these anti-inflammatory effects of SOD in related systems, but have often also found protective effects of catalase and other antioxidants, including the protein caeruloplasmin (Chapter 4). For example, both SOD and catalase showed inhibitory effects against inflammation induced by the implantation of carrageenan-soaked sponges beneath the skin of rats. Injection of xanthine oxidase into the hind-feet of rats produced a swelling that could be partially inhibited by SOD, catalase, or the hydroxyl radical scavenger, mannitol.

One criticism that can be levelled at many, but not all, of these experiments, is the lack of suitable controls. Thus controls with inactivated SOD and catalase should always be performed to rule out 'non-specific' anti-inflammatory effects of proteins.

8.3.2 *Oxidants and human rheumatoid disease*

Synovial fluid from the knee joints of human rheumatoid patients contains increased levels of 'diene conjugates' and TBA-reactive material, *suggestive* of increased lipid peroxidation *in vivo* (it must be pointed out that the significance of UV-absorbing diene conjugates and TBA reactive material as indices of lipid peroxidation in human body fluids is unclear, and that prostaglandin endoperoxides, formed in increased amounts during inflammation, are TBA reactive; see Chapter 4, Section 4.6 for further discussion of these points). Nevertheless, the TBA reactivity of synovial fluid correlates with disease severity as measured both by clinical and by laboratory-based parameters such as acute-phase-protein concentration and erythrocyte sedimentation rate. McCord in the USA showed that the breakdown of hyaluronic acid observed in synovial fluid from rheumatoid patients could be reproduced by exposing purified hyaluronic acid to a O_2^--generating system *in vitro*, whence it can be shown to be due to O_2^--dependent formation of OH˙ radicals that attack the hyaluronic acid (these observations do not, of course, prove that OH˙ is responsible for hyaluronic-acid degradation *in vivo*: there is some evidence that rheumatoid synovium synthesizes hyaluronic acid in a form of lower molecular mass than normal). Human synovial fluid contains little, if any catalase, glutathione peroxidase or GSH, and only traces of SOD activity (in the form of the glycoprotein EC-SOD; Chapter 4, Section 4.7.6). Thus O_2^- and H_2O_2 generated by phagocytes in the inflamed rheumatoid joint would not be efficiently scavenged, and OH˙ could form provided that a source of iron was available. Ascorbate concentrations in both synovial fluid and plasma of patients with rheumatoid arthritis are markedly sub-normal; and most of the ascorbate that is present is in the oxidized (dehydroascorbate) form. It may be that ascorbate is being rapidly oxidized by O_2^- (Chapter 3, Section 3.2.1) and hypochlorous acid (Chapter 7, Section 7.3) produced by activated neutrophils.

Iron and rheumatoid arthritis

That rheumatoid arthritis is accompanied by abnormalities in body iron metabolism has been known for many years. A rapid fall in the 'total iron' content of blood plasma at the onset of inflammation (Table 8.2) is followed by a drop in haemoglobin concentration and increased deposition of iron proteins in the synovial membranes. The drop in plasma iron correlates closely with the activity of the inflammatory process. The iron in the rheumatoid synovial membrane is largely present within ferritin. In early rheumatoid disease, the presence of ferritin and haemosiderin iron implies a poor prognosis. Attempts to reverse the 'anaemia of rheumatoid arthritis' by giving oral iron salts to patients are usually ineffective in the absence of improvement in their disease, and such oral treatment has been shown to worsen the symptoms of at least one patient. Intravenous iron therapy, e.g. by injection of iron dextran, frequently causes problems. Although this has often been attributed to the dextran, it may be the iron that is at fault. For example, Winyard *et al.* in England reported that exacerbations of joint inflammation in rheumatoid patients given iron-dextran infusions occurred at exactly the time that plasma transferrin became saturated with iron and non-transferrin-bound iron was present in body fluids. Indeed, iron-overloaded patients suffering from idiopathic haemochromatosis frequently show joint inflammation (Chapter 2, Section 2.4.3). When subjected to oxidant stress, both ferritin and (to a lesser extent) haemosiderin can release iron in a form capable of stimulating radical reactions such as $OH^{\cdot}$ production and lipid peroxidation. This suggests that the inflamed synovial membrane could

Table 8.2. Protein and iron concentrations in plasma and synovial fluid from rheumatoid patients. Data abstracted from Gutteridge (1986). *Biochim. Biophys. Acta*, **869**, 119–27

	Normal control plasma (8 samples)	Rheumatoid plasma (8 samples)	Rheumatoid knee-joint synovial fluid (9 samples)
Total non-haem iron [μM] $\pm$ SD	17.9 $\pm$ 6.7	7.5 $\pm$ 5.7	8.9 $\pm$ 4.1
Transferrin g/l $\pm$ SD	2.92 $\pm$ 0.38	2.76 $\pm$ 4.2	1.67 $\pm$ 0.31
% Transferrin saturation with iron $\pm$ SD	29.9 $\pm$ 10.2	16.1 $\pm$ 12.8	26.3 $\pm$ 11.7
Albumin g/l $\pm$ SD	50.3 $\pm$ 4.4	36.3 $\pm$ 7.9	17.2 $\pm$ 6.3
Caeruloplasmin g/l $\pm$ SD	0.259 $\pm$ 0.079	0.469 $\pm$ 0.085	0.256 $\pm$ 0.077

easily be damaged by externally-generated H_2O_2, which might penetrate into the tissue and lead to internal OH˙ production.

Ferritin, containing some iron, is also present in synovial fluid. Superoxide is capable of mobilizing iron from ferritin (Chapter 3, Section 3.5.4) and the released iron can cause OH˙ formation from H_2O_2. Thus ferritin might be a source of iron that could mediate formation of OH˙ and hyaluronic acid degradation in synovial fluid. Another possible source of iron is haemoglobin: an inflamed rheumatoid joint easily suffers mild traumatic injury, resulting in bleeding. Haemoglobin liberated from lysed erythrocytes can easily be degraded by H_2O_2 to liberate iron ions (Chapter 3, Section 3.5.4). Also, engulfment of erythrocytes by macrophage-like cells in the rheumatoid synovium could contribute to the increased iron deposition.

Application of the bleomycin assay for 'loosely-bound iron' (Chapter 3) to samples of synovial fluid from knee joints of patients with active joint disease gives a positive result in a minority of cases. The bleomycin assay presumably measures iron, released from proteins such as ferritin or haemoglobin under oxidative stress, that has not yet bound to transferrin or to the iron-binding protein lactoferrin secreted by phagocytic cells. Values recorded in the bleomycin assay on synovial fluids range from 0–5 μM. Thus, even at sites of intense inflammation, the availability of iron to stimulate radical reactions is very small.

Are oxidants important in rheumatoid arthritis?
There seems little doubt that oxidant damage, resulting from neutrophil activation, occurs in the synovial fluid of the inflamed rheumatoid joint. However, the major question is how important is this damage in relation to the major pathology of the disease such as the cartilage destruction and erosive bone damage? Events occurring at the interface of cartilage and pannus may be very different from those happening in the bulk synovial fluid. There are many other potential agents of damage, including proteolytic enzymes released from phagocytes (such as elastase), prostaglandins, and other products derived from arachidonic acid.

Two approaches have been used to assess the importance of free radicals. Some scientists have claimed that injection of SOD into the inflamed knee joint of rheumatoid patients produces a beneficial effect, thus implying that oxygen radicals are important. The experiments published to date have not been adequately controlled, however. As discussed previously, SOD has a well-established anti-inflammatory effect in several animal model systems of inflammation, but it does not appear to act only by bulk-scavenging of O_2^- in free solution. For example, Michelson *et al.* in France found that rat CuZnSOD, or MnSOD, did not suppress carrageenan-induced paw inflammation in the rat, whereas human and bovine CuZnSOD did. Thus the SOD may have to bind to certain specific sites to exert its anti-inflammatory activity (different SOD proteins have different surface charge distributions).

D. R. Blake and the authors in England adopted an alternative approach, based on the key role proposed for iron ions in causing OH˙ formation and lipid peroxidation. The action of desferrioxamine, an iron-chelating agent that suppresses iron-dependent radical reactions (Chapter 3, Section 3.5.4), was tested on various animal modes of inflammation. It was found that low doses of desferrioxamine aggravate some acute rat models of inflammation, but larger doses are anti-inflammatory. In Glynn–Dumonde synovitis in guinea-pigs, desferrioxamine (100 mg/kg of body weight) aggravated the acute phase of the inflammation, but repeated administration depressed the chronic phase. A similar effect is seen in the rat allergic air pouch model of acute to chronic inflammation (subcutaneous injection of sterile air in rodents results in formation of an 'air pouch' with a lining that superficially resembles synovium; this model is very useful in testing pro- and anti-inflammatory agents). Desferrioxamine treatment also decreases the severity of adjuvant-induced joint inflammation in rats. Interestingly, a similar reduction in severity was observed if the rats were made slightly anaemic, by feeding them on an iron-deficient diet, before performing the experiments. It is not clear why desferrioxamine may sometimes make inflammation worse during the acute phase; by stopping OH˙ production it might protect neutrophils against destruction by their own oxygen-derived products. It is also possible that some OH˙ production is necessary for the effective regulation of inflammation, e.g. by inactivation of arachidonic-acid metabolites (Chapter 7, Section 7.3.2). Neither should it be assumed that the action of desferrioxamine in suppressing chronic inflammation is necessarily related to its effect on radical reactions; it may block the proliferation of inflammatory cells by chelating essential iron, and iron may be involved in controlling lymphocyte traffic in inflamed areas (for an explanation of lymphocytes, see Appendix II).

The suppressive action of desferrioxamine on chronic inflammation in animals was sufficiently encouraging for preliminary trials with rheumatoid patients to be carried out. Giordano *et al.* in Italy injected 1 g of the chelator intra-muscularly and observed an abrupt rise in haemoglobin concentrations. No fall in the acute-phase response was observed, but the speed of the change in haemoglobin suggests that the effect might have been mediated by suppressing inflammation. No ill-effects were reported. Of seven rheumatoid patients given larger doses of desferrioxamine (up to 3 g/day for 5 days each week for 1–3 weeks), four developed ocular abnormalities that reversed on drug withdrawal. Two patients who received the sedative prochlorperazine to combat nausea during desferrioxamine therapy became unconscious for 48–72 hours, possibly because this combination of drugs mediates iron transfer across the blood–brain barrier and achieves removal of iron essential to the functioning of the nervous system. It is clear that doses of desferrioxamine suitable for treatment of iron overload are not necessarily safe in other disease states, and the combination of this chelator with phenothiazine drugs should be strictly avoided. Although desferrioxamine is of potential use in a number of human diseases (on the basis of animal experiments) caution should be employed in its use, and patients receiving it should be carefully checked for ocular and auditory abnormalities. Whether these side-effects are peculiar to desferrioxamine or are a general consequence of iron chelation is still under investigation.

Could oxidants be involved in causing cartilage damage in the inflamed rheumatoid joint? Pannus is rich in macrophages. As macrophages and other phagocytes adhere to cartilage surfaces, it is likely that a microenvironment sealed off from the bulk synovial fluid can be created. Direct measurements of pH beneath isolated adherent macrophages suggest that the pH in this microenvironment can fall to less than 5, sometimes as low as pH 3.6. Oxygen radicals, proteolytic enzymes, myeloperoxidase (from neutrophils), and other products are released into this microenvironment. The low pH could cause damage to cartilage by:

1. Favouring formation of the more-reactive $HO_2^{\cdot}$ from O_2^{-}.

2. Assisting release of iron from proteins and thus $OH^{\cdot}$ formation.

3. HOCl inactivating a_1-antiprotease in synovial fluid and thus assisting proteolytic attack upon the cartilage. Indeed, oxidatively-inactivated a_1-antiprotease can be demonstrated in synovial fluid from inflamed rheumatoid joints. It has also been shown that H_2O_2-generating systems significantly inhibit proteoglycan synthesis by cultured bovine articular cartilage. Thus it is possible that a combined attack by $OH^{\cdot}$ and proteolytic enzymes, as well as an inhibition of proteoglycan resynthesis, facilitates cartilage damage. Indeed, it has been shown that injection of a modified form of the H_2O_2-producing enzyme glucose oxidase (rendered cationic to assist retention within the joint) into the knee joints of mice causes severe cartilage damage, including the death of chondrocytes.

It must be realized, however, that much of the cartilage degradation taking place in rheumatoid arthritis may not be due to external attack by oxidants or proteases, but to increased release of proteolytic enzymes such as collagenase and neutral proteases *from the chondrocytes themselves* into the cartilage matrix. The high content of macrophages in rheumatoid synovial lining causes significant production of interleukin 1, which is released from these cells (Chapter 4, Section 4.7.2.5). This inflammatory mediator stimulates both synoviocytes and chondrocytes to release proteolytic enzymes, and it can also promote bone resorption. These actions are unrelated to oxidant production. However, the additional presence of oxidants might still facilitate cartilage degradation by endogenous proteases. For example, HOCl can activate latent collagenase and gelatinase enzymes (Chapter 7, Section 7.3)

Thus, although it seems likely that oxidants contribute significantly to damage in the inflamed rheumatoid joint, many more experiments need to be done to investigate this and to evaluate antioxidants, and some of the newer iron-chelating agents (Chapter 3), for therapeutic use. One illustration of the potential pro-inflammatory action of oxidants may be provided by *alcaptonuria*. This disease is an inborn defect in an enzyme (*homogentisate oxidase*) involved in the metabolic degradation of the amino acid tyrosine. Alcapto-

nuric patients accumulate homogentisic acid, They show pigmen-

tation of cartilage and connective tissue and may gradually develop severe inflammatory arthritis. Martin and Batkoff in the USA showed that the oxidation of homogentisic acid generates O_2^- and H_2O_2, and also OH· if iron ions are present. They therefore made the interesting suggestion that the arthritis of alcaptonuria is induced by increased oxygen-radical production. Autoxidation can be expected to produce melanin-like pigments, accounting for the pigmentation observed.

8.3.3 *Are oxidants important in autoimmune diseases?*

It would seem that oxygen radicals and released enzymes, such as myeloperoxidase and hydrolases, might play an important role in the autoimmune diseases, and that therapy directed against them might prove clinically useful. Some evidence consistent with this suggestion has been provided by Emerit *et al.* in France. They have shown that the serum of patients with lupus erythematosus contains a 'clastogenic factor' that induces chromosome damage when it is added to cultures of lymphocytes isolated from normal human donors. The chemical identity of clastogenic factor is unknown, but they have reported that inclusion of SOD in the culture medium prevents its effect. Lupus patients are light-sensitive and in about 40 per cent of cases develop a characteristic 'butterfly' rash across the cheeks and the bridge of the nose when exposed to sunlight. Lymphocytes from lupus patients are damaged by exposure to light, in the wavelength range 360–400 nm, in the presence of clastogenic factor: an effect again prevented by SOD.

Other evidence for the importance of oxygen radical reactions in autoimmune disease has been provided by Harman in the USA. He has studied New Zealand Black (NZB) mice, a strain that spontaneously develops an autoimmune disease that has some of the features of human systemic lupus. He finds that addition of antioxidants such as tocopheryl acetate or Santoquin (ethoxyquin, Table 4, Chapter 4) to their diet decreases the manifestations of this disease, and prolongs the average lifespan of the mice. The serum of NZB mice also contains a clastogenic factor whose action is decreased by SOD.

A few individual case-reports have appeared as to the benefit of oxygen radical scavengers in the treatment of human autoimmune disease, although proper controlled trials have not been done. SOD and simple copper chelates that can react with O_2^- (Chapter 3) have been reported as helpful in lupus and in dermatomyositis, and SOD encapsulated in liposomes (Chapter 4) has been used in the treatment of Crohn's disease. Large doses of vitamin E have sometimes been reported as helpful in lupus cases.

8.3.4 *Anti-inflammatory drugs and oxidant damage*

Possible mediators of inflammation include prostanoids, proteases, leuko-trienes, and oxidants produced by activated phagocytic cells. Relatively little attention has been given to oxidant scavengers as anti-inflammatory agents (Section 8.2.3), although some studies with Ebselen® (Chapter 4, Section 4.7.5) and with iron chelators have been performed. Most anti-inflammatory drugs developed have been directed at pathways of eicosanoid synthesis.

Corticosteroids inhibit phospholipase A_2 activity, which plays a role in the release of arachidonic acid within cells. Effective naturally-occurring corti-costeroids include cortisol; synthetic ones include prednisolone and dexa-methasone (Fig. 8.3). Corticosteroids decrease the formation of prostaglan-dins, thromboxanes, and leukotrienes. Corticosteroids act indirectly by causing release of a phospholipase-inhibitory protein known as *lipocortin*. They can also decrease formation of platelet-activating factor and of the powerful inflammatory mediator interleukin-1 (Chapter 4, Section 4.7.6). Steroids are powerful and effective therapeutic agents in rheumatoid arthri-tis, but side-effects limit their use.

In view of the powerful therapeutic actions of corticosteroids, several non-steroid drugs that can inhibit both cyclo-oxygenase and lipoxygenase are undergoing development. BW755C and benoxaprofen (Fig. 8.3) (which inhibits lipoxygenase more effectively than it does cyclo-oxygenase) are prototypes of this class of drug, although neither is suitable for human use. Attempts are also being made to develop selective inhibitors of lipoxygenase, such as *piriprost*. We have already seen that some anti-oxidants, such as NDGA and propyl gallate, are lipoxygenase inhibitors (Chapter 7, Section 7.6.2).

However, the drugs most commonly used in the day-to-day treatment of rheumatoid arthritis were developed as cyclo-oxygenase inhibitors, blocking prostaglandin production. The first drug of this type to be synthesized was aspirin (Fig. 8.3). In 1763, the Reverend Edmund Stone of Oxford, England, read a report to the Royal Society on the anti-fever action of willow bark.

Free radicals, ageing, and disease

STEROIDS

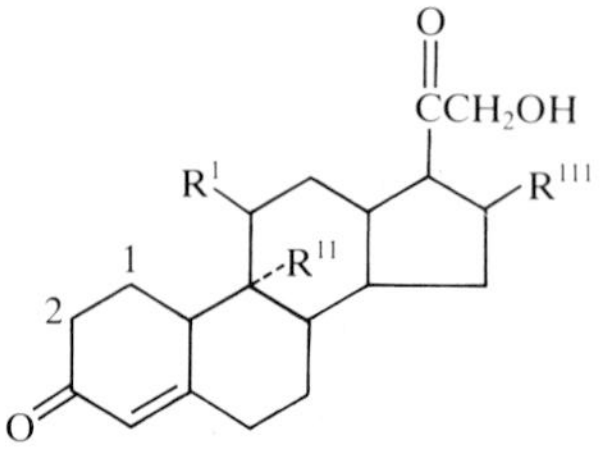

Cortisol $\quad$ (R^I=O; R^{II}–H; R^{III}–H; C 1–2 saturated)

Prednisolone $\quad$ (R^I–OH; R^{II}–H; R^{III}–H; C 1–2 unsaturated)

Dexamethasone $\quad$ (R^I–OH; R^{II}–F; R^{III}–CH$_3$; C 1–2 unsaturated)

ANTI-MALARIALS

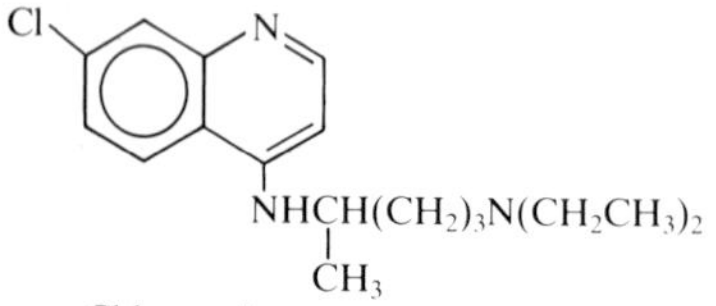

NHCH(CH$_2$)$_3$N(CH$_2$CH$_3$)$_2$·2HCl

Chloroquine

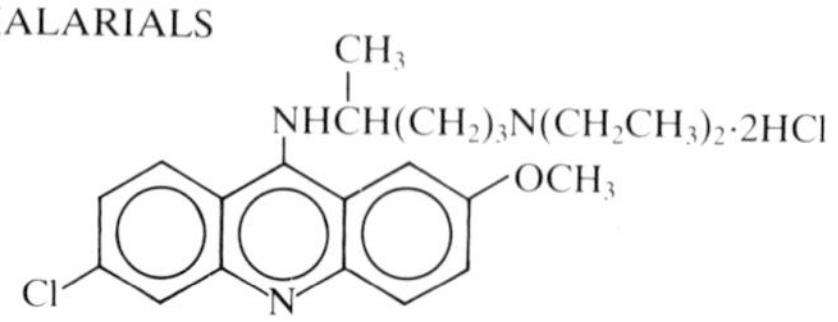

Quinacrine hydrochloride

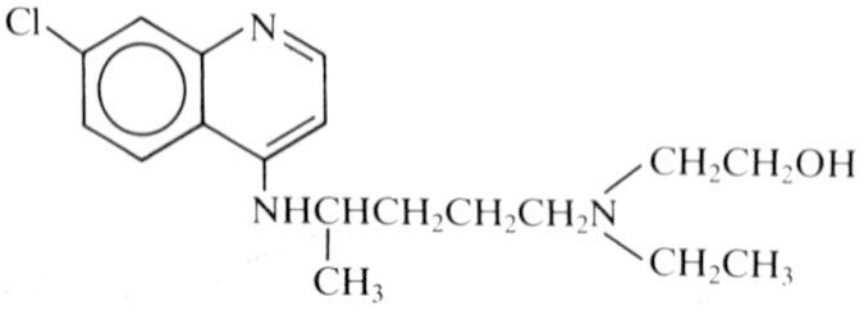

Hydroxychloroquine

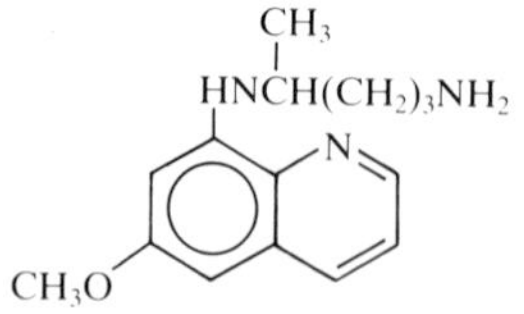

Primaquine

MISCELLANEOUS

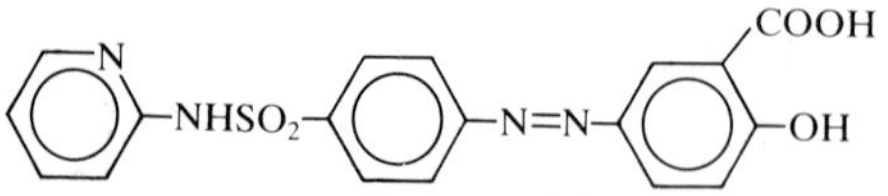

Sulphasalazine

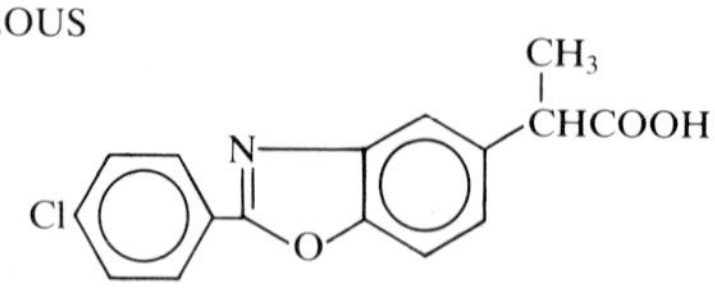

Benoxaprofen

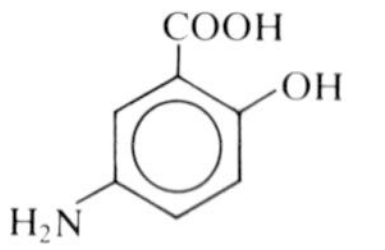

5–Aminosalicylic acid

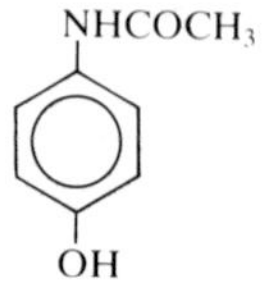

Paracetamol

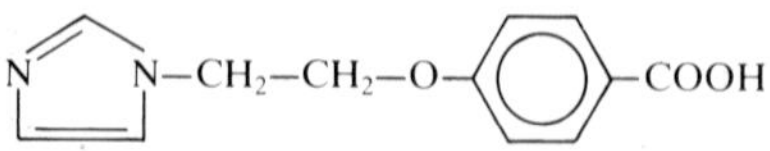

Dazoxiben

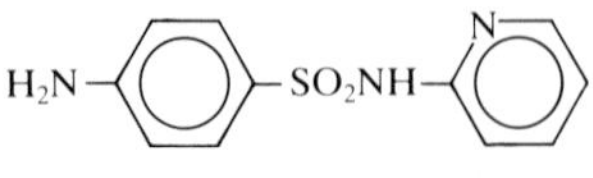

Sulphapyridine

Fig. 8.3. Structures of some drugs used in the treatment of rheumatoid arthritis.

NON-STEROIDALS

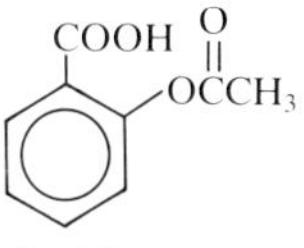

Aspirin
(acetylsalicylic acid)

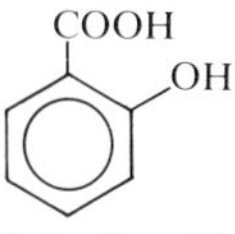

Salicylic acid
(2-hydroxybenzoic acid)

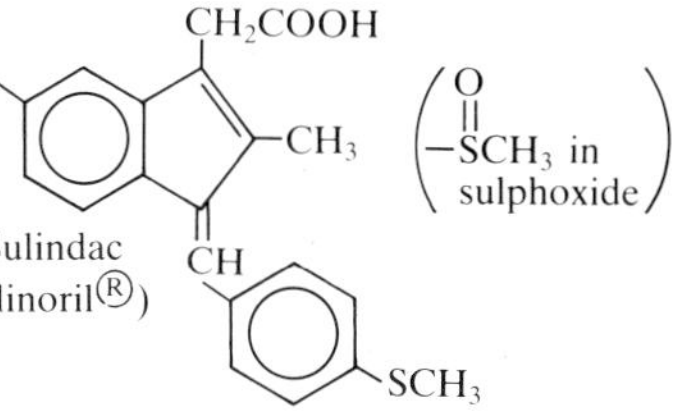

Sulindac
(Clinoril®)

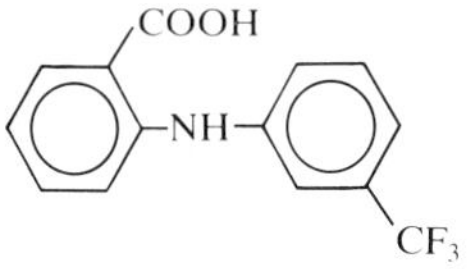

$\left(\begin{array}{c} O \\ \| \\ -SCH_3 \text{ in} \\ \text{sulphoxide} \end{array}\right)$

Flufenamic acid

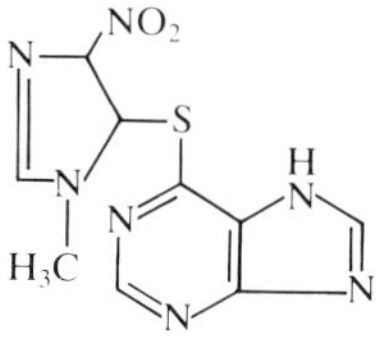

Azathioprine

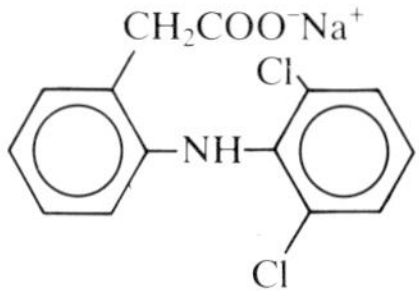

Diclofenac sodium

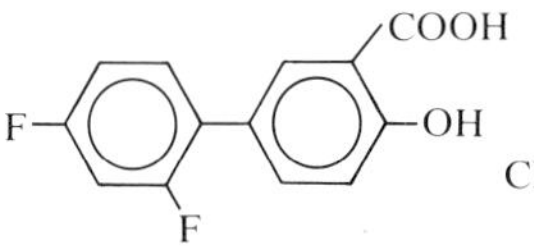

Diflunisal

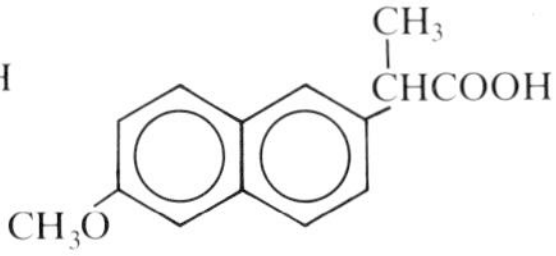

Naproxen

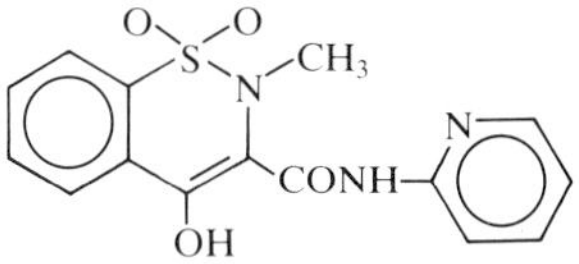

Piroxicam

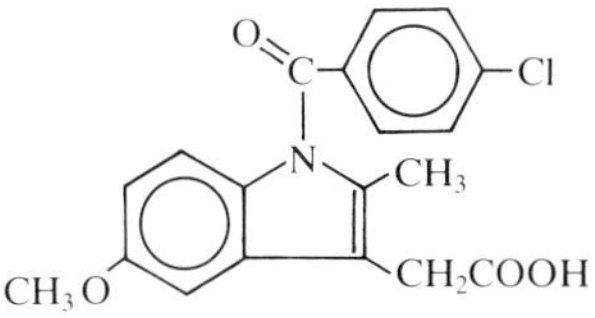

Indomethacin

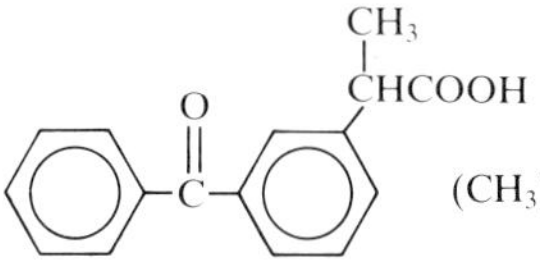

Ketoprofen

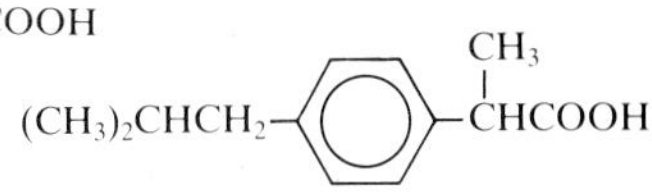

Ibuprofen

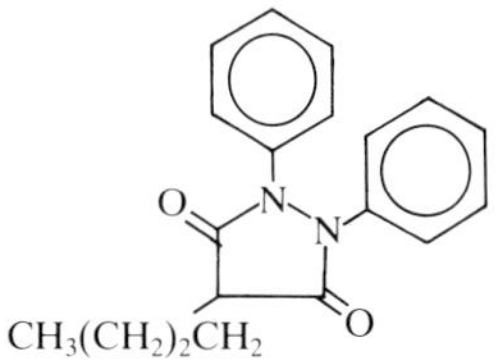

Phenylbutazone

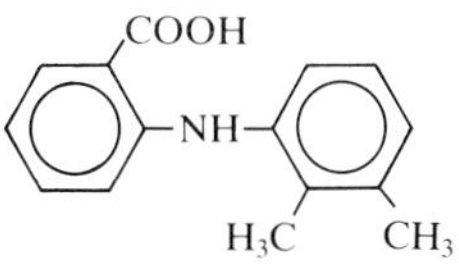

Mefenamic acid

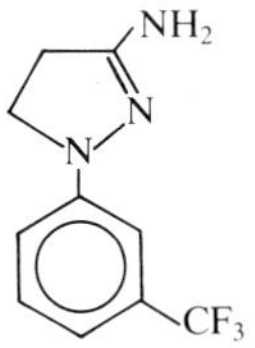

BW755C

THIOLS

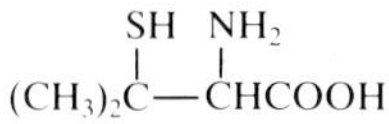

Penicillamine

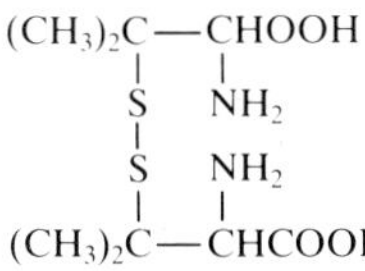

Penicillamine disulphide

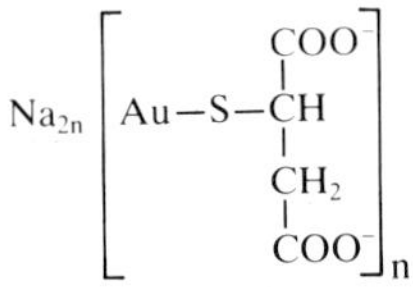

Gold sodium thiomalate

The active component was later extracted and shown to be salicylate (Fig. 8.3); in 1899 the Bayer Company in Germany introduced acetylsalicylate (aspirin). It is interesting to note that aspirin and salicylate are equally potent as anti-inflammatory agents, but aspirin is a more potent cyclo-oxygenase inhibitor. Aspirin ingested by humans is quickly hydrolysed to salicylate by esterases in the digestive tract and in liver. Several other non-steroidal anti-inflammatory drugs (NSAIDs) have been developed as cyclo-oxygenase inhibitors: they include diclofenac sodium, indomethacin, ibuprofen, phenyl-butazone, piroxicam, and mefenamic acid (Fig. 8.3).

Aspirin and other NSAIDs have three major pharmacological properties:

(i) reduction of swelling and redness associated with inflammation;
(ii) ability to decrease elevated body temperature;
(iii) an analgesic effect.

Major side-effects of NSAIDs are gastrointestinal disturbances, irritation of the stomach mucosa, and renal injury. In 1974, Ferreira and Vane in England proposed that all NSAIDs act by inhibition of prostaglandin biosynthesis. Much evidence supports this proposal, but there are still unexplained variations in the sensitivity of different biological systems to NSAIDs. These might be caused by subtle variations in the structure of cyclo-oxygenase in different tissues, by different rates of penetration of drugs to the correct site, or by differential rates of drug metabolism. However, it is also possible that NSAIDs may have multiple mechanisms of action. In general, none of the NSAID drugs decreases cartilage damage or bone erosion in rheumatoid arthritis, and some may even enhance it, perhaps by encouraging interleukin-1 formation by macrophages in the pannus or (in the case of salicylate) by inhibiting synthesis of cartilage proteoglycans.

Some thiol ($-SH$) compounds have been claimed to modify the underly-ing disease processes in rheumatoid arthritis; they include penicillamine (often given as its disulphide, which is reduced to penicillamine *in vivo*; Fig. 8.3), and gold–thiol complexes such as aurothiomalate (Fig. 8.3), and aurothioglucose. How much 'disease modification' they exert is a matter of debate. Gold complexes may act by depressing the activity of macrophages and lymphocytes, as well as interfering with eicosanoid production.

Several drugs developed for other diseases have been found to be of use in treatment of rheumatoid arthritis. These include the anti-malarials *chloro-quine* and *hydroxychloroquine* (Fig. 8.3), the antibiotic *rifamycin*, and the drug *sulphasalazine*. Anti-malarials such as hydroxychloroquine have been reported to decrease interleukin-1-induced cartilage degradation *in vitro*. Sulphasalazine consists of 5-aminosalicylic acid linked by an azo ($-N = N-$) bond to sulphapyridine. *In vitro*, it inhibits both lipoxygenase and 15-hydroxyprostaglandin dehydrogenase activities; the latter enzyme catalyses the first step in biological inactivation of E, F, and D-type prostaglandins. It is not clear whether these effects account for the clinical actions of sulphasalazine. It is particularly effective in the treatment of ulcerative colitis, but evidence exists that the true anti-colitic agent is 5-aminosalicylate, produced as a result of hydrolysis of sulphasalazine by bacteria in the human colon. Sulphasalazine has been reported to give some benefit in rheumatoid arthritis, although the effective agent here is probably the sulphasalazine molecule itself. Several immunosuppressive agents have been used in the treatment of rheumatoid arthritis, since it is believed that immunological activity contributes to joint damage. Those tried include cyclophosphamide, azathioprine, and methotrexate (Section 8.8.4).

Anti-inflammatory drugs as antioxidants?
The possibility that NSAIDs and other anti-inflammatory drugs have multiple mechanisms of action has led to the question as to whether they could have antioxidant effects.

Anti-inflamatory drugs could conceivably affect oxidant damage at sites of inflammation in several ways. First, they might directly scavenge such reactive oxidants as hydroxyl radical (OH$^\cdot$) and hypochlorous acid (HOCl). It has been shown that most, if not all, anti-inflammatory drugs are capable of reacting quickly with OH$^\cdot$ (rate constants in the range 5.10^9–$10^{10}\text{M}^{-1}\text{s}^{-1}$). This would be expected from their chemical structure (Fig. 8.3) since aromatic compounds and thiol compounds such as penicillamine are known to react very rapidly with OH$^\cdot$ (Chapter 2). Hence, in terms of OH$^\cdot$ scavenging, any anti-inflammatory drug present at a site of inflammation would remove OH$^\cdot$ if the drug concentration were high enough; at least millimolar concentrations would be required. Most drugs do not achieve such concentrations. However, in rheumatoid patients given aspirin, salicylate concentrations in plasma and synovial fluid reach almost 1 mM and might be capable of intercepting OH$^\cdot$. One of the products of OH$^\cdot$ attack on salicylate is 2,3-dihydroxybenzoate (Chapter 2, Section 2.5.2). Concentrations of this substance in plasma from aspirin-treated rheumatoid patients are greater than in control subjects consuming aspirin, and it is *possible* that some or all of the 2,3-dihydroxybenzoate arises as a product of OH$^\cdot$ attack on the aromatic ring structure.

The authors have also examined the ability of anti-inflammatory drugs to scavenge HOCl. Most drugs tested are able to react with HOCl, but only in a few cases was the reaction found to be fast enough for HOCl scavenging to be feasible at the drug concentrations found at sites of inflammation *in vivo*. Potential HOCl scavengers include thiol compounds (penicillamine, gold sodium thiomalate), phenylbutazone, and primaquine. It is also interesting to note that 5-aminosalicylate, the active component of sulphasalazine in the treatment of ulcerative colitis, is a powerful scavenger of HOCl whereas sulphasalazine itself is not. 4-Aminosalicylate, a newer drug apparently effective in the treatment of ulcerative colitis, is also an excellent scavenger of HOCl.

In considering the likelihood of scavenging of OH˙ or HOCl by a drug at a site of inflammation, one must also think about whether the drug-derived radical might itself do biological damage. For example, reaction of HOCl and OH˙ with penicillamine could conceivably produce toxic products that might contribute to the side-effects of this drug. A particular feature of penicillamine therapy is the large number of autoimmune reactions it produces, such as lupus-like syndromes. RS˙, RSO_2˙ and RSO˙ radicals formed from penicillamine might combine with proteins and alter their antigenicity.

The production of hydroperoxides by cyclo-oxygenase is followed by their oxidation by the peroxidase activity of this enzyme, during which a haem-associated radical species is generated that can destroy the enzyme (Chapter 7, Section 7.6.1). Some drugs might be able to scavenge this species, being oxidized by it and so protecting cyclo-oxygenase and *increasing* prostanoid production. 5-aminosalicylate and the sulphide form of the anti-inflammatory drug sulindac (Fig. 8.3) are able to do this *in vitro*.

Another way in which anti-inflammatory drugs might mediate oxidant reactions is by altering oxidant production by neutrophils, monocytes, and macrophages. Corticosteroids, high concentrations of antimalarials, and some NSAIDs (e.g. piroxicam) have been reported to diminish O_2^- production by activated phagocytes *in vitro*. In some cases, e.g. with piroxicam, the effect has been demonstrated in cells isolated from patients taking the drug, and so it might be of biological relevance. Inhibition of O_2^- production should lead to the diminished formation of H_2O_2, and hence to decreased production of HOCl and, possibly, OH˙.

Some NSAID drugs are able to *stimulate* oxidant formation under certain conditions. In particular, naproxen and benoxaprofen have a photosensitizing action and are able to stimulate singlet O_2 formation (Chapter 2, Section 2.6). This effect was bad enough in the case of benoxaprofen to cause its withdrawal from general clinical use.

8.4 Ischaemia/reoxygenation injury

8.4.1 *Intestinal injury*

Damage to the heart or brain by completely depriving a portion of the tissue of O_2 (*ischaemia*) is a major cause of death in Western society. Atherosclerosis, leading to the blockage of an essential artery, is usually the culprit (Section 8.2). Severe restriction of blood flow, leading to O_2 concentrations lower than normal (*hypoxia*), but not complete O_2 deprivation, can also result if the blocked artery is the major, but not the only, source of blood to the tissue in question.

Tissues made hypoxic or ischaemic survive for a variable time, depending on the tissue in question. Thus skeletal muscle is fairly resistant to hypoxic injury, whereas the brain is very sensitive and does not usually survive ischaemia for more than a few minutes. However, any tissue made ischaemic for a sufficient period will be irreversibly injured. Tissues respond to ischaemia in a number of ways. Early responses may include increased glycogen degradation and anaerobic glycolysis, leading to lactate production and acidosis. ATP levels begin to fall, and AMP is degraded to cause an accumulation of hypoxanthine (Fig. 8.4). Eventually, glycolysis slows and membrane damage becomes visible under the microscope.

If the period of ischaemia or hypoxia is insufficiently long to irreversibly injure the tissue, most of it can be salvaged by reperfusing the tissue with blood and re-introducing O_2 and nutrients. In this situation, reperfusion is a beneficial process overall. However, McCord, Parks, Granger *et al.* in the USA showed in the early 1980s that re-introduction of O_2 to an ischaemic or hypoxic tissue could cause an additional insult to the tissue (*reoxygenation injury*) that is, in part, mediated by oxygen radicals. The relative importance of reoxygenation (often called reperfusion) injury depends on the time of ischaemia/hypoxia. If this is sufficiently long, the tissue is irreversibly injured and will die, so reoxygenation injury is not important (one cannot further

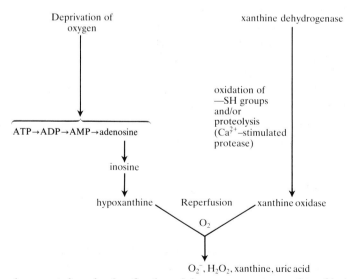

Fig. 8.4. A suggested mechanism for tissue injury upon reoxygenation of ischaemic or hypoxic tissues. Modified from McCord, J. M., *Fed. Proc.* **46,** 2402–2406 (1987).

injure a dying tissue). However, for a relatively-brief period of ischaemia/ hypoxia, the reoxygenation injury component may become more important and the amount of tissue remaining undamaged can be significantly increased by including oxidant scavengers in the reoxygenation fluid. The meaning of 'relatively brief' in this context depends on the tissue in question, whether one is dealing with ischaemia or hypoxia and, if the latter, what degree of hypoxia is actually achieved.

Although the enzyme xanthine oxidase is frequently used as a source of superoxide (O_2^-) in experiments *in vitro*, almost all the xanthine-oxidizing activity present in normal tissues is a dehydrogenase enzyme that transfers electrons not to O_2, but to NAD^+, as it oxidizes xanthine or hypoxanthine into uric acid (Chapter 3, Section 3.1.2). When tissues are disrupted, the dehydrogenase becomes converted into the oxidase enzyme by oxidation of essential —SH groups or by limited proteolysis. Hence, on purification, the enzyme appears as an oxidase, producing O_2^- and H_2O_2 when xanthine or hypoxanthine are oxidized. McCord's group suggested that the dehydrogenase to oxidase conversion also occurs in ischaemic/hypoxic tissue, by proteolytic attack (although conversion by oxidation of enzyme thiol groups is also possible). The depletion of ATP in hypoxic tissue causes hypoxanthine accumulation. This hypoxanthine can be oxidized by the xanthine oxidase when the tissue is oxygenated, causing rapid generation of O_2^- and H_2O_2, which might lead to severe tissue damage. The hypothesis of McCord *et al.* is summarized in Fig. 8.4.

The first evidence supporting the hypothesis came from studies upon the intestine. Partial arterial occlusion of a segment of cat small intestine (hypoxia), followed by reperfusion, causes gross histologically-observable damage to the tissue and increases intestinal vascular permeability. Intravenous administration of SOD, or oral administration of allopurinol (an inhibitor of xanthine oxidase) to the animals before removal of the arterial occlusion, offered protection against damage. Infusion of a mixture of hypoxanthine and xanthine oxidase into the arterial supply of a segment of normal cat intestine greatly increased vascular permeability, an effect that was decreased by the presence of SOD or dimethylsulphoxide (DMSO) in the infusion. The effect of DMSO, a powerful scavenger of OH· (Chapter 2), perhaps suggests that this radical was involved in the damage, although inhibition by a single scavenger is insufficient evidence to prove OH· involvement. Desferrioxamine, which suppresses iron-dependent OH· generation (Chapter 2), also protected against the injury, suggesting that iron ions are required. Iron ions able to catalyse OH· production could arise as a result of cellular injury and metal ion release from intracellular storage sites, proteolytic digestion of metalloproteins, and/or breakdown by H_2O_2 of haemoglobin liberated as a result of bleeding upon tissue reperfusion. Other scientists found that regional intestinal ischaemia in cats resulted in an accumulation of hypoxanthine in the tissue; the hypoxanthine disappeared quickly on reperfusion and both lipid peroxidation and the formation of oxidized glutathione could be measured in the reperfused intestine. It has also been reported that SOD diminishes the lethal effect of bowel ischaemia in rats. Finally, it is known that intestine is very rich in xanthine dehydrogenase activity, and that conversion of this enzyme to oxidase upon ischaemia/hypoxia does occur.

As far as intestinal ischaemia/reoxygenation is concerned, the essential features of the proposal of McCord *et al.* (Fig. 8.4) seem to be supported by experimental evidence. Although allopurinol is not a specific inhibitor of xanthine oxidase (Section 8.4.2), other evidence for the involvement of this enzyme has been obtained. Pterinaldehyde (2-amino-4-hydroxypteridine-6-carboxaldehyde), a powerful xanthine-oxidase inhibitor that often contaminates commercial preparations of the vitamin folic acid, also offered protection. Feeding animals on a diet rich in tungsten decreases tissue concentrations of xanthine dehydrogenase because the tungsten is 'accidentally' incorporated into the enzyme in place of molybdenum at the active site. Tungsten-containing enzyme is inactive. Intestinal segments from rats pretreated in this way showed much less re-oxygenation injury after hypoxia than segments from normal rats.

The hypoxanthine–xanthine oxidase system may not be the only source of radicals to which reoxygenated intestine (and other tissues) are subjected *in vivo*. Generation of oxygen-derived species by neutrophils entering reoxygenated intestine is another potential source of damage. Entering neutrophils (as well as neutrophils trapped within the hypoxic tissue) can adhere to endothelium and release products (O_2^-, H_2O_2, eicosanoids, proteolytic enzymes) that can worsen injury. Neutrophil depletion of animals, or pre-treatment of them with an antibody that prevents neutrophil adherence to endothelium, has been reported to diminish reoxygenation injury to intestine in whole animal studies. Mitochondria damaged by ischaemia may also 'leak' more electrons than usual from their electron transport chain (Chapter 3), thus forming more O_2^-.

A role of oxidants in reoxygenation injury has been proposed for some other parts of the gastrointestinal tract. Severe bleeding can cause a rapid fall in blood pressure, leading to tissue hypoxia, so that when blood volume is restored a 'shock' syndrome (*haemorrhagic shock*) can result. Pre-treatment of rats or cats with allopurinol or SOD before haemorrhagic shock has been reported to minimize the gastric lesions observed. Both allopurinol and desferrioxamine have been claimed to increase survival in dogs subjected to severe haemorrhagic shock. Oxidants and eicosanoids may also play some role in shock induced by bacterial endotoxin e.g. in sepsis caused by Gram-negative bacteria. The effects of endotoxin may be largely mediated by increased formation of *tumour necrosis factors*, (sometimes called *cachectin*), polypeptide hormones released from monocytes and macrophages. TNFs have many biological effects, including an ability to sensitize neutrophils and monocytes so that they respond to stimuli by increased oxidant formation. Nathan in the USA has reported that TNFs can stimulate oxidant production by neutrophils that are adhering to surfaces, as can happen in ischaemic/reperfused tissues.

8.4.2 Cardiac injury

The pioneering studies of Parks, Granger, McCord *et al.* naturally attracted the attention of cardiologists, since ischaemic/hypoxic injury to heart muscle is a major cause of death. Again, after relatively short periods of ischaemia or hypoxia, reoxygenation gives an additional insult to the tissue that can be diminished by SOD (manganese or copper–zinc enzymes), by catalase and scavengers of the OH˙ radical such as mannitol, by desferrioxamine, or by allopurinol. For example, reperfusion of heart after periods of ischaemia too short to produce any necrosis results in prolonged depression of contractile function (*stunning*) and sometimes generation of arrhythmias. Many studies showing protective effects of the above antioxidants have been done with isolated perfused hearts from rats, rabbits, dogs, and pigs. Some studies have also been performed with organs *in situ*. For example, in open-chest dogs (chest opened, coronary-artery branch partially or completely ligated, clamp removed after various periods) it was shown that pre-treatment of animals with SOD, catalase, or allopurinol could decrease the size of the infarction produced by occlusion of a coronary-artery branch, followed by reoxygenation. The extent of the protection by antioxidants has been very variable in different experiments. At least three reasons can account for this. Firstly, the importance of reoxygenation injury as a fraction of total tissue damage declines as the period of ischaemia increases, so that in experiments with long periods of ischaemia/hypoxia so much tissue injury has been done in that phase of the experiment that radical scavengers are unlikely to offer much protection (a tissue irreversibly injured by O_2-deprivation will die however it is reperfused). In dog heart, for example, coronary occlusion for less than about 20 minutes produces stunning on reperfusion, but essentially no infarct. Antioxidants are very effective in protecting against stunning. Occlusion for more than 4 hours followed by reperfusion gives an infarct about the same size as that resulting from permanent occlusion (i.e. damage is maximum and cannot be further increased), and antioxidants have no effect. Second, in many animal studies (especially with open-chest dogs), the extent of O_2 deprivation during arterial occlusion can vary considerably because of collateral vessels. Thus 'ischaemia' is often hypoxia, to variable extents. If flow does not cease completely, not only does some O_2 enter the 'ischaemic' tissue, but also some metabolic products, such as H^+, are removed and acidosis is less severe. Third, studies by Hearse in England and McCord *et al.* in the USA have shown that many of the 'antioxidants' used in these studies show bell-shaped dose–response curves. For example, in studies upon production of arrhythmias in rat hearts after a brief period of ischaemia, there is an optimal concentration of SOD for protection, and higher concentrations give smaller protective effects. The reason for these effects remains to be established. For example, a deleterious effect of impurities in some of the antioxidants studied might become significant at very high antioxidant concentrations.

The large number of successful reports using *in-vitro* and *in-situ* studies upon animals naturally led to proposals that the model shown in Fig. 8.4 could account for reoxygenation damage in heart. Experiments *in vitro* found that exposure of myocytes or whole hearts to oxidant-generating systems produces severe injury, including inactivation of the ATP-dependent Ca^{2+}-sequestering system of cardiac sarcoplasmic reticulum. Ca^{2+} has also been suggested to activate the protease that converts xanthine dehydrogenase to the oxidase form (Fig. 8.4). Of course, *in-vivo* sources of radicals other than xanthine oxidase, such as mitochondria (Section 8.4.1) and infiltrating neutrophils, may also be important. Neutrophils play little role in the stunning seen after brief ischaemia followed by reperfusion, but they may be important after longer ischaemic periods. Thus depleting animals of neutrophils, or injecting them with an antibody that prevents neutrophil adherence to endothelium, before performing ischaemia/reperfusion studies *in situ* (using an open-chest system) can decrease the size of the infarct zone produced.

However, some fundamental problems have arisen. Rat and dog heart contain xanthine dehydrogenase activity, but the rate of proteolytic conversion of dehydrogenase to oxidase during ischaemia is very slow (e.g. in rat heart it has been reported that about 20 per cent of enzyme is present as oxidase to start with and it takes about 4 hours of ischaemia for this to increase to 30 per cent). The activity of xanthine dehydrogenase activity in rat heart has been reported to increase with the age of the animals. Even worse, rabbit, pig, and human hearts have been reported not to contain any xanthine oxidase activity at all. If these reports are correct, how then can one explain the protective effects of allopurinol reported in rabbit heart by several groups? It may be that allopurinol inhibits another enzyme involved in oxidant generation. It might also protect by preventing the depletion of purine nucleotides, that can be used as substrates for resynthesis of ATP on reoxygenation. The authors have pointed to the fact that pre-treatment of animals with allopurinol caused formation of oxypurinol (Chapter 3). For example, administration of allopurinol to humans produces plasma concentrations of oxypurinol of 40–300 μmoles/litre. Oxypurinol is a good scavenger of $OH^{\cdot}$ and of the myeloperoxidase-derived oxidant HOCl. It has also been questioned as to whether the established protective action of mannitol is really due to $OH^{\cdot}$ scavenging. Both mannitol and glucose react with $OH^{\cdot}$ with about the same rate-constant (Chapter 2), but mannitol is far more protective against ischaemia/reperfusion-induced arrhythmias in isolated rat heart than is equimolar glucose. If $OH^{\cdot}$ is really formed in the reperfused heart, then a source of iron ions must be identified. This iron might be released as a result of cell injury (Section 8.4.1) or from bleeding caused by reperfusion of myocardium that has undergone vascular damage during ischaemia. Another possible source is myoglobin. The authors have shown that incubation of cardiac oxymyoglobin with excess H_2O_2 can lead to breakdown of the haem and liberation of iron ions in a form capable of catalysing $OH^{\cdot}$ production.

Despite these problems, the effectiveness of SOD, desferrioxamine, and other antioxidants in many of the studies performed cannot be doubted. In 'classical' human myocardial infarction, it has been suggested that occlusion of a branch of the coronary artery is maintained for such a long period that ischaemic/hypoxic injury is the sole cause of tissue death, reperfusion injury being insignificant. However, the early use of thrombolytic agents (such as streptokinase or tissue plasminogen activator) to produce clot dissolution is now becoming commonplace in many countries. Combined use of an antioxidant (such as desferrioxamine or SOD) and a thrombolytic agent might be expected to give enhanced benefit. Controlled trials using human recombinant CuZnSOD are underway to test this possibility. Reperfusion injury might also play a significant role in the depressed myocardial function sometimes seen after open-heart surgery and transplantation (Section 8.4.4), and in exercise-induced angina pectoris.

8.4.3 *Cerebral injury*

We have already seen that traumatic injury to the brain or spinal cord can lead to tissue degeneration that may well involve free-radical reactions such as lipid peroxidation (Chapter 4, Section 4.9). Cerebral ischaemia or hypoxia, followed by reperfusion, should also stimulate lipid peroxidation, again probably because of metal–ion release from injured cells or from bleeding in the reperfused area. The acidosis caused by lactate accumulation may accelerate lipid peroxidation and $OH^{\cdot}$ formation. It is not yet clear if ischaemia in brain causes the xanthine dehydrogenase to oxidase conversion (Fig. 8.4). Rat and gerbil brain have been reported to contain low activities of xanthine dehydrogenase, but the activity of this enzyme in human brain does not appear to have been clearly established. In any case, several groups have reported that SOD, desferrioxamine, or other iron chelators (e.g. the lazaroids; Section 4.9) have some beneficial effects in minimizing brain or spinal-cord injury after ischaemia/reperfusion studies in animals. Few of the

experiments with enzymes have been accompanied by controls with inactivated enzymes, however.

Another mechanism that may contribute to cerebral injury after hypoxia/
ischaemia is the generation within the tissue of excitatory amino-acid
neurotransmitters such as glutamate or aspartate. These excitatory neurotransmitters cause neurones to 'fire off' continuously until they are damaged.
Synthetic excitatory neurotoxins of this type include *kainic acid* and *quino-
linic acid*. Dykens, Stern, and Trenkner, in the USA, reported that the death
of cultured mouse cerebellar neurones caused by kainic-acid treatment could
be prevented by pre-treating cells with allopurinol or by adding SOD plus
catalase to the culture medium. They therefore suggested that excitotoxin-
induced neural damage may be mediated by O_2^- generated by xanthine
oxidase (presumably oxidizing hypoxanthine accumulating as a result of
excessive ATP degradation during repeated neuronal 'firing'). These observations provide an interesting link between two apparently-different mechanisms of brain injury during ischaemia/reperfusion: free-radical reactions and
the generation of excitatory amino acids.

8.4.4 *Preservation of organs for transplantation: kidney, liver, and skin*

The sensitivity of kidney to oxidant damage is well established. For example,
kidney damage by the antibiotic gentamicin may involve increased oxidant
production (Chapter 6, Section 6.11). Deposition of antigen–antibody complexes in the glomeruli can lead to complement activation and infiltration of
neutrophils. Resident phagocytic (mesangial) cells in the glomeruli can also
respond to complement activation by secreting eicosanoids, O_2^- and H_2O_2.
Ward *et al.* in the USA found that infusion of an antibody against
glomerular basement membranes into rats caused severe glomerular inflammation involving both neutrophils entering the tissue and complement
activation. Infusion of catalase offered protection, but infusion of SOD had
only a small protective effect. Both H_2O_2 and HOCl (from the action of
myeloperoxidase) might be involved in the damage, as might the action of
proteases such as elastase and collagenase. Several other groups have
reported that injection into rats of 'nephrotoxic serum' containing antibody
to rat glomerular basement membrane produces renal injury that can be
ameliorated by treating the animals with SOD. The acute phase of the
glomerular injury produced by injecting antibody against glomerular basement membrane into rabbits was reported to be suppressed by desferrioxamine. These variable results indicate the complexity of this model system of
immune injury to kidney.

Reoxygenation of animal kidneys after a brief period of ischaemia or
hypoxia produces a reoxygenation injury that can again contribute to tissue
damage. (After prolonged periods of ischaemia, the ischaemic damage itself

may be so extensive that reperfusion injury is insignificant; Section 8.4.3.) Several groups have reported that pre-treatment of whole animals or isolated organs with SOD, catalase, desferrioxamine, or allopurinol diminishes reperfusion injury.

There are two clinical situations in which the kidney is subjected to temporary hypoxia; shock due to severe blood loss (when all body organs are affected to variable degrees) and renal transplantation. Some degree of hypoxia in organs stored for transplantation is difficult to avoid; transplanted kidneys would thus be subjected to a limited degree of reperfusion injury after removal of the renal artery clamp. Rat kidney contains xanthine dehydrogenase, although the rate of its irreversible (proteolytic) conversion to xanthine oxidase in ischaemic tissues is slow (with a half-life of about 6 hours). There appears to be much less xanthine oxidase in dog or human kidney. Hence it is uncertain whether the model in Fig. 8.4 accounts for reperfusion injury in kidney.

Transplanted kidneys are frequently infiltrated by host phagocytic cells, and it is possible that oxidants produced by these might sometimes contribute to transplant rejection. Thus antioxidants may have a beneficial role not only in allowing organ preservation for longer periods, but also in diminishing graft rejection. These comments could equally well be applied to other transplanted organs, such as liver (see below), heart (Section 8.4.3), and to transplants of corneas or of skin flaps. Thus several groups have reported that the survival time of hypoxic skin flaps is increased by antioxidants such as SOD, desferrioxamine, or allopurinol.

Several groups have been interested in the possibility of reperfusion injury in transplanted livers, although no clear demonstration of the effectiveness of antioxidants in minimizing tissue damage has yet been achieved. Rat liver contains xanthine dehydrogenase, but again this is only very slowly converted to oxidase during ischaemia. Dog or human liver have been reported to contain very little of either form of the enzyme.

8.4.5   *Rheumatoid arthritis*

The involvement of neutrophils in reoxygenation injury in gut (Section 8.4.1), kidney (Section 8.4.4), and heart (Section 8.4.2) implies that 'inflammatory' responses such as neutrophil infiltration can contribute to hypoxic/ischaemic injury. It is also possible that hypoxia/reperfusion injury might contribute to tissue damage at sites of inflammation. Blake *et al.* in England have shown that tensing an inflamed human rheumatoid knee joint can generate intra-articular pressures in excess of capillary perfusion pressure, resulting in sharp drops in O_2 concentration within the joint. Upon relaxing the joint, the O_2 concentration gradually returns to normal. Thus a reperfusion injury is certainly possible. Blake *et al.* found some xanthine-oxidizing activity in both normal and rheumatoid synovial membranes, and the concentration of hypoxanthine has been reported to be increased in synovial fluid from patients with rheumatoid arthritis. These observations show that injury by the mechanism in Fig. 8.4 might be possible.

However, O_2 concentrations might also affect oxidant production by activated neutrophils, which are abundant in the inflamed rheumatoid joint (Section 8.1). Edwards *et al.* (1983) in Wales pointed out that the K_m for O_2 of the respiratory burst oxidase of rat neutrophils is within the range of physiological O_2 concentrations in body fluids. If the same is true of human neutrophils, it follows that local O_2 concentrations at sites of inflammation could modulate O_2^- and H_2O_2 production.

8.4.6 *Pancreas*

Inflammation of the pancreas can be initiated by several means, including excess alcohol intake, pancreatic-duct obstruction by a migrating gallstone, or by a period of ischaemia, e.g. as a result of haemorrhagic shock. Bulkley *et al.* in the USA found that the injury induced by fatty-acid infusion, ischaemia, or partial duct obstruction in isolated, perfused, dog pancreas could in each case be diminished by including SOD and catalase in the perfusing medium (SOD or catalase given separately had little effect). Conflicting results have been obtained as to whether SOD and catalase are effective against pancreatitis in *in-vivo* models. The issue is important because current therapies for the treatment of pancreatitis in humans are inadequate.

8.4.7 *Limb ischaemia*

Limb ischaemia or hypoxia occurs during several cardiovascular conditions, such as atherosclerosis, embolism, arterial injury resulting in rapid blood loss, and (to a controlled extent) in the use of tourniquets to provide a 'bloodless field' in orthopaedic surgery. Modern microsurgical techniques for re-attachment of digits or limbs often result in the re-attachment of tissue that has been ischaemic for a considerable time. Skeletal muscle is, compared with most tissues, fairly resistant to ischaemic injury. However, after several hours of ischaemia, an ischaemic/reperfusion injury to muscle can be demonstrated. Experiments with rat hindlimbs have shown some protective effects of SOD plus catalase against the injury, and other experiments with isolated dog gracilis muscle have shown some protection by SOD, allopurinol, or dimethylsulphoxide.

8.5 Lung damage and the adult respiratory distress syndrome

We began this book by discussing the toxic effects of elevated O_2 concentrations upon organisms and the proposal that free radicals are involved (Chapter 1). In Chapter 3 (Section 3.5.1) we reviewed the evidence that exposure of lung to elevated O_2 concentrations can result in increased intracellular generation of O_2^- (and hence of H_2O_2) by promoting the leakage of electrons onto O_2 from the electron transport chains of mitochondria and endoplasmic reticulum.

A second source of oxygen-derived species within lung *in vivo* is the accumulation and activation of neutrophils. Few neutrophils are present normally in lung, but they accumulate during the later stages of O_2-induced lung damage, perhaps because lung macrophages damaged by hyperoxia release chemotactic factors that attract neutrophils. Neutrophil involvement may be particularly important in some forms of the so-called 'adult respiratory distress syndrome'. ARDS, acute respiratory failure due to pulmonary oedema, is a frequent result of severe shock, aspiration of stomach contents, and tissue damage due to burns, accidents, or massive infections. Several reports document the dramatic appearance of neutrophils or neutrophil-derived products in lung-lavage fluids of patients early in the course of the syndrome. This neutrophil accumulation may well be mediated by activation of the complement system, and it has been proposed that oxygen radicals produced by activated neutrophils may be of importance in producing the lung damage seen in some forms of ARDS. For example, Ward *et al.* in the USA found that the acute lung injury produced in rats after severe skin burns or injection of cobra-venom factor (both cause complement activation, and neutrophil infiltration into lung) could be minimized by treating the animals with SOD, catalase, desferrioxamine, or apolactoferrin, and worsened by infusion of iron(III) salts. These results imply that iron-dependent formation of OH˙ radicals from O_2^- and H_2O_2 generated by activated neutrophils plays a significant part in the lung damage. Platelets may also be involved, perhaps because platelet-derived factors can accelerate neutrophil O_2^- generation. Activated neutrophils additionally produce HOCl by the action of myeloperoxidase, and they secrete elastase. Lung elastic fibres can be degraded by elastase if a_1-antiprotease is inactivated by oxidants such as HOCl (Chapter 6, Section 6.4). Severe burns can lead to extensive fluid loss and a shock syndrome (Section 8.4), and administration of allopurinol or SOD to mice before thermal skin injury has been reported to increase survival times.

Alveolar macrophages might also contribute to lung damage in ARDS. Instillation of immunoglobulin A into the lungs of rats, followed by intravenous injection of antigen, produces a lung injury in which neutrophils are not involved. Complement activation may instead accelerate oxidant production by alveolar macrophages. In this model, SOD, catalase, or desferrioxamine again diminish the lung injury observed. Macrophages release an elastase that is not inhibitable by a_1-antiprotease (Chapter 6, Section 6.4).

The relevance of these observations to ARDS in humans is supported by observations that ARDS patients may exhale H_2O_2 at an unusually high rate, and may show abnormally low plasma ascorbate/dehydroascorbate ratios. However, it must be realized that ARDS is a blanket term for several closely-related conditions provoked by different stimuli, and it is likely that the precise role played by oxidant damage differs in each. Thus ARDS has been observed to develop in neutropenic human leukaemia patients, i.e. it can occur without significant neutrophil infiltration into the lung. Studies on the protective action of catalase in animal systems must also be interpreted with care, since Gordon *et al.* (1987) in the USA have shown that contaminants in the protein can account for the protective effects observed in some systems. This underlines the point that all studies of the antioxidant action of enzymes must be accompanied by controls using inactivated enzyme.

Xanthine dehydrogenase/oxidase has been reported to be present in rat lung, and the possibility of reoxygenation injury (Fig. 8.4) to lung during shock caused by excessive blood loss, or during heart–lung transplantation, has been raised. How much xanthine dehydrogenase/oxidase activity is present in human lung remains to be established. As stressed previously (Chapter 3, Section 3.5.3), the histological complexity of lung means that a low overall activity of an enzyme in lung homogenates is compatible with a high activity in one or two cell types, such as endothelial cells.

8.6 Exercise-induced oxidant damage

During exercise, bodily O_2 consumption is greatly increased and it seems likely that more O_2^- and H_2O_2 form *in vivo*, since O_2^- can be a product of electron leakage from mitochondrial electron transport chains (Chapter 3, Section 3.5.1). Thus Davies *et al.* in the USA found that severe forced physical exercise in rats results in muscle damage, seen as a decrease in mitochondrial respiratory control, loss of structural integrity of sarcoplasmic reticulum, and increased levels of some markers of lipid peroxidation. Vitamin E-deficient rats have markedly lower endurance capacity for exercise. Protection against oxidative damage might be offered by careful endurance training: Quintanilha and Packer in the USA found that such training increased the activities of glutathione peroxidase, glutathione reductase, catalase, and SOD in rat heart and skeletal muscle, although it seemed to decrease the content of vitamin E in muscle mitochondria.

Do these observations have relevance in humans? Prolonged exercise, especially in untrained individuals, produces muscle damage, as demonstrated by microscopic studies or by release of muscle enzymes into the circulation. Some myoglobin can also leak into plasma. Exercise increases the number of circulating neutrophils and may produce some features of an acute-phase response, e.g. falls in plasma zinc and iron, and rises in C-reactive protein. Losses of iron and zinc in sweat may contribute to these changes. The authors found that trained athletes have higher concentrations of caeruloplasmin in plasma than normal controls; caeruloplasmin is part of the extracellular antioxidant defences (Chapter 4, Section 4.7.6). Dillard *et al.* in the USA found that some human subjects responded to exercise by increased exhalation of pentane (possibly arising from lipid peroxidation), an effect that could be diminished by pre-treating the subjects with excess oral vitamin E. By contrast, other scientists have found no increase in serum TBA reactivity after maximal exercise by long-distance runners.

It is certain that exercise can induce muscle damage. Damaged tissues often peroxidize more rapidly than normal (Chapter 4, Section 4.9). Thus the extent of muscle damage may be a simple determinant of whether or not lipid peroxidation end-products can be detected in the plasma of exercising humans. In any case, TBA reactivity is an unspecific method to look for these end-products. Hence oxidant formation during exercise may be a consequence of muscle injury rather than a cause of it.

8.7 Ageing

As an organism ages, its chance of death increases, so that all individuals of a given species are dead by some age, characteristic of that species. It seems likely that the maximum lifespan is around 90–115 years for humans. Theories to explain the mechanism of ageing are currently grouped under two main headings, *developmentally-programmed ageing* and *damage accumulation ageing*.

In primitive populations, few individuals reach their maximum lifespan because of infectious diseases or lack of an adequate food supply. In Western societies, such deaths are rare. Those few people who die under the age of 35 usually do so as a result of accidents, whereas older people more often die of diseases such as ischaemic damage to the heart (Section 8.4.2) or cancer (Section 8.8). The overall incidence of cancer varies strikingly with age; for most cancers the rate of incidence rises approximately proportionally to the fourth to sixth power of age. This is seen both in short-lived species such as rats and mice (about 30 per cent have cancer at age 2–3 years) and long-lived species such as humans (about 30 per cent have cancer at age 85). Thus the average lifespan (or *mean lifespan*) of 'advanced' societies is greater than that of primitive ones, but the *maximum* lifespan is probably no different. As we have seen, many of these fatal diseases, which cluster in the latter third of our lifespan, have some free-radical component.

The nature of the ageing process has been the subject of considerable speculation. Any theories proposed must explain that, although maximum lifespan may be fixed for a species, the lifespan actually achieved by a member of that species can be altered by environmental conditions. Thus restriction of food intake during the early growth phase of life has been shown to produce statistically significant increases in mean lifespan in several species including rats, *Daphnia*, rotifers, and *Drosophila*. The results are complicated and the age at which underfeeding is introduced is critical. The composition of the diet is another important factor. The most dramatic example of the effect of dietary intake on lifespan can be seen with the queen honeybee (*Apis mellifera*). The queen can live for as long as 6 years whereas worker bees live from 3 to 6 months. The difference in lifespan cannot be a fundamental difference in genetic make up of the insects since their genomes are identical. The difference in lifespan is simply the result of whether or not the insects were fed royal jelly during their larval stage, which increases lifespan by some twentyfold.

Lifespan can also be affected by temperature. Cold-blooded animals, such as insects and reptiles, live longer at lower temperatures. For example, *Drosophila* has a mean lifespan of 120 days at 10 °C, but only 14 days at 30 °C. It seems likely that this effect is due to greater metabolic activity and oxygen consumption at the higher temperatures. Indeed, there is evidence for an inverse correlation between the basal metabolic rates of animals and their lifespan in that, in general, larger animals consume less oxygen per unit of body mass than do smaller ones, and they live longer. We saw in Chapter 1 that cold-blooded animals are much more resistant to the toxic effects of elevated oxygen concentrations at lower environmental temperatures. Insects consume much more oxygen when flying than when they are at rest. Prevention of houseflies from flying by removing their wings, or by confining them in small bottles, has been shown by Sohal in the USA to produce a marked increase in lifespan (Fig. 8.5).

8.7.1 *Programmed ageing*

The programmed theory of ageing proposes a purposeful sequence of events encoded into the genome. Evidence in support of ageing changes under genetic control can be seen in isolated examples, the most dramatic of which is the Pacific Coast Kokanee salmon (*Oncorhynchus nerka kennerlyi*). The life cycle of this fish culminates in a migration to spawning grounds, followed by spawning and the death of the fish within a few weeks. During this period distinct morphological changes occur to the snout, jaws, back, and skin of the fish. These changes, resulting in widespread tissue degeneration, appear

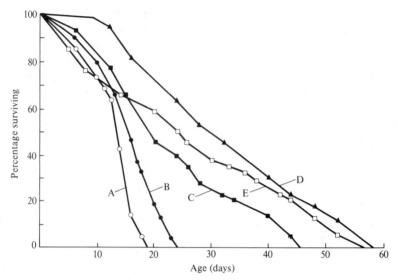

Fig. 8.5. Effect of activity on the lifespan of the housefly, *Musca domestica*. The survival curves of male houseflies under different experimental conditions. A: 50 flies in a large cage. B: 50 de-winged flies in a large cage. C: Each fly in a large cage. (The absence of other flies means that the insect is less disturbed and flies less.) D: Each normal fly in a small bottle to prevent flying. E: Each de-winged fly in a small bottle. Data by courtesy of Dr R. S. Sohal.

to be under hormonal control. Castration of immature fish can markedly delay the onset of degenerative changes.

The human progeroid ('premature ageing') syndromes (*Hutchinson– Gilford syndrome* and *Werner's syndrome*) also provide some evidence that a few genes can have major effects on ageing. Hutchinson–Gilford syndrome (progeria) is a rare disease (1 in every 4–8 million births) of genetic origin. The progeroid syndromes individually, however, only show some features of ageing and cannot be cited as truly representative of 'accelerated human ageing'. They are further discussed in Section 8.7.6.

Against a genetic theory of ageing is the observation that few wild animals ever survive to old age (they usually get killed first). Therefore, nature is unlikely to have favoured the development of genetically-programmed ageing. Extending this observation to humans, it should be noted that since life expectancy in pre-historic times was about 18 years, the human species has survived for a much longer period of time with an 18-year life expectancy than with the present 75-year expectancy. Natural selection for a genetically-programmed ageing seems unlikely to have occurred under these circumstances. An alternative hypothesis is that it is *longevity* that has evolved, not an ageing mechanism, and this idea could explain the remarkably constant maximum lifespans of individuals in each species. Thus close relatives of longer-lived humans also tend to live longer than average. A strong positive correlation between the efficiency of DNA repair processes (as measured by DNA excision repair activities) and species longevity has been claimed. These and many additional observations have led to the suggestion that there are 'longevity-assurance' or 'longevity-determining' genes. The latter concept has been explored by Cutler in the USA, who concludes that longer-lived species have better antioxidant protective mechanisms in relation to rates of radical generation than do shorter-lived species. For example, the activities of superoxide dismutase in different mammals were found to correlate with the metabolic rate multiplied by the maximum lifespan of the species (Table 8.3). Thus the superoxide dismutase activity of human liver, expressed per unit metabolic rate, is higher than that of other primate species and much higher than that of other mammals. Since SOD appears to have a similar structure and enzymic activity in all animals (Chapter 3, Section 3.1.2), Cutler has suggested that a change in gene regulation has allowed synthesis of higher cellular SOD concentrations in humans, contributing to longevity.

Table 8.3. Relationship of antioxidants to longevity. Results are abstracted from the article by Cutler, R. G. (1984). In *Free radicals in biology* (W. A. Pryor, Ed.), volume VI, pp. 371–428. Academic Press, New York.

Excellent positive correlation to maximum life span
  SOD activity (expressed per unit metabolic rate) in liver, brain, and heart of
    primate and rodent species.
  Plasma uric acid (per unit metabolic rate) in primates.

Weak to moderate positive correlation to maximum life span
  Ratio of MnSOD to total SOD activity in brain, but not liver, of primates.
  Serum carotenoids in range of mammals.
  Plasma vitamin E (expressed per unit metabolic rate) in a range of mammals.
  Plasma caeruloplasmin in primates.

Poor or no positive correlation to maximum life span
  Serum vitamin A in a range of mammals.
  *Plasma ascorbic acid in a range of mammals.
  *GSH concentrations in lens and whole blood in a range of mammals.
  *Activity of glutathione peroxidase or glutathione-S-transferase in brain, liver,
    or blood of a range of mammals.

* Some suggestion of negative correlations.

8.7.2   *Damage accumulation ageing*

Damage accumulation theories of ageing assume some progressive accumulation of damage because repair and maintenance is always less than that required for 'indefinite' survival. Faulty macromolecules can accumulate through 'wear and tear', and through errors in their synthesis.

Error catastrophe theory of ageing
The error catastrophe theory of ageing was introduced by Orgel in the USA in 1963. He proposed that errors in transcription of RNA and its translation into protein would lead to accumulation of altered non-functional proteins with age in cells, eventually reaching levels that ultimately lead to an 'error catastrophe' of complete failure to function. Some altered non-functional or less active enzymes have been isolated from aged cells (e.g. rat liver SOD, Section 8.7.3). In humans, development of cataract (lens opacity) is common after the sixth decade. The crystallins, which constitute more than 90 per cent of the lens proteins, become cross-linked and aggregate, by mechanisms that involve free-radical reactions (Chapter 5, Section 5.2). Abnormal proteins are found in old human erythrocytes. However, in both these cases, it does not seem that transcription and translation are at fault, but rather that proteins do not turn over quickly and so are not replaced when damaged (the mature human erythrocyte has no protein synthesis machinery and that of the lens is very limited).

The evidence for accumulation of altered proteins to high concentrations in most tissues is weak. Oliver *et al.* in the USA found little accumulation of altered proteins with age in isolated human fibroblasts, although fibroblasts isolated from patients with Werner's syndrome or progeria seemed to contain higher concentrations of abnormal proteins than did control fibroblasts. Further, the error rate (fidelity) of protein synthesis does not appear to change with age. Another factor that could be involved is if the cellular proteolytic mechanisms that degrade abnormal proteins (Chapter 3, Section 3.4.3) also decline with age. A little evidence consistent with this has been provided in some systems. Thus the capacity of erythrocytes or isolated lung fibroblasts to degrade abnormal proteins decreases with age. Again, however, one must point to the observation that there is no marked accumulation of abnormal proteins with age in most human tissues.

8.7.3   *Free-radical theory of ageing*

The free-radical theory of ageing was introduced in 1956 by Denham Harman in the USA. He proposed that normal ageing results from random deleterious damage to tissues by free radicals produced during normal aerobic metabolism. In support of this concept we know that radiation produces damaging free radicals and accelerates some features of the ageing process, although radiation-induced 'ageing' does not exactly mimic normal ageing. Oxygen free radicals are produced during normal metabolism and may escape scavenging to cause damage. For example, Ames in the USA has calculated that in excess of 1000 oxidative 'hits' per cell per day take place upon DNA in the human body (Chapter 2, Section 2.5.3). The effects of various toxins (Chapter 6) show that the antioxidant defences of many tissues are sufficient only to protect them against normal rates of radical generation; they cannot cope with increased rates. It seems reasonable to suggest that perhaps they can only cope with, say 99.9 per cent of normal radical generation, so that there is a very slow progressive damage, undetectable in the short term for humans. In addition, failure of the normal 'self-recognition' mechanisms with increasing age may allow autoantibodies to form (Section 8.3.1) and provoke increased attack by phagocytes. That free radicals are somehow involved in the ageing process is suggested by the accumulation of 'age pigments' (Section 8.7.4), and by the observations of Cutler on the correlation of SOD activity with lifespan. The involvement of radicals could also explain the approximate inverse correlation between the basal metabolic rates of animals and their lifespan, and the effect of temperature on lifespan and oxygen toxicity as described above (fewer radicals should be made at lower temperatures). The faster O_2 is consumed by an organism, the more radicals it is likely to make.

Cutler also found some correlation between lifespan and the concentrations of uric acid, carotenoids, and vitamin E in animals (Table 8.3). However, such important antioxidants as glutathione peroxidase and GSH are not positively correlated with lifespan—in fact, correlations tend to be negative. The negative correlation of glutathione-S-transferase (a 'detoxifying' system) with lifespan may relate to other observations showing that levels of cytochrome P-450 per unit body mass also tend to be smaller in long-lived animals than in species such as the rat and mouse. Hence the oxygenation of environmental toxins, which can result in formation of oxygen radicals (Chapter 3, Section 3.5.1) as well as sometimes generating cytotoxic and carcinogenic products (e.g. Chapter 6 Section 6.1 and 6.9; Chapter 8, Section 8.8), will be lower (per unit body mass) in longer-lived animals. However, the negative correlation of lifespan with glutathione peroxidase, a key enzyme in the metabolism of H_2O_2, might be held to cast some doubt on the validity of the whole analysis.

Is there evidence that antioxidant protective mechanisms fail with age? A negative answer to this question does not rule out the free-radical theory of ageing, but a positive answer would support it. Gershon in Israel observed that SOD purified from the livers of old rats is less stable to heating, and of lower activity, than the enzyme from younger rats. The SOD activities of mouse brain, rat lens, and rat erythrocytes have been observed to decrease slightly with age, as have the GSH concentrations and glutathione reductase activities of mouse tissues. Unfortunately, these positive results are matched by an equal number of negative ones. Thus no change in SOD activity with age has been observed in human cells in culture or in human erythrocytes.

Do free-radical reactions increase in tissues with age? Some evidence consistent with this is provided by studies of lipid peroxidation. For example, the membranes of the oldest erythrocytes in the human bloodstream show decreased fluidity and increased cross-linking of membrane proteins, although erythrocytes are not good tissue models of 'whole body' ageing; studies on nucleated post-mitotic cells are more relevant. Older rats have been claimed to exhale more ethane and pentane than do younger ones. However, if tissues deteriorate with age, then an increased rate of lipid peroxidation might be expected anyway, since lipid peroxidation can probably increase as a consequence of tissue degeneration (Chapter 4, Section 4.9). Ames *et al.* studied the excretion of thymine and thymidine glycol, as an index of oxidative damage to DNA *in vivo* (Chapter 2, Section 2.5.3), in human urine and found no change with age. However, they did report that, per unit body weight, monkeys, rats, and mice excrete more of these products than do humans. Urinary output of the glycols was found to be approximately correlated to the metabolic rate of these organisms. Overall, one may conclude that the evidence for an increase in intracellular free-radical reactions with age is not strong.

Attempts have been made to test directly the role of free radicals in the ageing process by supplying antioxidants to various organisms, and examining their effect on longevity. Thiol compounds (such as glutathione and mercaptoethylamine), butylated hydroxytoluene, a-tocopherol, and Santoquin (ethoxyquin; Chapter 4, Section 4.7.4) have been used. The most striking results have been obtained with lower organisms. Thus vitamin E prolongs the average lifespan of several fairly-simple organisms such as *Drosophila*, nematode worms, and the rotifer *Philodina* (Fig. 8.6). The antioxidant NDGA (Chapter 4, Section 4.7.4) increases the mean lifespan of *Drosophila* by 20 per cent, but the singlet oxygen scavengers, DABCO (Chapter 2, Section 2.8.1) and β-carotene, have no effect on the lifespan of this insect. NDGA also increases the lifespan of the yellow-fever mosquito *Aedes aegypti*, as does raising tissue GSH concentrations by giving thiazolidine-4-carboxylate (Chapter 3).

However, the effects of administered antioxidants on the lifespan of mammals are small or zero. Thus vitamin E has no significant effect on lifespan in mice, although it has been reported to diminish the decline in immune response with age in one strain of female mice. Claims that antioxidants such as mercaptoethylamine and Santoquin raise the mean lifespan of mice by up to 18 per cent have been challenged; apparently some of the control animals did not live as long as they should have done and so

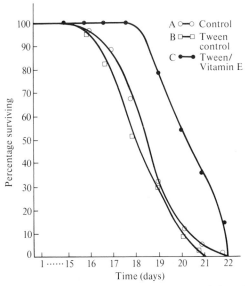

Fig 8.6. Survival of the rotifer *Philodina* in the presence of vitamin E. Three groups, each of 32 rotifers, were used. Group A is the control group. In group C, vitamin E dissolved in the detergent Tween was added to the growth medium. Group B received Tween alone. Rotifers were counted as dead if they did not move when prodded with a pipette. Data from the article by Enesco, H. and Verdone-Smith, C., *Exp. Gerontol.* **15**, 335–8 (1980), with permission.

the apparent increase in lifespan might have been a diminution of some environmental stress by the antioxidants rather than an effect on the ageing process itself. Thus the antioxidants could act to diminish tissue damage produced by radical reactions resulting from toxins in the food supply or in the surrounding air, or from excessive amounts of polyunsaturated fatty acids or deficiencies in antioxidants in the diet. Background exposure to ionizing radiation in some laboratories could be another factor.

The small (or zero) effects of large doses of antioxidants on the lifespan of mammals would tend to argue against the free-radical theory of ageing— scavenging of radicals would be expected to give much greater increases in lifespan. Cutler has approached this objection by proposing that excessive dosing of animals with an antioxidant can produce decreases in the rate of synthesis of 'natural' antioxidants, so that the total 'tissue antioxidant potential', which he claims is a major determinant of lifespan, remains unaltered. Consistent with this, Sohal *et al.* in the USA found that inhibition of catalase in houseflies (by feeding them aminotriazole) or inhibition of CuZnSOD (by diethyldithiocarbamate) produced increases in glutathione concentrations; neither aminotriazole nor diethyldithiocarbamate decreased the lifespan of the insects. One treatment that did decrease lifespan was giving the insects an excess intake of iron, which is particularly interesting in view of the stimulatory role played by low-molecular-mass iron complexes in such free-radical reactions as lipid peroxidation and OH˙ generation. Another point that must be made is that antioxidant inhibitors of lipid peroxidation, such as a-tocopherol and Santoquin, might not necessarily protect against free-radical damage to DNA and non-membrane-bound proteins, nor could any molecule acting only as a scavenger protect against 'site-specific' damage by radicals such as OH˙. Thus it is possible to argue this point in many ways, but it must be stressed that there is no clear experimental proof for the free-radical theory of ageing.

An involvement of oxygen radicals in the process of differentiation has also been suggested. Thus Sohal *et al.* observed striking increases in MnSOD activity during differentiation of the slime-mold *Physarum polycephalum*. Differentiation appeared to be accompanied by falls in GSH concentration.

The evidence most often quoted for increased free-radical formation during ageing is the increased accumulation of fluorescent pigments in tissues with age. Some interest has also been expressed in amyloid as a marker of ageing. Let us review what is known about fluorescent pigments; amyloid is considered in Section 8.7.7.

8.7.4 *Fluorescent tissue pigments*

The first description of the intracellular pigment known as *lipofuscin* was made in 1842 by Hannover, who reported its presence in neurones. The fluorescence properties of the pigment were described in 1911, and it has been found to accumulate, in amounts increasing with age, in many tissues, both in humans and in a wide variety of animals, including rats, nematodes, *Drosophila*, and houseflies. It is also found in several fungi and in cells in culture. In general, the most metabolically-active tissues show most lipofuscin deposition. In houseflies, the lipofuscin content of muscles increases with flight activity: flies that are more active live for shorter times (Fig. 8.5) and accumulate pigment faster. There is little or no lipofuscin in human heart muscle up to the second decade of life, but it then accumulates at about 0.3 per cent of the total heart volume per further decade of life. On the other hand, lipofuscin appears earlier in some tissues, such as the spinal cord.

Deposition of lipofuscin is not an irreversible process. For example, the drug *centrophenoxine*, also known as meclophenoxate, lucidril, or dimethyl-aminoethyl-*p*-chlorophenoxyacetate, has been claimed to decrease the number of fluorescent granules in the central nervous system of animals *in vivo*. Its mode of action is unclear, although it has sometimes been prescribed for elderly patients as a tonic, and even its 'lipofuscinolytic' effect has been disputed.

The terms *age pigment* and *lipofuscin* are frequently used as synonyms for the fluorescent pigment that accumulates in older tissues. However, the term 'age pigment' (if it is to be used at all) should, strictly speaking, only be applied to the pigment that accumulates in neurones and in cardiac muscle fibres where, because of their inability to divide, pigment accumulation is roughly correlated with age. In other tissues, the normal cell turnover can result in lipofuscin removal.

The colour of lipofuscin varies from red, through yellow, to dark-brown, and it occurs intracellularly as granules bounded by a single membrane, their diameters being in the range 1–5 μm. Both the number of granules, and their size, increase with age. Many different types of lipid are present in the pigments, including triglyceride, phospholipid, and cholesterol, and an equally wide variety of proteins as judged from the amino-acid composition of hydrolysed pigments. Lipofuscin contains a high concentration of metal ions such as zinc, copper, and especially iron.

Extraction of lipofuscin granules with a mixture of organic solvents (chloroform plus methanol) solubilizes part of the material, and the solution so obtained shows fluorescence characteristics quite similar to those of the conjugated Schiff bases, dihydropyridine dicarbaldehydes and aldehyde polymers formed during lipid peroxidation (Chapter 4, Section 4.6.7). However, all of these extracted fluorophores display blue fluorescence, whereas microscopists observe green to yellow fluorescence from lipofuscin *in situ*. This suggests that the extracted material is not truly representative of the tissue fluorophores.

On the basis of the fluorophores extracted by organic solvents, it is widely thought that lipofuscin represents the end-product of the oxidative destruction of lipids, and their cross-linking with proteins and other compounds bearing amino groups. However, work by Porta in the USA and by Jolly *et al.* in New Zealand suggests it may be misleading to emphasize a central role for lipid peroxides in the composition of lipofuscin (see Section 8.7.6). For example, Jolly *et al.* found that the pigment accumulating in the livers of sheep with the disease ceroid lipofuscinosis contained two-thirds protein. As already noted in Chapter 4 (Section 4.6.7) fluorescent material can be generated *in vitro* by free-radical damage to carbohydrates, DNA, and proteins, as well as to unsaturated lipids.

There is a stronger case for implicating increased lipid peroxidation as a key event in formation of the fluorescent pigment *ceroid*, a term first introduced by Lillie in 1942. The amount of this pigment produced in animal tissues is greatly increased if they are fed on diets deficient in vitamin E or abnormally rich in polyunsaturated fatty acids. Such feeding induces a 'yellow fat disease' in, for example, pigs and minks that can be prevented by feeding excess vitamin E or other antioxidants such as N,N'-diphenyl-p-phenylenediamine (Chapter 4, Section 4.7.3). Mitchinson in England has shown that macrophages in cell culture accumulate ceroid when they are exposed to complexes of albumin and lipids rich in polyunsaturated fatty acids (Fig. 8.7). Ceroid formation does not take place, however, when the lipids contain monounsaturated fatty acids or a mixture of cholesterol and albumin, nor if antioxidants such as BHA or BHT (Chapter 4, Section 4.7.4) are added to the culture medium together with the lipids. Peroxidation of lipids by macrophages has been implicated in the pathogenesis of athero- sclerosis (Section 8.2); it is interesting to note that ceroid within macrophages and extracellular ceroid (presumably released from lysing cells) are present in human atherosclerotic lesions. Collins in Sweden exposed human glial cells in culture to rat liver mitochondria, some of which could be taken up by phagocytosis. The result was an increased accumulation of fluorescent pigment in the glial cells.

Thus *ceroid* and *lipofuscin* must not be thought of as chemically identical, and the amount of lipofuscin accumulated does not necessarily reflect the rate of free-radical reactions. Hence accumulation of fluorescent pigments should not necessarily be regarded as a 'fingerprint' of free-radical damage to biological systems, nor is there any clear evidence that 'age pigments' are deleterious to cells. Another reason for caution in interpreting the signifi- cance of fluorescent pigments comes from a consideration of the way that these pigments arise. Histochemical studies strongly suggest that they are derived from lysosomes. The cellular lysosomes are continually digesting parts of the cell cytoplasm, a process known as *autophagy*. Proteins and lipids are taken into the lysosomes and degraded. It could be that lysosomes have a special affinity for peroxidized lipids and so gradually accumulate them, but it seems much more likely that lysosomes accumulate 'normal' lipids that peroxidize more rapidly than usual once they are inside these organelles. Disruption of lipid organization by lysosomal hydrolytic enzymes should aid lipid peroxidation, as would the high internal concentrations of copper and iron salts within lysosomes. These metal ions are probably derived from ingested metalloproteins, including transferrin (Chapter 2). The acidic pH of lysosomes should aid iron-dependent radical reactions, in part by helping to keep iron ions in solution. Lipofuscin and ceroid deposition are promoted by a number of abnormalities of fat metabolism, including *abetalipoprotein- aemia*. In this disease, dietary fat is digested and absorbed, but held up in the intestinal mucosal cells (Chapter 4, Section 4.7.3). Presumably more lipid than usual is degraded within lysosomes in such diseases. The association of fluorescent pigments with lysosomes is seen especially in *Chediak–Higashi syndrome*. This is a rare, human, autosomal-recessive disease in which 'giant granules' of lysosomal origin are present in many tissues. A similar syndrome has been described in Aleutian mink, albino whales, beige mice, and partial albino Hereford cattle. The lysosomal granules frequently enclose large amounts of pigmented lipid material. Their presence in neutrophils is

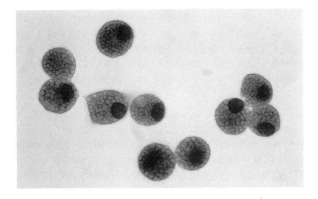

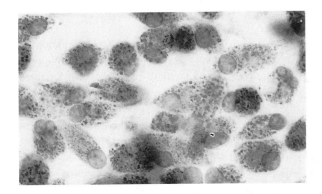

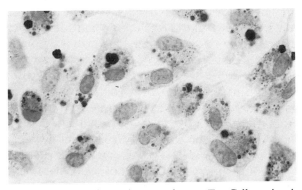

Fig. 8.7. Ceroid in mouse peritoneal macrophages. *Top* Cells maintained for 3 days in presence of cholesterol linoleate ester. 'Rings' of ceroid are present with unstained centres, also some granules × 1100. *Middle* as top, using cholesterol arachidonate ester. All cells contain innumerable small granules of ceroid rather than rings. *Bottom* as top, trilinolein (no cholesterol). Cells contain small granules and larger lumps of ceroid. Photographs by courtesy of M. J. Mitchinson and K. L. H. Carpenter. [Also see *Br. J. Exp. Pathol.* **68**, 427 (1987)]. No ceroid was found with cholesterol alone or a cholesterol oleate ester, i.e. ceroid accumulated only when the cells were incubated in the presence of a peroxidizable lipid material.

correlated with impaired bacterial capacity, causing recurrent infections. Other symptoms include defective skin pigmentation and nervous disorders. The primary cause of Chediak–Higashi syndrome is unknown.

8.7.5   *Progeroid syndromes*

Progeria (Hutchinson–Gilford syndrome) is extremely rare, affecting 1 in 4–8 million births. It is almost certainly genetic in origin being dominant rather than recessive. Affected individuals appear normal at birth but at about 12 months of age severe growth retardation is observed. Hair loss and loss of subcutaneous fat make the skin appear aged; pigmented 'age spots' are also present. The patients have an average weight of 25–30 lbs and a height of around 40 inches as teenagers. They have normal to above average intelligence with a median age of death at 12 years. Most deaths are due to heart failure. Many features of normal ageing are not present in progeria; ones absent include the increased frequency of malignant tumours, cataracts, and bone demineralization.

Werner's syndrome is an autosomal-recessive mutation with some striking differences to progeria. Affected individuals are usually normal during childhood but stop growth during their teens. Greying and whitening of the hair occurs at an early age, the skin appears old with a scaly appearance. The patients develop early cataracts, tumours, bone demineralization, and diabetes and show peripheral muscular atrophy, poor wound healing, poor gonad development, and atherosclerosis. They usually die in their 40s from atherosclerosis-related conditions.

Together the progeroid syndromes have most of the features associated with accelerated ageing, suggesting that normal human ageing can be greatly affected by simple mutations in one or both of a pair of key genes. It is interesting to note that fluorescent pigments are not markedly increased in the progeroid syndromes, nor have any defects in antioxidant protective enzymes (SOD, catalase, glutathione peroxidase) been reported.

By contrast, in *Huntington's chorea*, a dominantly-inherited condition that appears in middle age, there is a massive accumulation of fluorescent pigment in the brain, accompanying degeneration of striatal tissue. For example, the content of fluorescent pigment in the brain of one such patient at death was equivalent to the level found in human brain extrapolated to 195 years of age. Huntington's chorea appears as involuntary muscular movements plus mental changes. There is then progressive mental and physical degeneration, eventually leading to death. The cause of this unpleasant condition is unknown.

8.7.6 *Neuronal ceroid lipofuscinosis (NCL)*

Neuronal ceroid lipofuscinosis refers to a series of recessively-inherited disorders that occur world-wide in about 1 per 100,000 people, but more frequently than this in Scandinavia. Some patients develop symptoms during infancy (infantile NCL), others in early childhood (late infantile NCL), others in late childhood (juvenile NCL), and yet others after adolescence (adult NCL). Juvenile NCL is sometimes known as *Batten's disease.*

The onset of the disease is marked by behavioural abnormalities which worsen to include disturbances of vision and speech, as well as muscular and mental deterioration, associated with seizures. In the terminal stages the brain is severely damaged and patients assume a contracted position. A similar disease has been observed in English red setter dogs (Fig. 8.8) and in sheep in New Zealand. There is a rapid accumulation of fluorescent pigments (Fig. 8.9), not only in the brain but also in other tissues. Hence the 'neuronal' part of the name is slightly misleading. The cause of NCL disease is unknown. Koppang in Norway has shown that in the red setter the genetic mode of transmission is autosomal recessive and provides a good model for the human juvenile type of NCL. Pigment formation in the dog's tissues begins at birth, if not *in utero.* Maximal accumulation in the brain is reached after 1 year of age, although at this time there is no clinical evidence of neuronal dysfunction. As pigment formation continues, accumulation is offset by an increasing rate of nerve-cell death upon which the pigment granules are removed. In the sheep model, studies by Jolly in New Zealand have shown that the isolated fluorescent pigment contains about 70 per cent of a protein with fairly-low molecular mass, the rest being mainly lipid. However, no evidence has yet been found to support defects in lipid metabolism or antioxidant protection mechanisms in NCL (catalase, SOD, and glutathione peroxidase are normal in the human tissues studied). It has been proposed that the pigment is derived from the abnormal turnover and catabolism of lysosomal and other membranes as a result of some defect in proteolytic systems. Indeed, injection of inhibitors of lysosomal proteases into the brains of young rats has been reported to induce the rapid formation of lipofuscin-like particles.

In collaboration with Westermarck in Finland, the authors have observed an increased content of iron in the cerebrospinal fluids of patients with NCL, as well as an increased absorption of iron from the gut. Whether these changes in iron metabolism are a cause or a consequence of the disease remains to be established. In a regime pioneered by Dr Westermarck, patients with NCL appear to respond to treatment with selenium and other antioxidants such as vitamin E, BHT, and the iron chelator desferrioxamine, although full double-blind controlled clinical trials have not been carried out (and are very difficult to do in these conditions).

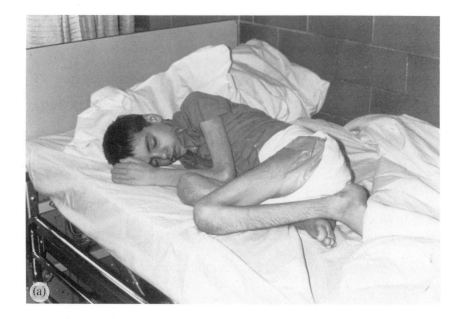

Fig. 8.8. Neuronal ceroid lipofuscinoses. (a) Shows the terminal clinical condition of a patient with juvenile NCL. (b) Shows a similar condition in an English setter dog. The dog, having been placed in such an awkward position, will not move, indicating the extensive damage suffered by the brain. Photograph by courtesy of Prof. D. Armstrong.

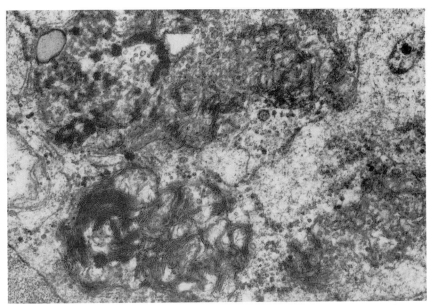

Fig. 8.9. Neuronal ceroid lipofuscinosis. The figure shows fluorescent pigment arranged in a 'fingerprint' manner in brain tissue from a 7-year-old child with the juvenile form of the disease. Photograph by courtesy of Prof. D. Armstrong.

8.7.7 *Amyloid*

'Amyloid' is a chemically-unspecific term that refers to tissue deposits of rigid, proteinaceous fibrils, 5–10 nm in diameter, which stain with Congo red dye and are birefringent under polarized light. Amyloid deposits vary widely in their protein composition and are the end-stage products of a variety of disease processes (e.g. Alzheimer's disease; Chapter 6, Section 6.13); the protein fibrils usually adopt a twisted β-pleated sheet conformation. Amyloids show whitish birefringence in unstained sections and a green birefringence under polarized light after staining with Congo red.

Diseases associated with amyloid deposition include tuberculosis, syphilis, rheumatoid arthritis, and multiple myeloma. Several rare inherited types of severe amyloid deposition have been described, such as *familial Mediterranean fever*, found principally in Mediterranean Jews and Armenians. This disease is eventually fatal, often because of kidney damage. Amyloid is also often deposited to a small extent in normal tissues during the ageing process (e.g. it is found in the seminal vesicles of 21 per cent of men over the age of 75). Organs severely affected by amyloid deposition are enlarged, abnormally firm, and their cut surfaces are waxy in appearance. Severe amyloid deposition interferes with the normal passage of water and solutes across the walls of blood vessels. In amyloidosis secondary to multiple myeloma, the amyloid deposit appears to be partially derived from antibody molecules, but this is not the case in other diseases. The *N*-terminal fragment of serum amyloid A, an acute-phase protein synthesized in increased amounts during tissue injury (Chapter 4, Section 4.7.6), may be an important component of other amyloid deposits but different proteins have been found in some deposits. Harman, in the USA, observed that the antioxidant Santoquin, prevented spontaneous amyloidosis in two strains of mice. The mechanism of this action is unknown.

8.7.8   *Model systems*

Munkres in the USA has described a 'rapidly-ageing' mutant of the fungus *Neurospora crassa*. The mutant ages, and accumulates fluorescent pigment, much more rapidly than does the wild-type strain. However, accumulation of pigment is decreased, and lifespan increased, by including in the culture medium antioxidants such as NDGA, thiol compounds, or *a*-tocopherol. This mutant has *increased* activities of superoxide dismutase, catalase, and glutathione peroxidase when compared to the wild-type fungus. Perhaps the primary defect of the mutant is an increased rate of radical generation, which stimulates the synthesis of the above protective enzymes, although not to a sufficiently great extent. Munkres has suggested that the mutant is unable to 'organize' its cell membranes properly, leading to an increased rate of lipid peroxidation. Indeed, a *N. crassa* mutant that was unable to synthesize inositol (Section 4.1) for incorporation into its membrane lipids was observed to show increased lipid peroxidation when placed on a medium deficient in this substance. Antioxidants offered some protection against the damage.

8.8   Cancer

A tumour may be defined as an abnormal lump or mass of tissue, the growth of which exceeds, and is unco-ordinated with, that of the normal tissue, continuing after the stimuli that initiated it have ceased. Most tumours form discrete masses, but in the *leukaemias* (tumours of myeloid or lymphoid cells) the tumour cells are spread through the bone marrow or lymphoid tissues, and also circulate in the blood. Tumours vary widely in their growth rates. The most important classification of tumours is that of *benign* or *malignant*. The cells of benign tumours remain at the site of origin, forming a cell mass. When growing in a solid tissue, they usually become enclosed in a layer of fibrous material, the capsule, formed by compression of the surrounding tissues. Benign tumours rarely kill, unless they press on a vital structure or secrete abnormal amounts of hormones.

Most fatal tumours are *malignant*, or *cancerous*. The cells of malignant tumours invade locally, and also pass through the bloodstream and lymphatic system to form secondary tumours (*metastases*) at other sites. The rate of growth and of metastasis formation differs from tumour to tumour. For example, breast cancers often grow very slowly, as do many rodent ulcers. Rapidly-growing malignant tumours usually lose their histological resemblance to their tissue of origin. Basal cell cancer of the epidermis rarely metastasizes, whereas melanoma frequently does.

The process of conversion of a normal cell to the malignant state is called *carcinogenesis*, and agents that induce it are called *carcinogens*. Carcinogenesis is a complicated multi-stage process: factors predisposing to malignancy include both inherited traits and environmental factors. Some people are much more sensitive to environmental carcinogens than are others, for example. The development of cancer in humans takes many years; progressive tissue and cellular changes are seen during the latent period. Often new cell populations appear that represent stages in evolution from normal cells through pre-neoplastic and pre-malignant to malignant cells.

Studies upon the action of carcinogens in mouse skin revealed that two distinct stages of carcinogenesis can be defined, and later work identified a third stage. This three-stage model seems to apply to many cancers. Initiation, the first stage, seems to relate to DNA alteration, e.g. reaction of carcinogens with DNA. Thus a major protection against cancer initiation may reside in efficiency of DNA repair. Successful initiation may require both DNA alteration and some degree of cell proliferation to allow the change to be 'fixed' in the DNA. Thus a successful initiator might not only cause changes in DNA, but also stimulate cell proliferation (this is especially necessary in tissues whose cells are normally non-dividing). Cell proliferation could be stimulated directly, or as a result of the need to replace some cells killed by the initiator.

Initiation is followed by *promotion*. Many carcinogens, given in large doses, are both initiators and promoters. However, a low dose of a carcinogen, itself too small to induce a tumour, can be effective if supplied together with certain non-carcinogenic substances known as *promoters*. For example, croton oil, a non-carcinogen, promotes the development of cancer by sub-carcinogenic doses of the hydrocarbon methylcholanthrene, but only if the oil is given simultaneously with, or after, the methylcholanthrene. Croton oil is obtained from seeds of the plant *Croton tiglium*. Fractionation of the oil has shown that the most powerful tumour promoter present is phorbol myristate acetate (PMA, Fig. 8.10), a compound that induces a respiratory burst in phagocytic cells (Chapter 7, Section 7.3.1). It is interesting to note in this context that many tumour promoters, including PMA, are

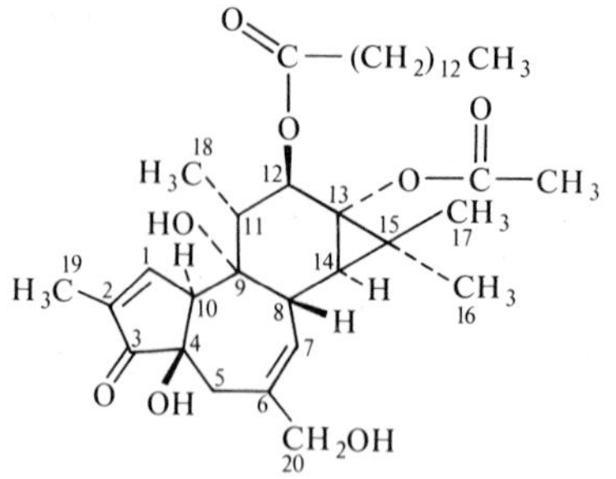

Fig. 8.10. Structure of PMA (phorbol myristate acetate). The proper chemical name of this compound is 12-*O*-tetradecanoylphorbol-13-acetate.

potent inducers of inflammation, but there is as yet no single unifying principle that explains the action of all the different promoting agents. During promotion, the latent altered genetic material of the initiated cell is thought to become expressed, perhaps by changes in the expression of genes that regulate cell differentiation and growth. Promoters can produce reversible changes in cell proliferation and phenotypic expression.

The final stage in carcinogenesis is the development of a pre-malignant lesion into a malignant one. This tumour *progression* may involve a second inheritable event involving additional attack upon DNA. Transposition of genetic material to a site where it can be transcribed may also be important.

The importance of changes in DNA to carcinogenesis is shown by several observations. About fifty viruses have been shown to induce tumours in infected animals after short latency periods. From many of these viruses *oncogenes* have been isolated, specific genes responsible for cell transformation. Nucleic acid hybridization studies showed that normal mammalian cells contain genes (*proto-oncogenes*) that are similar in base sequence to the viral oncogenes. Proto-oncogenes appear to encode growth factors or cell-surface receptors for such factors, or they may be involved in transferring the growth 'signal' from surface receptors into the nucleus (Table 8.4). The precise role of proto-oncogenes in relation to initiation, promotion, and progression of carcinogenesis is not entirely clear; activation of proto-oncogenes (Table 8.4) could play a role in all three processes, although it appears that activation of a single proto-oncogene cannot alone produce malignant transformation.

A few human tumours develop mainly, or wholly, as a result of an inherited genetic anomaly. *Familial polyposis coli*, in which colon cancer almost invariably develops, is one example; it is associated with a gene defect on chromosome 5. Defects in DNA repair mechanisms also predispose to the development of cancer. In *Xeroderma pigmentosum*, an inborn recessive defect in the ability to repair UV-induced lesions in DNA, there is severe skin

Table 8.4. A classification of proto-oncogenes. Proto-oncogenes can be broadly classified into five groups, depending on the function of the gene product

(1)	Genes coding for growth factors
(2)	Genes coding for receptors that recognize growth factors
(3)	Genes coding for cytoplasmic protein kinases
(4)	Genes coding for intracellular signal transducers, i.e. molecules associated with the cytoplasmic face of the plasma membrane, involved in transducing growth at stimulating signals.
(5)	Genes coding for proteins that act within the nucleus.

Information abstracted from James *et al.*, *Br. J. Cancer* **57**, 440 (1988), and Bell, *Cancer Lett.* **40**, 1 (1988). Proto-oncogenes can be activated to the oncogenic state by point mutation, gene amplification, deletion, chromosomal translocation, or viral insertion.

damage and a high risk of skin cancer on exposure to sunlight or to other sources of ultraviolet light. (The nature of UV-induced damage to DNA is discussed in Chapter 6, Section 6.17.) Patients with *Fanconi's syndrome*, in which DNA repair is also defective, show an increased incidence of leukaemia. This is also seen in *Bloom's syndrome* and in *ataxia telangiectasia*. Bloom's syndrome is an autosomal-recessive condition in which about 20 per cent of patients develop cancer before the age of 20. Ataxia telangiectasia occurs in about 1 per 40,000 births, and about 10 per cent of patients develop cancers at an early age. In all these conditions, DNA repair mechanisms are defective. In *Down's syndrome* (mongolism), which is usually caused by the presence of three copies of chromosome 21, an increased risk of cancer has also been noted.

8.8.1 *Oxidants and carcinogenesis*

In view of the importance of DNA damage in carcinogenesis, it is conceivable that any agent capable of reacting with DNA and chemically modifying it could be carcinogenic. Thus it is safest to think of all mutagens as carcinogens until they are proved otherwise.

Exposure of organisms to ionizing radiation has long been known to lead to the development of cancer. It appears that radiation-induced carcinogenesis involves both initiation and promotion. Some DNA damage by radiation occurs by direct absorption of energy by this molecule, but most is mediated by ionization of water and formation of highly-reactive species such as hydroxyl radical, OH'. Hydroxyl radical attack upon DNA generates a whole series of modified purine and pyrimidine bases (Chapter 2, Section 2.5.3). Attack of OH' upon deoxyribose also yields a multiplicity of products. However, there has been no identification of any specific types of OH'-

mediated damage to DNA that participate in carcinogenesis; it may be that faulty repair of OH'-induced lesions is responsible. The ability of OH' to damage DNA fits in well with many observations upon bacteria, and upon animal or plant cells in culture, that show a mutagenic effect of elevated O_2 concentrations. Indeed, exposure of bacteria or cultured mammalian cells to activated human neutrophils or to hypoxanthine plus xanthine oxidase is known to produce DNA damage. Catalase usually protects against the DNA damage in these experiments; the reported effects of SOD are more variable. It may be that H_2O_2 penetrates into the cell, reaches the nucleus and reacts with transition metal ions bound at or close to the DNA to form OH' (neither O_2^- nor H_2O_2 alone react at significant rates with DNA). The DNA within neutrophils themselves is fragmented during phagocytosis, perhaps by the same mechanism. Thus exposing cells and tissues to oxidant stress might tend to promote DNA damage. An *in-vivo* example of this is perhaps provided by the observation that the sperm produced by male rats in which epididymal inflammation has been induced by treatment with chloromethane (CH_3Cl), are genetically abnormal, even though chloromethane itself does not attack DNA directly. Peroxy radicals, cytotoxic aldehydes, and alkoxy radicals, generated by decomposition of lipid peroxides, might also be able to cause DNA damage. DNA damage by oxidants could account for the increased risk of cancer development in chronically-inflamed tissues.

Thus there is little doubt that oxidant stress can cause DNA damage in cells. Conceivably therefore (unless the oxidant stress is severe enough to kill the cell), oxidants could be involved in initiation, promotion, or progression. The available literature has tended to concentrate on the promotional aspects of oxidants. For example, it has been known for many years that several organic peroxides, such as benzoyl and lauryl peroxides, are skin irritants and tumour promoters in mice. Tumour promotion by PMA may be related to its ability to accelerate production of O_2^- and H_2O_2 by activated phagocytic cells (Chapter 7, Section 7.3.1); many tumour promoters are powerful stimulators of inflammation. There have been several reports that oxidants can lead to cell transformation. For example, Cerutti in Switzerland reported that the xanthine–xanthine oxidase system could promote transformation in initiated mouse embryo fibroblasts, or in mouse epidermal cells. Weitzman *et al.* in the USA found that a mouse fibroblast line could be transformed by the same radical-generating system, or by activated neutrophils. Much more detailed experiments are necessary to ascertain exactly how the oxidants are acting in these systems.

Another possibility that has been raised is that tumour promoters might depress cellular antioxidant defences, such as SOD or catalase. For example, it has been reported that treatment of mouse skin with tumour promoters leads to a fall in SOD activity. This might lead to increased oxidative damage to DNA *in vivo*. Indeed, mitochondria from several malignant animal tumours, and from tumour cells in culture, are deficient in manganese SOD activity when compared to mitochondria from normal tissues. Lowered activities of CuZnSOD are often seen in tumours, but not always. These observations led Oberley and Buettner in the USA to propose that decreased antioxidant protection, and concomitant increased radical generation, might explain many of the properties of cancer cells. Many animal tumour cell lines are also low in catalase activity (Table 8.5). Oberley further observed that growth of a tumour model system in mice was decreased by injection of a low-molecular-mass compound, copper(II) (3,5-diisopropylsalicylate)$_2$, that has superoxide-scavenging activity.

If the hypothesis of Oberley *et al.* were valid, one might expect to see evidence of increased free-radical reactions in malignant cells. This has been looked at in the context of oxidative damage to lipids, and the results are not as expected. It has been known for three decades that subcellular fractions prepared from malignant cells in culture, or from samples of animal cancers, show abnormally *low* rates of lipid peroxidation *in vitro*. The precise reasons for this depend on the cells studied, but they include a decreased content of polyunsaturated fatty-acid side-chains in membrane lipids, an increased accumulation of *a*-tocopherol, and drops in cellular activities of cytochrome P-450 and NADPH-cytochrome-P450 reductase. Pre-malignant cells may also show decreased rates of lipid peroxidation. Slater *et al.* have claimed that dividing hepatocytes in partially-hepatectomized rats accumulate *a*-tocopherol during DNA replication, suggesting that 'decreased peroxidizability', at least as accounted for by this mechanism, may be a feature of normal cell division rather than a feature of the cancerous state. Damage to DNA during replication has more serious consequences than when the DNA is not replicating, so it is possible that cells take steps to increase their antioxidant defences at this time.

Several groups have attempted to detect radical reactions by subjecting healthy control tissue and malignant tissue to electron-spin-resonance studies (Chapter 2, Section 2.5.1). For example, Slater, in England, and Benedetto *et al.*, in Italy, observed that frozen, ground, samples of normal human cervix give strong ESR signals; these are almost *absent* in cancerous cervical tissue. The signal may be due to a peroxy radical, perhaps generated from a peroxide produced on grinding of the frozen tissue, by the action of a lipoxygenase enzyme.

These studies thus suggest that free-radical reactions are *decreased* in malignant cells, although it must be pointed out that oxidant stress is not always reflected in lipid peroxidation (Chapter 6, Section 6.1). A further problem in assessing the hypothesis of Oberley *et al.* is that there is no good evidence for decreases in SOD, catalase, or glutathione peroxidase activities in human cancerous tissue. Surgical biopsy samples of most human cancers do not seem deficient in these enzymes. Nathan and Cohn, in the USA, studied several human malignant cells, and found no evidence that they were abnormally sensitive to H_2O_2. It should be borne in mind that areas within a large tumour mass often have a poor blood supply and so are often anoxic, which may lead to decreases in SOD activity (Chapter 6, Section 6.10). For example, Petkau *et al.* in Canada, studied a rat breast cancer and found that its SOD activity was 54 ($\pm$ 10) $\mu g/g$ tissue at the centre of the tumour, but 117 ($\pm$ 38) $\mu g/g$ at the edge. Exposure of the tumour-bearing rats to elevated O_2 concentrations raised both values, to 162 ($\pm$ 73) and 286 ($\pm$ 103) $\mu g/g$, respectively. This variation, and the fact that rapidly-growing tumours frequently lose histological resemblance to their tissue of origin, make it extremely difficult to obtain valid comparisons of tumour enzyme activities with the 'normal' state.

The increased risk of cancer in Down's syndrome, Bloom's syndrome, ataxia telangiectasia, and Fanconi's anaemia has prompted investigations of cellular antioxidant defences in these conditions. Joenje *et al.*, in Holland, reported that lymphocytes, taken from Fanconi patients and cultured *in vitro*, show much smaller numbers of chromosomal aberrations if the O_2 concentration in the culture medium is decreased, suggesting some abnormality in antioxidant defence. Indeed, the SOD activity of erythrocytes in Fanconi patients has been reported to be decreased by 30–40 per cent, whereas catalase and glutathione peroxidase are normal. The erythrocyte SOD present was reported to have normal specific activity and electrophoretic mobility, i.e. there is no evidence for production of a mutant enzyme as a result of a defect in the gene encoding SOD. Cultures of fibroblasts from patients with Bloom's syndrome have been reported to release a 'clastogenic factor' that can induce chromosome breaks if it is added to cultures of normal lymphocytes. This action of clastogenic factor could be suppressed by including SOD in the culture medium. It is possible that clastogenic factors are cytotoxic lipid peroxides. It is, of course, essential in such work to bear in mind the ease with which peroxides develop on storage or mishandling of biological fluids or other lipid-containing media (Chapter 4).

On the other hand, in Down's syndrome the tissue content of CuZnSOD is increased by approximately 50 per cent. This is because the gene coding for CuZnSOD is located on chromosome 21. There have been reports that mitochondrial manganese SOD is slightly ($\sim$ 30 per cent) decreased in platelets from Down's patients; whether such a small decrease could be biologically meaningful is unclear. Erythrocyte glutathione peroxidase activity has been reported to be increased by 30–50 per cent in erythrocytes from Down's syndrome patients, possibly because the increased SOD activity leads to more H_2O_2 generation.

Iron and cancer
DNA damage within cells by O_2^- and H_2O_2, perhaps resulting in carcino-

genesis, might be mediated by formation of OH˙ close to the DNA. Metal ions, probably those of iron, are required for this process. It is therefore interesting to note that malignant disease, like chronic inflammation (Section 8.3.2), produces changes in body iron distribution. In most cases, iron is lost from the blood (e.g. the percentage iron saturation of transferrin drops markedly) and it accumulates in liver, spleen, and bone marrow. Perhaps this serves to withhold iron from the tumour and so slow its growth, since, in general, tumours seem to contain less 'total iron' and have a lower degree of iron saturation in ferritin than do normal cells. This is not always the case, however, since human breast tumours appear to accumulate iron, and in Hodgkin's disease heavy deposits of iron and ferritin are seen surrounding the tumour nodules. In some cancers, including Hodgkin's disease, breast cancer, and leukaemia, the normally-low concentrations of ferritin protein present in the blood are greatly increased. In some patients with acute leukaemia, plasma-transferrin saturation is close to 100 per cent and non-transferrin-bound iron may be present. Shires, in the USA, observed that incubation of rat liver nuclei with iron(II) salts causes extensive O_2-dependent DNA damage. Perhaps this may account for the increased risk of hepatoma in iron-overloaded patients with idiopathic haemochromatosis, since the liver takes up low-molecular-mass iron complexes from the circulation in this disease (Chapter 2, Section 2.4.3). It has also been suggested (Chapter 7, Section 7.3) that the iron content of asbestos may relate to its ability to produce inflammation and cancer, since asbestos will promote formation of OH˙ from H_2O_2 *in vitro*, a reaction inhibitable by desferrioxamine. Asbestos fibres can also accelerate lipid peroxidation; this again is inhibitable by desferrioxamine.

Free-radical reactions have been claimed to be involved not only in the promotion of cancer, but also in 'immune surveillance', in causing the DNA damage produced by some carcinogens, and in the mechanism of action of antitumour drugs. Let us now look at the evidence available for the role of radicals in these processes. Their importance in cancer treatment using radiation combined with hypoxic cell sensitizers, has already been discussed (Chapter 6, Section 6.10).

8.8.2 *Immune surveillance*

The concept of immune surveillance postulates that malignant tumours are recognizable as 'foreign tissue', and that one role of the immune system is to identify and destroy them. The popularity of the concept has waxed and waned at various times during the past 20 years. Many cancers develop without any apparent defect in the immune system, and do not provoke any immune response. Nevertheless, there is *some* evidence that the immune response of the host influences the development of a few malignant tumours.

For example, there is an increased risk of malignancy after immunosuppressive therapy; the overall increased risk of skin cancer in kidney transplant patients, treated with immunosuppressive drugs to prevent rejection of the foreign kidney, has been estimated as over seven-fold. However, there appears to be no increase in the incidence of cancers of the lung, gut, or breast. Infection of patients by human immunodeficiency virus (HIV, formerly called HTLV-III; Fig. 8.11) can lead to severe dysfunction of the immune system, accompanied by life-threatening infections by organisms that are normally easily dealt with (such as *Pneumocystis carinii*, which leads to pneumonia). Patients in whom HIV infection leads to this 'acquired immune deficiency syndrome' (AIDS), show a marked tendency to develop certain 'unusual' cancers, such as Kaposi's sarcoma.

In 1973, Edelson and Cohn in the USA demonstrated that the myeloperoxidase–H_2O_2–halide system can kill tumour cells *in vitro*. Monocytes, macrophages, and neutrophils can also kill malignant cells, provided that the 'respiratory burst' of the phagocytes is activated. The killing mechanism can involve released proteolytic enzymes, H_2O_2, O_2^- and HOCl (from neutrophil

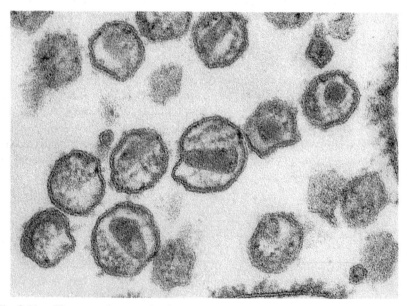

Fig. 8.11. Electron micrograph of sectioned human immunodeficiency virus (HIV). Each virus particle has a lipid bilayer envelope which is derived from the plasma membrane of the host cell and in which are inserted viral glycopeptides. Enclosed within the envelope is the typical elongated nucleoid of HIV which contains the viral RNA. Magnification × 270 000. Photograph by courtesy of Dr David Hockley. The lipid bilayer envelope of the virus makes it sensitive to lipid-disrupting agents, including agents that stimulate lipid peroxidation.

myeloperoxidase). Many animal tumour cells are especially sensitive to H_2O_2 because of their low catalase activities (Table 8.5), but such a lack of H_2O_2-metabolizing enzymes may not be a feature of human malignant cells *in vivo*, as discussed previously. Drapier and Hibbs in the USA found that murine cytotoxic activated macrophages could attack tumour cells and cause loss of intracellular iron from them, in part by damage to iron–sulphur proteins such as the enzyme aconitase (an enzyme of the Krebs cycle). This release of iron could provide the necessary transition metal for release of highly-reactive OH⁻ from H_2O_2.

It is, of course, necessary to activate the phagocyte respiratory burst, either by coating the tumour cells with antibody, or by supplying another inducer. Thus non-immunogenic malignant tumours will, in general, not be attacked by phagocytes. Almost a century ago, Coley in New York observed that cancer patients repeatedly administered 'bacterial broths' could show hae-morrhagic necrosis of their tumours. Later work showed that the active agent was bacterial endotoxin, which stimulates production of tumour necrosis factor (TFN) by macrophages (Section 8.4.1). The precise role of TFN as an antitumour agent *in vivo* is not clear, but it is known to 'prime' phagocytic cells to respond to stimuli by increased oxidant formation (Section 8.4.1).

Mammals also show some kind of 'natural immunity' towards a wide range of tumour cells. This poorly-understood ability appears to be mediated by 'natural killer' (NK) cells, a species of lymphocyte (Appendix II). NK cells can apparently lyse a range of malignant cells, with variable efficiency and without prior 'activation' of the NK cells. NK cells also attack certain types of cell in the normal bone marrow and thymus gland, as well as cells infected by some viruses or parasites. Little is known about the mechanism by which NK cells recognize and kill their targets. Claims that superoxide or hydroxyl radicals produced by NK cells are responsible for target cell killing have not

Table 8.5. Activities of 'antioxidant' enzymes and GSH concentrations in some tumour cell lines

Cell line	Catalase (units mg⁻¹ protein)	SOD (µg mg⁻¹ protein)	Glutathione peroxidase (units mg⁻¹ protein)	Glutathione reductase (units mg⁻¹ protein)	G6PDH[1] (units mg⁻¹ protein)	GSH (µg mg⁻¹ protein)
A	< 0.02	0.25	11	0.04	0.05	6.6
B	< 0.02	0.53	36	0.18	0.11	18.1
C	< 0.02	0.50	13	0.06	0.16	24.4

[1] Glucose 6-phosphate dehydrogenase, the first enzyme of the pentose phosphate pathway.
Data were abstracted from Bozzi *et al.*, *Cancer Biochem. Biophys.* **3**, 135–41 (1979). Line A was Ehrlich ascites cells; B was Yoshida ascites cells; C, Novikoff ascites cells.

Free radicals, ageing, and disease

been substantiated; proteolytic enzymes secreted by the NK cells may be involved, and/or proteins that form 'pores' in the membrane of the target cell.

8.8.3 *Mechanism of action of carcinogens*

In 1775, the English physician and surgeon Percival Pott reported that the occurrence of cancer of the scrotum in chimney-sweeps could be correlated with their exposure to soot and tar. Further work showed that repeated application of coal tar to the skin of animals eventually produces malignant tumours at the site of application. Later, aromatic hydrocarbons such as *benzpyrene* (Fig. 8.12) were isolated from coal tar and shown to be carcinogenic. Carcinogenic hydrocarbons are present in combusted organic matter (e.g. soot) and in cigarette tar. In the late 1800s, the German physician Rehn noticed an association between exposure of dye workers to aromatic amines such as benzidine (Fig. 8.12) and the development of bladder cancer. Not until the 1930s was it shown that several aromatic amines would directly induce bladder cancer in animals.

A wide range of chemicals has since been shown to be carcinogenic; Table 8.6 lists some of them. About one-third of all cancer cases in Europe and North America can be related to the presence of carcinogens in cigarettes and other tobacco products (Chapter 6, Section 6.4). Other carcinogens have been identified from epidemiological studies (e.g. vinyl chloride), from routine bioassays (e.g. nitrosamines), and from investigations into the cause of other diseases in man (e.g. the plant product cycasin; asbestos) or in animals (e.g. the hepatotoxic aflatoxins).

Repeated exposure to carcinogens is usually necessary to produce a malignant tumour, and this often takes a considerable time because of the complex multistep nature of carcinogenesis. Many carcinogens, e.g. benzpyrene, are both initiators and promoters of carcinogenesis (*complete carcinogens*), whereas others are initiators only (*incomplete carcinogens*) and will produce malignant transformation only if a promoter is present. Low doses of a complete carcinogen, themselves too small to produce carcinogenesis, can be caused to do so by a promoter. Thus urethane is an incomplete carcinogen to mouse skin. How a carcinogen behaves may also depend on the tissue treated with it, e.g. urethane has been suggested to act as a complete carcinogen in fetal animal lung.

Carcinogens are believed to act by attacking DNA—indeed, several tumours produced by administration of carcinogens to animals contain activated proto-oncogenes. Some chemical carcinogens, such as highly-reactive *N*-methyl-*N'*-nitrosoguanidine or *N*-methyl-*N'*-nitrosourea (Fig. 8.12) can chemically modify DNA directly. Most carcinogens, however, have to be converted into the active forms. Conversion of a *pro-carcinogen* to the

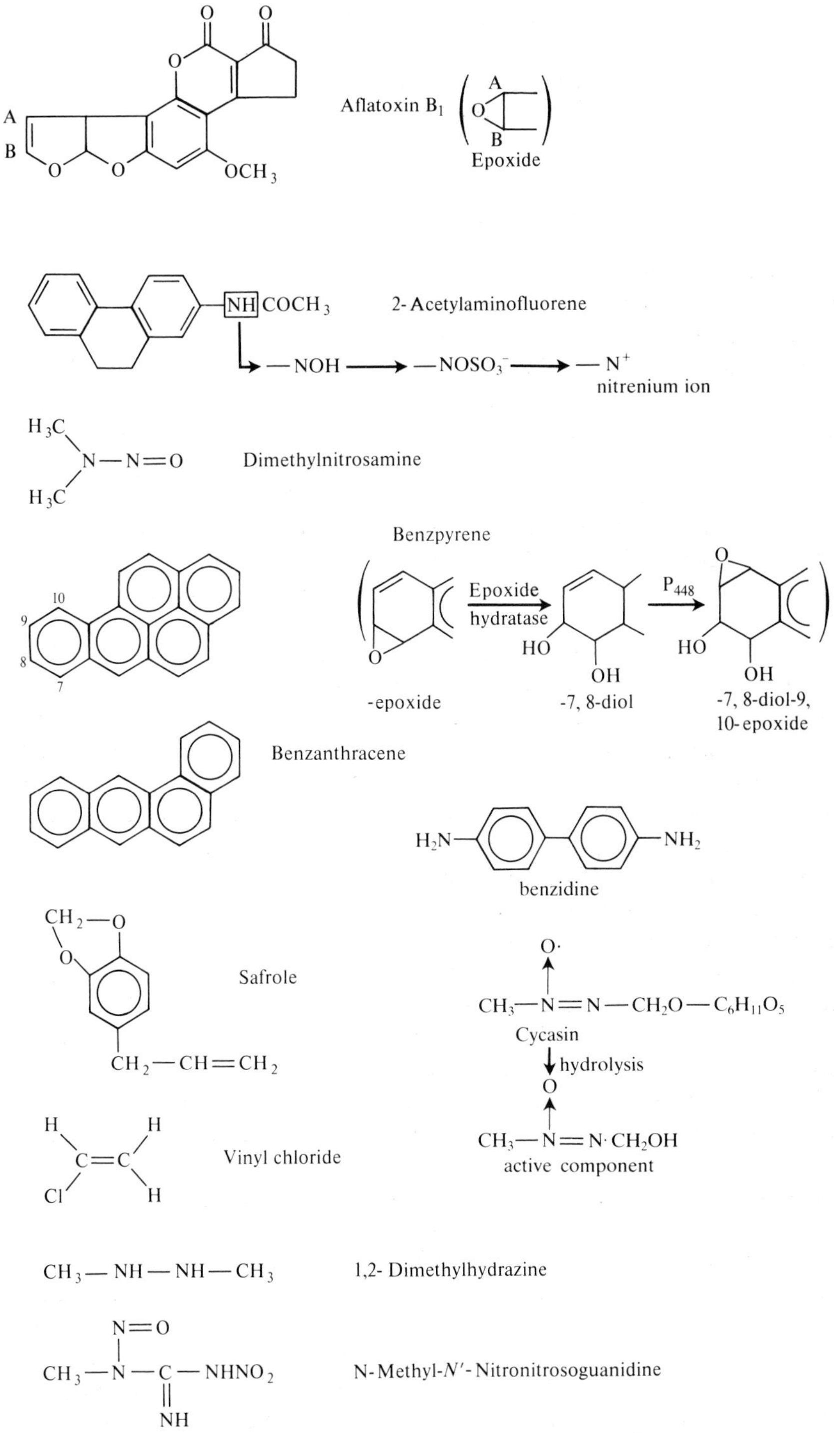

Aflatoxin B1
A
B
O
Epoxide
OCH3
NH COCH3
2- Acetylaminofluorene
—NOH
—NOSO3⁻
— N⁺
nitrenium ion
H3C
N—N=O
Dimethylnitrosamine
Benzpyrene
10
9
8
7
Epoxide
hydratase
P448
HO
OH
-epoxide
-7, 8-diol
-7, 8-diol-9,
10-epoxide
Benzanthracene
H2N
NH2
benzidine
CH2—O
Safrole
CH2—CH=CH2
O·
CH3—N=N—CH2O—C6H11O5
Cycasin
hydrolysis
CH3—N=N·CH2OH
active component
H
C=C
Cl
Vinyl chloride
CH3— NH— NH— CH3
1,2- Dimethylhydrazine
N=O
CH3—N —C — NHNO2
NH
N-Methyl-N′-Nitronitrosoguanidine

Fig. 8.12. Structure and metabolism of some carcinogens.

ultimate carcinogen can proceed in one or more steps (if more than one step is required, the intermediates may be called *proximate carcinogens*). The necessary metabolism is sometimes carried out by gut bacteria, e.g. the hydrolysis of cycasin (Table 8.6) to the proximate carcinogen methylazoxy-methanol (Fig. 8.12), which is then metabolically activated to a DNA-damaging product. More often, enzymes of the body tissues are involved. The first example of metabolic activation to be discovered was the conversion of the aromatic amine 2-acetylaminofluorene (Fig. 8.12) to an N-hydroxy-lated product. This type of metabolism is common to several aromatic amines, including 2-naphthylamine, which induces cancer of the bladder in humans. The N-OH product then undergoes several reactions, including combination with sulphate to give an N-O-sulphate ester that is probably the ultimate carcinogen; loss of sulphate produces a reactive *nitrenium ion* that combines with guanine residues in DNA. Peroxidase enzymes can also oxidize the N-OH product to an N-O˙ radical, but whether this plays any role in carcinogenesis is unclear.

N-hydroxylation of 2-acetylaminofluorene is catalysed by the microsomal mixed-function oxidase system (Chapter 3, Section 3.5.1). This system catalyses activation of many other pro-carcinogens, including dimethylnitro-samine (to highly-reactive methyl radicals; Chapter 3, Section 3.5.1) and

Table 8.6. Some chemical carcinogens

Industrial chemicals

Chemical	Site of cancer	Chemical	Site of cancer
Aromatic amines	Bladder	Arsenic	Skin, bronchus
Asbestos*	Bronchus, pleura	Benzene	Bone-marrow
Tars, oils	Skin, lungs	Vinyl chloride	Liver
Diethylstilboestrol	Vagina	Aromatic hydrocarbons	Lung

Naturally occurring chemicals

Chemical	Source	Site of cancer
Cycasin	Cycad plants	Liver, kidney, intestine (rats)
Ptaquiloside	Bracken fern	Bladder, intestine (rats, cows)
Not yet purified	Betel nut	Mouth (betel nut and leaves are often chewed in Southeast Asian countries)
Safrole	Oil of sassafras	Liver (rats)
Aflatoxins (several known, most active is B₁)	*Aspergillus flavus* (fungus)	Liver (rats)

* The carcinogenicity of asbestos is considered in Chapter 7, Section 3. The purification and structure of ptaquiloside can be found in Niwa *et al.*, *Tetrahedron Lett.* **24,** 4117 and 5371 (1983).

carbon tetrachloride, CCl_4. Formation of the free radicals $\dot{C}Cl_3$ and $\dot{O}_2CCl_3$ is known to be important in the hepatotoxicity of CCl_4 (Chapter 6, Section 6.9) and it might also contribute to their carcinogenicity. The cytochrome P-450 system is additionally involved in formation of the ultimate carcinogen, a 7,8-diol-9,10-epoxide (Fig. 8.12), from benzpyrene. Cytochrome P-450, or probably special forms of it known collectively as cytochrome P-448, convert benzpyrene into a 7,8-epoxide. This is acted upon by the enzyme *epoxide hydratase* to form a diol, that is then a substrate for further epoxidation (Fig. 8.12). The resulting 7,8-diol-9,10-epoxide combines with guanine residues in DNA. Epoxidation of the benzpyrene molecule occurs at other positions on the ring as well, although the products that result are less carcinogenic. It is interesting to note that phenobarbital, an inducer of cytochrome P-450 (Chapter 3, Section 3.5.1) is a promoter of carcinogenesis in liver. The activity of microsomal mixed-function oxidase systems in animals is approximately inversely correlated to body weight. Thus, for a given dose per unit body weight, small mammals such as mice will tend to convert more of a given dose of a pro-carcinogen into active intermediates than will large animals such as humans. This must be borne in mind in toxicity testing.

The peroxidase activity associated with prostaglandin synthetase (Chapter 7, Section 7.6.1) can co-oxidize several pro-carcinogens *in vitro*. It can convert benzpyrene into 6-hydroxybenzpyrene, the 7,8-diol into a 7,8-diol-9,10 epoxide, and several other pro-carcinogens into more active forms; 6-hydroxybenzpyrene rapidly oxidizes into a mixture of quinones, with production of O_2^- radical. Other substrates for the peroxidase action of prostaglandin synthetase include benzidine, 2-aminofluorene, 2-naphthylamine and 2,5-diaminoanisole. It is unclear at the moment how important these prostaglandin synthetase-catalysed metabolic transformations are *in vivo*. Peroxidizing microsomes, or irradiated food lipids, have been reported to convert benzpyrene into mutagenic products, including 6-hydroxybenzpyrene, and sometimes to form diol epoxides from benzpyrene-7,8-diol. Recently, Reed *et al.* in the USA found that oxidizing sulphite ions (which generate a large number of radical species; Chapter 6, Section 6.5) could also produce diol epoxides from the 7,8-diol, perhaps by the oxidizing action of the peroxy radical $\dot{O}_2SO_3^-$, formed by reaction of sulphite radical ($\dot{S}O_3^-$) with oxygen. This observation is interesting because SO_2 has been claimed to be a co-carcinogen for the development of lung cancer in animals treated with benzpyrene.

As well as forming diols, epoxides are substrates for glutathione-S-transferase enzymes (Chapter 3), being converted into more soluble glutathione derivatives that can then be excreted. This, and other metabolic pathways such as conjugation with glucuronic acid, often represent detoxification mechanisms, and so the overall effect of most carcinogens *in vivo* depends upon the balance between activation and detoxification mechanisms, as well as upon the efficiency with which any DNA lesions produced by the ultimate carcinogen are repaired.

Several 'antioxidants', such as BHA and BHT, have been observed to decrease the carcinogenicity of benzpyrene and other pro-carcinogens, usually by promoting increased synthesis of the enzymes involved in detoxification pathways. Ascorbic acid can diminish the carcinogenicity of nitroso-compounds by reducing the nitroso-group to an inactive product. (Claims that ascorbic acid prolongs the survival time of human cancer patients have not been validated by clinical trials, however.) GSH is not only a substrate for transferase enzymes, but can also often combine directly with the ions or radicals that attack DNA. Other dietary agents, including vitamin A, ß-carotene, and selenium, have been suggested to exert some anti-cancer effect.

Selenium has some general protective effect against carcinogenesis induced in animals by chemicals that require metabolic activation, such as dimethyla-minoazobenzene. Selenium-deficient animals are also more sensitive to certain carcinogens, especially if the animals are fed a diet rich in polyunsa-turated fatty acids. It should be noted that the experimental results obtained in all these studies vary depending on the species used, the doses of selenium and carcinogen administered, and on the exact times of administration. It has been *suggested* that excessive consumption of polyunsaturated fatty acids by humans might predispose to cancer, but there is no clear evidence for this. Since many carcinogens are lipid-soluble, increased fat intake might lead to increased carcinogen intake. As we have seen, some organic hydroperoxides can act as tumour promoters, although claims that malondialdehyde is a carcinogen have been disputed.

Peroxisome proliferators

The ability of oxidants such as H_2O_2 to induce DNA damage in cells, perhaps by site-specific hydroxyl radical formation, is of particular interest in relation to the mechanism of action of several carcinogens that have been called 'peroxisome proliferators'. A wide range of compounds, including trichlor-oethylene, drugs that lower blood lipid concentrations (e.g. clofibrate [Atromid-S]), plasticizers (e.g. di-(2-ethylhexyl)phthalate), and phenoxyace-tic acid herbicides (e.g. 2,4-dichlorophenoxyacetic acid), produce enlarge-ment of the liver, accompanied by marked increases in the number of hepatic peroxisomes, in several animal species. Peroxisomes not only increase in number, but their overall balance of enzyme activities changes (for a basic description of these organelles, see Chapter 3). Thus the activity of the peroxisomal system for ß-oxidation of fatty acids often increases more than that of catalase.

Some peroxisome proliferators increase the incidence of malignant tu-mours in rat liver. Reddy *et al.* in the USA have suggested that tumours result because the excess H_2O_2 in peroxisomes (e.g. due to increased -

oxidation of fatty acids) cannot be fully metabolized by the catalase present and leaks out of these organelles. If some of the H_2O_2 survives to reach the nucleas, DNA damage could be expected. Indeed, perfusion of lauric acid, a substrate for peroxisomal ß-oxidation, into isolated livers from control rats produced no increase in efflux of oxidized glutathione from the liver. However, if the rats had been pre-treated with the peroxisome proliferator nafenopin, livers isolated from them responded to infusion of lauric acid by greatly-increased GSSG efflux. This observation is consistent with an escape of H_2O_2 from the peroxisomes in treated animals; its metabolism by glutathione peroxidase in the cytosol would account for the increased GSSG efflux (Chapter 3, Section 3.1.1). Hypolipidaemic agents also cause proliferation of hepatic endoplasmic reticulum and its associated cytochromes, which might also be sources of O_2^- and H_2O_2 *in vivo* (Chapter 3, Section 3.5.1).

8.8.4 *Antitumour drugs and oxidant production*

The object of cancer chemotherapy is to kill cancer cells with as little damage as possible to normal cells. In malignant tumours, many cells are dividing and so most drugs used are designed to interfere with cell proliferation, often by blocking synthesis of DNA, RNA, or protein. The first reported use of chemotherapy was in 1942, when the DNA-modifying agent nitrogen mustard was used in a patient with lymphoma. One major class of chemotherapeutic agents is thus provided by the alkylating agents (Table 8.7); they or their metabolites bind to DNA, and chemically-modifying it, interfering with replication and transcription. A second major class of drugs is designed to interfere with metabolic reactions. Thus methotrexate, a folic acid antagonist, inhibits the enzyme dihydrofolate reductase and prevents transfer of methyl groups in several biosynthetic reactions, including synthesis of deoxythymidine in DNA. 5-Fluorouracil is an analogue of thymine and prevents DNA synthesis by inhibiting thymidylate synthetase. Cytosine arabinoside contains an arabinose sugar instead of ribose, and it interferes with DNA polymerase activity. Cisplatin (*cis*-diamminedichloroplatinum(II), see Fig. 8.13) probably acts by cross-linking guanine residues in DNA. Hydroxyurea inhibits ribonucleoside diphosphate reductase (Chapter 7, Section 7.1). Vincristine and vinblastine interfere with spindle formation during mitosis, and anti-steroidal agents have been developed to treat cancers whose growth depends on the continued availability of steroid hormones. Of course, any normal cells undergoing division will also be damaged by most of these drugs; dividing cells are found in the intestinal epithelium, hair follicles, gonads, and bone marrow. These side-effects limit the dose frequency, and size of dose that can be administered. Most, if not all, of the agents that chemically modify DNA are themselves mutagenic, and

Table 8.7.   Major families of drugs used in cancer chemotherapy

Alkylating agents (modify DNA) nitrogen mustard chlorambucil melphalan cyclophosphamide nitrosoureas (BCNU†, CCNU, methyl CCNU) busulphan	Antimetabolites (interfere with DNA synthesis) 5-fluorouracil methotrexate cytosine arabinoside 6-mercaptopurine 6-thioguanine
Natural products (several mechanisms of action) doxorubicin (Adriamycin)* daunorubicin* actinomycin D* mitomycin C* vinblastine/vincristine bleomycin*	Other agents (several mechanisms of action) Tamoxifen cisplatin* dicarbazine* procarbazine*

* Probably act, at least in part, at the DNA level.
† An inhibitor of glutathione reductase (Chapter 3, Section 3.1.1).

thus can be regarded as potentially carcinogenic, meaning that they must be handled with care. A number of antitumour antibiotics exist, and they can be harvested after microbiological fermentation. Usually they act, at least in part, by binding to DNA and interfering with unwinding, replication and/or transcription or causing strand breakage. The antitumour antibiotics can be grouped under several distinct chemical types. These include the anthracylines and other quinone-containing drugs (such as mitomycins, streptonigrin, daunomycin, and doxorubicin, (otherwise known as Adriamycin), the metal-chelators (such as tallysomycin and the bleomycins), the protein antitumour antibiotics (such as macromomycin and neocarzinostatin), and aureolic-acid-based antibiotics (such as mithramycin, chromomycins, and olivomycins). Free-radical reactions have been suggested to be important in the mechanism of

Fig. 8.13.   Structures of some antitumour drugs. (a) Streptonigrin; (b) adriamycin; (c) daunomycin; (d) AD32; (e) mitomycin C; (f) actinomycin D (thr, threonine; val, valine; pro, proline; sar, sarcosine; NMeval, *N*-methylvaline), (g) *cis*-diamminedichloroplatinum(II). (*Cis* means that similar groups are next to each other, i.e. the two —NH₂ groups and the two —Cl groups are adjacent.) The *trans*-isomer [in brackets] is not an effective antitumour drug.

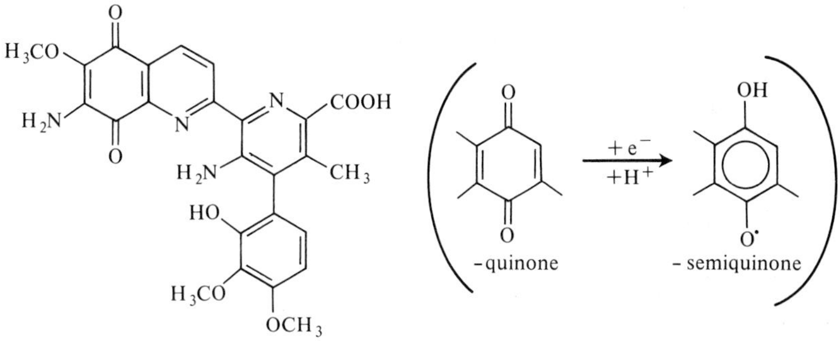

(a)

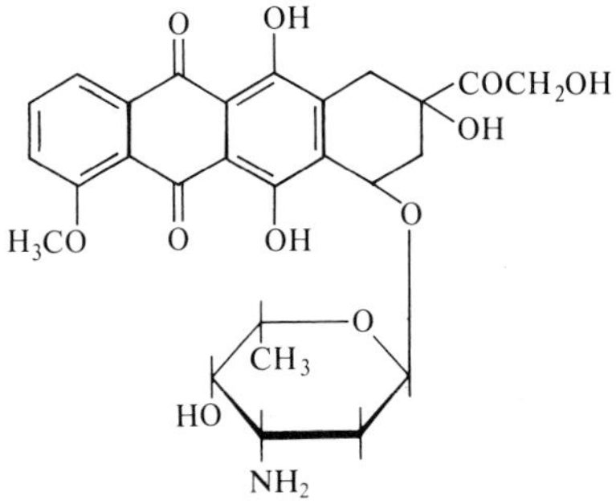

(b)

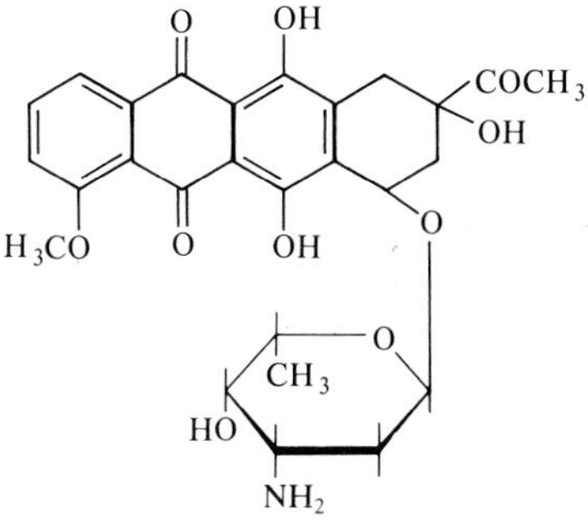

(c)

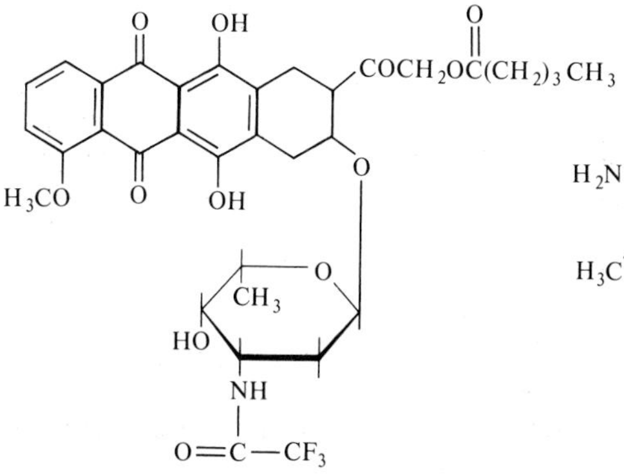

(d)

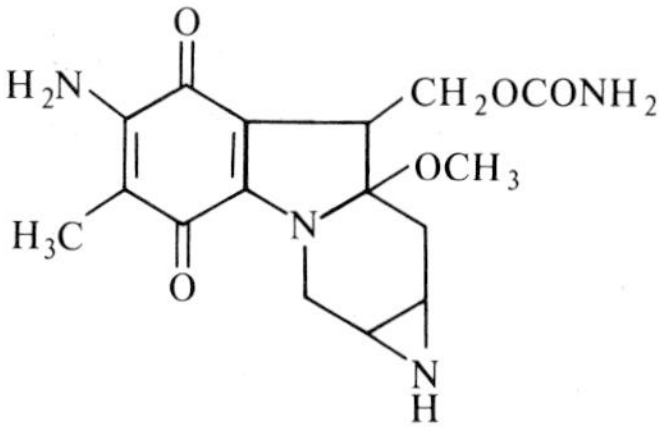

(e)

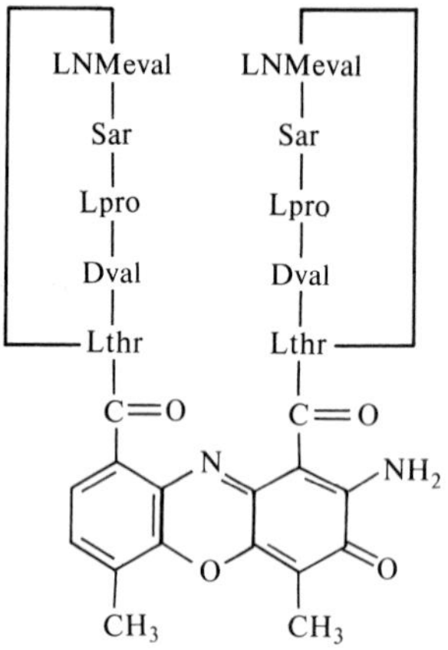

(f)

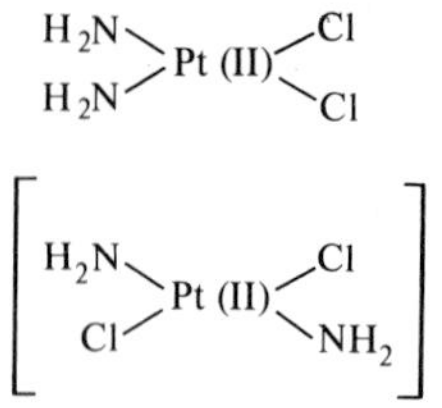

(g)

action of several antitumour antibiotics. We will therefore concentrate our discussion on these drugs.

Bleomycin

Bleomycin is the collective name given to a family of glycopeptide antibiotics produced by *Streptomyces verticillus*. It was first isolated by Umezawa in Japan. The clinical preparation commonly used, Blenoxane, is a mixture of several bleomycins that differ slightly in structure, although bleomycin A_2 is the major component (Fig. 8.14). Bleomycins are active against several human cancers, including Hodgkins's disease and cancer of the testis.

Bleomycins produce their action by binding to DNA, especially adjacent to guanine residues. They cause single-strand, and some double-strand, breaks and degradation of the deoxyribose sugar to form, among other products, *base propenals*, which further break down to release the three carbon aldehyde malondialdehyde (MDA). Since this reacts with thiobarbituric acid to give a pink chromogen (Chapter 4, Section 4.6.8), the TBA test can be used to follow DNA breakdown by bleomycin. Some of the base propenals are themselves cytotoxic.

Bleomycins are powerful chelators of transition metal ions, such as Cu^{2+}, Co^{2+}, Zn^{2+}, Fe^{2+}, Fe^{3+}, or Ni^{2+}, by donation of electrons from nitrogen atoms and from a $\diagdown C = O$ group to the metal ion (fig. 8.14). Indeed, degradation of DNA by bleomycin absolutely requires that a reduced transition metal ion is present within the DNA–bleomycin complex. O_2 is also needed. *In vitro*, complexes of bleomycin with iron(II), copper(I), cobalt(II), manganese(II), and vanadium(IV) have all been claimed to be

Bleomycin A$_2$

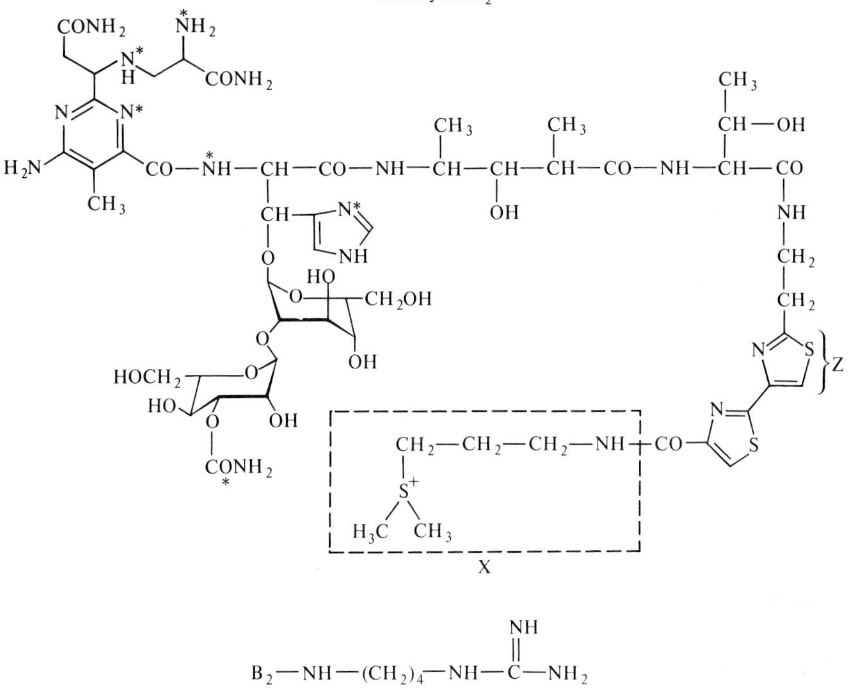

Fig. 8.14. Structure of bleomycin A$_2$. In bleomycin B$_2$ the substituent shown replaces the terminal group (X) of bleomycin A$_2$. The asterisks denote the atoms that probably interact with bound transition-metal ions. The *phleomycin* antibiotics differ from the bleomycins only in the absence of one of the double bonds in the ring marked Z. Phleomycins have not so far been used clinically because they cause kidney damage. Bleomycin is usually supplied by the manufacturers as a sulphate salt.

capable of causing DNA degradation. However, only iron (and possibly copper) ions seem likely to mediate bleomycin-induced DNA damage *in vivo*.

A DNA-degrading bleomycin–Fe(II) complex can be generated by adding Fe(II) salts to bleomycin or by reducing a bleomycin–Fe(III) complex with such biological reducing agents such as ascorbate, GSH, or a system generating superoxide radicals (O_2^-). A bleomycin–Fe(III) complex can also be reduced by the microsomal NADPH-cytochrome P-450 reductase enzyme and by a similar reducing system in the nuclear membrane. Bleomycin–Fe(III) can also be activated to degrade DNA in the presence of H_2O_2 although, in this case, there is no evidence for reduction of ferric ions to the ferrous state.

Incubation of bleomycin alone (i.e. in the absence of DNA) with an Fe(II) salt in aqueous solution causes formation of superoxide (O_2^-) and hydroxyl

(OH˙) radicals, detected by spin-trapping experiments (Chapter 2, Section 2.5.1). In such incubations, the drug itself undergoes chemical modification (presumably by OH˙ attack) that destroys its activity, but it can be protected by addition of DNA. By contrast, oxidant scavengers such as superoxide dismutase, low-molecular-mass scavengers of O_2^-, catalase, caeruloplasmin, methanol, vitamin E, thiourea, mannitol, or dimethylsulphoxide offer little, if any, protection to DNA against damage by a bleomycin–Fe(II) complex bound to it. Indeed, propyl gallate and some other phenolic antioxidants can make the damage worse *in vitro* by reducing Fe(III) in the complex back to Fe(II). DNA degradation can most easily be inhibited by preventing iron binding to the bleomycin, using such chelating agents as EDTA, DETAPAC, or desferrioxamine. However, if the iron has already bound to the bleomycin and DNA damage has begun, much higher concentrations of chelators may be required to stop it, since it takes time for the iron to transfer to the chelator molecule.

The chemistry of DNA damage by activated bleomycin may involve ferryl species; a mechanism proposed by Sugiura in Japan is illustrated, with some modifications, in Figure 8.15. Reactants leading to formation of the DNA-attacking species can be generated by complexation of Fe(II)–bleomycin with O_2, or by reaction of Fe(III)–bleomycin with H_2O_2.

The ability of bleomycin to attack the DNA of tumour cells *in vivo* presumably means that it is able to 'pick up' the necessary transition metal

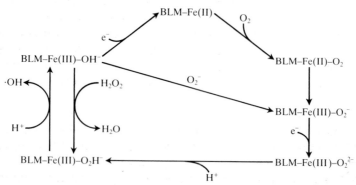

Fig. 8.15. Mechanism of bleomycin (BLM) action. The species that attacks DNA is thought to be a ferric peroxide, BLM–Fe(III)–O₂H⁻. It can be formed by direct reaction of ferric bleomycin with H_2O_2, or from a BLM–iron(III)–O_2^- complex. Diagram adapted from Sugiura *et al.*, *Biochem. Biophys. Res. Commun.* **105** 1511–18 (1982). Under appropriate conditions, BLM–Fe(III)–O_2^- might decompose to release O_2^-, and BLM–Fe(III)–O₂ H⁻ to release OH˙, explaining why these radicals have been detected (e.g. by spin-trapping) in some bleomycin-containing systems (e⁻ represents a reducing system, e.g. thiols, ascorbic acid, NAPDH-cytochrome P-450 reductase).

ions (presumably iron) within the body. (Indeed, the authors have used the iron-requirement of bleomycin as a means of measuring 'available iron' in human body fluids; Chapter 2, Section 2.4.3). Perhaps iron is chelated from the cellular low-molecular-mass iron pool. DNA damage can lead to poly(ADP-ribose) synthetase activation and lethal NAD depletion (Chapter 6).

A major side-effect of therapy with bleomycins is lung damage. The lung appears to have only low activities of *bleomycin hydrolase*, an enzyme which breaks down the bleomycin molecule into an inactive form. This enzyme is present in most body tissues; the content of it in malignant tumours is one factor that affects their sensitivity to bleomycin. Lung damage by bleomycin may again involve radical reactions, since microsomal fractions from rat lungs can reduce Fe(III)–bleomycin in the presence of NADPH, forming Fe(II)–bleomycin and hence oxygen radicals. Consistent with this, bleomycin-induced lung damage in hamsters is increased by exposure to elevated oxygen concentrations. Another important factor might be the availability of iron within the tissue. Hay *et al.* in England reported that intravenous injection of bleomycin into rats produced no lung injury, yet simultaneous exposure to hyperoxia or simultaneous tracheal installation of ferric ions led to marked bleomycin-dependent lung damage. However, attempts to use desferrioxamine to prevent bleomycin-induced lung fibrosis in experimental animals have given negative or marginally-positive results. This may be because desferrioxamine does not penetrate easily into lung cells, especially as induction of mild iron deficiency in hamsters has been reported to decrease lung fibrosis induced by subsequent treatment with bleomycin.

The toxicity of bleomycin to some strains of bacteria might also involve oxygen radicals, since Umezawa in Japan, has reported that strains of *E. coli* K12 with increased SOD and catalase activities, due to previous exposure to paraquat, are more resistant to the toxic effects of bleomycin.

Quinone antitumour agents
Several antitumour antibiotics are quinones. The possibility therefore arises that redox cycling and/or reaction of semiquinones with —SH groups could participate in their activity against malignant cells. Generation of O_2^- and H_2O_2 by redox cycling could produce DNA damage if they led to generation of highly-reactive $OH^.$ radical close to the DNA molecule. Reduction of Fe(III) to Fe(II), which facilitates $OH^.$ generation, could be achieved not only by O_2^-, but also directly by semiquinones ($SQ^{.-}$). Thus two pathways of Fe(II) generation are feasible:

$$A \quad SQ^{.-} + O_2 \longrightarrow SQ + O_2^-$$
$$O_2^- + Fe(III){-}chelate \longrightarrow O_2 + Fe(II){-}chelate$$
$$B \quad SQ^{.-} + Fe(III){-}chelate \longrightarrow Fe(II){-}chelate + SQ$$

The balance between pathways A and B will depend on the rate constants for the appropriate reactions and on the concentrations of O_2 and biological Fe(III)–chelate. In general, route B is not favoured until O_2 concentrations become very low.

Let us now look at some examples of quinone antitumour agents.

Naphthol

The compound 1-naphthol (Fig. 8.16) was reported by Cohen *et al.* in England to be selectively toxic to isolated colon cancer cells, as opposed to isolated normal colonic cells. It has been proposed that 1-naphthol may exert its toxic effects by metabolism to form 1,2- or 1,4-naphthoquinones (Fig. 8.16). These could then attack critical cellular —SH groups and/or undergo redox cycling to form O_2^- and H_2O_2 (Chapter 6, Section 6.6.3). The selectivity of the toxicity to tumour cells remains to be explained, however.

Streptonigrin

Streptonigrin (Fig. 8.13), an antibiotic produced by *Streptomyces flocculus* was one of the first quinone antibiotics whose action *in vivo* has been shown to involve oxygen radicals in some organisms. It does have antitumour effects in Man, but it is not generally used clinically because of serious side-effects. The toxic action of streptonigrin on *E. coli* requires the presence of oxygen and is decreased if the superoxide dismutase activity of the cells is raised (Chapter 3, Section 3.5.3). Increased uptake of iron salts by *E. coli*, following the addition of citrate to the growth medium, enhances the bacteriocidal action of streptonigrin. This effect can be inhibited by desferrioxamine, suggesting a role for iron in the oxygen-radical reactions leading to bacterial damage. Similarly, the killing of *Neisseria gonorrhoea* by streptonigrin seems to depend on the availability of iron to promote Fenton-type reactions. Indeed, streptonigrin is known to bind ions of iron, copper, zinc, cobalt, cadmium, and manganese.

The quinone part of streptonigrin (Fig. 8.13) can be reduced by bacterial enzymes to a semiquinone form that can then reduce oxygen to form O_2^-. Whether or not radicals such as OH˙ are responsible for the observed toxic

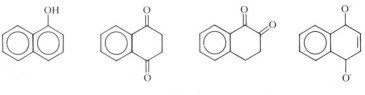

1-naphthol   1,4–naphthoquinone   1,2–naphthoquinone   1,4–naphthosemiquinone

Fig. 8.16. Structures of 1-naphthol and its metabolites.

effects of streptonigrin in cancer cells (which include DNA damage) is hard to determine, since the semiquinone itself is a reactive molecule (the toxicity of quinones and semiquinones is discussed in Chapter 6, Section 6.6.3). However, the toxicity of streptonigrin to mouse mammary tumour cells is decreased under hypoxic conditions, suggesting that oxygen radical generation might be involved.

Actinomycin D
Actinomycin is a peptide antibiotic, produced by *Streptomyces* species, (Fig. 8.13) that is used mainly in the treatment of childhood tumours. Its antitumour activity is generally attributed to its intercalation between the bases of DNA, so preventing RNA synthesis. Intercalation occurs adjacent to guanine bases. This may not be the whole story, however, since actinomycin D, incubated with microsomal fractions in the presence of NADPH, is slowly reduced into a radical intermediate that can convert oxygen into O_2^-. DNA-bound actinomycin D cannot be reduced, however. Oxygen-radical production could be relevant *in vivo* since the toxicity of actinomycin D to mouse mammary tumour cells is decreased at low oxygen concentrations. O_2 concentrations might also influence metabolism of the drug, however.

Mitomycin C
Mitomycin C, isolated from *Streptomyces caespitosus*, is used, in combination with other agents, in the palliative treatment of a large number of advanced human cancers. *In vitro*, it can be reduced by microsomal or nuclear electron transport chains at the expense of NADPH. Reduction forms a semiquinone (fig. 8.13) that can react with oxygen to generate O_2^-. The P-450 system also converts mitomycin C into products that attack DNA directly, and produce cross-linking of the strands. Usually the toxicity of mitomycin C is greater under conditions of low oxygen concentration, suggesting that its reductive activation by the NADPH-cytochrome P-450 system to a product that attacks DNA is far more important in damaging its target cells than is its ability to redox cycle with production of O_2^- and H_2O_2. The damaging action of mitomycin C against large 'solid' tumours, which often contain hypoxic areas (Chapter 6, Section 6.10) is consistent with this conclusion.

The anthracycline antibiotics
Anthracyclines are tetracyclic (i.e. have four joined ring structures) antibiotics produced by various strains of *Streptomyces*. The tetracyclic structure is linked to carbohydrate; complete removal of carbohydrate from anthracyclines usually leads to loss of antitumour activity. Anthracyclines are widely used in the treatment of acute leukaemia, breast cancer, Hodgkin's disease, and sarcomas. The best known are *daunorubicin* (sometimes called *daunomy-*

cin) and *doxorubicin* (often called *Adriamycin*). In both these compounds, the attached sugar is *daunosamine* (Fig. 8.13). Like all antitumour drugs, the anthracyclines produce a number of side-effects, the most serious being damage to the heart. An acute cardiotoxicity occurs within minutes of drug administration and is revealed as arrhythmias and non-specific electrocardiographic changes. More serious than this is a chronic irreversible congestive heart failure, which becomes a significant problem once a certain total dose of anthracycline has been administered. It is this cardiotoxicity that limits the doses of doxorubicin and daunorubicin that can safely be given to cancer patients.

The mechanism of action of anthracyclines is not entirely clear, and may involve multiple processes. Doxorubicin and daunorubicin can intercalate between DNA bases, interfering with DNA replication and the transcription of RNA. DNA strand breakage is also observed, perhaps as a result of intercalation and/or as a result of oxidant generation (see below). Membrane damage, and interference with ion transport, can also occur. Thus doxorubicin is toxic to at least one tumour cell line, even if it is prevented from entering the cells. It binds to cell membranes and alters their permeability to ions such as Ca^{2+}. Indeed the doxorubicin derivative AD32 (*N*-trifluoroacetyladriamycin-14-valerate; see Fig. 8.13) will not bind to DNA, and yet it is still an effective antitumour agent. Increased membrane permeability to Ca^{2+} ions may in part explain the cardiotoxic effects of adriamycin. Recent reports suggest that doxorubicinol, a major metabolite of doxorubicin, formed by NADPH-dependent reduction of the side-chain keto ($\gtrless C = O$) group to —CHOH, is a much more powerful inhibitor of membrane-associated ion pumps than doxorubicin. Doxorubicin also interferes with electron transport in cardiac mitochondria and administration of ubiquinone (Chapter 3, Section 3.5.1) has been reported to protect experimental animals against doxorubicin-induced heart damage.

Redox cycling of anthracyclines

The pioneering work of Handa and Sato in Japan showed that quinone drugs can be reduced by the microsomal electron transport chain to form reactive semiquinones that can combine directly with cellular components, or react with oxygen to form superoxide. Bachur, Gee, and Friedman in the USA suggested that oxygen radicals could be responsible for the DNA damage caused by anthracyclines. Reduction of these drugs can also be catalysed by several enzymes, such as xanthine oxidase or ferredoxin reductase, by the electron transport chain located in the nuclear envelope, and by the mitochondrial NADH–dehydrogenase complex (Chapter 3, Section 3.5.1), hence interfering with the normal flow of electrons during mitochondrial electron transport.

Formation of O_2^- and its conversion to more-reactive radicals such as OH˙

could be involved in causing DNA damage, but the OH˙ would have to be formed very closely to the DNA. Unfortunately, binding of anthracyclines to DNA prevents their reduction by enzymes (because of inaccessibility). Oxidant damage could also be involved in the cardiotoxicity of anthracyclines, since several new anthracycline analogues with less cardiotoxicity (such as *mitoxantrone*) appear to undergo less redox cycling *in vitro*. Treatment of various animals with ascorbic acid, vitamin E, or the thiol compound *N*-acetylcysteine has shown some protective effects against the acute toxicity of doxorubicin, but little action against the chronic cardiotoxicity, a more serious clinical problem. Of course, some of these scavengers (e.g. thiols) might react directly with semiquinones. Heart tissue is low in catalase activity and its glutathione peroxidase activity may decrease after doxorubicin treatment of animals. Mixtures of an α-tocopherol ester and dimethylsulphoxide have been observed to offer some protection against ulceration induced by the direct application of doxorubicin to animal tissues, but none of these scavengers has yet been shown to protect against the side-effects of doxorubicin in human patients.

Reductive metabolism of anthracyclines can lead to loss of sugar from the molecules, giving inactive *aglycones*. For example, if doxorubicin is incubated with NADPH and liver microsomes *in vitro*, NADPH is consumed as NADPH-cytochrome P-450 reductase catalyses formation of doxorubicin semiquinone. The semiquinone reacts rapidly with O_2 to give O_2^-. As O_2 concentration falls to low levels, however, there is rapid formation of aglycone. NADPH-dependent reduction of anthracyclines at low O_2 concentrations can also generate reactive products that can combine directly with DNA, proteins, and membrane lipids.

Any damage done by conversion of O_2^- and H_2O_2 into more reactive species, such as OH˙, would require the presence of transition metal ions. (Claims that doxorubicin semiquinone reacts directly with H_2O_2 to form OH˙ have not been substantiated.) It is therefore interesting to note that doxorubicin and daunorubicin form complexes with a number of metal ions including Fe(II), Fe(III), and Cu(II). Complexes of doxorubicin and Fe(II) can oxidize in air to generate O_2^- and OH˙. Doxorubicin forms a strong complex with Fe(III), in which three doxorubicin molecules associate with each Fe^{3+} ion. In the resulting complex, the Fe(III) is slowly reduced to Fe(II). *In vitro*, the doxorubicin–Fe(III) complex is a powerful stimulator of lipid peroxidation, possibly because it efficiently decomposes traces of lipid hydroperoxides (Chapter 4, Section 4.2.2); the observed peroxidation is not inhibited by scavengers of OH˙. However, evidence for increased rates of lipid peroxidation in animals treated with doxorubicin is very weak. Thiol compounds such as cysteine or GSH can rapidly reduce the doxorubicin–Fe(III) complex, leading to increased formation of O_2^- and H_2O_2. The doxorubicin semiquinone is sufficiently reducing to mobilize iron, as Fe(II), from ferritin; this mobilization is faster than that produced by O_2^-. Indication of the biological relevance of interactions of anthracyclines and metal ions is perhaps provided by the observation that the chelating agent ICRF-187, a derivative of EDTA, has been reported to decrease the cardiotoxicity of doxorubicin to rabbits, mice, pigs, dogs, and humans. A combination of doxorubicin and iron, called *quelamycin*, has been found to be highly cytotoxic to heart and other tissues in preliminary clinical trials.

Thus, although there is some evidence that the cardiotoxicity of anthracyclines may be mediated by oxidant formation, it is not fully clear if damage to cancer cells *in vivo* is achieved by the same mechanism. Some isolated tumour cells exposed to doxorubicin *in vitro*, in the presence of air, can be injured by mechanisms that appear to involve O_2^-, and OH·, and metal ions. The toxicity of the anthracyclines to some isolated tumour cells is decreased under hypoxic conditions, *suggesting* a role for oxygen radicals. This is not always the case, however. For example, the toxicity of doxorubicin to mouse mammary tumour is greater under hypoxic conditions.

Carminic acid
Carminic acid is a major component of cochineal, a red food colouring obtained from the insect *Napalea coccinellifera*. As can be seen in Fig. 8.17, carminic acid has close structural similarities to the anthracycline doxorubicin. Indeed, carminic acid has been reported to inhibit the growth of tumours in rats but does not appear to bind to DNA. In the authors' laboratories it has been shown to redox cycle as effectively as the other anthracycline drugs and to produce hydroxyl radicals able to degrade the sugar deoxyribose. Thus serious questions are raised about its safety.

Protein antitumour drugs

Neocarzinostatin. Neocarzinostatin is a protein, 109 amino acids long, secreted by a mutant strain of *Streptomyces carzinostaticus*. It has actions against a number of tumour cells *in vitro* and *in vivo*. A small fluorescent

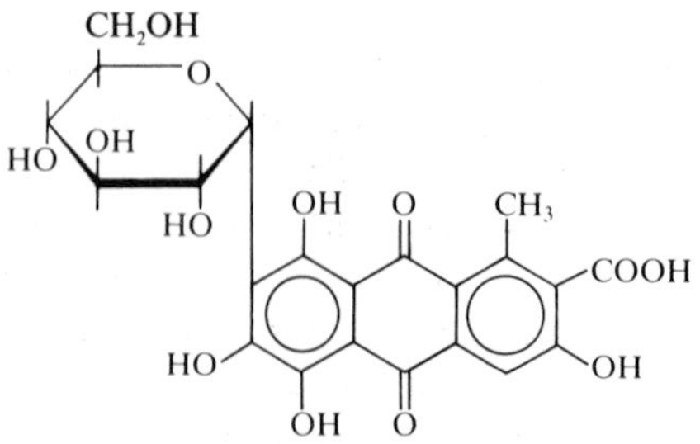

Fig. 8.17. Structure of carminic acid.

molecule bound to the protein is actually responsible for its biological activity. This chromophore becomes detached from the protein and binds to the DNA of the target cells, mainly adjacent to adenine and thymine. It promotes damage to deoxyribose, leading to strand scission and formation of TBA-reactive material. DNA damage by the neocarzinostatin chromophore *in vitro* requires the presence of both oxygen and a reducing agent, such as a thiol compound. The thiol reduces the chromophore to a free-radical species (detectable by ESR) that abstracts a hydrogen atom from C-5 of deoxyribose. The deoxyribose radical reacts with O_2 to give a peroxy radical, whose decomposition leads to the strand breakage. Depleting the GSH content of tumour cells decreases the DNA damage observed on subsequent treatment with neocarzinostatin, suggesting that this mechanism is relevant *in vivo*. Indeed, incubation of the neocarzinostatin chromophore with thiols in the absence of DNA inactivates it, perhaps by self-destruction of the radical species. DNA degradation by neocarzinostatin *in vitro* is not prevented by SOD, catalase, scavengers of hydroxyl radical, or metal-ion-chelating agents. Elucidation of the exact mechanism of the damage must await final determination of the structure of the chromophore. The protein part of neocarzinostatin does not attack DNA, and probably serves only as a vehicle for transporting the chromophore into cells.

Macromomycin and auromomycin. Macromomycin, a protein 112 amino acids long, with an amino-acid sequence similar to that of the neocarzinostatin protein, has been isolated from *Streptomyces macromomyceticus*. It has some antitumour effects, and induces DNA strand breakage. During purification of macromomycin, a small fluorescent molecule is released. Readdition of this greatly increases the antibacterial and antitumour activities of the protein, the complex of macromomycin and the fluorescent chromophore being known as *auromomycin*. The chromophore itself can induce DNA damage *in vitro*, again probably by some type of free-radical mechanism.

8.8.5 *Footprinting of DNA by using hydroxyl radical and other active complexes*

The binding of various proteins to DNA regulates the process of gene expression. Thus it is often important to identify the DNA sequence to which a given regulatory protein can bind. The problem can be approached by the technique of 'footprinting', which creates images of areas of protein binding to DNA. Essentially, the DNA–protein complex is treated with a DNA-degrading reagent such as a deoxyribonuclease. The bound protein protects the part of the DNA to which it is attached, while the rest of the DNA is hydrolysed. The protected part can then be isolated by polyacrylamide-gel electrophoresis and studied.

The high reactivity of OH˙ towards DNA has led to a development of DNA footprinting in which this radical is used. Essentially, the reaction of Fe(II)–EDTA with H_2O_2 is used to produce OH˙. Fe(II)–EDTA has an overall negative charge at physiological pH, and so will not bind to DNA (which is also negatively charged). Any OH˙ radicals produced by reaction of Fe(II)–EDTA with H_2O_2 that escape scavenging by EDTA itself enter 'free solution' and randomly damage sites in DNA. A bound protein blocks the cleavage of DNA by OH˙, allowing identification of the base sequence to which the protein binds. Since OH˙ has no significant base or sequence preference for reaction with DNA, it does not have the specificity problems encountered in the use of nuclease enzymes.

Another agent that has been used in DNA footprinting is *methidium propyl EDTA-Fe(II)*. Methidium intercalates between the DNA bases with no particular site specificity; attachment of EDTA–Fe(II) to it leads to generation of OH˙ and essentially random breakage of the DNA strands. By contrast, attachment of EDTA-Fe(II) to the intercalating compound *distamycin* results in DNA strand breakage at the sites to which distamycin binds. This approach allows cleavage of DNA at *specific* sites by attaching an agent generating OH˙ [Fe(II)–EDTA] to a compound that intercalates at specific sites. The copper–phenanthroline complex, which degrades DNA (Chapter 2, Section 2) has also been used in footprinting studies.

8.9 Further reading

Adelson, R., Saul, R. L., and Ames, B. N. (1988). Oxidative damage to DNA: relation to species metabolic rate and life span. *Proc. Natl. Acad. Sci. USA* **85**, 2706.

Adler, S. *et al.* (1986). Complement membrane attack complex stimulates production of reactive oxygen metabolites by cultured rat mesangial cells. *J. Clin. Invest.* **77**, 762.

Allan, G. *et al.* (1986). The protective action of allopurinol in an experimental model of haemorrhagic shock and reperfusion. *Br. J. Pharmacol.* **89**, 149.

Andrews, F. J. *et al.* (1987). Effect of iron chelation on inflammatory joint disease. *Ann. Rheum. Dis.* **46**, 327.

Andrews, F. J. *et al.* (1987). Effect of nutritional iron deficiency on acute and chronic inflammation. *Ann. Rheum. Dis.* **46**, 859.

Angel, M. F. *et al.* (1986). Free radicals: basic concepts concerning their chemistry, pathophysiology and relevance to plastic surgery. *Plastic Reconstruct. Surg.* **79**, 990.

Anzueto, L. *et al.* (1987). A role for iron in oxidant-mediated ischemic injury to intestinal microvasculature. *Am. J. Physiol.* **253**, G49.

Aruoma, O. I. and Halliwell, B. (1988). The iron-binding and hydroxyl radical scavenging action of anti-inflammatory drugs. *Xenobiotica* **18**, 459.

Aruoma, O. I. *et al.* (1987). The scavenging of oxidants by sulphasalazine and its metabolites. A possible contribution to their anti-inflammatory effects? *Biochem. Pharmacol.* **36**, 3739.

Aruoma, O. I. *et al.* (1988). Iron, copper and zinc concentrations in human sweat and plasma during exercise. *Clin. Chim. Acta* **177**, 81.

Axford, J. S. *et al.* (1987). Reduced B-cell galactosyltransferase activity in rheumatoid arthritis. *Lancet* **ii**, 1486.

Bachur, N. R., Gee, M. V., and Friedman, R. (1982). Nuclear catalyzed antibiotic free radical formation. *Cancer Res.* **42**, 1078.

Badylak, S. F. *et al.* (1987). Protection from reperfusion injury in the isolated rat heart by postischemic deferoxamine and oxypurinol administration. *Cardiovasc. Res.* **21**, 500.

Baldwin, S. R. *et al.* (1986). Oxidant activity in expired breath of patients with adult respiratory distress syndrome. *Lancet* **i**, 11.

Ball, R. Y. *et al.* (1987). Ceroid accumulation by murine peritoneal macrophages exposed to artificial lipoproteins. *Br. J. Exp. Pathol.* **68**, 427.

Baret, A. *et al.* (1984). Pharmacokinetic and anti-inflammatory properties in the rat of superoxide dismutases (CuSODs and MnSOD) from various species. *Biochem. Pharmacol.* **33**, 2755.

Bates, E. J., Johnson, C. C., and Lowther, D. A. (1985). Inhibition of proteoglycan synthesis by hydrogen peroxide in cultured bovine articular cartilage. *Biochim. Biophys. Acta* **838**, 221.

Baud, L. and Ardaillou, R. (1986). Reactive oxygen species: production and role in the kidney. *Am. J. Physiol.* **251**, F765.

Begleiter, A. (1985). Studies on the mechanism of action of quinone antitumour agents. *Biochem. Pharmacol.* **34**, 2629.

Benedetti, A. *et al.* (1984). Loss of lipid peroxidation as a histochemical marker for preneoplastic hepatocellular foci of rats. *Cancer Res.* **44**, 5712.

Bennett, R. M. and Skosey, J. L. (1977). Lactoferrin and lysozyme levels in synovial fluid. Differential indices of articular inflammation and degradation. *Arthritis Rheum.* **20**, 84.

Beraldo, H. *et al.* (1985). Iron(III)–adriamycin and iron(III)–daunorubicin complexes: physicochemical characteristics, interaction with DNA, and antitumour activity. *Biochemistry* **24**, 284.

Bernier, M., Hearse, D. J., and Manning, A. S. (1986). Reperfusion-induced arrhythmias and oxygen-derived free radicals. Studies with 'anti-free radical' intervention and a free radical-generating system in the isolated perfused rat heart. *Circ. Res.* **58**, 331.

Betz, A. L. (1985). Identification of hypoxanthine transport and xanthine oxidase activity in brain capillaries. *J. Neurochem.* **44**, 574.

Beutler, B. and Cerami, A. (1988). The history, properties and biological effects of cachectin. *Biochemistry* **27**, 7575

Bickers, D. R., Dixit, R., and Mukhtar, H. (1984). Enhancement of bleomycin-mediated DNA damage by epidermal microsomal systems. *Biochim. Biophys. Acta* **781**, 265.

Biemond, P. *et al.* (1986). Intraarticular ferritin-bound iron in rheumatoid arthritis. A factor that increases oxygen free radical-induced tissue destruction. *Arthritis Rheum.* **29**, 1187.

Biemond, P. *et al.* (1986). Superoxide production by polymorphonuclear leucocytes in rheumatoid arthritis and osteoarthritis: *in vivo* inhibition by the antirheumatic drug piroxicam due to interference with the activation of the NADPH-oxidase. *Ann. Rheum. Dis.* **45**, 249.

Blackett, A. D. and Hall, D. A. (1980). The action of vitamin E on the ageing of connective tissue in the mouse. *Mech. Age Dev.* **14**, 305.

Blake, D. R. *et al.* (1981). The importance of iron in rheumatoid disease. *Lancet* **ii**, 1142.

Blake, D. R. *et al.* (1983). Effect of a specific iron chelating agent on animal models of inflammation. *Ann. Rheum. Dis.* **42**, 89.

Blake, D. R. *et al.* (1989). Hypoxic-reperfusion injury in the inflamed rheumatoid joint. *Lancet* **i**, 289.

Bodmer, W. F. *et al.* (1987). Localization of the gene for familial adenomatous polyposis on chromosome 5. *Nature*, **328**, 614.

Bolli, R. (1988). Oxygen-derived free radicals and postischemic myocardial dysfunction ('stunned myocardium'). *J. Am. Coll. Cardiol.* **12**, 239.

Borello, S. *et al.* (1986). Restoration of hydroperoxide-dependent lipid peroxidation by 3-methylcholanthrene induction of cytochrome P-448 in hepatoma microsomes. *FEBS Lett.* **209**, 305.

Bowern, N. *et al.* (1984). Inhibition of autoimmune neuropathological process by treatment with an iron-chelating agent. *J. Exp. Med.* **160**, 1532.

Boyce, N. W. and Holdsworth, S. R. (1986). Hydroxyl radical mediation of immune renal injury by desferrioxamine. *Kidney Intl.* **30**, 813.

Bragt, P. C., Bansburg, J. I., and Bonta, I. L. (1980). Antiinflammatory effects of free radical scavengers and antioxidants. *Inflammation* **4**, 289.

Braude, S. *et al.* (1985). Adult respiratory distress syndrome after allogeneic bone-marrow transplantation: evidence for a neutrophil-independent mechanism. *Lancet* **i**, 1239.

Braunwald, E. and Kloner, R. A. (1985). Myocardial reperfusion: a double-edged sword? *J. Clin. Invest.* **76**, 1713.

Bredehorst, R. *et al.* (1987). Doxorubicin enhances complement susceptibility of human melanoma cells by extracellular oxygen radical formation. *J. Biol. Chem.* **262**, 2034.

Breed, J. G. S. *et al.* (1980). Failure of the antioxidant vitamin E to protect against adriamycin-induced cardiotoxicity in the rabbit. *Cancer Res.* **40**, 2033.

Brock, J. H. and Mainou-Fowler, T. (1983). The role of iron and transferrin in lymphocyte transformation. *Immunol. Today* **4**, 347.

Brooks, P. M., Kean, W. F., and Buchanan, W. W. (1986). *The clinical pharmacology of anti-inflammatory drugs.* Taylor and Francis, London.

Brown, W. T. (1985). Genetics of human ageing. In *Review of biological research in ageing* (M. Rothstein, Ed.) vol. 2, p. 105. A. R. Liss Inc, New York.

Burkhardt, H. *et al.* (1986). Oxygen radicals as effectors of cartilage destruction. Direct degradative effect on matrix components and indirect action via activation of latent collagenase from polymorphonuclear leukocytes. *Arthritis Rheum.* **29**, 379.

Cameron, J. S. and Simmonds, H. A. (1987). Use and abuse of allopurinol. *Br. Med. J.* **294**, 1504.

Campbell, W. W. and Anderson, R. A. (1987). Effects of aerobic exercise and training on the trace minerals chromium, zinc and copper. *Sports Med.* **4**, 9.

Carew, T. E., Schwenke, D. C., and Steinberg, D. (1987). Antiatherogenic effect of probucol unrelated to its hypercholesterolemic effect: evidence that anti-oxidants *in vivo* can selectively inhibit low density lipoprotein degradation in macrophage-rich fatty streaks and slow the progression of atherosclerosis in the Watanabe heritable hyperlipidemic rabbit. *Proc. Natl. Acad. Sci. USA* **84**, 7725.

Castano, E. M. and Frangione, B. (1988). Human amyloidosis, Alzheimer disease and related disorders. *Lab. Invest.* **58**, 122.

Cebellos, I. *et al.* (1988). Expression of transfected human CuZn superoxide dismutase gene in mouse L cells and N520Y neuroblastoma cells induces enhancement of glutathione peroxidase activity. *Biochim. Biophys. Acta* **949**, 58.

Cerchiari, E. L. *et al.* (1987). Protective effects of combined superoxide dismutase and deferoxamine on recovery of cerebral blood flow and function after cardiac arrest in dogs. *Stroke* **18**, 869.

Chandler, D. B. *et al.* (1988). Effect of iron deficiency on bleomycin-induced lung fibrosis in the hamster. *Am. Rev. Resp. Dis.* **137**, 85.

Chaudri, G. *et al.* (1986). Effect of antioxidants on primary alloantigen-induced T cell activation and proliferation. *J. Immunol.* **137**, 2646.

Cheeseman, K. H. *et al.* (1986). Lipid peroxidation in regenerating rat liver. *FEBS Lett.* **209**, 191.

Chellman, G. J., Bus, J. S., and Working, P. K. (1986). Role of epididymal inflammation in the induction of dominant lethal mutations in Fischer 344 rat sperm by methyl chloride. *Proc. Natl. Acad. Sci. USA* **83**, 8087.

Conway, J. G. *et al.* (1987). Role of fatty acyl coenzyme A oxidase in the efflux of oxidized glutathione from perfused livers of rats treated with the peroxisome proliferator nafenopin. *Cancer Res.* **47**, 4795.

Crawford, D. *et al.* (1988). Antioxidant enzymes in Xeroderma Pigmentosum fibroblasts. *Cancer Res.* **48**, 2132.

Crosti, N. *et al.* (1985). Coordinate expression of Mn-containing superoxide dismutase and Cu,Zn-containing superoxide dismutase in human fibroblasts with trisomy 21. *J. Cell Sci.* **79**, 95.

Cuperus, R. A., Muijsers, A. O., and Wever, R. (1985). Antiarthritic drugs containing thiol groups scavenge hypochlorite and inhibit its formation by myeloperoxidase from human leukocytes. *Arthritis Rheum.* **28**, 1228.

Cutler, R. G. (1985). Peroxide-producing potential of tissues: inverse correlation with longevity of mammalian species. *Proc. Natl. Acad. Sci. USA* **82**, 4798.

Dalsing, M. C. *et al.* (1983). Superoxide dismutase: a cellular protective enzyme in bowel ischaemia. *J. Surg. Res.* **34**, 589.

D'Arcy Doherty, M., Cohen, G. M., and Smith, M. T. (1984). Mechanisms of toxic injury to isolated hepatocytes by 1-naphthol. *Biochem. Pharmacol.* **33**, 543.

Dean, R. T., Roberts, C. R., and Forni, L. G. (1984). Oxygen-centred free radicals can efficiently degrade the polypeptide of proteoglycans in whole cartilage. *Biosci. Rep.* **4**, 1017.

Deguchi, Y., Negoro, S., and Kishimoto, S. (1987). C-FOS expression in human skin fibroblasts by reperfusion after oxygen deficiency; a recovery change of human skin fibroblasts after oxygen deficiency stress. *Biochem. Biophys. Res. Commun.* **149**, 1093.

Dillard, C. J. *et al.* (1978). Effects of exercise, vitamin E and ozone on pulmonary function and lipid peroxidation. *J. Appl. Physiol.* **45**, 927.

Dingle, J. T. and Gordon, J. L. (Eds) (1981). *Cellular interactions*, Research monographs in cell and tissue physiology, volume 6. Elsevier/North-Holland, Amsterdam.

Dorner, R. W., Alexander, R. L., and Moore, T. L. (1987). Rheumatoid factors. *Clin. Chim. Acta* **167**, 1.

Doroshow, J. H. (1986). Prevention of doxorubicin-induced killing of MCF-7 human breast cancer cells by oxygen radical scavengers and iron chelating agents. *Biochem. Biophys. Res. Commun.* **135**, 330.

Doroshow, J. H. and Davies, K. J. A. (1983). Comparative cardiac oxygen radical metabolism by anthracycline antibiotics, mitoxantrone, bisantrene, 4'-(9-acridinyl-amino)-methanesulfon-*m*-anisidide, and neocarzinostatin. *Biochem. Pharmacol.* **32**, 2935.

Doroshow, J. H. and Davies, K. J. A. (1986). Redox cycling of anthracyclines by cardiac mitochondria. Formation of superoxide anion, hydrogen peroxide and hydroxyl radical. *J. Biol. Chem.* **261**, 3068.

Doroshow, J. H., Locker, G. Y., and Myers, C. E. (1980). Enzymatic defenses of the mouse heart against reactive oxygen metabolites. Alterations produced by doxorubicin. *J. Clin. Invest.* **65**, 128.

Drapier, J. C. and Hibbs, J. R. Jr (1986). Murine cytotoxic activated macrophages inhibit aconitase in tumour cells. *J. Clin. Invest.* **78**, 790.

Dykens, J. A., Stern, A. and Trenkner, E. (1987). Mechanism of kainate toxicity to cerebellar neurons *in vitro* is analogous to reperfusion tissue injury. *J. Neurochem.* **49**, 1222.

Eddy, L. J. *et al.* (1987). Free radical-producing enzyme, xanthine oxidase, is undetectable in human hearts. *Am. J. Physiol.* **253**, H709.

Edwards, S. W. *et al.* (1983). Decrease in apparent K_m for oxygen after stimulation of respiration of rat polymorphonuclear leukocytes. *FEBS Lett.* **161**, 60.

Ehrenfeld, G. M. *et al.* (1987). Copper-dependent cleavage of DNA by bleomycin. *Biochemistry* **26**, 931.

Engerson, T. D. *et al.* (1987). Conversion of xanthine dehydrogenase to oxidase in ischemic rat tissues. *J. Clin. Invest.* **79**, 1564.

Fahl, W. E. *et al.* (1984). DNA damage related to increased hydrogen peroxide generation by hypolipidemic drug-induced liver peroxisomes. *Proc. Natl. Acad. Sci. USA* **81**, 7827.

Farber, E. (1982). Chemical carcinogenesis. A biologic perspective. *Am. J. Pathol.* **106**, 271.

Fleckenstein, A. *et al.* (1987). The role of calcium in the pathogenesis of experimental arteriosclerosis. *Trends Pharm. Sci.* **8**, 496.

Frank, M. M. (1987). Complement in the pathophysiology of human disease. *New Engl. J. Med.* **316**, 1525.

Gibson, B. W. *et al.* (1984). A revised primary structure for neocarzinostatin based on fast atom bombardment and gas-chromatographic mass spectrometry. *J. Biol. Chem.* **259**, 10801.

Giordano, N. *et al.* (1984). Increased storage of iron and anaemia in rheumatoid arthritis: usefulness of desferrioxamine. *Br. Med. J.* **289**, 961.

Gohil, K. *et al.* (1987). Effect of exercise training on tissue vitamin E and ubiquinone content. *J. Appl. Physiol.* **63**, 1638.

Gordon, T. *et al.* (1987). Apparent effect of catalase on airway edema in guinea pigs. Role of endotoxin contamination. *Am. Rev. Resp. Dis.* **135**, 854.

Gower, J. D. (1988). A role for dietary lipids and antioxidants in the activation of carcinogens. *Free Rad. Biol. Med.* **5**, 95.

Grisham, M. B. *et al.* (1986). Xanthine oxidase and neutrophil infiltration in intestinal ischemia. *Am. J. Physiol.* **251**, G567.

Grollman, A. P. *et al.* (1985). Origin and cytotoxic properties of base propenals derived from DNA. *Cancer Res.* **45**, 1127.

Grootveld, M. and Halliwell, B. (1988). 2,3-Dihydroxybenzoic acid is a product of human aspirin metabolism. *Biochem. Pharmacol.* **37**, 271.

Grum, C. M. *et al.* (1986). Absence of xanthine oxidase or xanthine dehydrogenase in the rabbit myocardium. *Biochem. Biophys. Res. Commun.* **141**, 1104.

Gutteridge, J. M. C. (1984). Lipid peroxidation and possible hydroxyl radical formation stimulated by the self-reduction of a doxorubicin-iron(III) complex. *Biochem. Pharmacol.* **33**, 1725.

Gutteridge, J. M. C. (1984). Streptonigrin-induced deoxyribose degradation: inhibition by superoxide dismutase, hydroxyl radical scavengers and iron chelators. *Biochem. Pharmacol.* **33**, 3059.

Gutteridge, J. M. C. (1987). Bleomycin-detectable iron in knee-joint synovial fluid from arthritic patients and its relationship to the extracellular antioxidant activities of caeruloplasmin, transferrin and lactoferrin. *Biochem. J.* **245**, 415.

Gutteridge, J. M. C. and Fu Xiao-Chang (1981). Enhancement of bleomycin–iron free radical damage to DNA by antioxidants and their inhibition of lipid peroxidation. *FEBS Lett.* **123**, 71.

Gutteridge, J. M. C. and Halliwell, B. (1987). Radical-promoting loosely-bound iron in biological fluids and the bleomycin assay. *Life Chem. Rep.* **4**, 113.

Gutteridge, J. M. C. and Quinlan, G. J. (1986). Carminic-acid promoted oxygen radical damage to lipid and carbohydrate. *Food Additives and Contaminants* **3**, 289.

Gutteridge, J. M. C. *et al.* (1985). The behaviour of caeruloplasmin in stored human extracellular fluids in relation to ferroxidase II activity, lipid peroxidation and phenanthroline-detectable copper. *Biochem. J.* **230**, 517.

Gutteridge, J. M. C., Beard, A. P. C., and Quinlan, G. J. (1985). Catalase enhances damage to DNA by bleomycin-iron(II): the role of hydroxyl radicals. *Biochem, Intl.* **10**, 441.

Halliwell, B. (1982). Production of superoxide, hydrogen peroxide and hydroxyl radicals by phagocytic cells: a cause of chronic inflammatory disease? *Cell Biol. Intl. Rep.* **6**, 529.

Halliwell, B. (1989). Lipid peroxidation *in vivo* and *in vitro* in relation to atherosclerosis: some fundamental questions. *Agents Actions* suppl. **26**, 223.

Halliwell, B. and Hoult, J. R. S. (1988). Oxidants, inflammation and anti-inflammatory drugs. *FASEB J.* **2**, 2867.

Harada, R. N., Vatter, A. E., and Repine, J. E. (1984). Macrophage effector function in pulmonary oxygen toxicity: hyperoxia damages and stimulates alveolar macrophages to make and release chemotaxins for polymorphonuclear leukocytes. *J. Leukocyte Biol.* **35**, 373.

Harman, D. (1981). The aging process. *Proc. Natl. Acad. Sci. USA.* **78**, 7124.

Harman, D., Eddy, D. E., Noffsinger, J. (1976). Free radical theory of aging: inhibition of amyloidosis in mice by antioxidants; possible mechanism. *J. Am. Geriat. Soc.* **24**, 203.

Hart, R. W. and Setlow, R. B. (1974). Correlation between deoxyribonucleic acid excision repair and life-span in a number of mammalian species. *Proc. Natl. Acad. Sci. USA* **71**, 2169.

Hassan, H. M. and Fridovich, I. (1977). Enzymatic defense against the toxicity of oxygen and of streptonigrin in *Escherichia coli. J. Bacteriol.* **129**, 1574.

Hassett, D. J. *et al.* (1987). Bacteria form intracellular free radicals in response to paraquat and streptonigrin: demonstration of the potency of hydroxyl radical. *J. Biol. Chem.* **262**, 13404.

Hay, J. G. *et al.* (1987). The effects of iron and desferrioxamine on the lung injury produced by intravenous bleomycin and hyperoxia. *Free Radical Res. Commun.* **4**, 109.

Hazelton, G. A. and Lang, C. A. (1985). Glutathione peroxidase and reductase activities in the aging mouse. *Mech. Age Dev.* **29**, 71.

Hearse, D. J., Humphrey, S. M., and Bullock, G. R. (1978). The oxygen paradox and the calcium paradox: two facets of the same problem? *J. Mol. Cell. Cardiol.* **10**, 641.

Hecht, S. M. (1986). DNA strand scission by activated bleomycin group antibiotics. *Fed. Proc.* **45**, 2784.

Hemilä, H., Roberts, P., and Wikström, M. (1984). Activated polymorphonuclear leucocytes consume vitamin C. *FEBS Lett.* **178**, 25.

Henderson, B. and Edwards, J. C. W. (1987). *The synovial lining in health and disease.* Chapman and Hall, London.

Henson, P. M. *et al.* (1984). Resolution of pulmonary inflammation. *Fed. Proc.* **43**, 2799.

Hewitt, S. D. *et al.* (1987). Effect of free radical altered IgG on allergic inflammation. *Ann. Rheum. Dis.* **46**, 866.

Hill, C. E. and Olson, M. S. (1987). Stimulation of uric acid release from the perfused rat liver by platelet activating factor or potassium. *Biochem. J.* **247**, 207.

Hirono, I. (1981). Natural carcinogenic products of plant origin. *CRC Crit. Rev. Toxicol.* **8**, 235.

Hoffstein, S. T., Gennaro, D. E., and Meunier, P. C. (1988). Cytochemical demonstration of constitutive H_2O_2 production by macrophages in synovial tissue from rats with adjuvant arthritis. *Am. J. Pathol.* **130**, 120.

Hoult, J. R. S. (1986). Pharmacological and biochemical actions of sulphasalazine. *Drugs,* **32** (suppl. 1), 18.

Huang, L. *et al.* (1987). Increased survival of skin flaps by scavengers of superoxide radical. *FASEB J.* **1**, 129.

Hurst, N. P. *et al.* (1986). Differential effects of mepacrine, chloroquine and hydroxychloroquine on superoxide anion generation, phospholipid methylation and arachidonic acid release by human blood monocytes. *Biochem. Pharmacol.* **35**, 3083.

Ip, C. (1985). Selenium inhibition of chemical carcinogenesis. *Fed. Proc.* **44**, 2573.

Ivy, G. O. *et al.* (1984). Inhibitors of lysosomal enzymes: accumulation of lipofuscin-like dense bodies in the brain. *Science* **226**, 985.

Jackson, M. J., Wagenmakers, A. J. M., and Edwards, R. H. T. (1987). Effect of inhibitors of arachidonic acid metabolism on efflux of intracellular enzymes from skeletal muscle following experimental damage. *Biochem. J.* **241**, 403.

Jain, S. K. (1988). Evidence for membrane lipid peroxidation during the *in vivo* aging of human erythrocytes. *Biochim. Biophys. Acta* **937**, 205.

Jansson, G. (1985). Formation of antibodies to native DNA in rats after administration of native DNA treated with the xanthine–xanthine oxidase system. *Free Radical Res. Commun.* **1**, 119.

Jasin, H. E. (1988). Oxidative cross-linking of immune complexes by human polymorphonuclear leukocytes. *J. Clin. Invest.* **81**, 6.

Ji, L. L., Stratman, F. W., and Lardy, H. A. (1988). Enzymatic down regulation with exercise in skeletal muscle. *Arch. Biochem. Biophys.* **263**, 137.

Joenje, H. *et al.* (1981). Oxygen-dependence of chromosomal aberrations in Fanconi's anaemia. *Nature* **290**, 142.

Joenje, H. *et al.* (1987). Oxygen toxicity and chromosomal breakage in ataxia telangiectasia. *Carcinogenesis* **8**, 341.

Johnson, K. J., Rehan, A., and Ward, P. A. (1988). The role of oxygen radicals in kidney disease. In *Proceedings of the Upjohn Symposium on Oxidants and Disease* (B. Halliwell, Ed.). Allen Press, Kansas.

Jürgens, G. *et al.* (1987). Modification of human serum low density lipoprotein by oxidation-characterization and pathophysiological implications. *Chem. Phys. Lipids* **45**, 315.

Kaplan, H. B. *et al.* (1984). Effect of non-steroidal anti-inflammatory agents on human neutrophil functions *in vitro* and *in vivo*. *Biochem. Pharmacol.* **33**, 371.

Kappen, L. S. and Goldberg, I. H. (1985). Activation of neocarzinostatin chromophore and formation of nascent DNA damage do not require molecular oxygen. *Nucl. Acid. Res.* **13**, 1637.

Kappen, L. S., Ellenberger, T. E., and Goldberg, I. H. (1987). Mechanism and base specificity of DNA breakage in intact cells by neocarzinostatin. *Biochemistry* **26**, 384.

Kedziora, J. and Bartosz, G. (1988). Down's syndrome: a pathology involving the lack of balance of reactive oxygen species. *Free Radical Biol. Med.* **4**, 317.

Keller, T. J. and Oppenheimer, N. J. (1987). Enhanced bleomycin-mediated damage of DNA opposite charged nicks. A model for bleomycin-directed double strand scission of DNA. *J. Biol. Chem.* **262**, 15144.

Kellogg, E. W. III and Fridovich, I. (1976). Superoxide dismutase in the rat and mouse as a function of age and longevity. *J. Gerontol.* **31**, 405.

Kennedy, K. A. *et al.* (1983). Effects of anthracyclines on oxygenated and hypoxic tumor cells. *Cancer Res.* **43**, 54.

Kensler, T. W. and Taffe, B. G. (1986). Free radicals in tumor promotion. *Adv. Free Radical Biol. Med.* **2**, 347.

Kimber, I. (1985). Natural killer cells. *Med. Lab. Sci.* **42**, 60.

Korthuis, R. J. *et al.* (1985). The role of oxygen-derived free radicals in ischemia-induced increases in canine skeletal muscle vascular permeability. *Circ. Res.* **57**, 599.

Kowall, N. W., Ferrante, R. J., and Martin, J. B. (1987). Patterns of cell loss in Huntington's disease. *Trends Neurosci.* **10**, 24.

Kukreja, R. C. *et al.* (1988). Oxygen radical-mediated lipid peroxidation and inhibition of Ca^{2+}-ATPase activity of cardiac sarcoplasmic reticulum. *Arch. Biochem. Biophys.* **261**, 447.

Kuwahara, J., Suzuki, T., and Sugiura, Y. (1985). Effective DNA cleavage by bleomycin–vanadium(IV) complex plus hydrogen peroxide. *Biochem. Biophys. Res. Commun.* **129**, 368.

Lasley, R. D. *et al.* (1988). Allopurinol enhanced adenine nucleotide repletion after myocardial ischemia in the isolated rat heart. *J. Clin. Invest.* **81**, 16.

Laughrea, M. (1982). On the error theories of aging. A review of the experimental data. *Exp. Gerontol.* **17**, 305.

Levander, O. A. (1987). A global view of human selenium nutrition. *Ann. Rev. Nutr.* **7**, 227.

Lewis, J. G., Hamilton, T., and Adams, D. O. (1986). The effect of macrophage development on the release of reactive oxygen intermediates and lipid oxidation products, and their ability to induce oxidative DNA damage in mammalian cells. *Carcinogenesis* **7**, 813.

Lim, K. H. *et al.* (1986). Prevention of reperfusion injury of the ischemic spinal cord: use of recombinant superoxide dismutase. *Ann. Thorac. Surg.* **42**, 282.

Linas, S. L., Whittenburg, D., and Repine, J. E. (1987). O_2 metabolites cause reperfusion injury after short but not prolonged renal ischemia. *Am. J. Physiol.* **253**, F685.

Link, E. M. and Riley, P. A. (1988). Role of hydrogen peroxide in the cytotoxicity of the xanthine/xanthine oxidase system. *Biochem. J.* **249**, 391.

Lorentzen, R. J. *et al.* (1979). Toxicity of metabolic benzo(*a*)pyrenediones to cultured cells and the dependence upon molecular oxygen. *Cancer Res.* **39**, 3194.

Macleod, A. M. and Catto, G. R. D. (1988). Cancer after transplantation. The risks are small. *Br. Med. J.* **297**, 4.

Mandel, H. G. *et al.* (1987). Metabolic basis for the protective effect of the antioxidant ethoxyquin on aflatoxin B_1 hepatocarcinogenesis in the rat. *Cancer Res.* **47**, 5218.

Marklund, S. L. *et al.* (1982). Copper- and zinc-containing superoxide dismutase, manganese-containing superoxide dismutase, catalase, and glutathione peroxidase in normal and neoplastic human cell lines and normal human tissues. *Cancer Res.* **42**, 1955.

Marnett, L. J. (1987). Peroxyl free radicals: potential mediators of tumor initiation and promotion. *Carcinogenesis* **8**, 1365.

Martin, J. P. Jr. and Batkoff, B. (1987). Homogentisic acid autoxidation and oxygen radical generation: implications for the etiology of alkaptonuric arthritis. *Free Radical Biol. Med.* **3**, 241.

Masotti, L., Casali, E., and Galeotti, T. (1988). Lipid peroxidation in tumour cells. *Free Radical Biol. Med.* **4**, 377.

Massie, H. R. and Williams, T. R. (1980). Singlet oxygen and aging in *Drosophila*. *Gerontology* **26**, 16.

Matsuda, Y. *et al.* (1982). Correlation between level of defense against active oxygen in *Escherichia coli K12* and resistance to bleomycin. *J. Antibiot.* **35**, 931.

Mavelli, I. *et al.* (1982). Superoxide dismutase, glutathione peroxidase and catalase in oxidative hemolysis. A study of Fanconi's anaemia erythrocytes. *Biochem. Biophys. Res. Commun.* **106**, 286.

McCord, J. M. (1987). Oxygen-derived radicals: a link between reperfusion injury and inflammation. *Fed. Proc.* **46**, 2402.

McCully, K. K. (1986). Exercise-induced injury to skeletal muscle. *Fed. Proc.* **45**, 2933.

Meldrum, B. (1985). Excitatory amino acids and anoxic/ischaemic brain damage. *Trends Neurosci.* **8**, 47.

Mickel, H. S. *et al.* (1987). Breathing 100% oxygen after global brain ischemia in mongolian gerbils results in increased lipid peroxidation and increased mortality. *Stroke* **18**, 426.

Mirvish, S. S. (1983). The etiology of gastric cancer. Intragastric nitrosamide formation and other theories. *J. Natl. Cancer Inst.* **71**, 631.

Moore, D. E. and Chappius, P. P. (1988). A comparative study of the photochemistry of the non-steroidal anti-inflammatory drugs, naproxen, benoxaprofen and indomethacin. *Photochem. Photobiol.* **47**, 173.

Moorhouse, C. P. *et al.* (1987). Allopurinol and oxypurinol are hydroxyl radical scavengers. *FEBS Lett.* **213**, 23.

Muehlmatter, D., Larsson, R., and Cerutti, P. (1988). Active oxygen induced strand breakage and poly ADP-ribosylation in promotable and non-promotable JB6 mouse epidermal cells. *Carcinogenesis* **9**, 239.

Muliawan, H., Scheulen, M. E., and Kappus, H. (1980). Acute adriamycin treatment of rats does not increase ethane expiration. *Res. Comm. Chem. Pathol. Pharmacol.* **30**, 509.

Mullane, K. M., Salmon, J. A., and Kraemer, R. (1987). Leukocyte-derived metabolites of arachidonic acid in ischemia-induced myocardial injury. *Fed. Proc.* **46**, 2422.

Munro, J. M. and Cotran, R. S. (1988). The pathogenesis of atherosclerosis: atherogenesis and inflammation. *Lab. Invest.* **58**, 249.

Myers, C. E. *et al.* (1982). Oxidative destruction of erythrocyte ghost membranes catalyzed by the doxorubicin–iron complex. *Biochemistry* **21**, 1707.

Näslund, U. *et al.* (1986). Superoxide dismutase and catalase reduce infarct size in a porcine myocardial occlusion-reperfusion model. *J. Mol. Cell. Cardiol.* **18**, 1077.

Nathan, C. F. (1987). Neutrophil activation on biological surfaces. Massive secretion of hydrogen peroxide in response to products of macrophages and lymphocytes. *J. Clin. Invest.* **80**, 1550.

Nathan, C. F. and Cohn, Z. A. (1981). Antitumor effects of hydrogen peroxide *in vivo. J. Exp. Med.* **154**, 1539.

Nishino, T. and Tsushima, K. (1986). Interaction of milk xanthine oxidase with folic acid. Inhibition of milk xanthine oxidase by folic acid and separation of the enzyme into two fractions on Sepharose 4B/folate gel. *J. Biol. Chem.* **261**, 11242.

Nomoto, K. I. *et al.* (1987). Early loss of myocardial myoglobin detected immunohistochemically following occlusion of the coronary artery in rats. *Exp. Mol. Pathol.* **47**, 390.

Oberley, L. W. (1984). The role of superoxide dismutase and gene amplification in carcinogenesis. *J. Theor. Biol.* **106**, 403.

O'Donnell-Tormey, J., De Boer, C. J., and Nathan, C. F. (1985). Resistance of human tumor cells in vitro to oxidative cytolysis. *J. Clin. Invest.* **76**, 80.

Oliver, C. N. *et al.* (1987). Age-related changes in oxidized proteins. *J. Biol. Chem.* **262**, 5488.

Olson, R. D. *et al.* (1988). Doxorubicin cardiotoxicity may be caused by its metabolite doxorubicinol. *Proc. Natl. Acad. Sci. USA* **85**, 3585.

Paller, M. S. and Hebbel, R. P. (1986). Ethane production as a measure of lipid peroxidation after renal ischemia. *Am. J. Physiol.* **251**, F839.

Palmer, D. N. *et al.* (1986). Ceroid lipofuscinosis in sheep. II. The major component of the lipopigment in liver, kidney, pancreas, and brain is low molecular weight protein. *J. Biol. Chem.* **261**, 1773.

Parekh, R. B. *et al.* (1985). Association of rheumatoid arthritis and primary osteoarthritis with changes in the glycosylation pattern of total serum IgG. *Nature* **316**, 452.

Patt, A. *et al.* (1988). Xanthine oxidase-derived hydrogen peroxide contributes to ischemia reperfusion-induced edema in gerbil brains. *J. Clin. Invest.* **81**, 1556.

Perkins, W. E. *et al.* (1982). Effect of ICRF-187 on doxorubicin-induced myocardial effects in the mouse and guinea pig. *Br. J. Cancer* **46**, 662.

Perry, M. A. *et al.* (1986). Role of oxygen radicals in ischemia-induced lesion in the cat stomach. *Gastroenterology* **90**, 362.

Peskin, A. V. *et al.* (1987). An unusual NAD(P)H-dependent O_2^--generating redox system in hepatoma 22a nuclei. *Free Radical Res. Commun.* **3**, 47.

Petkau, A. *et al.* (1977). Modification of superoxide dismutase in rat mammary carcinoma. *Res. Comm. Chem. Pathol. Pharmacol.* **17**, 125.

Pitkänen, P. *et al.* (1983). Amyloid of the seminal vesicles. A distinctive and common localized form of senile amyloidosis. *Am. J. Pathol.* **110**, 64.

Poot, M. *et al.* (1986). *De novo* synthesis of glutathione in human fibroblasts during *in vitro* ageing and in some metabolic diseases as measured by a flow cytometric method. *Biochim. Biophys. Acta* **883**, 580.

Potmesil, M., Israel, M., and Silber, R. (1984). Two mechanisms of adriamycin–DNA interaction in L1210 cells. *Biochem. Pharmacol.* **33**, 3137.

Pryor, W. A. (1987). The free-radical theory of ageing revisited: a critique and a suggested disease-specific theory. In *Modern biological theories of aging* (R. N. Butler *et al.*, Eds), p. 89. Raven Press, New York.

Puppo, A. and Halliwell, B. (1988). Formation of hydroxyl radicals in biological systems. Does myoglobin stimulate hydroxyl radical formation from hydrogen peroxide? *Free Radical Res. Commun.* **4**, 415.

Rana, R. S. and Munkres, K. D. (1978). Ageing of *Neurospora crassa*. V. Lipid peroxidation and decay of respiratory enzymes in an inositol auxotroph. *Mech. Age Develop.* **7**, 241.

Ratych, R. E. and Bulkley, G. B. (1986). Free radical-mediated postischemic reperfusion injury in the kidney. *J. Free Radical Biol. Med.* **2**, 311.

Reed, G. A. *et al.* (1986). Epoxidation of (±)-7,8-dihydroxy-7,8-dihydrobenzo[a]pyrene during (bi)sulfite autoxidation: activation of a procarcinogen by a co-carcinogen. *Proc. Natl. Acad. Sci. USA* **83**, 7499.

Reiss, U. and Gershon, D. (1976). Rat-liver superoxide dismutase. Purification and age-related modifications. *Eur. J. Biochem.* **63**, 617.

Richie, J. P. Jr., Mills, B. J., and Lang, C. A. (1987). Correction of a glutathione deficiency in the aging mosquito increases its longevity. *Proc. Soc. Exp. Biol. Med.* **184**, 113.

Riva, S., Manning, A. S., and Hearse, D. J. (1987). Superoxide dismutase and the reduction of reperfusion-induced arrhythmias: *in vivo* dose–response studies in the rat. *Cardiovasc. Drugs Ther.* **1**, 133.

Roberts, C. R., Mort, J. S., and Roughley, P. J. (1987). Treatment of cartilage proteoglycan aggregate with hydrogen peroxide. Relationship between observed degradation products and those that can occur naturally during aging. *Biochem. J.* **247**, 349.

Roberts, G. P. and Gibbons, R. A. (1966). The action of neutral hypochlorite on epithelial mucopolysaccharides. *Biochem. J.* **98**, 426.

Rodell, T. C. *et al.* (1987). Xanthine oxidase mediates elastase-induced injury to isolated lungs and endothelium. *J. Appl. Physiol.* **63**, 2159.

Rotrosen, D. and Gallin, J. I. (1987). Disorders of phagocyte function. *Ann. Rev. Immunol.* **5**, 127.

Rowley, D. A. *et al.* (1984). Lipid peroxidation in rheumatoid arthritis: thiobarbituric acid-reactive material and catalytic iron salts in synovial fluid from rheumatoid patients. *Clin. Sci.* **66**, 691.

Roxin, L. E., Hedin, G., and Venge, P. (1986). Muscle cell leakage of myoglobin after long-term exercise and relation to the individual performance. *Intl. J. Sports Med.* **7**, 259.

Saéz, J. C. *et al.* (1984). Superoxide radical involvement in the pathogenesis of burn shock. *Circ. Shock* **12**, 229.

Sagai, M. and Ichinose, T. (1980). Age-related changes in lipid peroxidation as measured by ethane, ethylene, butane and pentane in respired gases of rats. *Life Sci.* **27**, 731.

Sanfey, H., Bulkley, G. B., and Cameron, J. L. (1984). The role of oxygen-derived free radicals in the pathogenesis of acute pancreatitis. *Ann. Surg.* **200**, 405.

Sartorelli, A. C. (1986). The role of mitomycin antibiotics in the chemotherapy of solid tumors. *Biochem. Pharmacol.* **35**, 67.

Saugstad, O. D. (1988). Hypoxanthine as an indicator of hypoxia: its role in health and disease through free radical production. *Pediat. Res.* **23**, 143.

Sausville, E. A., Peisach, T., and Horwtiz, S. B. (1978). Effect of chelating agents and metal ions on the degradation of DNA by bleomycin. *Biochemistry* **17**, 2740.

Schalkwijk, J. *et al.* (1985). Cationization of catalase, peroxidase and superoxide dismutase. Effect of improved intraarticular retention on experimental arthritis in mice. *J. Clin. Invest.* **76**, 198.

Schalkwijk, J. *et al.* (1986). An experimental model for hydrogen peroxide-induced tissue damage. Effects of a single inflammatory mediator on (peri)articular tissues. *Arthritis Rheum.* **29**, 532.

Schisler, N. J. and Singh, S. M. (1985). Tissue-specific developmental regulation of superoxide dismutase (SOD-1 and SOD-2) activities in genetic strains of mice. *Biochem. Genet.* **23**, 291.

Schlote, W. and Boellard, J. W. (1983). Role of lipopigment during aging of nerve and glial cells in the human central nervous system. In *Brain ageing: neuropathology and neuropharmacology* (J. Cervós-Navarro and H. I. Sarkander, Eds), p. 27. Raven Press, New York.

Scott, D. L. *et al.* (1987). Long-term outcome of treating rheumatoid arthritis: results after 20 years. *Lancet* **i**, 1108.

Sebti, S. M., De Leon, J. C., and Lazo, J. S. (1987). Purification, characterization and amino acid composition of rabbit pulmonary bleomycin hydrolase. *Biochemistry* **26**, 4213.

Selevan, S. G. *et al.* (1985). A study of occupational exposure to antineoplastic drugs and fetal loss in nurses. *New Engl. J. Med.* **313**, 1173.

Shasby, D. M. *et al.* (1985). Reversible oxidant-induced increases in albumin transfer across cultured endothelium: alterations in cell shape and calcium homeostasis. *Blood* **65**, 605.

Shires, T. K. (1982). Iron-induced DNA damage and synthesis in isolated rat liver nuclei. *Biochem. J.* **205**, 321.

Siesjö, B. K. *et al.* (1985). Influence of acidosis on lipid peroxidation in brain tissues *in vitro*. *J. Cerebral Blood Flow Metab.* **5**, 253.

Sigman, D. S. (1986). Nuclease activity of 1,10-phenanthroline-copper ion. *Acc. Chem. Res.* **19**, 180.

Silver, I. A., Murrills, R. J., and Etherington, D. J. (1988). Microenvironment beneath adherent macrophages and osteoclasts. *Exp. Cell. Res.* **175**, 266.

Silverman, D. J. and Santucci, L. A. (1988). Potential for free radical-induced lipid peroxidation as a cause of endothelial cell injury in Rocky Mountain Spotted Fever. *Infect. Immun.* **56**, 3110.

Simpson, P. J. and Lucchesi, B. R. (1987). Free radicals and myocardial ischemia and reperfusion injury. *J. Lab. Clin. Med.* **110**, 13.

Sinha, B. K. *et al.* (1984). Enzymatic activation and binding of adriamycin to nuclear DNA. *Cancer Res.* **44**, 2892.

Sinha, B. K. *et al.* (1987). Differential formation of hydroxyl radicals by adriamycin in sensitive and resistant MCF-7 human breast tumor cells: implications for the mechanism of action. *Biochemistry* **26**, 3776.

Sohal, R. S. and Allen, R. G. (1986). Relationship between oxygen metabolism, aging and development. *Adv. Free Radical Biol. Med.* **2**, 117.

Sohal, R. S. *et al.* (1985). Iron induces oxidative stress and may alter the rate of aging in the housefly, *Musca domestica*. *Mech. Age Dev.* **32**, 33.

Southard, J. H. *et al.* (1987). Oxygen-derived free radical damage in organ preservation: activities of superoxide dismutase and xanthine oxidase. *Surgery* **101**, 566.

Stone, E. (1763). An account of the success of the bark of the willow in the cure of the agues. *Phil. Trans. Roy. Soc. Lond.* **53**, 195.

Strachan, A. F. *et al.* (1984). C reactive protein concentrations during long distance running. *Br. Med. J.* **289**, 1249.

Sugiyama, H. *et al.* (1988). Chemistry of the alkali-labile lesion formed from iron(II) bleomycin and d(CGCTTTAAAGCG). *Biochemistry* **27**, 58.

Svingen, B. A. *et al.* (1981). Protection against adriamycin-induced skin necrosis in the rat by dimethyl sulfoxide and α-tocopherol. *Cancer Res.* **41**, 3395.

Taylor, J. S., Schultz, P. G., and Dervan, P. B. (1984). DNA affinity cleaving. Sequence specific cleavage of DNA by distamycin.EDTA.Fe(II) and EDTA.distamycin.Fe(II). *Tetrahedron* **40**, 457.

Teicher, B. A., Lazo, J. S., and Sartorelli, A. C. (1981). Classification of antineoplastic agents by their selective toxicities towards oxygenated and hypoxic tumor cells. *Cancer Res.* **41**, 73.

Terkeltaub, R *et al.* (1984). Lipoproteins containing apoprotein B are a major regulator of neutrophil responses to monosodium urate crystals. *J. Clin. Invest.* **73**, 1719.

Thaw, H. H., Collins, V. P., and Brunk, U. T. (1984). Influence of oxygen tension, pro-oxidants and antioxidants on the formation of lipid peroxidation products (lipofuscin) in individual cultivated human glial cells. *Mech. Age Dev.* **24**, 211.

Theofilopoulous, A. N. and Dixon, F. B. (1982). Autoimmune diseases. Immunopathology and etiopathogenesis. *Am. J. Pathol.* **108**, 321.

Thomas, C. E. and Aust, S. D. (1986). Release of iron from ferritin by cardiotoxic anthracycline antibiotics. *Arch. Biochem. Biophys.* **248**, 684.

Till, G. O. *et al.* (1985). Lipid peroxidation and acute lung injury after thermal trauma to skin. Evidence of a role for hydroxyl radicals. *Am. J. Pathol.* **119**, 376.

Tomasi, A. *et al.* (1984). Studies on human uterine cervix and rat uterus using S-, X- and Q-band electron-spin-resonance spectroscopy. *Biochem. J.* **224**, 431.

Tritton, T. R. and Yee, G. (1982). The anticancer agent adriamycin can be actively cytotoxic without entering cells. *Science* **217**, 248.

Tsuchida, M., Miura, T., and Aibara, K. (1987). Lipofuscin and lipofuscin-like substances. *Chem. Phys. Lipids* **44**, 297.

Tullius, T. D. (1988). DNA footprinting with hydroxyl radical. *Nature* **332**, 663.

Uetrecht, J. P. (1988). Mechanisms of drug-induced lupus. *Chem. Res. Toxicol.* **1**, 133.

Umezawa, H. (1983). Studies of microbial products in rising to the challenge of curing cancer. *Proc. Roy. Soc. Lond. ser. B.* **217**, 357.

Usui, T. *et al.* (1982). Possible prevention from the progression of cardiotoxicity in adriamycin-treated rabbits by coenzyme Q_{10}. *Toxicol. Lett.* **12**, 75.

Van Hinsbergh, V. W. M. (1984). LDL cytotoxicity. The state of the art. *Atherosclerosis* **53**, 113.

Van Kessel, K. P. M. *et al.* (1987). Further evidence against a role for toxic oxygen products as lytic agents in NK cell-mediated cytotoxicity. *Immunology* **62**, 675.

Vandré, D. D., Shepherd, V. L., and Montgomery, R. (1979). Effects of macromomycin on the ultrastructure and biological properties of cultured mammalian cells. *Cancer Res.* **39**, 4091.

Vane, J. and Botting, R. (1987). Inflammation and the mechanism of action of anti-inflammatory drugs. *FASEB J.* **1**, 89.

Viinikka, L., Vuori, J., and Ylikorkala, O. (1984). Lipid peroxides, prostacyclin and thromboxane A_2 in runners during acute exercise. *Med. Sci. Sports Exercise* **16**, 275.

Ward, P. A. *et al.* (1983). Evidence for role of hydroxyl radical in complement and neutrophil-dependent tissue injury. *J. Clin. Invest.* **72**, 789.

Ward, P. A. *et al.* (1988). Immune complexes, oxygen radicals and lung injury. In *Proceedings of the Upjohn Symposium on Oxidants and Disease* (B. Halliwell, Ed.). Lawrence Press, Kansas.

Wasil, M. *et al.* (1987). Biologically-significant scavenging of the myeloperoxidase-derived oxidant hypochlorous acid by some anti-inflammatory drugs. *Biochem. Pharmacol.* **36,** 3847.

Weinberg, E. D. (1981). Iron and neoplasia. *Biol. Trace Element Res.* **3,** 55.

Werts, E. D. and Gould, M. N. (1986). Relationships between cellular superoxide dismutase and susceptibility to chemically induced cancer in the rat mammary gland. *Carcinogenesis* **7,** 1197.

Wharton, S. A. and Hipkiss, A. R. (1985). Degradation of peptides and proteins of different sizes by homogenates of human MRC 5 lung fibroblasts. Aged cells have a decreased ability to degrade shortened proteins. *FEBS Lett.* **184,** 249.

Wilmer, J. and Schubert, J. (1981). Mutagenicity of irradiated solutions of nucleic acid bases and nucleosides in *Salmonella typhimurium. Mutat. Res.* **88,** 337.

Winyard, P. G. *et al.* (1987). Mechanism of exacerbation of rheumatoid synovitis by total-dose iron-dextran infusion: *in vivo* demonstration of iron-promoted oxidant stress. *Lancet* **i,** 69.

Wolf, G. (1982). Is dietary β-carotene an anti-cancer agent? *Nutr. Rev.* **40,** 257.

Woodruff, T. *et al.* (1986). Is chronic synovitis an example of reperfusion injury? *Ann. Rheum. Dis.* **45,** 608.

Yeowell, H. N. and White, J. R. (1982). Iron requirement in the bactericidal mechanism of streptonigrin. *Antimicrob. Agents Chemother.* **22,** 961.

Yokode, M. *et al.* (1988). Cholesterylester accumulation in macrophages incubated with low density lipoprotein pretreated with cigarette smoke extract. *Proc. Natl. Acad. Sci. USA* **85,** 2344.

Yoshino, S. *et al.* (1985). Effect of blood on the activity and persistence of antigen induced inflammation in the rat air pouch. *Ann. Rheum. Dis.* **44,** 485.

Younes, M. *et al.* (1987). Inhibition of lipid peroxidation by superoxide dismutase following regional intestinal ischemia and reperfusion. *Res. Exp. Med.* **187,** 9.

Young, J. D. E. and Cohn, Z. A. (1986). Cell-mediated killing: a common mechanism? *Cell* **46,** 641.

Zimmerman, R. and Cerutti, P. (1984). Active oxygen acts as a promoter of transformation in mouse embryo C3H/10T$\frac{1}{2}$/C18 fibroblasts. *Proc. Natl. Acad. Sci. USA* **81,** 2085.

Appendix I

A consideration of atomic structure and bonding

A.1 Atomic structure

For the purposes of this book it will be sufficient to consider a simple model of atomic structure in which the atom consists of a positively charged nucleus that is surrounded by one or more negatively charged electrons. The nucleus contains two types of particle of approximately equal mass, the positively charged proton and the uncharged neutron. By comparison with these particles, the mass of the electron is negligible so that virtually all of the mass of the atom is contributed by its nucleus. The *atomic number* of an element is defined as the number of protons in its nucleus, the *mass number* as the number of protons plus neutrons. In the neutral atom, the atomic number also equals the number of electrons surrounding the nucleus. The simplest atom is that of the element hydrogen, containing one proton (atomic number equals one, mass number equals one) and one electron. All other elements contain neutrons in the nucleus.

Some elements exist as *isotopes*, in which the atoms contain the same number of protons and electrons, but different numbers of neutrons. These isotopes can be stable or unstable, the unstable ones undergoing radioactive decay at various rates. In this process, the nucleus of the radioactive isotope changes, and a new element is formed. For example, an isotope of the element uranium (atomic number 92) with a mass number of 238 undergoes nuclear disintegration to produce two fragments, one with two protons and two neutrons and the other with 90 protons and 144 neutrons, in fact an isotope of the element thorium. Fortunately the elements with which we are largely concerned in this book, carbon, hydrogen and oxygen, exist almost exclusively as one isotopic form in nature (Table A.1).

The electrons surrounding the atomic nucleus possess a negative charge. Since they do not spiral into the nucleus, they must possess energy to counteract the attractive electric force tending to pull them in. In 1900, Planck suggested that energy is quantized, i.e. that energy changes only occur in small, definite amounts known as 'quanta'. Application of Planck's quantum theory to the atom, by Bohr, produced a model in which the electrons exist in specific orbits, or 'electron shells', each associated with a

Table A.1. Isotopes of some common elements

Element	Isotope	Number of protons in nucleus	Number of neutrons in nucleus	Comments
Chlorine	$^{35}_{17}Cl$	17	18	Both isotopes are stable and occur
	$^{37}_{17}Cl$	17	20	naturally, ^{35}Cl being more abundant
Carbon	$^{12}_{6}C$	6	6	Over 90% of naturally occuring
	$^{13}_{6}C$	6	7	carbon is $^{12}_{6}C$. Small amounts of
	$^{14}_{6}C$	6	8	the radioactive isotope $^{14}_{6}C$ are formed by the bombardment of atmospheric CO_2 with cosmic rays (i.e. streams of neutrons arising from outer space). This isotope undergoes slow radioactive decay (50% decay after 5600 years)
Oxygen	$^{16}_{8}O$	8	8	Over 90% of naturally occuring
	$^{17}_{8}O$	8	9	oxygen is the isotope $^{16}_{8}O$
	$^{18}_{8}O$	8	10	
Hydrogen	$^{1}_{1}H$	1	0	Over 99% of hydrogen is $^{1}_{1}H$.
	$^{2}_{1}H$	1	1	Deuterium ($^{2}_{1}H$) is a stable
	$^{3}_{1}H$	1	2	isotope, whereas tritium ($^{3}_{1}H$) is radioactive. Deuterium oxide is known as heavy 'water', and is used in detecting the presence of singlet oxygen in biological systems.

The superscript number on the left of the symbol for the element represents the mass number, and the subscript the atomic number. All atoms of a given element have the same number of protons, but sometimes have different numbers of neutrons, giving rise to isotopes.

particular energy level. The 'K'-shell electrons, lying closest to the nucleus, have the lowest energy, and the energy successively increases as one proceeds outwards to the so-called L-, M-, and N-shells. The K-shell can hold a maximum of two electrons, the L-shell, 8, M-shell, 18, and N-shell, 32. Table A.2 shows the location of electrons in each of these shells for the elements up to atomic number 36.

Subsequent developments of atomic theory have shown that an electron has some of the properties of a particle, and some of the properties of a wave motion. As a result, the position of an electron at a given time cannot be

Table A.2. Location of electrons in shells for the elements with atomic numbers 1 to 36

Atomic number of element	Element	Symbol	Shell K	Shell L	Shell M	Shell N
1	Hydrogen	H	1			
2	Helium	He	2			
3	Lithium	Li	2	1		
4	Beryllium	Be	2	2		
5	Boron	B	2	3		
6	Carbon	C	2	4		
7	Nitrogen	N	2	5		
8	Oxygen	O	2	6		
9	Fluorine	F	2	7		
10	Neon	Ne	2	8		
11	Sodium	Na	2	8	1	
12	Magnesium	Mg	2	8	2	
13	Aluminium	Al	2	8	3	
14	Silicon	Si	2	8	4	
15	Phosphorus	P	2	8	5	
16	Sulphur	S	2	8	6	
17	Chlorine	Cl	2	8	7	
18	Argon	Ar	2	8	8	
19	Potassium	K	2	8	8	1
20	Calcium	Ca	2	8	8	2
21	Scandium	Sc	2	8	9	2
22	Titanium	Ti	2	8	10	2
23	Vanadium	V	2	8	11	2
24	Chromium	Cr	2	8	13	1
25	Manganese	Mn	2	8	13	2
26	Iron	Fe	2	8	14	2
27	Cobalt	Co	2	8	15	2
28	Nickel	Ni	2	8	16	2
29	Copper	Cu	2	8	18	1
30	Zinc	Zn	2	8	18	2
31	Gallium	Ga	2	8	18	3
32	Germanium	Ge	2	8	18	4
33	Arsenic	As	2	8	18	5
34	Selenium	Se	2	8	18	6
35	Bromine	Br	2	8	18	7
36	Krypton	Kr	2	8	18	8

precisely located, but only the region of space where it is most likely to be. These regions are referred to as *orbitals*. Each electron in an atom has its energy defined by four so-called quantum numbers. The first, or *principal quantum number* (n) defines the main energy level the electron occupies. For the K-shell, $n = 1$; for L, $n = 2$; for M, $n = 3$; and for N, $n = 4$. The second, or *azimuthal quantum number* (l) governs the shape of the orbital and has values from zero up to ($n - 1$). When $l = 0$, the electrons are called 's' electrons; when $l = 1$, they are 'p' electrons; $l = 2$, 'd' electrons; and $l = 3$ gives 'f' electrons. The third quantum number is the *magnetic quantum number* (m) and, for each value of l, m has values of l, ($l - 1$), ..., 0, $- 1$, ..., $- l$. Finally, the fourth quantum number, or *spin quantum number*, can have values of either $\frac{1}{2}$ or $- \frac{1}{2}$ only. Table A.3 shows how various combinations of these four quantum numbers can fill the electron shells, and (hopefully!) makes the above explanation a bit clearer. *Pauli's principle* states that 'no two electrons can have the same four quantum numbers'. Since the spin quantum number has only two possible values ($\pm \frac{1}{2}$), it follows that an orbital can hold only two electrons at most (Table A.3).

In filling the available orbitals electrons will enter the orbitals with the lowest total energy content (*Aufbau principle*). The order of filling is:

1s   2s   2p   3s   3p   4s   3d   4p   5s   4d   5p   6s   4f   5d   6p   7s   5f

lowest energy   increasing energy ⟶   highest energy

Table A.4 gives the electronic energy configurations of the elements with atomic numbers from 1 to 32. When the elements are arranged in the *Periodic Table* (Fig. A.1), elements with similar electronic arrangements fall into similar 'groups' (vertical rows), e.g. the group II elements all have two electrons in their outermost electron shell, and the group IV elements have four. Since the 4s-orbital is of lower energy than the 3d-orbitals, these latter orbitals remain empty until the 4s-orbital is filled (e.g. see the elements potassium and calcium in Table A.4). In subsequent elements the five 3d-orbitals receive electrons, creating the first row of the so-called *d-block* in the Periodic Table (Fig. A.1). Some of these d-block elements are called *transition elements*, meaning elements in which an inner shell of electrons is incomplete (in this case these are electrons in the fourth shell, but all the d-orbitals of the third shell are not yet full). The term transition element, as defined above, applies to scandium and subsequent elements as far as nickel, although it is often extended to include the whole of the first row of the d-block.

If orbitals of equal energy are available, e.g. the three 2p-orbitals in the L-shell, or the five 3d-orbitals in the M-shell (Table A.3), each is filled with one electron before any receives two (*Hund's rule*). Hence one can further break

Table A.3. Orbitals available in the principal electron shells

Shell	Principal quantum number	Value of l (azimuthal quantum number)	Electron type	Value of m (magenetic quantum number)	Value of s (spin quantum number)	Maximum number of electrons in shell	
K	1	0	s	0	$\pm\frac{1}{2}$	2 (1s-orbital)	
L	2	0	s	0	$\pm\frac{1}{2}$	2 (2s-orbital)	8
		1	p	1, 0, −1	$\pm\frac{1}{2}$	3 × 2 (three 2p-orbitals)	
M	3	0	s	0	$\pm\frac{1}{2}$	2 (3s-orbital)	
		1	p	1, 0, −1	$\pm\frac{1}{2}$	3 × 2 (three 2p-orbitals)	18
		2	d	2, 1, 0, −1, −2	$\pm\frac{1}{2}$	5 × 2 (five 3d-orbitals)	
N	4	0	s	0	$\pm\frac{1}{2}$	2 (4s-orbital)	
		1	p	1, 0, −1	$\pm\frac{1}{2}$	3 × 2 (three 4p-orbitals)	32
		2	d	2, 1, 0, −1, −2	$\pm\frac{1}{2}$	5 × 2 (five 4d-orbitals)	
		3	f	3, 2, 1, 0, −1, −2, −3	$\pm\frac{1}{2}$	7 × 2 (seven 4f-orbitals)	

Table A.4. Electronic configuration of the elements

Element	Atomic number	Symbol	Configuration	Place in periodic table
Hydrogen	1	H	$1s^1$	uncertain
Helium	2	He	$1s^2$	Group 0 (inert gases)
Lithium	3	Li	$1s^2 2s^1$	Group I (alkali metals)
Beryllium	4	Be	$1s^2 2s^2$	Group II (alkaline-earth metals)
Boron	5	B	$1s^2 2s^2 2p^1$	Group III
Carbon	6	C	$1s^2 2s^2 2p^2$	Group IV
Nitrogen	7	N	$1s^2 2s^2 2p^3$	Group V
Oxygen	8	O	$1s^2 2s^2 2p^4$	Group VI
Fluorine	9	F	$1s^2 2s^2 2p^5$	Group VII (halogen elements)
Neon	10	Ne	$1s^2 2s^2 2p^6$	Group 0
Sodium	11	Na	$1s^2 2s^2 2p^6 3s^1$	Group I
Magnesium	12	Mg	$1s^2 2s^2 2p^6 3s^2$	Group II
Aluminium	13	Al	$1s^2 2s^2 2p^6 3s^2 3p^1$	Group III
Silicon	14	Si	$1s^2 2s^2 2p^6 3s^2 3p^2$	Group IV
Phosphorus	15	P	$1s^2 2s^2 2p^6 3s^2 3p^3$	Group V
Sulphur	16	S	$1s^2 2s^2 2p^6 3s^2 3p^4$	Group VI
Chlorine	17	Cl	$1s^2 2s^2 2p^6 3s^2 3p^5$	Group VII
Argon	18	Ar	$1s^2 2s^2 2p^6 3s^2 3p^6$	Group 0
Potassium	19	K	$1s^2 2s^2 2p^6 3s^2 3p^6 4s^1$	Group I
Calcium	20	Ca	$1s^2 2s^2 2p^6 3s^2 3p^6 4s^2$	Group II
Scandium	21	Sc	$1s^2 2s^2 2p^6 3s^2 3p^6 4s^2 3d^1$	d-block
Titanium	22	Ti	$1s^2 2s^2 2p^6 3s^2 3p^6 4s^2 3d^2$	d-block
Vanadium	23	V	$1s^2 2s^2 2p^6 3s^2 3p^6 4s^2 3d^3$	d-block
Chromium	24	Cr	$1s^2 2s^2 2p^6 3s^2 3p^6 4s^1 3d^5$	d-block
Manganese	25	Mn	$1s^2 2s^2 2p^6 3s^2 3p^6 4s^2 3d^5$	d-block
Iron	26	Fe	$1s^2 2s^2 2p^6 3s^2 3p^6 4s^2 3d^6$	d-block
Cobalt	27	Co	$1s^2 2s^2 2p^6 3s^2 3p^6 4s^2 3d^7$	d-block
Nickel	28	Ni	$1s^2 2s^2 2p^6 3s^2 3p^6 4s^2 3d^8$	d-block
Copper	29	Cu	$1s^2 2s^2 2p^6 3s^2 3p^6 4s^1 3d^{10}$	d-block
Zinc	30	Zn	$1s^2 2s^2 2p^6 3s^2 3p^6 4s^2 3d^{10}$	d-block
Gallium	31	Ga	$1s^2 2s^2 2p^6 3s^2 3p^6 4s^2 3d^{10} 4p^1$	Group III
Germanium	32	Ge	$1s^1 2s^2 2p^6 3s^2 3p^6 4s^2 3d^{10} 4p^2$	Group IV

down the electronic configurations shown in Table A.4. The element boron has two 1s, two 2s, and one 2p electrons. Three 2p-orbitals of equal energy are available (Table A.3), and they are often written as $2p_x$, $2p_y$, and $2p_z$. If we represent each orbital as a square box and an electron as an arrow:

Fig. A.1. The Periodic Table.

Groups

	I	II	III	IV	V	VI	VII	O

s-block

Group I	Group II
1 H	
3 Li	4 Be
11 Na	12 Mg
19 K	20 Ca
37 Rb	38 Sr
55 Cs	56 Ba
87 Fr	88 Ra

d-block

21 Sc	22 Ti	23 V	24 Cr	25 Mn	26 Fe	27 Co	28 Ni	29 Cu	30 Zn
39 Y	40 Zr	41 Nb	42 Mo	43 Tc	44 Ru	45 Rh	46 Pd	47 Ag	48 Cd
57 La	72 Hf	73 Ta	74 W	75 Re	76 Os	77 Ir	78 Pt	79 Au	80 Hg
89 Ac									

p-block

III	IV	V	VI	VII	O
				1 H	2 He
5 B	6 C	7 N	8 O	9 F	10 Ne
13 Al	14 Si	15 P	16 S	17 Cl	18 Ar
31 Ga	32 Ge	33 As	34 Se	35 Br	36 Kr
49 In	50 Sn	51 Sb	52 Te	53 I	54 Xe
81 Tl	82 Pb	83 Bi	84 Po	85 At	86 Rn

f-block

Lanthanides

58 Ce	59 Pr	60 Nd	61 Pm	62 Sm	63 Eu	64 Gd	65 Tb	66 Dy	67 Ho	68 Er	69 Tm	70 Yb	71 Lu

Actinides

90 Th	91 Pa	92 U	93 Np	94 Pu	95 Am	96 Cm	97 Bk	98 Cf	99 Es	100 Fm	101 Md	102 No	103 Lr

For the next element, carbon, the extra electron enters another 2p-orbital in obedience to Hund's rule:

	1s	2s	2p
C	↑↓	↑↓	↑ ↑ ☐

And for nitrogen we have:

	1s	2s	2p
N	↑↓	↑↓	↑ ↑ ↑

Further electrons will now begin to 'pair-up' to fill the 2p-orbitals, e.g. for the oxygen atom:

	1s	2s	2p
O	↑↓	↑↓	↑↓ ↑ ↑

Table A.5 Electronic configuration of the elements scandium to zinc in the first row of the d-block of the Periodic Table.

		3d					4s
Scandium	Ar	↑					↑↓
Titanium	Ar	↑	↑				↑↓
Vanadium	Ar	↑	↑	↑			↑↓
Chromium	Ar	↑	↑	↑	↑	↑	↑
Manganese	Ar	↑	↑	↑	↑	↑	↑↓
Iron	Ar	↑↓	↑	↑	↑	↑	↑↓
Cobalt	Ar	↑↓	↑↓	↑	↑	↑	↑↓
Nickel	Ar	↑↓	↑↓	↑↓	↑	↑	↑↓
Copper	Ar	↑↓	↑↓	↑↓	↑↓	↑↓	↑
Zinc	Ar	↑↓	↑↓	↑↓	↑↓	↑↓	↑↓

'Ar' is used as an abbreviation for the argon configuration. $1s^2 2s^2 2p^6 3s^2 3p^6$, to simplify the table, i.e. each element begins with the argon configuration. The 'unusual' electronic configurations of chromium and copper seem to be due to the increased relative stability of atoms in which each 3d-orbital contains either one or two electrons.

Hund's rule is particularly important in the d-block elements, e.g. Table A.5 uses the same 'electrons-in-boxes' notation for the elements in the first row of this block. Each of the five 3d-orbitals receives one electron, before any receives two.

We shall now consider how atoms join together to form molecules in chemical reactions.

A.2   Bonding between atoms

A.2.1   *Ionic bonding*

As with our consideration of atomic structure, the account of chemical bonding that follows is the simplest possible model consistent with the requirements of this book.

Essentially two types of chemical bond can be distinguished. The first is called *ionic bonding*. This tends to happen when so-called electropositive elements combine with electronegative ones. Electropositive elements, such as those in groups I and II of the Periodic Table (Fig. A.1), tend to lose their outermost electrons easily, whereas electronegative elements (group VII, and oxygen and sulphur in group VI) tend to accept extra electrons. By doing so, they gain the electronic configuration of the nearest inert gases, which seems to be a particularly stable configuration in view of the relative lack of reactivity of these elements! Consider, for example, the combination of an atom of sodium with one of chlorine. Sodium, an electropositive group I element, has the electronic configuration $1s^2 2s^2 2p^6 3s^1$. If a sodium atom loses one electron, it then has the configuration $1s^2 2s^2 2p^6$, that of the inert gas, neon. It is still the element sodium because its nucleus is unchanged, but the loss of one electron leaves the atom with a positive charge, forming an ion or, more specifically, a *cation* (positively charged ion). For chlorine, configuration $1s^2 2s^2 2p^6 3s^2 3p^5$, acceptance of an electron gives the argon electron-configuration $1s^2 2s^2 2p^6 3s^2 3p^6$, and produces a negatively charged ion (*anion*) Cl^-.

In the case of a group II element such as magnesium, it must lose two electrons to gain an inert-gas electron-configuration. Thus one atom of magnesium can provide electrons for acceptance by two chlorine atoms, giving magnesium chloride a formula $MgCl_2$, i.e.

$$Mg \longrightarrow Mg^{2+} + 2e^-$$
$$1s^2 2s^2 2p^6 3s^2 \qquad 1s^2 2s^2 2p^6$$

(neon configuration).

An atom of oxygen, however, can accept two electrons and combine with magnesium to form an oxide MgO:

$$O + 2e^- \longrightarrow O^{2-}$$
$$1s^2 2s^2 2p^4 \qquad 1s^2 2s^2 2p^6$$
(neon configuration).

Once formed, anions and cations are held together by the electric attraction of their opposite charges. Each ion will exert an effect on each other ion in its vicinity, and these effects cause the ions to pack together into an *ionic crystal lattice*, as shown in Fig. A.2 for NaCl. Each Na$^+$ ion is surrounded by six Cl$^-$ ions, and vice versa. Once the lattice has formed, it cannot be said that any one Na$^+$ ion 'belongs' to any one Cl$^-$ ion, nor can 'molecules' of sodium chloride be said to exist in the solid. The formula of an ionic compound merely indicates the combining-ratio of the elements involved. A considerable amount of energy is required to disrupt all the electrostatic forces between the many millions of ions in a crystal of an ionic compound, so such compounds are usually solids with high melting-points. Ionic compounds are mostly soluble in water, and the solutions conduct electricity because of the presence of ions to carry the current. The properties of an ionic compound are those of its constituent ions.

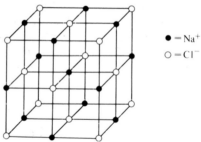

Fig. A.2.   Crystal structure of sodium chloride. The exact type of lattice formed by an ionic compound depends on the relative sizes of the ions. NaCl forms a cubic lattice, as shown.

A.2.2   Covalent bonding

The covalent bond involves a sharing of a pair of electrons between the two bonded atoms. In 'normal' covalent bonding, each atom contributes one electron to the shared pair; but in *dative covalent bonding*, one atom contributes both of the shared electrons. The element hydrogen is usually found in nature as *diatomic* molecules, H$_2$. Two hydrogen atoms are sharing a pair of electrons. If we represent the electron of each hydrogen atom by a cross (×) we can write:

$$H \times + {}^\times H \rightarrow H \, {}^\times_\times \, H$$

where ${}^\times_\times$ is the shared pair of electrons. Many other gaseous elements, including oxygen and chlorine, exist as covalently bonded diatomic molecules.

When chlorine combines with hydrogen, the covalent compound, hydrogen chloride, is formed. If we represent the outermost electrons of the chlorine atom as circles, we can write:

$$\overset{\text{OO}}{\underset{\text{OO}}{\text{O}}} \overset{}{\text{Cl}}{}^{\text{O}} + \text{H} \times \longrightarrow \text{H} \overset{\text{OO}}{\underset{\text{OO}}{\overset{\times}{\text{Cl}}}}\text{O}$$

hydrogen chloride

where $\overset{\times}{\underset{\text{o}}{}}$ is the shared pair of electrons. In reality, of course, one electron is the same as any other, in that once the bond is formed, the electron originating from the chlorine cannot be distinguished from that which came from the hydrogen. Similarly for the covalent compound ammonia, NH_3:

$$\overset{\text{OO}}{\underset{\text{O}}{\text{O}}}\text{N} + 3\text{H} \times \longrightarrow \begin{array}{c} \text{H} \\ \overset{\text{O}\times}{\underset{\text{O}\times}{\text{O}\text{N}\text{O}}}\,\text{H} \\ \text{H} \end{array}$$

In all the above cases, each atom has contributed one electron to the covalent bond. The compound ammonia also undergoes dative covalent bonding using the spare pair (*'lone-pair'*) of electrons on the nitrogen atom. For example, it forms a covalent bond with a proton (H^+), formed by the loss of one electron from a hydrogen atom and thus possessing no electrons of its own.

$$\begin{array}{c} \text{H} \\ \overset{\text{O}\times}{\underset{\text{O}\times}{\text{O}\text{N}\text{O}}}\,\text{H} + \text{H}^+ \\ \text{H} \end{array} \longrightarrow \left[\begin{array}{c} \text{H} \\ \text{H}\overset{\text{O}\times}{\underset{\text{O}\times}{\text{O}\text{N}\text{O}}}\text{H} \\ \text{H} \end{array}\right]^+$$

ammonia   ammonium ion, NH_4^+

Once formed, each of the four covalent bonds in NH_4^+ is indistinguishable from the others.

Covalent compounds do not conduct electricity and are usually gases, liquids, or low-melting-point solids at room temperature, because the forces of interaction between the molecules are weak (by contrast, covalent bonds themselves are usually very strong). Covalent bonds, unlike ionic bonds, have definite directions in space, and so their length, and the angles between them, can be measured and quoted.

The orbital theory applied to atomic structure (Section A.1) can also be applied to covalent compounds, the bonding electrons being considered as occupying *molecular orbitals* formed by interaction of the atomic orbitals in which the electrons were originally located. The various possible interactions produce molecular orbitals of different energy levels, each of which can hold a maximum of two electrons with opposite values of the spin quantum number (i.e. Pauli's principle is obeyed). In the simplest case, the hydrogen molecule, there are two possible molecular orbitals formed by interaction of the 1s atomic orbitals. The lowest energy orbital is the *bonding molecular orbital* (often written as $\sigma 1s$) in which the electron is most likely to be found between the two nuclei. There is also an *antibonding molecular orbital* (written as $\sigma*1s$) of higher energy in which there is little chance of finding an electron between the two nuclei. A bonding molecular orbital is more stable than the atomic orbitals that might give rise to it, whereas an antibonding molecular orbital is less stable. The two electrons in the hydrogen molecule have opposite spin, and both occupy the bonding molecular orbital. Hence H_2 is much more stable than are the isolated H atoms. By contrast, helium atoms have the electron configuration $1s^2$, and if they combined to give He_2, both the bonding and antibonding molecular orbitals would contain two electrons, and there would be no effective gain in stability. Hence He_2 does not form.

The combination of p-type atomic orbitals can produce two types of molecular orbital by overlapping in different ways. These are known as σ and π. Hence, for one of the 2p-orbitals (say $2p_x$) combining with another such orbital, there will be two bonding molecular orbitals, $\sigma 2p_x$ and $\pi 2p_x$, and two antibonding molecular orbitals, $\sigma*2p_x$ and $\pi*2p_x$. Energy increases in the order:

$$\sigma 2p_x < \quad \pi 2p_x \quad < \quad \pi*2p_x \quad < \quad \sigma*2p_x.$$

With this in mind, we can consider bonding in three more-complicated cases: the gases, nitrogen, oxygen, and fluorine. The nitrogen atom has the configuration $1s^2 2s^2 2p^3$. If two atoms join together to form a diatomic molecule N_2, the four 1s-electrons (two from each atom) fully occupy both a $\sigma 1s$ bonding and a $\sigma*1s$ antibonding orbital, and so there is no net bonding. The four 2s-electrons similarly occupy $\sigma 2s$ and $\sigma*2s$ molecular orbitals, and again no net bond results. Six electrons are left, located in two $2p_x$, two $2p_y$, and two $2p_z$ atomic orbitals. If the axis of the bond between the atoms is taken to be that of the $2p_x$ orbitals, they can overlap along this axis to produce a bonding $\sigma 2p_x$ molecular orbital that can hold both electrons. The $2p_y$ and $2p_z$ atomic orbitals cannot overlap along their axes, but they can overlap laterally to give bonding $\pi 2p_y$ and $\pi 2p_z$ molecular orbitals, each of which holds two electrons with different values of the spin quantum number. The 2p antibonding orbitals are not occupied; and the net result is a triple covalent bond $N \equiv N$, i.e. one σ covalent bond and two π covalent bonds. N_2 is thus far more stable than are the individual N-atoms.

The oxygen atom (configuration, $1s^2 2s^2 2p^4$) has one extra electron, and so when O_2 is formed there are two more electrons to consider. These must occupy the next highest molecular orbital in terms of energy. In fact, there are two such orbitals of equal energy, $\pi*2p_y$ and $\pi*2p_z$. By Hund's rule, each must receive one electron. Since the presence of these electrons in antibonding orbitals energetically cancels out one of the $\pi2p$ bonding orbitals, the two oxygen atoms are effectively joined by a double bond, i.e. $O=O$.

The fluorine molecule contains two more electrons than does O_2, and so the $\pi*2p_y$ and $\pi*2p_z$ orbitals are both full. Since three bonding and two antibonding molecular orbitals are occupied, the fluorine molecule effectively contains a single bond, $F-F$.

A.2.3 Non-ideal character of bonds

The discussion so far has implied an equal sharing of the bonding electrons between two atoms joined by a covalent bond. However, this is only the case when both atoms have a similar attraction for the electrons, i.e. are equally electronegative. This is often not the case. Consider, for example, the water molecule, which contains two oxygen–hydrogen covalent bonds:

$$2H \times + \;\overset{\circ\circ}{\underset{\circ\circ}{O}}\; \longrightarrow \;\overset{H}{\underset{\circ\circ}{\overset{\times\circ}{O}}}\!\overset{\times}{O}\,H$$

Oxygen is more electronegative than hydrogen, and so takes a slightly greater 'share' of the bonding electrons than it should, giving it a slight negative charge (written as δ^-). The hydrogen similarly has a slight positive charge i.e.

$$\overset{\delta^+}{H}\diagdown\qquad\diagup\overset{\delta^+}{H}$$
$$O_{\delta^-}$$

where the dash between the atoms represents a covalent bond.

The existence of these charges gives water many of its properties. They attract water molecules to each other, so raising the boiling point to 100 °C at normal atmospheric pressure e.g.

$$H\diagdown\quad\diagup\overset{\delta^+}{H}\cdots\cdots\overset{\delta^-}{O}\diagup\overset{\cdot\cdot}{H}$$
$$\underset{O}{}\qquad\qquad\qquad H$$

weak electrostatic bond

These weak electrostatic bonds are called *hydrogen bonds*. The small charges also allow water to hydrate ions; water molecules cluster around ions and help to stabilize them, e.g. for A^+ and B^- ions;

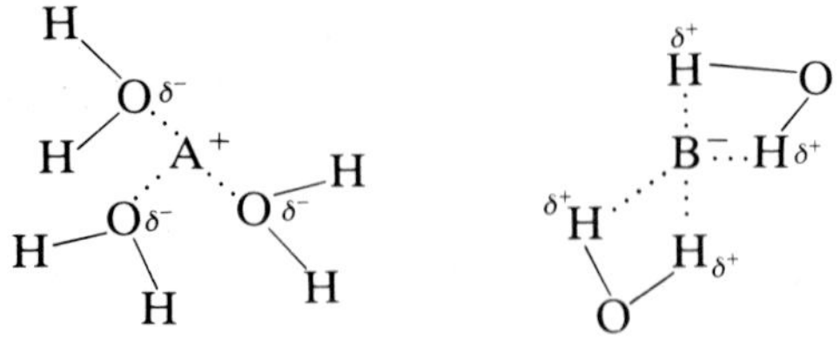

The energy obtained when ions become hydrated is what provides the energy to disrupt the crystal lattice when ionic compounds dissolve in water. In those cases where the energy of hydration would be much smaller than the energy needed to disrupt the lattice, then the ionic compound will not dissolve in water.

A.2.4 *Hydrocarbons and electron delocalization*

The element carbon has four electrons in its outermost shell (Table A.4), and normally forms four covalent bonds. Carbon atoms can covalently bond to each other to form long chains. For example, the compound butane, used as a fuel in cigarette lighters, has the structure

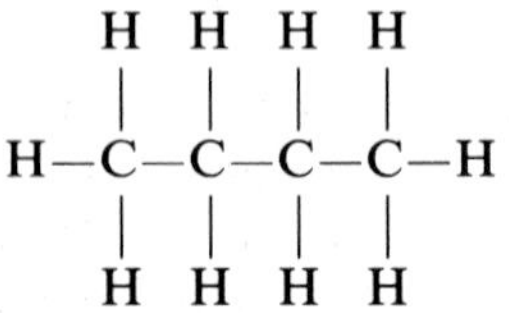

$$
\begin{array}{ccccc}
 & H & H & H & H \\
 & | & | & | & | \\
H- & C- & C- & C- & C-H \quad (C_4H_{10}) \\
 & | & | & | & | \\
 & H & H & H & H
\end{array}
$$

each dash (—) representing a covalent bond. Butane is referred to as a *hydrocarbon*, since the molecule contains carbon and hydrogen only. Two other hydrocarbon gases, ethane and pentane, are released during the peroxidation of membrane lipids (Chapter 4). They have the structures:

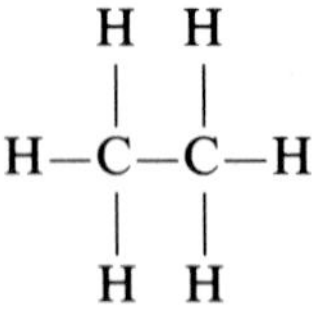

$$
\begin{array}{cc}
H & H \\
| & | \\
H-C-C-H \quad \text{(ethane)} \\
| & | \\
H & H
\end{array}
\qquad
\begin{array}{ccccc}
H & H & H & H & H \\
| & | & | & | & | \\
H-C-C-C-C-C-H \quad \text{(pentane)} \\
| & | & | & | & | \\
H & H & H & H & H
\end{array}
$$

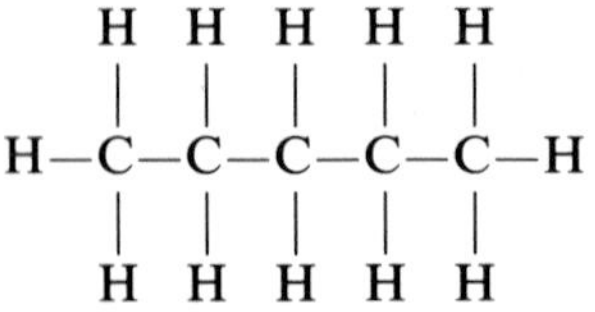

Carbon atoms can also form double covalent bonds (written as $>C{=}C<$) and triple covalent bonds (—C≡C—) with each other. A double bond consists of four shared electrons (two pairs), and a triple bond has six shared electrons (three pairs). The simplest hydrocarbon containing a double bond is the gas *ethene*, otherwise known as ethylene.

It has the structure

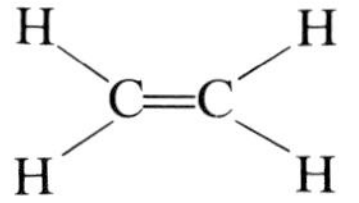

$$\underset{H}{\overset{H}{\diagdown}}C=C\underset{H}{\overset{H}{\diagup}}$$

Ethene is produced in several assays for the detection of hydroxyl radicals (Chapter 2).

Ethyne, otherwise known as acetylene, contains a triple bond and has the structure H—C≡C—H.

Organic compounds containing carbon–carbon double or triple bonds are often said to be *unsaturated*. Many constituents of membrane lipids are of this type (see Chapter 4).

The organic liquid *benzene* has the overall formula C_6H_6. Given that carbon forms four covalent bonds, the structure of benzene might be drawn as containing three carbon–carbon single bonds, and three double bonds, i.e.

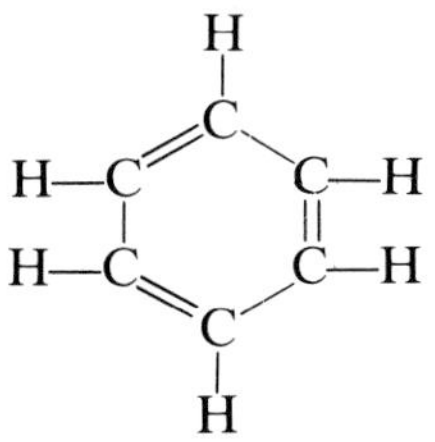

This structure cannot be correct, however, since benzene does not show the characteristic chemical reactions of compounds containing double bonds. A carbon–carbon single bond is normally 0.154 nm long (one nanometre, nm, is 10^{-9} metre), and a carbon–carbon double bond, 0.134 nm; yet all the bond lengths between the carbon atoms in benzene are equal at 0.139 nm, i.e. intermediate between the double and single bond lengths. The six electrons, that should have formed three double bonds, appear to be 'spread around' all six bonds. This is often drawn as:

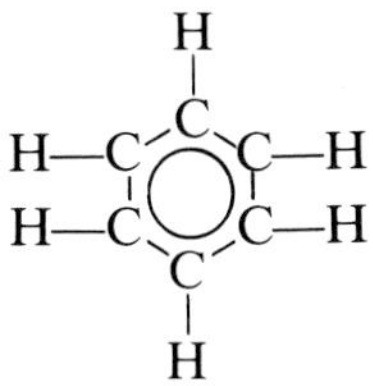

or, in abbreviated form,

This abbreviated form is used in this book. Compounds containing the benzene ring are called *aromatic compounds*. This delocalization of electrons over several bonds greatly increases the stability of a molecule. Other examples can be seen in haem rings (Chapter 2), which show extensive delocalization of electrons, and in several ions such as nitrate (NO_3^-) and carbonate (CO_3^{2-}). In each case the negative charge is spread between each of the bonds, i.e.

$$\begin{bmatrix} & O & \\ & \overset{\cdot\cdot}{\underset{\cdot\cdot}{N}} & \\ O & & O \end{bmatrix}^{-}$$

(each O has, on average, one-third of the negative charge)

$$\begin{bmatrix} & O & \\ & C & \\ O & & O \end{bmatrix}^{2-}$$

(each O has, on average, two-thirds of a negative charge).

A.3   Further reading

Harrison, P. M. and Hoare, R. J. (1980). *Metals in biochemistry.* Chapman and Hall, London.

Hughes, M. N. (1981). *The inorganic chemistry of biological processes.* Wiley and Sons, London.

Liptrot, G. F. (1978). *Modern inorganic chemistry.* Mills and Boon, London.

Appendix II: A simple guide to lymphocytes

All the cells of the immune system originate from stem cells in the bone marrow, as do erythrocytes and platelets. Monocytes and neutrophils mediate protection against invading foreign organisms; they ingest, and can kill, anything that they recognize as foreign. Monocytes invading a site of inflammation develop into macrophages.

The *lymphocytes* are responsible for specific immunity, and two main classes exist. Bone-marrow-derived (B) lymphocytes mature in the marrow and migrate directly to the spleen and lymph nodes. They are then ready to react to foreign substances (antigens). Thymus-derived (T) lymphocytes migrate from the bone marrow to the thymus where they mature before migrating to the peripheral lymphoid organs (spleen and lymph nodes). These two types of lymphocyte, although morphologically very similar, can be readily distinguished by their different cell-surface components. Several subsets of T cells exist.

The principal function of B lymphocytes is to synthesize and secrete antibodies (*immunoglobulins*). When these cells are stimulated by encountering a foreign antigen, they first go through several cycles of cell division and then differentiate into specialized antibody-secreting *plasma cells*. The immune system can produce specific antibodies (i.e. that bind with high affinity) to almost any antigen encountered. When an animal comes into contact with an antigen, many B cells are stimulated. Each B cell carries on its surface membrane immunoglobulin molecules with different antigen binding sites. These act as the receptors for antigen and the secreted immunoglobin of each cell has the same antigen binding site (specificity) as the membrane-bound immunoglobulin. Each B cell will divide to produce a *clone* of daughter cells, all producing immunoglobulin with the same specificity. Since many B cells bind the antigen the serum of the immunized animal will contain a mixture of the antibodies produced by many clones; *polyclonal antiserum*. (A monoclonal antibody is produced when a normal lymphocyte is fused with an immortalized cancer cell, producing a single clone which secretes antibody.) A second contact with the same antigen elicits a more rapid and extensive antibody response (*immunological memory*).

T lymphocytes do not secrete a single major protein product such as immunoglobulin but carry out their functions either by direct contact with other cells or by producing *lymphokines* (secreted proteins that have powerful biological effects on other cells at very low concentrations). T cells always recognize suitably-presented antigen in association with 'self'. The nature of this phenomenon can be made clearer by describing one of the earliest experiments that demonstrated it. If an animal is immunized with a virus (e.g. influenza A), immune T cells from the animal are able to kill influenza A-infected cells, but not uninfected target cells or cells infected with a different virus. The immune T cells are therefore specific for influenza A virus. The experiment, however, has a second part. The infected target cells must come from the same animal (or inbred strain) as the immune T cells or they cannot be killed. Thus the T cells do not recognize virus only; they appear to 'see' virus + self. Further genetic experiments showed that the 'self' components required were membrane glycoproteins coded in a region of the genome called the *major histocompatibility complex*. This name derives from earlier experiments which showed that the same gene products are responsible for the rejection of organ grafts, a reaction also known to be initiated by T lymphocytes. Thus, in both the response to foreign pathogens and to tissue grafts, T lymphocytes recognize and respond to MHC antigens, although in one case foreign MHC and in the other self MHC + antigen. Exactly how T cells 'see' (self + antigen) is still unclear. It has been reported that H_2O_2 interferes with T-cell function, an interesting observation in view of the production of H_2O_2 by activated phagocytes.

Some of the functions of T cells can be carried out by these cells alone, e.g. graft rejection. Other functions involve different cell types as well. Very often these responses are immunoregulatory; the T cells control the responses of other cell types. Thus *helper T cells* stimulate antigen-exposed B lymphocytes to secrete antibody and *suppressor T cells* can switch off antibody production. This latter response is important in terminating a response to a foreign pathogen but may also prevent inappropriate immune responses to self-antigens (*autoimmunity*).

Many immunoregulatory functions are mediated by lymphokines and cytokines. They have powerful effects on the growth and differentiation of their target cells and act as local hormones within the immune system. Lymphokines and cytokines include *interleukins I and II, tumour necrosis factors, interferons α and γ and lymphotoxins*. For example, interleukin II is a powerful stimulator of division in both the T cells producing it and in other T cells. Only T cells which have been first stimulated by contact with antigen express receptors for interleukin II. *Large granular lymphocytes* are another subset of lymphocytes that have cytolytic activity; they include natural killer (NK) cells (Chapter 8, Section 8.8.2), lymphokine-activated killer (LAK) cells, and K cells, which mediate antibody-dependent cell-mediated cytotoxicity (ADCC).

Index

Where a topic is discussed in several places in the text, those pages containing the most detailed account of it are printed in **bold type**. The term *def* in parentheses after a page number indicates that that page contains a definition of the term indexed.